Strategies in Regenerative Medicine

Matteo Santin
Editor

Strategies in Regenerative Medicine

Integrating Biology with Materials Design

 Springer

Editor

Matteo Santin
School of Pharmacy
and Biomolecular Sciences
University of Brighton
m.santin@brighton.ac.uk

ISBN 978-0-387-74659-3   e-ISBN 978-0-387-74660-9
DOI 10.1007/978-0-387-74660-9

Library of Congress Control Number: 2008940958

Printed on acid-free paper

springer.com

I would like to dedicate this book to those who have always stood by my side with patience and understanding

my parents

And

my best friend and light of my life, my wife Chiara

I would like also to express my deepest gratitude to Ms Rachel Steer for her great support during the book editing.

Preface

The profound transformations occurred in our modern age have been made possible by the unique combination of new technologies. Among them, medicine has completely changed our perception of life. Longevity has been significantly extended and linked to new lifestyles. The negative impact that pathologies and ageing have always had on the quality of our life is now mitigated by the availability of treatments daily applied to many individuals worldwide. For many years, pharmacological and surgical treatments have been supported by the introduction of biomedical devices. Biomedical implants have played a key role in the development of these treatments and achieved the objective of replacing tissue and organ structures and functionalities. Gradually, the scientific and clinical communities have understood that replacement could be improved by materials able to interact with the tissues and to participate in their metabolism and functions. This approach soon led to biomedical implants with improved clinical performances, but also to a new aspiration; rather than replacing damaged tissues and organs scientists and clinicians nowadays aim at their partial or complete regeneration. As a consequence of this ambition, the disciplines of tissue engineering and regenerative medicine have recently emerged. It is the dawn of a fascinating era where scientists from various disciplines, clinicians, and industry will need to intensify their collaborative efforts to provide our society with new and affordable solutions. The individual perception of life as well as societal issues such as longer working life and healthcare costs demand prompt responses. It is responsibility of scientists, clinicians, entrepreneurs, and politicians to respond in a coordinated manner. To this purpose, new knowledge and multi-disciplinary language have been emerging which have to permeate the thinking, activity, and communications of the experts and be acquired by the young students. Educational programs and book such the present one are therefore fundamental to prepare the next generations of scientists while triggering new thoughts and research activities among the most experienced. Through its objective assessment of the state-of-the-art in tissue engineering and regenerative medicine, this book serves our communities and its future technological aspirations. The flexible character of this book allows the reader to face problems from different perspectives and into different depths. The authors who have accepted to collaborate with Dr. Matteo Santin to the drafting of this book have uniquely combined information and highlighted literature sources which can pave the way towards a personalized study of the field. Indeed, the presentation of scientific, technological, and clinical knowledge has been organized at different levels to leave to the reader the choice of building up his/her intellectual pathway. New insights in the fields of biomaterials and biocompatibility, of molecular and cell biology, and of tissue and organ regeneration are combined with a critical assessment of clinical applications in regenerative medicine thus setting the scene in which our achievements and future challenge reside.

Naples, Italy   Luigi Nicolais

Profile. *Professor Luigi Nicolais has been one of the first material scientists to develop biomaterials with physico-chemical properties tuned to the specific clinical applications. His outstanding academic curriculum is demonstrated by the many scientific publications, patent applications and by his long-standing teaching experience and is underpinned by numerous national and international collaborations with research and industrial organizations. In recent years, Prof. Nicolais has made his experience available to society by serving for the local government of the Campania Region, Italy, as minister of Research and Technology Transfer and for the Italian Government (year 2006 elections) as Minister of the Public Sector and Innovation.*

Contents

1 Introduction: History of Regenerative Medicine   1
  Stephen F. Badylak, Alan J. Russell, and Matteo Santin

2 Soft Tissues Characteristics and Strategies for Their
  Replacement and Regeneration 15
  Maurizio Ventre, Paolo A. Netti, Francesco Urciuolo,
  and Luigi Ambrosio

3 Biomaterials for Tissue Engineering of Hard Tissues 55
  Elisabeth Engel, Oscar Castaño, Emiliano Salvagni,
  Maria Pau Ginebra, and Josep A. Planell

4 Biomimetic and Bio-responsive Materials in Regenerative
  Medicine 97
  Jacob F. Pollock and Kevin E. Healy

5 Clinical Approaches to Skin Regeneration 155
  S. E. James, S. Booth, P. Gilbert, I. Jones, and R. Shevchenko

6 Angiogenesis in Development, Disease, and Regeneration 189
  Rakesh K. Jain and Dai Fukumura

7 Tissue Engineering of Small– and Large– Diameter
  Blood Vessels 231
  Dörthe Schmidt and Simon P. Hoerstrup

8 Pancreas Biology, Pathology, and Tissue Engineering 261
  Wendy M. MacFarlane, Adrian J. Bone, and Moira Harrison

9 The Holy Grail of Hepatocyte Culturing and Therapeutic Use 283
  Andreas K. Nussler, Natascha C. Nussler, Vera Merk,
  Marc Brulport, Wiebke Schormann, Ping Yao,
  and Jan G. Hengstler

10   **Peripheral Nerve Injury, Repair, and Regeneration**   321
     Rudolf K. Potucek, Stephen W.P. Kemp, Naweed I. Syed,
     and Rajiv Midha

11   **Therapeutic Strategies in Ocular Tissue Regeneration: The Role**
     of Stem Cells   341
     K. Ramaesh, N. Stone, and B. Dhillon

12   **Cartilage Development, Physiology, Pathologies,**
     and Regeneration   367
     Xibin Wang, Lars Rackwitz, Ulrich Nöth, and Rocky S. Tuan

13   **Basic Science and Clinical Strategies for Articular Cartilage**
     Regeneration/Repair   395
     Barry W. Oakes

14   **Bone Biology: Development and Regeneration Mechanisms**
     in Physiological and Pathological Conditions   431
     Hideki Yoshikawa, Noriyuki Tsumaki, and Akira Myoui

15   **Clinical Applications of Bone Tissue Engineering**   449
     Silvia Scaglione and Rodolfo Quarto

16   **Conclusions: Towards High-Performance and Industrially**
     Sustainable Tissue Engineering Products   467
     Matteo Santin

Index   495

Contributors

Luigi Ambrosio Institute of Composite and Biomedical Materials, C.N.R, Naples, Italy, e-mail: ambrosio@unina.it

Stephen F. Badylak Department of Surgery, University of Pittsburgh,100 Technology Drive, Suite 200, Pittsburgh, PA 15219-3130, e-mail: badylaks@upmc.edu

Adrian J. Bone School of Pharmacy & Biomolecular Sciences, University of Brighton, Cockcroft Building Lewes Road, Brighton BN2 4GJ, UK, e-mail: a.j.bone@brighton.ac.uk

Booth S. McIndoe Burns Centre, Queen Victoria Hospital, East Grinstead, Sussex, UK.

Marc Brulport University of Dortmund, Leibniz Research Center for Working Environment and Human Factors, Ardeystr. 67, D-44139 Dortmund and Center for Toxicology, University of Leipzig, Haertelstr. 16-18, 04107 Leipzig.

Oscar Castaño Biomaterials, Biomechanics and Tissue Engineering Research Group. Institut de Bioenginyeria de Catalunya, IBEC. Universitat Politècnica de Catalunya. Baldiri Reixac, 13 08028 Barcelona, Spain.

B. Dhillon Tennent Institute of Ophthalmology Gartnavel General Hospital, Glasgow, Princess Alexandra Eye Pavilion Chalmers Street, Edinburgh.

Elisabeth Engel Biomaterials, Biomechanics and Tissue Engineering Research Group. Institut de Bioenginyeria de Catalunya, IBEC. Universitat Politècnica de Catalunya. Baldiri Reixac, 13 08028 Barcelona, Spain.

Dai Fukumura Edwin L. Steele Laboratory Department of Radiation Oncology, Massachusetts General Hospital 100 Blossom Street, Cox-736 Boston, MA 02114, e-mail: dai@steele.mgh.harvard.edu

P. Gilbert McIndoe Burns Centre, Queen Victoria Hospital, East Grinstead, Sussex, UK, e-mail: p.gilbert@qvh.nhs.uk

Maria Pau Ginebra Biomaterials, Biomechanics and Tissue Engineering Research Group. Institut de Bioenginyeria de Catalunya, IBEC. Universitat Politècnica de Catalunya. Baldiri Reixac, 13 08028 Barcelona, Spain.

Moira Harrison School of Pharmacy & Biomolecular Sciences, University of Brighton, Cockcroft Building Lewes Road, Brighton BN2 4GJ, UK, e-mail: m.harrison@brighton.ac.uk

Kevin E. Healy Department of Materials Science and Engineering, University of California, Berkeley.

Jan G. Hengstler University of Dortmund, Leibniz Research Center for Working Environment and Human Factors, Ardeystr. 67, D-44139 Dortmund and Center for Toxicology, University of Leipzig, Haertelstr. 16-18, 04107 Leipzig, e-mail: Jan.hengstler@medizin.uni-leipzig.de

Simon P. Hoerstrup Division of Regenerative Medicine (Tissue Engineering and Cell Transplantation), Department of Surgical Research and Clinic for Cardiovascular Surgery, University and University Hospital Zurich, Rämistrasse 100, 8091 Zurich, Switzerland, e-mail: simon_philipp.hoerstrup@usz.ch

Rakesh K. Jain Harvard Medical School and Edwin L. Steele Laboratory for Tumor Biology Department of Radiation Oncology Massachusetts General Hospital 100 Blossom St, Cox 7 Boston, MA 02114, e-mail: jain@steele.mgh.harvard.edu

S.E. James School of Pharmacy and Biomolecular Sciences, University of Brighton, Sussex, UK, e-mail: s.e.james@brighton.ac.uk

I. Jones Blond McIndoe Research Foundation, Queen Victoria Hospital, East Grinstead; Sussex, UK.

Stephen W.P. Kemp Hotchkiss Brain Institute, Faculty of Medicine, University of Calgary, Calgary, Alberta, Canada.

Wendy M. MacFarlane School of Pharmacy & Biomolecular Sciences, University of Brighton, Cockcroft Building Lewes Road, Brighton BN2 4GJ, UK, e-mail: w.m.macfarlane@brighton.ac.uk

Vera Merk Universitätsmedizin Berlin, Campus Virchow-Klinikum, Department of General, Visceral and Transplantation Surgery, Augustenburger Platz 1, 13353 Berlin, Germany.

Rajiv Midha Hotchkiss Brain Institute, Faculty of Medicine, University of Calgary, Calgary, Alberta, Canada.

Akira Myoui Department of Orthopaedic Surgery, Osaka University Graduate School of Medicine, 2-2 Yamadaoka, Suita 565-0871, Japan.

Paolo A. Netti Interdepartmental Centre on Biomaterials, University of Naples Federico II, Italy, e-mail: nettipa@unina.it

Ulrich Noth Orthopaedic Center for Musculoskeletal Research, König-Ludwig-Haus, Julius-Maximilians-University, Würzburg, Germany.

Andreas K. Nussler Technical University Munich, Department of Traumatology, Ismaningerstr. 11, 81829 Munich, Germany, e-mail: andreas.nuessler@gmail.com

Natascha C. Nussler Universitätsmedizin Berlin, Campus Virchow-Klinikum, Department of General, Visceral and Transplantation Surgery, Augustenburger Platz 1, 13353 Berlin, Germany.

Barry W. Oakes Mercy Tissue Engineering and Department of Anatomy and Cell Biology, Monash University, Melbourne, Australia, 2 Reeve Court, Cheltenham, 3192, Australia, e-mail: barry.oakes@optusnet.com.au

Josep A. Planell Biomaterials, Biomechanics and Tissue Engineering Research Group. Institut de Bioenginyeria de Catalunya, IBEC. Universitat Politècnica de Catalunya.. Baldiri Reixac, 13 08028 Barcelona, Spain, e-mail: Josep.A.Planell@UPC.edu

Jacob F. Pollock Department of Bioengineering, University of California, Berkeley.

Rudolf K. Potucek Hotchkiss Brain Institute, Faculty of Medicine, University of Calgary, Calgary, Alberta, Canada.

Rodolfo Quarto Advanced Biotechnology Center, Genova, Italy; Department of Pharmaceutical and Food Chemistry and Technologies, University of Genova, Italy, e-mail: rodolfo.quarto@unige.it

Lars Rackwitz Cartilage Biology and Orthopaedics Branch, National Institute of Arthritis and Musculoskeletal and Skin Diseases, National Institutes of Health, Department of Health and Human Services, Bethesda, MD 20892, USA; [2]Orthopaedic Center for Musculoskeletal Research, König-Ludwig-Haus, Julius-Maximilians-University, Würzburg, Germany.

K. Ramaesh Tennent Institute of Ophthalmology Gartnavel General Hospital, Glasgow, Princess Alexandra Eye Pavilion Chalmers Street, Edinburgh.

J. Russell Alan McGowan Institute for Regenerative Medicine, University of Pittsburgh, Pittsburgh Tissue Engineering Initiative, 100 Technology Drive, Suite 200, Pittsburgh, PA 15219, e-mail: russellaj@upmc.edu

Emiliano Salvagni Biomaterials, Biomechanics and Tissue Engineering Research Group. Institut de Bioenginyeria de Catalunya, IBEC. Universitat Politècnica de Catalunya.. Baldiri Reixac, 13 08028 Barcelona, Spain.

Matteo Santin School of Pharmacy & Biomolecular Sciences, University of Brighton, Cockcroft Building Lewes Road, Brighton BN2 4GJ, UK, e-mail: m.santin@brighton.ac.uk

Silvia Scaglione Advanced Biotechnology Center, Genova, Italy; Department of Communication, Computer and System Sciences, University of Genova, Italy, e-mail: silvia.scaglione@unige.it

Dörthe Schmidt Division of Regenerative Medicine (Tissue Engineering and Cell Transplantation), Department of Surgical Research and Clinic for Cardiovascular Surgery, University and University Hospital Zurich, Rämistrasse 100, 8091 Zurich, Switzerland, e-mail: doerthe.schmidt@usz.ch

Wiebke Schormann University of Dortmund, Leibniz Research Center for Working Environment and Human Factors, Ardeystr. 67, D-44139 Dortmund and Center for Toxicology, University of Leipzig, Haertelstr. 16-18, 04107 Leipzig.

R. Shevchenko School of Pharmacy and Biomolecular Sciences, University of Brighton, Sussex, UK; Blond McIndoe Research Foundation, Queen Victoria Hospital, East Grinstead; Sussex, UK.

N. Stone Tennent Institute of Ophthalmology Gartnavel General Hospital, Glasgow, Princess Alexandra Eye Pavilion Chalmers Street, Edinburgh.

Naweed I. Syed Hotchkiss Brain Institute, Faculty of Medicine, University of Calgary, Calgary, Alberta, Canada, e-mail: nisyed@ucalgary.ca

Noriyuki Tsumaki Department of Orthopaedic Surgery, Osaka University Graduate School of Medicine, 2-2 Yamadaoka, Suita 565-0871, Japan.

Rocky S. Tuan Cartilage Biology and Orthopaedics Branch, National Institute of Arthritis and Musculoskeletal and Skin Diseases, National Institutes of Health, Department of Health and Human Services, Bethesda, MD 20892, USA, e-mail: tuanr@mail.nih.gov

Francesco Urciuolo Institute of Composite and Biomedical Materials, C.N.R, Naples, Italy, e-mail: urciuolo@unina.it

Maurizio Ventre Interdepartmental Centre on Biomaterials, University of Naples Federico II, Italy, e-mail: maventre@unina.it

Xibin Wang Cartilage Biology and Orthopaedics Branch, National Institute of Arthritis and Musculoskeletal and Skin Diseases, National Institutes of Health, Department of Health and Human Services, Bethesda, MD 20892, USA.

Ping Yao Universitätsmedizin Berlin, Campus Virchow-Klinikum, Department of General, Visceral and Transplantation Surgery, Augustenburger Platz 1, 13353 Berlin, Germany.

Hideki Yoshikawa Department of Orthopaedic Surgery, Osaka University Graduate School of Medicine, 2-2 Yamadaoka, Suita 565-0871, Japan, e-mail: yhideki@ort.med.osaka-u.ac.jp

The Editor's Profile

Dr. Matteo Santin was born in Cava de' Tirreni (Salerno) Italy. Currently, he is a Reader in Tissue Regeneration at the School of Pharmacy & Biomolecular Sciences, University of Brighton, UK.

Dr. Santin has achieved a BSc Hon and a PhD in Biomaterials at the University of Naples and a PhD in Biomedical Sciences at the University of Brighton.

Previously, he has spent his career at University of Naples and CNR (Italy), University of Connecticut (USA), Independent University of Eastern Piedmont (Italy).

He has been working in the field of Biomaterials and Tissue Engineering since 1991 working on the synthesis of bio-responsive biomaterials, protein adsorption and host response.

In his most recent work at the University of Brighton, he has developed novel biomaterials such as the soybean-based biomaterials and the nano-structured and biocompetent semi-dendrimeric systems. Dr. Santin's research interest also focusses on the activation and phenotype plasticity of monocytes/ macrophages and the differentiation of adult stem cells.

Since 1994, Dr. Santin has filed several patent applications and he has regularly published his work in international peer-reviewed journals. Among them, it is worth noting the first in vitro study about the inflammatory potential of the silk fibroin (1999), the first description of the role of the adsorbed alpha1-microglobulin on biomaterials surface and new insight on the mechanisms of biomaterials mineralization and in-stent restenosis.

Throughout the years, Dr. Santin has established many collaborative links with UK and overseas research groups.

In 2005, he has been awarded by the European Society for Biomaterials with the Jean Leray Award.

Chapter 1
Introduction: History of Regenerative Medicine

Stephen F. Badylak, Alan J. Russell, and Matteo Santin

Contents

1.1	Introduction	1
1.2	The Pioneers	2
	1.2.1 The Pioneers of Tissue Engineering and of Clinical Tissue Regeneration	2
	1.2.2 The Pioneers of the Biomimicking Biomaterials Science	4
	1.2.3 The Pioneers of the Tissue Engineering Science	7
1.3	The Birth of Regenerative Medicine	8
	1.3.1 Early Clinical Applications	9
	1.3.2 Commercialization Efforts	9
1.4	Challenges	11
Questions		11
References		12

1.1 Introduction

The majority of species on earth have the ability to regenerate body parts. Higher order mammals, including humans, have lost the ability to re-grow limbs and vital organs and have replaced tissue regeneration with the processes of inflammation and scar tissue formation (Metcalfe & Ferguson 2007). The human body does have the inherent ability, however, to regenerate selected cell populations and tissues on a routine basis: bone marrow, the liver, the epidermis, and the cells that constitute the intestinal lining among others. The dramatic idea that through medicine we may be able to minimize scar tissue formation and extend this regenerative capacity to all body parts has been an elusive dream since the times of Greek mythology when Prometheus was sentenced to eternal suffering as a bird ate his liver for eternity while the liver regenerated. It is fascinating to consider that even the Greeks seemed to predict the regenerative capacity of the liver.

S.F. Badylak (✉)
Department of Surgery, University of Pittsburgh, 100 Technology Drive, Suite 200
Pittsburgh, PA 15219-3130
e-mail: badylaks@upmc.edu

M. Santin (ed.), *Strategies in Regenerative Medicine*,
DOI 10.1007/978-0-387-74660-9_1, © Springer Science+Business Media, LLC 2009

1.2 The Pioneers

1.2.1 The Pioneers of Tissue Engineering and of Clinical Tissue Regeneration

In the 1930s, the Nobel laureate *Alexis Carrel* collaborated with *Charles Lindberg* at the Rockefeller Institute in New York in an effort to grow viable and functional tissues and organs in vitro for use as replacement body parts (www.charleslind-bergh.com/heart/carrel.asp; www.nobel.se/index.html).

Later, difficult clinical cases were treated by pioneering approaches which are still the basis for many surgical procedures in modern surgery. The work performed by *Sir Archibald McIndoe* (www.qvh.nhs.uk; www.surgical-tutor.org.uk) deserves a particular emphasis. Archibald McIndoe was a New Zealand civilian plastic surgeon who, during World War II, pioneered exceptional treatments for badly burned Allied aircrew – known as the "Guinea Pigs" – at the Queen Victoria Hospital, East Grinstead, UK (Bennett 1988) (Fig. 1.1).

Sir Archibald's ("Archie" as he was known at East Grinstead) pioneering surgical procedures were first applied to the fighter pilots who sustained horrific burns during the Battle of Britain. Later, as the bombing program intensified against the industrial heart of Germany, many burned bomber crews benefited from the same treatments; these patients representing 80% of all those who were under McInodoe's care. By the end of the war, 649 Guinea Pigs had been treated by McIndoe including British (65–70%), Canadian (27%), Australian (8%), and New Zealander (8%). McIndoe's techniques for the treatment of casualties disfigured in combat transformed reconstructive surgery procedures. This inspired surgeon ("The Boss" or

Fig 1.1 Sir Archibald McIndoe and some of the aircrew he treated (photo reprinted with permission by the Blond McIndoe Museum)

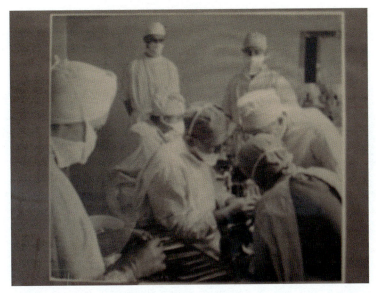

Fig 1.2 Sir McIndoe during a surgical procedure (photo reprinted with permission by the Blond McIndoe Museum)

"The Maestro" as he was also known) finely defined procedures of tissue resection, grafting, stitching, and reconstitution which to date are still at the basis of plastic surgery (Fig. 1.2). Alongside surgical restoration, McInode also fought a psychological battle in rebuilding the morale and self-confidence of his disfigured patients.

Among McIndoe's many intuitions, there was the awareness that, in the majority of the clinical cases, successful tissue regeneration depended on the re-establishment of vascularization (Fig. 1.3a and b).

He has been reported to confide to one of his colleagues: 'The next great era in surgery will be when we learn how to transplant tissue from one person to another. I foresee the day when whole limbs, kidneys, lungs and even hearts will be surgically replaced'. Indeed, his vision was later fulfilled when the Leverhulme Trust funded a project to transplant donor skin to severely burned patients. During the last 40 years, the Blond McIndoe Centre, established one year after McIndoe's death in 1960, has pioneered leading-edge surgical techniques including the first successful toe-to-thumb transplant, immunology of organ transplantation and tissue typing, matching grafts to recipients, and improving the success of corneal grafts (see Chapter 5 and Chapter 11).

However, it was soon recognized that tissue transplants suffered of limitations such as donors' availability and immunological reaction. As a consequence, in the late 1970s, *Eugene Bell* and colleagues developed strategies for the in vitro production of skin, glandular tissues, and blood vessels (Bell et al. 1981). Simultaneously, reports by Reinwald and Green (1975) showed the

Fig 1.3 An injured fighter pilot before **(a)** and after **(b)** plastic surgery by Sir Archibald McIndoe (photo reprinted with permission by the Blond McIndoe Museum)

feasibility and potential for the development of sustainable tissue equivalents through the use of co-culture systems and bioreactors, especially with tissues such as skin.

1.2.2 The Pioneers of the Biomimicking Biomaterials Science

Such in vitro strategies were paralleled by the concept of regenerating damaged tissues through the implantation of biomaterials able to interact with cells and surrounding tissues to encourage their growth. Indeed, the same concept was later applied to the in vitro culturing itself to produce bioreactor beds (see Chapters 2–4). At the beginning, synthetic and natural biomaterials used to replace damaged tissue were modified to support complete tissue regeneration. Work was performed by material scientists such as *Prof. Samuel Huang*, University of Connecticut, USA and *Prof. Luigi Nicolais*, University of Naples, Italy who sought for inspiration from natural materials and structures and tried to mimic them through the synthesis and engineering of new biomaterials (Fig. 1.4a and b). Biodegradable materials with engineered domains (e.g., hydrophilic/hydrophobic) and biodomains (e.g., amino acids, peptides and derivitized proteins) were synthesized to biomimic the biochemistry of living tissues and

Fig 1.4 Prof. Samuel Huang (**a**) and Prof. Luigi Nicolais (**b**)

make these biomaterials responsive to tissue remodeling (Huang et al. 1979, Huang & Leong 1979, Ramos & Huang 2002). Engineering solutions were offered to obtain composite materials with tuned physico-chemical properties and biomimicking architecture (Iannace et al. 1989, Mensitieri et al. 1994).

A decisive step forward towards the synthesis of biomaterials able to support tissue growth both in vitro and in vivo was made following the key findings of *Prof. Jim Anderson*, Case-Western Reserve University, *Prof. Buddy Ratner* and *Prof. Allan Hoffman*, University of Washington Engineered Biomaterials, USA. These scientists introduced a truly multi-disciplinary approach to the field by linking for the first time biological response to the physico-chemical properties of biomaterial surfaces (Anderson & Miller 1984, Marchant et al. 1984, Ratner et al. 1975). In Europe, among many, two distinguished scientists paralleled their American colleagues; *Prof. William Bonfield*, at that time at Queen Mary and Westfield College, London, UK, *Prof. David Williams*, University of Liverpool, UK and *Prof. Larry Hench*, Imperial College London offered new insights into the synthesis of bioactive biomaterials (polymers and ceramics) and into the assessment of their biological performances (Bonfield 1997, Remes & Williams 1992, Hench 2006) (Fig. 1.5).

The need for coordinated approaches towards the development of new biomaterials was almost simultaneously recognized across the world and led to the establishment of the American Society for Biomaterials (www.biomaterials.org) as well as to that of the European (www.esbiomaterials.eu), and Japanese (www.soc.nii.ac.jp/jsbm) societies; all these society were founded by the common effort of material scientists, surgeons, and biologists.

Indeed, the title of the Fifth Annual Biomaterials Symposium entitled "Prosthesis and Tissue – The Interface Problem" that was held at Clemson

Fig 1.5 Prof. Jim Anderson (*right*) with Prof. Luigi Ambrosio (*center*) and Prof. Paolo Netti (*left*) at the Biomaterials Consensus Meeting in Sorrento, September 2005

University, USA in April 1973 reflected the need for a multi-disciplinary approach exploring the interactions taking place at biomaterial/tissue interface. In 1976, the European Society for Biomaterials was founded by 31 scientists from different countries and disciplines and the first president and vice-president to be elected were *Prof. George Winter* and *Prof. Jean Leray*. Since then, the Society has evolved into a very multi-disciplinary Community (Fig. 1.6b).

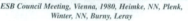
ESB Council Meeting, Vienna, 1980, Heimke, NN, Plenk, Winter, NN, Burny, Leray

Fig 1.6 ESB council meeting in **(a)** 1980 featuring Prof. George Winter (4th from the *left*) and Prof. Jean Leray (*right end*), **(b)** in 2007 featuring from left to right Dr Antonio Merolli, Prof. Liz Tanner, Prof James Kirckpatrick (ex-president of the ESB), Dr. Lucy DiSilvio, Prof Josep Plannell, Prof. Luigi Ambrosio (ESB President elected in 2007), Prof. Mina Kellomaki, Prof. Dirk Gijpma

1.2.3 The Pioneers of the Tissue Engineering Science

In the early 70s, Dr. W. T. Green, Children's Hospital, Boston, USA tried to generate cartilage by culturing chondrocytes on bone fragments and then implanting the construct in mice (Vacanti 2006). Although the experiment was unsuccessful, the author recognized that a better outcome could have been obtained by the use of better substrate materials. Indeed, it was only after few years that scaffolds based on biomaterials such as collagen were introduced for the tissue engineering of skin. However, the first breakthrough in tissue engineering was obtained by Langer (Millenium Technology Prize Winner, www.millenniumprize.fi) and Vacanti who successfully regenerated cartilage by mouse subcutaneous implantation of synthetic biodegradable scaffolds seeded with chondrocytes (Cao et al. 1997).

By the late 1980s, the concept of in vitro tissue growth as a solution to the shortage of available organs for tissue transplantation became a driving force for initiatives within the Department of Commerce (the Advanced Technology Program can claim much of the credit for launching the first generation of tissue engineering companies), the National Science Foundation and the National Institutes of Health in the United States. The notion of off-the-shelf replacement tissues and organs captured the imagination of clinicians, basic scientists, and engineers alike. The discipline of tissue engineering and later regenerative medicine emerged as a focused and directed interdisciplinary effort towards the restoration of functional tissues and organs. The potential commercial applications spawned the creation of the first tissue engineering companies.

The earliest attempts at commercialization of tissue engineering technology involved the generation of skin and cartilage. Organogenesis was established in 1985 followed by Advanced Tissue Sciences in 1987. Both companies targeted off-the-shelf skin substitutes as a viable strategy for commercializing tissue engineering technology. This thought process was driven by the willingness of the Food and Drug Administration to regulate their products as Class 3 medical devices rather than biologics. As these products moved from the laboratory into the clinical setting, issues such as cell source and upscale bioreactors and immunocompatibility were not nearly as problematic as reimbursement mechanics and a changing regulatory landscape. These same issues exist today as the science of regenerative medicine and the business regenerative medicine search for common ground (see Chapter 16, Section 16.4).

The approach that was most commonly used and identified with tissue engineering involved the use of biodegradable materials that were seeded with cells, cultured in a bioreactor for a period of time, and then followed by implantation into the recipient. This approach was based upon the concept that the biomaterial would degrade during the process of cell growth and differentiation and that the living material would become a vascularized living component of host tissue following implantation. In fact, the extent to which such cells survive the implantation process and become part of a neotissue remains an unanswered question.

In a study sponsored by the National Science Foundation, the World Technology Evaluation Center (WTEC) published a report indicating that US investors focused their tissue engineering efforts on cell therapies based on autologous cells (i.e., cells derived from the patient him/herself). The report, published in 2002, is available at www.wtec.org/loyola/te/final/te_final.pdf. There was significant concern regarding the immunosuppressive therapies that would be required with the use of allogeneic cell sources (cells derived from donors). Non-US investors were more focused upon allogeneic cell-based therapies. Both the autologous and allogeneic cell-based approaches have their advantages and disadvantages (Cancedda et al. 2003, Metcalfe and Ferguson 2007). It is generally accepted, however, that cell-seeded scaffolds can indeed induce significant alterations in the default mechanisms of wound healing and facilitate constructive remodeling and de novo functional tissue formation.

According to an article recently written by Prof. Charles Vacanti, the first record of the word tissue engineering can be found in the title of a paper published in 1991 in Surgical Technology International. The article was entitled "Functional Organ Replacement, The New Technology of Tissue Engineering".

The work of Langer and Joseph Vacanti was inspirational for many institutions and new research centers were established with a mission to develop tissue engineering for many biomedical applications. The Pittsburgh Tissue Engineering Initiative by Prof. Peter Johnson, the Cardiovascular Tissue Engineering Centre established by Dr. Robert Nerem at Georgia Tech, and the tissue engineering work performed by Dr. Charles Vacanti at the UMass Medical School in the States were paralleled by the efforts of scientists such as Dr. Julia Polak, Imperial College London in the UK and Prof. Koichi Tanaka and Prof. Minora Ueda in Japan. Indeed, the adolescent years seem to be in the past and we are now approaching tissue engineering's prime years with vigor and anticipation. The scientific community has been recently organized beneath a single umbrella in the form of the Tissue Engineering Regenerative Medicine Society International (TERMIS, www.termis.org). TERMIS unites the societies from the main continents; the Tissue Engineering founded in the States in 1994 by Dr. Charles and Joseph Vacanti, the European Tissue Engineering Society funded in the year 2000 by the joint efforts of German (Prof. Stark and Horch), British (Dr. Julia Polak), Italian (Prof. Cancedda), and Swedish (Prof. Hilborn) scientists, and the Japanese Tissue Engineering Society that was promoted by Dr. Ueda.

1.3 The Birth of Regenerative Medicine

In the late 1990s, the term "regenerative medicine" emerged from the interdisciplinary activities of the tissue engineering community and research in stem cell biology. The scientific boundaries between tissue engineering and regenerative medicine remain somewhat unclear. It is generally accepted, however, that tissue engineering provides a set of tools that can be used to perform regenerative medicine. Regenerative medicine encompasses methods and strategies that extend beyond the traditional tools used within the field of tissue engineering. Both disciplines have the same ultimate objective of providing functional tissue and organs for patients.

1.3.1 Early Clinical Applications

After the famous attempt by Dr. Charles Vacanti to regenerate an ear subcutaneously in a mouse (Cao et al. 1997), other clinical applications followed (Vacanti 2006). In 1991, a patient with a congenital malformation of the sternum was the first to be treated by Drs. Upton and C. Vacanti with polymeric scaffolds seeded with chondrocytes. In 1998, a coral-based scaffold, previously seeded with autologous bone cells (osteoblasts), was used by Drs. Shuffelberg and J. Vacanti to regenerate the distal phalanx of an amputated thumb, the construct being attached to the proximal end of the amputated finger. In 2001, a paper was published by Shin'oka et al. who shown the regeneration of the pulmonary artery.

1.3.2 Commercialization Efforts

The clinical need for effective therapies for chronic skin wounds such as diabetic ulcers and venostasis ulcers was a major driver in the strategic decisions made by early tissue engineering companies. One in seven Medicare dollars is spent for the treatment of diabetes induced disease in the United States and it was felt that a significant portion of this dollar expenditure could find its way into the pockets of successful regenerative medicine companies.

One of the first products to market was the Integra Dermal Regeneration Template (see Chapter 5). This acellular scaffold provided an environment for wound healing using the patient's own cells following in vivo implantation. The absence of a cellular component greatly simplified the commercialization efforts and this product is still widely used today.

Cultured skin substitute products that include the use of cells onto a biodegradable matrix combined with an ex vivo bioreactor culture phase, were extraordinarily difficult to bring to market successfully. Both ATS and Organogenesis were challenged by a changing regulatory landscape, a continuous struggle with reimbursement issues, and a highly complex requirement for the manufacture and shipment of a living product. That said, the current market for the Organogenesis product now exceeds $50 million per year.

Non-skin tissue engineered products have seen variable degrees of success in the clinics. In 1995, Genzyme Tissue Repair began commercialization efforts for tissue engineered cartilage. Autologous chondrocytes were harvested from patients, shipped to Genzyme where they were cultured for 4–8 weeks, and then returned to the surgeon for implantation. Carticell® was approved as a biologic by the Food and Drug Administration in 1997. The considerable infrastructure required to safely culture cells and tissue represented unique challenges that have been largely overcome by Genzyme. In a similar approach, corneal cell sheets have been developed by Okano at Tokyo Women's Hospital. The key advancement in this technology was a clever mechanism for removing cells from the culture dish and maintaining an intact sheet by the use of N-isopropylamide to the surface of the culture dish (Nishida et al. 2004). The commercialization efforts of this technology are still in the early stage.

The commercialization of acellular biologic scaffolds derived from a variety of tissues and from different species have seen considerable commercial success. LifeCell, Inc., incorporated in 1986, provided decellularized human dermis as a biologic scaffold for guided tissue regeneration. After struggling for several years, the company showed its first profits in the early 2000s and now has expanded the applications of its decellularized dermal substrate into the orthopedic and general surgery fields. Of note, the LifeCell product was regulated as a tissue transplant and not as a medical device. Although this classification provided some early advantages, the long-term and more widespread applications have prompted LifeCell to seek reclassification of their product. Porcine derived small intestinal submucosa (SIS) is widely used to promote constructive remodeling of skin, musculotendinous tissues, lower urinary tract, dura mater, and gastrointestinal tissues among others. The SIS scaffold has been extensively investigated with regard to mechanisms of remodeling and host response and has provided a model for commercialization of similar acellular scaffold materials. Several other decellularized tissues including porcine dermis and bovine pericardium have been introduced into the marketplace for orthopedic soft tissue applications. In short, the regulatory burden and commercialization pathway for these acellular biologic scaffolds has proven much simpler than other regenerative medicine approaches. More than one million patients have benefited from the use of these biologic scaffolds in the past 5–6 years alone.

Bone morphogenetic protein (BMP) is the most widely used compound for the induction of bone growth. Medtronic, Inc. developed a spine fusion device utilizing a collagen sponge infused with the BMP peptide. Now in clinical use, this regenerative medicine approach for bony reconstruction in the spine is generating significant income. *Dr. Yilin Cao* has made significant advancements in craniofacial reconstruction in Shanghai, China.

Recently, a regenerative medicine approach for the replacement of the lower urinary tract tissue has been developed by *Dr. Anthony Atala* through the company Tengion, Inc. Autologous cells isolated and cultured ex vivo are seeded onto a synthetic biodegradable material followed by an ex vivo culture phase. The cell seeded neoorgan construct has been successfully implanted into patients for functional remodeling as a neoorgan. This approach is challenging because of the need for extensive and complex ex vivo culture systems, the use of autologous cells, and specialized surgical expertise. However, it represents one of the earliest successful approaches to the regeneration of functional tissue, and because of the high cost of transplantation, it may well represent a key successful business model milestone for the field.

1.4 Challenges

The field of regenerative medicine is in its adolescence and while the efforts of companies like ATS and Organogenesis led to dramatic advances in science, the adolescent years were complicated by failed business models. Significant challenges remain including unanswered questions regarding the use of autologous versus allogenous cell sources, the establishment of reimbursement mechanisms that will allow for sustainable business strategies, regulatory guidelines that not only vary from country to country but are subject to frequent change within each country, and the undeveloped use of stem cells of adult mesenchymal tissues or embryos in regenerative medicine applications (see Chapter 16).

In spite of these obstacles, optimism for the potential solutions that regenerative medicine offers remains very high. Regenerative medicine efforts by large medical device companies are becoming evident through the introduction of new products and the field is now moving towards the regeneration and repair of increasingly complex tissues and whole organ replacement. The technology to develop effective therapies for previously untreatable diseases either exists or is being developed. These technological advancements must be balanced by a sustainable business strategy if tissue engineering and regenerative medicine are to become therapies widely available worldwide.

Questions

1. Highlight at least two main technological approaches characterizing the early pioneering attempts to ensure tissue viability in vitro and in clinical applications.
2. Critically evaluate the early tissue regeneration approaches pioneered by Archibald McIndoe and compare it with current surgical procedures as described in Chapter 5.
3. When and where was the term "tissue engineering" mentioned for the first time?
4. Which tissue was regenerated by a tissue engineering approach and implanted in clinics for the first time? Who did it and when?
5. Which type of biodegradable biomaterials was used for the first clinical application of a tissue engineering construct? Do the characteristics of this biomaterial reflect any biomimetic approach? If yes, describe the main biomimetic feature.

6. List the main Societies for biomaterials and tissue engineering. Can you identify their main mission?
7. What was the main scientific and technological progress made in biomaterial science in the 70s and 80s?
8. Are tissue engineering products classified as medical devices or biologics?
9. Provide examples of acellular and cellular products currently available on the markets. In the light of information provided in the following chapters critically discuss their commercial and clinical potential.
10. Provide a brief and critical overview of the commercialization strategies so far adopted by companies worldwide.

References

Anderson JM, Miller KM (1984) Biomaterial biocompatibility and the macrophage. Biomaterials 5:5–10

Bell E, Ehrlich HP, Buttle DJ, Nakatsuji T (1981) Living tissue formed in vitro and accepted as skin-equivalent tissue of full thickness. Science 211:1052–1054

Bennett JP (1988) A history of the Queen Victoria Hospital, East Grinstead. Br J Plast Surg 41:422–440

Bonfield W (1997) Tailor-making analogue biomaterials for skeletal implants. J Pathol 181:A59 Suppl S

Cancedda R, Dozin B, Giannoni P, Quarto R (2003) Tissue engineering and cell therapy of cartilage and bone. Matrix Biol 22:81–91

Cao Y, Vacanti JP, Paige KT, Upton J, Vacanti CA (1997) Transplantation of chondrocytes utilizing a polymer-cell construct to produce tissue-engineered cartilage in the shape of a human ear. Plastic Reconstr Surg 100:297–302

Hench LL (2006) The story of bioglass®. J Mater Sci: Mater Med 17:967–978

Huang SJ, Bansleben DA, Knox JR (1979) Biodegradable polymers – Chymotrypsin degradation of a low-molecular weight poly(ester-urea) containing phenylalanine. J Appl Polymer Sci 23:429–437

Huang SJ, Leong KW (1979) Biodegradable polymers – Polymers derived from gelatin and lysine esters. Abstracts of papers of the Am Chem Soc, 135

Iannace S, Nicolais L, Ambrosio G (1989) USA Italy workshop on polymers for biomedical applications – Capri, Italy. Biomaterials 10:640–641

Marchant RE, Anderson JM, Phua K, Hiltner A (1984) In vivo biocompatibility studies. 2. Biomer – Preliminary cell-adhesion and surface characterization studies. J Biomed Mater Res 18:309–315

Mensitieri M, Ambrosio L, Nicolais L, Balzano L, Lepore D (1994) The rheological behavior of animal vitreus and its comparison with vitreal substitutes. J Mater Sci: Mater Med 5:743–747

Metcalfe AD, Ferguson MWJ (2007) Tissue engineering of replacement skin: the crossroad of biomaterials, wound healing, embryonic development, stem cells and regeneration. J R Soc. Interface 4:413–437

Nishida K, Yamato M, Hayashida Y, Watanabe K, Maeda N, Watanabe H, Yamamoto K, Nagai S, Kikuchi A, Tano Y, Okano T (2004) Functional bioengineered corneal epithelial sheet grafts from corneal stem cells expanded ex vivo on a temperature-responsive cell culture surface. Transplantation 77:379–385

Nobile MR, Acierno D, Incarnato L, Amendola E, Nicolais L, Carfagna C (1990) Improvement of the processability of advanced polymers. J Appl Polym Sci 41:2723–2737

Ramos M, Huang SJ (2002) Functional hydrophilic-hydrophobic hydrogels derived from condensation of polycaprolactone diol and poly(ethylene glycol) with itaconic anhydride. In Functional Condensation Polymers, Carraher CE and Swift GG eds, Kluwer Academic/Plenum Publishers, New York, 185–198

Ratner BD, Horbett T, Hoffman AS, Hauschka SD (1975) Cell adhesion to polymeric materials – Implications with respect to biocompatibility. J Biomed Mater Res 9:407–422

Reinwald J, Green H (1975) Serial cultivation of strains of human epidermal keratinocytes: the formation of keratinizing colonies from single cells. Cell 6:331–344

Remes A, Williams DF (1992) Immune response in biocompatibility. Biomaterials 13:731–743

Vacanti CA (2006) History of tissue engineering and a glimpse into its future. Tissue Eng 12:1137–1142

Chapter 2
Soft Tissues Characteristics and Strategies for Their Replacement and Regeneration

Maurizio Ventre, Paolo A. Netti, Francesco Urciuolo, and Luigi Ambrosio

Contents

2.1   Composition-Structure-Properties of Soft Tissues: Introduction 16
2.2   Tendons and Ligaments 17
       2.2.1   Composition and Structure..................................... 17
       2.2.2   Mechanical Properties... 18
2.3   Skin.. 20
       2.3.1   Composition and Structure.................................... 20
       2.3.2   Mechanical Properties... 22
2.4   Arteries 25
       2.4.1   Composition and Structure.................................... 25
       2.4.2   Mechanical Properties... 27
2.5   Cartilage 28
       2.5.1   Composition and Structure.................................... 28
       2.5.2   Mechanical Properties... 31
2.6   Soft Tissue Replacements..................................... 34
       2.6.1   Introduction 34
2.7   Ligament Prostheses.. 34
2.8   Skin Replacements 38
       2.8.1   Skin Substitutes for Wound Cover 40
       2.8.2   Skin Substitutes for Wound Closure............................. 41
2.9   Vascular Grafts.. 42
2.10   Cartilage Replacement 45
       2.10.1 Joint Resurfacing 45
       2.10.2 Biological Autograft 45
       2.10.3 Total Joint Replacement 46
       2.10.4 Tissue Engineered Constructs 46
Questions/Exercises.. 49
References 50

P.A. Netti (✉)
Interdepartmental Centre on Biomaterials, University of Naples Federico II, Italy
e-mail: nettipa@unina.it

M. Santin (ed.), *Strategies in Regenerative Medicine*,
DOI 10.1007/978-0-387-74660-9_2, © Springer Science+Business Media, LLC 2009

2.1 Composition-Structure-Properties of Soft Tissues: Introduction

The extracellular matrix (ECM) of connective tissues is usually represented as a composite material, in which insoluble fibers are embedded in an aqueous matrix containing a wide variety of soluble macromolecules. The main characteristic of the ECM is to provide the tissue with specific physiological and biomechanical properties. These macroscopic properties are strictly related to the tissue composition and spatial arrangement of its microconstituents.

The most abundant proteins of the ECM of vertebrate body belong to the family of collagens. In particular, fibril forming collagens have been widely examined due to their property to form high ordered structure. Alongside collagens, elastin and proteoglycans play important mechanical and biological functions.

In collagen rich soft biological tissues exists a correlation between their inner microstructure and their macroscopic mechanical properties. The interest in the mechanical and rheological behavior of connective tissues involves several disciplines of biological and clinical research. Examples can be found in the field of plastic (Thacker et al. 1977), orthopedic (Harris et al. 2000, Woo et al. 2006) and cardiovascular surgery (van Andel et al. 2003), transplantations (Khan et al. 2005), and wound healing process (Hierner et al. 2005, http://www.uweb.engr.washington.edu/research/tutorials/mechproperties.html).

A general approach for the quantitative description of connective tissues mechanical properties is difficult to achieve. Connectives possess very different microstructures, despite having quite similar composition. We have chosen to describe four models not only because they can be considered as a representative population of soft tissues, but also because each of them experiences in vivo very different tensional states. The discussion aims at providing a picture of how composition and microstructure can affect the macroscopic mechanical response of soft tissues. The four tissues differ from one another by compositional and structural features. The soft tissues are presented at increasing levels of structural complexity. One of the simplest models, in terms of spatial configuration of its microconstituents, is represented by tendons and ligaments. Tendons in particular, are mainly subjected to uniaxial tension and are constituted by a close packed arrangement of collagen fibers oriented parallel to the tendon axis, which is the direction of the applied load. Ligaments possess a similar structure, but their collagen fibers are arranged in a less ordered fashion than in tendons. Skin structure is more complex. It is an entangled mass of fibrous elements, although the predominant fiber directions are parallel to the epidermal surface. Several layers can be discerned within skin structure. Each layer is characterized by specific features. Dermal components are arranged in a way that causes the skin to resist to biaxial stresses. Arteries are multilayered structures as well. They display axial symmetry. The microcomponents confer the tissue with elasticity to withstand pulsating inflation and bending. Cartilage is a highly hydrated three-dimensional structure, whose mechanical properties are governed by osmotic phenomena. It is rich in charged macromolecules which render the cartilage particularly resistant to compressive stresses.

2.2 Tendons and Ligaments

2.2.1 Composition and Structure

Tendons and ligaments are dense connective tissues. They act as load transferring agents between the anatomical elements that they connect. Usually tendons are referred to as connective structures which link muscle and bones together. They are able to transmit the forces generated by muscle contraction to bones thus allowing joint motion. Ligaments, in turn, connect the articular extremity of bones. Their main function is to guide the correct movement of joints. Tendons and ligaments are subjected to severe stress during physical activity. Their microarchitecture is assembled in a rope-like structure, which provides the tissues with high stiffness and flexibility. Collagen, which is the basic building block of both tendons and ligaments, is assembled in supramolecular structures with growing level of complexity during morphogenetic and remodeling events. The structure is composed of collagen molecules, fibrils, fibers, bundles, fascicles, and tendon units.

All the above mentioned hierarchical units develop along the tendon axis. Under polarized light, fibers, however, are not straight. They display a characteristic crimped pattern.

The most abundant constituents of tendons are the proteins which belong to the collagen superfamily, proteoglycans, elastin, fibroblasts, and water. Collagen is the main component. It constitutes about the 60–70% of the total dry weight. In particular the 95% of the collagen molecules belong to the type I family. The remaining 5% consists of type III and type V collagen. The expression pattern of type III collagen is not constant during tissue development. It has been shown that type III is more abundant in the fetal/neonatal stages and the molecule is bound to fibril surface during these stages (Birk and Mayne 1997). Collagen type III concentration decreases during development. In adult tissues, the presence of the protein is limited to the sheets, which wraps the fascicles around (endotenon) (Birk and Mayne 1997). Numerous works suggest that type III collagen molecules are involved in fibril assembly and regulate fibril diameter (Romanic et al. 1991). Type V is also found in tendons, in particular into the core of the heterotypic type I/type III fibrils. It has been demonstrated that type V collagen profoundly affects fibril diameter in corneal stroma (Birk 2001). However, its effect in tendon fibrils has not yet been unraveled. Other collagen types like type XII and XIV (which belong to the fibril associated collagens with interrupted triple helices family FACIT have been localized on fibril surface. Type XII surface is believed to mediate collagen-decorin and collagen-fibromodulin interactions (Font et al. 1996). Type XIV seems to have a role in regulation of fibrillogenesis (Young et al. 2002). Other collagen molecules including II, IX, and XI are present in trace quantities near the tendon-bone insertion. This transition zone has the function of mediating the abrupt mismatch of the mechanical properties (from the compliant tendon to the stiff bone). The transition zone is morphologically distinct from tendon and bone. It is rather a fibrocartilagineous tissue which becomes mineralized near the bone insertion.

Collagen molecules undergo to extensive enzymatic crosslink reactions during development (Bailey et al. 1998). Adult tendons are highly crosslinked and display increased stiffness, reduced strain at break, and high tensile strength.

A small amount of elastin is present in tendons. Elastin comprises about the 2% of the tendon mass. Elastic fibers, which are made of microfibrils and amorphous elastin, form a three dimensional network of thin fibers which interpenetrate the array of collagen fibers.

Beside collagen and elastin, protoeglycans are also present in small quantities. It has been shown that the proteoglycan content is strictly related to the loading condition. Tendons are usually subjected to uniaxial tension. However, this could not be the case in few anatomical sites. As an example, in bovine flexor tendon, which is mainly subjected to uniaxial tension, the proteoglycan content is very low ($<0.5\%$ of the dry weight) (Koob and Vogel 1987). In the compression portion of the flexor digitorium profundus bovine tendon, the proteoglycan is significantly higher ($\approx 3.5\%$) (Vogel and Koob 1989). Among proteoglycans, the small leucine rich proteoglycan decorin is one of the most abundant. It is located on the surface of collagen fibrils. Decorin takes part in regulating fibril diameter during morphogenesis and profoundly affects the viscoelastic behavior of the tissue (Watanabe et al. 2005, Pins et al. 1997).

There is no extracellular matrix component that can discriminate tendons from ligaments. Only the relative amount of ECM proteins together with their spatial assembly determines the biological and physical properties of tendons/ligaments. The main structural element of ligaments is collagen whose spatial assembly follows the same well defined hierarchical organization as displayed in tendons. However, collagen fibers in ligaments are not as polarized as in tendons. Elastin is also an important component of ligaments, but its content strongly depends on the anatomical site: in human ACL, the elastin comprises about the 5% of the total dry weight, while in bovine ligamentum nuchae, it is the predominant component. These differences in terms of both composition and structure discriminate the gross mechanical behavior of ligaments and tendons.

2.2.2 Mechanical Properties

Tendons and ligaments exert very important biomechanical functions since they are involved in joint motion and joint stability. The knowledge of their mechanical properties is of paramount importance in several disciplines, like orthopedics, transplantations, and tissue engineering. In vivo tests have in principle the advantage of taking into account loading conditions which could match those experienced during physical activity. However, there are several drawbacks connected to this kind of experiments. The quantification of the loads exerted during motion could not be straightforward as well as the determination of tissue deformation. In vitro tests, on the other hand, can be controlled and monitored more easily. Uniaxial tension is a useful test which provides basic information on the mechanical response. However, it is not sufficient to capture the overall anisotropic behavior of tendons and ligaments, which is a feature common to almost all soft connective tissues. Among all tendons and ligaments certainly Achilles' Tendon and Anterior Cruciate Ligament (ACL) are the most studied. Achilles' Tendon is often chosen as model of connective tissue to study owing to its very regular structure, composed of parallel collagen bundles which lie along tendon longitudinal axis. The mechanical properties of ACL are widely documented because of its relevance in clinics. ACL is indeed one of the most injured ligaments of the body.

As previously stated, the composition and structural assembly of ligaments and tendons depend on the anatomical site. This dependence produces differences in the macroscopic mechanical response of these tissues. However, from a simple uniaxial tension experiment it is possible to find general aspects which are shared by the majority of tendons and ligaments. The stress–strain diagram provided by a uniaxial tension test of tendons and ligaments is reported in Fig. 2.1.

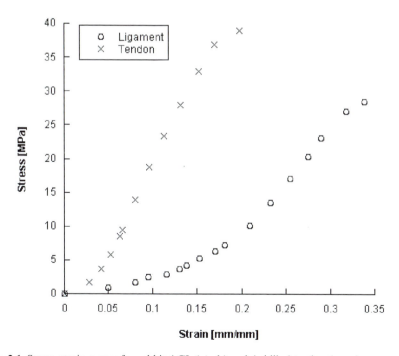

Fig. 2.1 Stress–strain curves for rabbit ACL (*circle*) and Achilles' tendon (*cross*)

Both curves display the characteristic J shape, which is a feature that is common to numerous collagen rich biological tissues. The first part of the diagram, which is usually referred to as the toe region involves small load values. It is generally accepted that the straightening of the crimped pattern of the collagen fibers occurs during the toe region. Once straight, the fibrils begin to stretch. Larger loads are required for further elongation. Of course, not all the fibers are crimped to the same extent. Different fiber groups become straight at different elongations: that is, the higher is the elongation, the more fibers are uncrimped and, therefore, are able to withstand loads. This process is also known as fiber recruitment and explains the upward concavity of the stress–strain plot reported in Fig. 2.1. The wider toe region displayed by ACL specimens reveals cues on the underpinning microstructure. ACL have a higher content of elastic tissue which probably confer the collagen fiber a more marked crimped pattern. Moreover, the collagen fibers of the ligament are assembled in a less ordered arrangement rather than tendons. Ligaments indeed are usually subjected to more complex stress states as tendons are (Netti et al. 1996). Stretching of aligned collagen fibers causes an almost linear stress–strain response. Microscopic failures of the collagenous network have been shown to occur in the linear region which can eventually lead to rupture for sufficiently high levels of stress.

Tendons and ligaments, like most biological tissues, exhibit time-dependent mechanical properties. The molecular basis of tendons and ligaments viscoelasticity is largely unknown. The high content of a hydrated matrix of proteoglycans is of course essential for the viscoelastic effect, as viscous shear stresses may arise from relative gliding of collagen fibers. The ground substance acts as an elastic matter which supports the fibrous elements of tendons and ligaments (Lanir 1978), during small strain rate experiments. According to the experiments of Jenkins and Little performed on bovine ligamentum nuchae (Jenkins and Little 1974), the stress relaxation phenomenon almost vanishes after removal of collagenous components. These data suggest that the main source of viscoelasticity arises from collagen bundles themselves and/or from interaction between collagen and viscous matrix.

2.3 Skin

2.3.1 Composition and Structure

An overview of skin is given in Chapter 5. Here, basic concepts are recalled and linked to the tissue mechanical properties. Skin is a multilayered structure which covers mammalian bodies. It could be considered as the largest organ: human skin has a total surface of approximately 2 m^2 and constitutes about 15–20% of the overall body weight. Three clearly visible strata are present in a skin cross section. The superficial epidermis, the dermis, and finally the hypodermis or subcutis (Fig. 2.2).

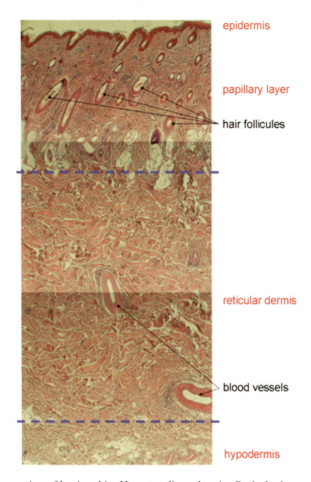

epidermis

papillary layer

hair follicules

reticular dermis

blood vessels

hypodermis

Fig. 2.2 Cross section of bovine skin. Hematoxylin and eosin. Optical micrograph

The epidermis is densely populated by keratinocytes. Closer inspection of the epidermal layer reveals the existence of four well defined cell layers: stratum basale, stratum spinosum, stratum granulosum, and stratum corneum. The deepest layer is the statum basale, in which cell divisions occur. Keratinocytes move from the stratum basale towards the surface. As the cells displace outward, they change morphology and become polyhedral. In the stratum granulosum, the degradation of the mitochondria and nuclei begins and the intracellular space is made up of the hydrophobic protein keratin. In the outmost layer, almost all cells are dead and form thin flat hexagonal squames. Besides keratinocytes, melano-cytes (which are pigment forming cells), and Langerhans cells (which are thought to have important immunological functions) have been found in the epidermis.

The dermis is a dense fibro-elastic connective tissue. It is mainly composed of collagen, arranged in fiber bundles, elastin, a viscous matrix named ground substance, and fat. The dermis is less cellularized than the epidermis. It is populated by fibroblasts which are responsible for the production and assembly of the dermal ECM. Collagen comprises about the 70–80% of the dermal fat free dry weight. It is assembled in the form of fibers which are in turn packed in collagen bundles. These bundles are textured in an irregular three dimensional network and they run almost parallel to the epithelial surface. Elastin is another key component of the dermis. It comprises 2–4% fat free dry weight and forms a network which wraps collagen bundles around. The ground substance is composed of water, proteoglycans, which can be bound to collagen fibril surface or can be present in a soluble form, and other soluble proteins and salts. The dermis can be roughly divided in two overlapped layers: the papillary layer and the reticular dermis. The papillary layer is the outmost layer which contains the sweat glands and hair follicles. Reticular dermis is predominantly composed of an entangled mass of collagen bundles. The epidermis is attached to the dermis through the basal lamina, which provides a physical barrier for cells and large macromolecules. Moreover, it determines a strong bonding between the layers owing to the presence of macromolecular attachments.

The hypodermis is the innermost part of skin. It is a fibrous-fatty layer, loosely connected to reticular dermis, whose thickness and composition depend on several factors suck as age, sex, race, endocrine status, nutrition. It acts as an insulating layer and it is present almost all over the body.

The most obvious function of skin is protection. Epidermis prevents the penetration of undesirable matter in the body, acts as a barrier against microbial attack, and prevents loss of fluids. The dermis is flexible, allowing joint motion but it is also robust since it has to protect the inner organs against external mechanical shocks. Besides these functions, the skin plays an important role in detecting external stimuli like pressure, heat, cold, and so forth. Moreover, skin contributes to the synthesis of D vitamin and body thermal regulation, owing to the presence of sweat glands and extensive blood supply.

In the next section, attention will be focused on the mechanical properties of skin and how these properties are affected by structure and composition.

2.3.2 Mechanical Properties

Mechanical behavior of skin is strictly related to its inner structure and to the relative amount of its components. Like other collagen rich connective tissues, mechanical properties of skin are non linear and anisotropic. Such properties vary from site to site and are age dependent.

Mechanical tests of skin have been performed both in vivo and in vitro. In vivo test have the great advantage to simulate stresses which can occur in physiological conditions. However, the influence of surrounding tissue on the specimen is unknown and, in general, in vivo tests are difficult to perform, that is, specimen clamping, difficulties in determining the imposed motion, etc.

Usually the preferred in vivo tests include traction, torsion, indentation, and suction. In vitro tests of course have the great advantage to separate the contribution of the individual skin layer on the overall mechanical response. Moreover, in vitro tests allow to study the gross mechanical behavior at large deformations. On the contrary, in vitro tests do not replicate the actual stresses which are experienced by the tissue in physiological conditions.

A typical stress–strain plot of skin samples, in two mutually orthogonal directions, is depicted in Fig. 2.3.

The diagrams can be divided in three regions. The first part is characterized by low stiffness. The distortion of the elastic and interfibrillar networks occurs in this region, which is usually referred to as toe region. It is believed that in this deformation range, the collagen network unravels and fibers align towards the direction of the applied load (mechanism also known as fiber recruitment). The third region is the linear region where the stress is mostly borne by collagen fibers and predominant mechanism of deformation is fiber stretch. Eventually the distorted collagen network brakes down leading to sample failure. The anisotropic mechanical behavior arises from anisotropies of the microstructure. In particular, anisotropy has to be ascribed to collagen fibers, which despite of being textured in a three dimensional network possess a preferential direction of run. The non-collagenous matrix exerts almost an isotropic contribution. The

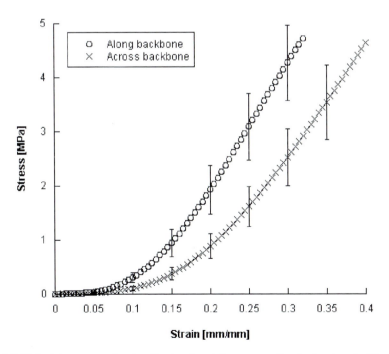

Fig. 2.3 Uniaxial stress–strain plot. Bovine (partially processed) reticular dermis. The tests have been carried out up to approximately 5 MPa

set of (preferential) collagen fibers direction constitutes the so called Langer's lines. Skin extensibility is minimum along the direction of a Langer's line and it is in maximum in the orthogonal direction. In other words, if a sample is uniaxially stretched along a Langer's line, its fibers are almost all aligned in the direction of deformation, and the resulting stress–strain response is characterized by a short toe region (i.e., lower extent of fiber recruitment). On the other hand, if the same specimen is stretched in the orthogonal direction, its fibers undergo extensive recruitment thus causing a larger delay in the fiber response to the applied load. This of course is evident in the stress–stain plot with a larger toe region.

No significant differences are observed in the slope of the linear region, which indicates that in this deformation range, collagen fibers are the actual load bearing elements.

Elastin network affects mechanical response at small deformations. Oxlund et al. (1988) demonstrated that the slope of the first stage of the stress–strain plot depends on the elastin content. Moreover, elastin holds the collagen network in an entangled and crimped form. Again, in the hypothesis of small deformation, (< 5%), elastin network drives the elastic recoil of the distorted collagen network.

Besides these structural heterogeneities which occur in the plane of the skin, other important heterogeneities occur through the thickness. As mentioned earlier, skin is indeed a multilayered structure. The epidermis little affects gross mechanical response in a full thickness skin. Papillary and reticular dermis possess different structure, being the former composed of close packed thin fibers and the latter is an entangled mass of larger bundles. It has been shown that, owing to its close packed structure, the papillary layer is stiffer than the reticular dermis (Ventre 2006). Keane et al. (1997) shown that the elastin network is assembled in different structures according to the layer where it resides. In the papillary dermis, elastin is present in the form of a network of thin microfibrils made of fibrillin. Larger elastic trunks rich of amorphous elastin have been found in the reticular dermis. This peculiar assembly is thought to have profound effect on the elastic recoil of dermis.

The macromolecules of the ground substance do not seem to affect ultimate tensile properties of dermis. However, Kronik and Sacks (1994) demonstrated that proteoglycans and interfibrillar proteins strongly influence the deformation mechanisms at fibrillar level. In particular, proteoglycan reticulum wraps collagen fibrils around, thus influencing relative gliding of individual fibrils. This profoundly affects the dynamical-mechanical response of the dermis. In fact the GAG-collagen interactions cause the fibril to move in concert.

Skin, like other collagen-rich soft tissues, exhibits preconditioning when subjected to cyclic deformations (Fig. 2.4).

Preconditioning has been recognized as a viscoelastic property and it basically is a gradual adaptation of the tissue to the applied load (Eshel and Lanir 2001). During cyclic (finite) deformations, irreversible rearrangements of the tissue microstructure occur. However, preconditioning does not involve

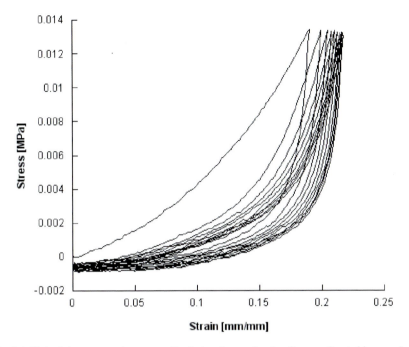

Fig. 2.4 Uniaxial stress–strain curves. Cyclic loading and unloading on Goat skin sample

configurational changes of the ECM alone. At large strain, the inner structure is severely damaged and the extent of damage presumably depends on the maximum stress previously experienced by the specimen (Gregersen et al. 1998), which is a phenomenon known in rubber elasticity as strain softening.

2.4 Arteries

2.4.1 Composition and Structure

Arteries are vessels of the circulatory system and are committed to transport the blood away from the heart (see Chapters 6 and 7). The main artery, the dorsal aorta, originates from the left ventricle and then divides and branches out into many smaller arteries so that each region of the body is provided with its own blood supply. Of course, larger arteries are located close to the heart, while they become thinner at the periphery. The former are usually called elastic and the latter muscular. However, it is possible to find arteries which display intermediate characteristics. Besides branched arteries, like bifurcation and pathological conditions, arteries usually display an axial symmetric, cylindrical shape, whose cross section is composed of three distinct layers: tunica intima, tunica media, and tunica adventitia. A clear study on the histological features and morphological characteristics of arteries has been made available by the College of Medicine of the University of Illinois Urbana Champaign (https://histo.life.uiuc.edu/histo/lab/lab3/lab3.htm).

The intima is the innermost part. It is composed of a single endothelial cell layer, which lines the media surface. Endothelial cells directly face the blood stream and, as a consequence of that, they are the primary responsible for the transduction of the mechanical stress which occurs at the lumen-blood interface (Chien 2006). The most common disease of arterial wall, the atherosclerosis, affects primarily the tunica intima and causes a significant alteration of the vessel mechanical properties (Nagaraj et al. 2005).

The tunica media is the middle layer of the artery. It is composed of collagen fibers, elastin, and smooth muscle cells. At closer inspection, the media is in turn composed of a series of concentrically packed elastic lamellae. The structural lamellar unit consists of an elastic tissue between interlamellar concentric layers of smooth muscle cells and fibrous tissue. The elastic core is composed of a layer rich of elastin fibers, associated with the adjacent interlamina of smooth muscle cell embedded in a collagen reticulum. Elastic fibers get across the interlamellar space and connect two adjacent elastic laminae. Wolinsky and Glagov (1967) showed that the characteristic dimension (thickness) of the lamellar unit remains constant at approximately 15 μm in almost all mammalian arteries. Thus the number of such lamellae decreases in small peripheral arteries, that is, the muscular arteries. In each lamina the elastic fibers, collagen bundles are arranged in a helical pattern. The smooth muscle cells follow this pattern and have their long axis parallel to collagen fibers. These helices have a small pitch so that the fibrous elements are almost circumferentially oriented. This peculiar structure provide the media with high strength in the circumferential direction yet allowing flexibility in the axial direction.

The outermost layer is the tunica adventitia. This layer is populated by fibroblasts which produce a collagen rich extracellular matrix. The thickness of the adventitia varies along the arterial tree. Also in this layer collagen fibers follow a helical pattern.

In order to better understand the mechanical properties of the artery, it is important to have a clear picture on which loads and deformations act on the arterial wall. Arteries, in physiological conditions, are mainly subjected to a pulsating inflation. The blood pressure can be affected by several factors: ventricular stroke volume, mechanical, and hydraulic properties of the arterial tree, physical characteristics of the blood fluid. The arterial pulse waveform itself can be decomposed in a steady component (mean arterial pressure) and a pulse pressure. Ideally, blood flow should be steady in order to optimize the energy requirements of the heart. A pulsating flow indeed requires the blood to be periodically accelerated and decelerated. This of course is costly, from an energetic point of view, since inertial forces become non-negligible. The elasticity of the vessel acts as a hydraulic rectifier which converts an intermittent output into a steady flow. This pulse inverting effect is caused by the periodic passive expansion and elastic contraction of the arterial wall, which balances the heart strokes: each time the heart contracts (systole) the elastic vessels expand and then recoil elastically (during the diastole) thus sustaining the flow in the peripheral arteries. In these latter, in particular, the blood flow is almost continuous because of the elastic reservoir effect of the upstream arterial tree (Shadwick 1999).

2.4.2 Mechanical Properties

Mechanical properties of arterial walls can be determined both in vivo and in vitro. In vivo tests of course provide a clear scenario of the mechanical response of the tissue under "on duty" physiological conditions. However, the effect of surrounding tissues, hormones, and nerve control cannot be evaluated accurately. Moreover, the utility of some of these techniques is limited by an intrinsic invasive nature, which usually requires the introduction of a device into the vessel lumen, making them risky for clinical studies. A complete mechanical characterization of arterial wall should account for a combination of inflation, extension, and twist, which can be performed and controlled more easily in vitro tests (Holzapfel et al. 2000).

In vitro tests usually involve inflation of straight arteries and uniaxial stretching performed on arterial stripes (Wolinsky and Glagov 1964, Holzapfel et al. 2005). The test stripes can be cut both longitudinally and circumferentially. Uniaxial tests on arterial patches give basic information on the gross mechanical behavior of arterial wall, but certainly they are not sufficient to provide a full picture of the anisotropic response. Thus, the potency of uniaxial tests still remains questionable.

It is known that the load free configuration of an artery is not a stress free configuration, that is, the excised arterial segment shortens after removal from their anatomical site (Rachev and Greenwald 2003). The in vivo pre-stretch in both longitudinal and circumferential directions must, therefore, be restored in in vitro tests, in order to reproduce a stress state which more closely mimics physiological conditions.

Besides the heterogeneity which arises from the multilayered structure of the arterial wall, the composition varies along the arterial tree (Cox 1978). Thus the mechanical response of arteries is position dependent. However, it is possible to sketch the general mechanical characteristic out. The typical stress–strain relationship in uniaxial test is non linear with a marked stiffening effect at large deformations which is a feature common to almost all biological soft tissues. The stiffening effect arises from the recruitment of collagen fibers.

Collagen fibers are mostly present in the tunica adventitia. Even though the adventitia is a three dimensional network, collagen fibers predominantly run parallel to the outer surface and are wound in a helical pattern. According to Driessen et al. (2005), collagen fibers form overlapped specular helices with preferential direction of winding oriented at approximately 50° with respect to the longitudinal axis. The spatial arrangement of collagen fibers within the medial layer is different form the one in the adventitia. Collagen bundles are packed in the interlamellar space (i.e., in the space between two adjacent elastic laminae) following a small pitch helical pattern. A network composed of elastic fibers goes across the interlamellar space and wraps collagen fibers around.

Tunica media and tunica adventitia are very different tissues in terms of both structure and composition. It is presumable that their mechanical response is different as well. A recent study of Holzapfel et al. (2005) points out the layer-specific mechanical properties of coronary arteries. They have shown that the response of the intima and the adventitia to longitudinal (uniaxial) loading is stiffer than the corresponding response in the circumferential direction. The media, in turn, exhibits an opposite behavior. These data are in agreement with the histological observations of Wolinsky and Glagov (1964) and suggest that the elastin rich medial layer governs the tissue mechanical response in the physiological pressure range, and confers the artery with the ability to recoil from the superimposed pulsating deformation. At larger distensions (i.e., at internal pressure above the systolic pressure), the thickness of the medial layer is considerably reduced. The adventitial collagen fibers rearrange circumferentially acting like a jacket reinforcement which prevents vessel rupture.

Arteries subjected to cyclic loading or inflations, display strain softening. The extent of strain softening is more pronounced during the first cycles, until the tissue reaches a reproducible cyclic response. The ability of an artery to exhibit viscoelastic behavior strictly depends on its composition. Thus arteries of the elastic type display almost immediately a nearly repeatable cyclic behavior, while muscular vessels display pronounced histeresis.

2.5 Cartilage

2.5.1 Composition and Structure

Cartilage is dense connective tissue (see Chapters 12 and 15). It is composed of a network of collagen and elastic fibers which are embedded in a firm gel-like matrix known as ground substance. Cellular component is constituted by chondrocytes that govern matrix turnover and remodeling. Cartilage is avascular, nutrient supply occurs by diffusion and convection through the matrix due to inflow and outflow of intersitial fluid during physiological loading. No nerves are observed. Cartilage does not heal readily after damage. It accomplishes several functions, like creating a template upon which bone deposition can occur; it covers the surface of joints, allowing bones to slide over one another, thus reducing friction and preventing damage; it also acts as a shock absorber (Grodzinsky et al. 2000). Cartilage is found in many places in the body including the joints, the rib cage, the ear, the nose, the bronchial tubes, and between intervertebral disc.

Cartilage can exist in the form of elastic cartilage, fibrocartilage, or hyaline cartilage, depending on the anatomical site and its different physical properties.

Hyaline cartilage is the most abundant type of cartilage. The name hyaline is derived from the Greek word *hyalos* meaning glass. This refers to the translucent matrix or ground substance. It is composed predominantly of type II collagen. Hyaline cartilage lines bones in joints (articular cartilage, or commonly, gristle) and is also present inside bones, serving as a center of ossification or bone growth. In addition, hyaline cartilage forms most of the skeleton during embryo development.

Elastic cartilage, besides collagen and ground substance, contains a large amount of elastin, which confers the tissue a distinctive yellow color. Elastic cartilage makes up the springy part of the outer ear and the epiglottis.

Fibrocartilge (also known as white cartilage) is a specialized type of cartilage found in areas requiring tough support or great tensile strength such as between intervertebral discs, the pubic, and other symphyses and in the transition region which links tendons/ligament to the bone. The fibrocartilage found in intervertebral disks contains more collagen compared to hyaline. In addition to the type II collagen found in hyaline and elastic cartilage, fibrocartilage contains type I collagen that assembles in the form of fiber bundles. When the hyaline cartilage at the end of long bones such as the femur is damaged, it is often replaced with fibrocartilage, which does not provide a weight bearing function.

The most studied kind of cartilage is Articular Cartilage (AC). It is a specialized connective tissue which covers the bony ends of synovial joints and it functions as a load bearing structure. AC is composed of an aqueous electrolyte solution (80%) and an organic matrix of collagen, proteoglycans, glycoproteins, lipids, and chondrocytes (Table 2.1).

It is believed that AC can protect bone from impact loads. The high content of proteoglycans (25% of dry weight) provides the tissue with a large osmotic

Table 2.1 Composition of AC. Data from Grodzinsky (1983) and Poole et al. (2001)

ECM	AC
Fibrous Protein (dry. wt.)	
Collagen	65–75 %
Elastin	–
Ground Substance (dry. wt.)	
Glycosamminoglycans (GAG)	15 % (CS, KS, HA)*
Proteoglycans	20 %
Glycoproteins and Others	–
Mineral	–
Water and Electrolytes (wet wt.)	65–80 %
Cells	Chondrocytes

*CS = chondroitin sulfates, KS = keratin sulfate, HA = hyaluronic acid.

swelling pressure making it able to withstand peak forces that can be several times higher than the body weight.

Proteoglycans are assemblies of disaccharidic chains chemically bound to a core protein.

The dominant disaccharidic GAG unit includes chondroitin 4- and 6- sulfate, keratan sulfate, dermatan sulfate, and hylauronic acid, which, however, is not covalently linked to any core protein. Representation of the assembly of disaccharidic units is shown in Fig. 2.5. Each repeating unit contains approximately 1–4 anionic charged groups, which may be a carboxyl group (hyaluronate, HA), a sulfate group (keratan sulfate, KS), or both of them (chondroitin sulfate, CS; dermatan sulfate; DS, heparin sulfate, HS). In articular cartilage, a typical subunit of proteoglycan is about 300–400 nm long, 40 nm in diameter, and has a molecular weight of 2×10^6. The subunit of proteoglycans aggregates with a long chain of hyaluronic acid (0.5×10^6 M.W.) and forms complexes of molecular weight of about 200–400 $\times 10^6$.

The aggregates of proteoglycans are linked to the collagen network. In physiological condition, they are able to develop a fixed charge density making the collagen network able to swell and retain a large amount of water and ionic constituent. A typical organization of the articular cartilage is shown in Fig. 2.6.

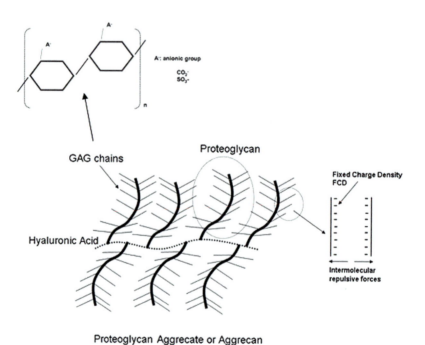

Fig. 2.5 Disaccharide units and proteoglycan aggregate

Fig. 2.6 Schematic representation of cell-matrix assembly in the articular cartilage

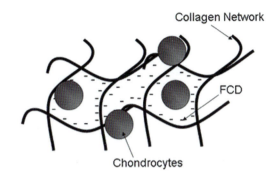

Collagen Network

FCD

Chondrocytes

2.5.2 Mechanical Properties

One of the most striking features of articular cartilage is that it normally exhibits little or no wear after the application of millions of cycles which can even overcome by an order of magnitude the body weight. Its unique mechanical and tribological properties, which are unparalleled in man-made load-bearing artifacts, have been attributed to the complex structure and composition of the extracellular tissue matrix. In response to an externally applied load, articular cartilage is subjected to a complex state of tensile, shear, and compressive stresses. Because of the large water content of the extracellular matrix 75–85%, mechanical loading also results in pressure gradients in the interstitial fluid. As the extracellular matrix is permeable to water and possesses a significant amount of fixed negative charge, pressure gradients cause movement and redistribution of the interstitial fluid. Fluid movement may also be accompanied by electrokinetic effects such as streaming potentials and currents as various ions are moved through the charged matrix (Grodzinsky et al. 2000). It is now well accepted that the primary mechanism of viscoelasticity in cartilage results from frictional interactions between the solid and fluid phases, although there is evidence that the solid matrix exhibits intrinsic viscoelasticity. Cartilage also exhibits highly non-linear mechanical properties such as strain-dependent moduli, strain-dependent hydraulic permeability (Holmes and Mow 1990, Cohen 1992, Ateshian et al. 1997, Grodzinsky 1983), and a difference of nearly two orders of magnitude in tensile and compressive moduli. These properties are also anisotropic, particularly in tension, and vary significantly with distance from the tissue surface and with joint area. More complex but equally important mechanical behaviors include the presence of internal swelling pressures that give rise to inhomogeneous residual stresses within normal articular cartilage. Finally, cartilage possesses important geometric and material characteristics that endow it with unique frictional properties. The low friction coefficient, coupled with fluid-dependent mechanisms of load support, allow for minimal tissue wear under a relatively harsh mechanical environment.

Influence of Charged Groups on Macroscopic Properties

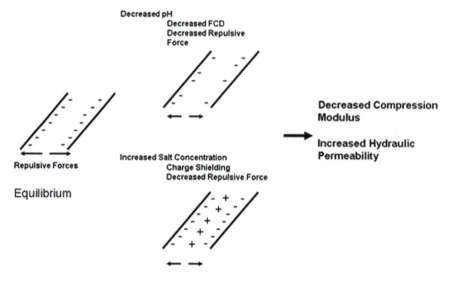

Fig. 2.7 Influence of charged groups on macroscopic properties

The dependence of physical properties of articular cartilage on its "electric" phase has been studied in several issues (Grodzinsky 1983). Elastic modulus and hydraulic permeability depends by fixed charge density (Fig. 2.7). Fixed charge density is varied by placing sample of articular cartilage in contact with an external solution at different salt concentrations. At each concentration, the pK of charged groups of GAG changes, resulting in a different fixed charge density (Donnan effect). By increasing salt concentration, the decreased elastic modulus and increased permeability is due to the electrostatic shielding of proteoglycan charge groups by counterions (Comper and Laurent 1978).

The electromechanical behavior of articular cartilage can be tested in an experimental apparatus as the one depicted in Fig. 2.8, used by Frank and Grodzinsky (1987). The effects of fluid movement-induced viscoelasticity are also evident: the application of a periodic sinusoidal displacement produces an out of phase response in terms of mechanical stress and electric potential being the latter caused by ions movements within the tissue.

Articular cartilage has been successfully modeled as a macrocontinuum of molecules, that is a gel with uniform charge density. The presence of such charge density is responsible for particular bio-mechanical properties of articular cartilage as depicted by Comper and Laurent (1978). They suggested that polysacaride-based natural tissue 'transduce what are perhaps normally regarded as unspecific forces or physicomechanical phenomena of their

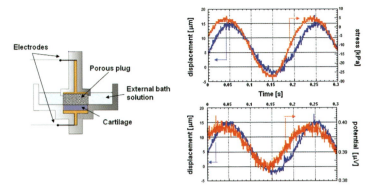

Fig. 2.8 Confined compression testing apparatus

environment, such as mechanical phenomena, osmotic pressure, diffusion, exclusion, electrostatic phenomena, and fluid flow, to ultimately influence the various cellular and molecular properties of their tissue in a specific way'. Understanding these coupled phenomena is essential for the normal, healthy function, as well as for the success of tissue engineered replacements. Mathematical models for articular cartilage are derived from the merging of two main theories: thermodynamic of polyelectrolyte gel and poroelastic theory. Their union gives rise to the multiphasic theory. This mathematical formulation is able to capture the mechanisms involved in articular cartilage during external solicitations: swelling, deformation, fluid flow, electrical potential, and solute transport. This has been developed in an independent manner by several authors (Huyghe and Janssen 1997, Levenston et al. 1998, Mauck et al. 2003). Cartilage is modeled as a continuum where the representative elementary volume (REV) is composed of a solid, fluid, ionic, and macromolecular phase each of them weighed by their own volume fraction. The macroscopic physical parameters are elastic modulus (H), Darcy's permeability (k), electrokinetic coupling coefficient (k_{ij}), molecular diffusion coefficient (D). The solid phase is modeled both as elastic or viscoelastic matrix (Netti et al. 2002). A typical experiment used to measure physical parameters of articular cartilage is the confined compression test (Fig. 2.9). Under constant deformation, cartilage undergoes stress relaxation due to fluid and pressure redistribution within the sample. By using multiphasic theory, it is possible to measure strain dependent elastic modulus and hydrauilic permeability by curve fitting of stress relaxation test.

Using this approach, it has been possible to measure physical properties of articular cartilage as a function of age and its main constituents: water content, collagen, and GAG. The results indicate a significant dependence σ_{ULT} (stress at failure in tension) and Et (Elastic modulus in tension) on collagen/ volume, while GAG/volume was a significant factor for H (Elastic Modulus in compression), kp (hydraulic permeability) (Williamson et al. 2001).

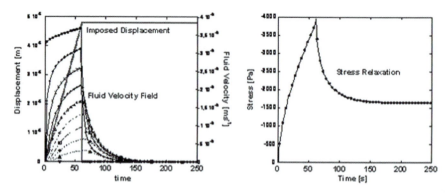

Fig. 2.9 Schematic of typical plots of a stress relaxation test

2.6 Soft Tissue Replacements

2.6.1 Introduction

Biomedical implants find application to treat those pathologies which cannot heal spontaneously or via surgical procedures or when the normal healing process is impaired by pathological conditions. Prosthetic implants must be biocompatible and have to fulfill specific functional tasks, which can be biological, physical, mechanical. Biocompatibility is the ability of the implant to avoid an adverse response of the body once in contact with it. Biomedical implants are in contact with body fluids. The biomaterial surface invariably adsorbs biological molecules which trigger cellular responses. Such responses can eventually determine the failure of the prosthesis. Thus biocompatibility of an implant is strictly related to the physical and chemical properties of its surface and, more importantly, to the characteristics of the thin layer of the biomacromolecules adsorbed on it.

As far as performance parameters are concerned, we are particularly interested in the mechanical behavior of prosthetic implants. Mismatch of the mechanical properties between the implant and the surrounding tissue can cause severe complications, like implant loosening, detachment, rupture, and abnormal tissue ingrowth.

The structures, properties, and main features of the prosthetic devices that have been applied in clinics or that are currently under investigation will be presented in the following sections.

2.7 Ligament Prostheses

ACL is a very important ligament of the knee joint. The ACL keeps the articular surfaces of the bones in compression and it contributes in defining the range of movements of the joint. In particular, it prevents anterior displacement of the tibia and with respect of the femur. Besides these mechanical functions, the sensory role of ACL has now been recognized. The presence of mechanoreceptors within the human ACL has been demonstrated (Zimny et al. 1986). The mechanoreceptors are believed to provide the central nervous system with information on movement and spatial position of the articulation. Unfortunately, the ACL is the most totally ruptured ligament of the knee. If treated conservatively, ACL rupture leads to joint instability, osteoarthritis, and meniscal tears (Noyes et al. 1983). Suturing the ACL remnants does not produce satisfactory results. Thus ACL replacement is the preferred therapeutic treatment. A wide variety of transplants, both autologous and allogenous, have been proposed for therapeutic treatment so far. Autografts like patellar tendon and semitendineous tendon graft are appealing because of their limited immunogenic potential and infection transmission. The greatest concern is related to the morbidity of the donor site. Use of allografts of course avoids the problems related to donor site morbidity and allows the harvesting of larger grafts. However, their use is strongly limited by their potential for transmission of pathogens and the sterilization procedures they are subjected can alter their structure and mechanical response. Moreover, both autograft and allografts undergo a dramatic decrease in mechanical properties between 2 weeks and 4 months after transplantation. This weakening effect is believed to arise from the revascularization and remodeling processes which affect the graft postoperatively (Fowler 1993). These observations stimulated the design and production of grafts based on synthetic and biological materials.

The commercially available prosthetic devices have been mainly implanted in North America and in Europe during the last decades.

Prosthetic devices are usually grouped in three types: permanent, stent (agumentation device), scaffold.

Permanent prostheses are designed in order to replace on a permanent basis the injured ACL. Therefore, they are characterized by high strength and increased resistance to cyclic loads. For example Gore Tex ligament, Stryker-Dacron belong to this category. Stents are temporary prostheses which have to protect the main graft during the remodeling phase that occurs postoperatively. Kennedy Ligament Agumentation Device (LAD) and Trevira Hochfest have been marketed as stent. Scaffolds are designed in a way that their structure and materials allow the infiltration of autogenous tissue within the articular portion of the scaffold device. Leeds-Keio and LARS (Ligament Advanced Reinforcement System) belong to this type of prosthetic.

Carbon fibers have been used owing to their chemical inertia and high strength. Integraft, Jenkins, and Biocarb are carbon-fiber based prostheses which have been commercialized. Despite early promising results, the poor shear stress behavior of carbon fibers caused a high failure rate of these prostheses. Moreover, the presence of carbon debris in the synovial fluid often caused severe inflammations which eventually required the removal of the device. Polyester PET and polytetrafluoro ethylene (PTFE) are well characterized biomaterials that have been widely used in cardiovascular surgery. Two of the most successfully synthetic ligaments, mainly implanted in Europe (the LARS and the Trevira), are constituted by PET fibers (www.LARSliaments.com). The Stryker-Dacron prostheses is another PET based device (www.stryker.com). It has been approved by the Food and Drug Administration (FDA) in 1989 with restrictions. Also the Leeds-Keio, which is one of the most popular scaffold is made of PET fibers. PTFE has been proposed as material for ligament replacement owing to its intrinsic chemical inertia and biocompatibility. The Gore Tex ligament is composed of microporous PTFE fibers. It gained FDA approval in 1986 with some limitations. Despite encouraging preliminary results Gore Tex artificial ligaments displayed unacceptable degradation in longer follow up studies (> 5 years) (Roolker et al. 2000). The use of polypropylene fibers is limited to Kennedy LAD. The LAD gained approval in 1986 to augment autogenous grafts.

Other synthetic materials found limited applications in the production of artificial ligaments prostheses. As example, we quote the polyaramidic fibers (Nomex) which have been used as portion of the composite ligament Prolast. Resorbable polymers like polylactid acid (PLA) and polyglycolic acid (PGA) are characterized by weak mechanical properties and fast degradation rate. So far, their application as structural element in ligament production is limited (Duval and Chaput 1997).

Owing to the peculiar biomechanics of the ACL, the synthetic materials have mainly been used in the form of filaments and fibers. The spatial assembly of the fibers as well as their fineness determines the macroscopic mechanical response of the prosthesis. The multifilament strand is the easiest geometry available. It is constituted by the juxtaposition of several hundreds of filaments. This particular structure allows adjacent fibers to move past each other, minimizing friction and, therefore, wear particle release under cyclic loading. The Integraft, Jenkins, and the articular portion of the LARS have been built according to this geometry.

Woven ligaments are obtained by the intertwining of two groups of threads, the warp and the weft. Usually in artificial ligament design and production, the density of the weft threads is higher than the warp one, being the weft direction parallel to the axis of the in vivo applied load. Warp filaments hold the structure together and prevent the disruption of the textile. The core of the Stryker-Dacron prosthesis is a set of woven narrow ribbons, while the Trevira Ligament is a simple woven ribbon.

In the braided structure, the two families of threads do not necessary follow two perpendicular directions. The breading angle can vary in the 20°–170° range. The Gore Tex ligament is an example of cylindrical braid and the Kennedy LAD is a flat narrow braided ribbon.

Knitted fabrics have also been used. Flat textiles can be either warp knitted or weft knitted. Knitted structures allow a better design of the anisotropic mechanical properties of the prosthesis. As a general rule, the mechanical properties are determined by the density of stitches in a particular direction. The outer shell of the Stryker is composed of weft knitted patches, and the intraosseous parts of the LARS are made of warp knits reinforced by straight multifilaments.

Despite the large variety of polymeric materials used and structures proposed, all synthetic ligaments did not provide satisfactory long-term results. The autologous repair with the bone-patellar tendon autograft still remains the gold-standard solution. The poor results displayed by synthetic grafts during follow-up studies are somehow surprising since in vitro testing of the same devices showed incredibly high ultimate tensile strength (UTS) and fatigue resistence. It should be noted, however, that the graft can experience in vivo harsh environment (Ambrosio et al. 1994) and multiaxial mechanical stresses which can dramatically decrease their mechanical stability. Guidoin et al. (2000) analyzed the mechanism of failure of a broad variety of explanted prostheses. They invariably ascribed the cause of failure to one of these three causes: (1) yarn-on-bone and yarn-on-yarn abrasion; (2) frexural and rotational fatigue; (3) detrimental collagen infiltration. Moreover, they pointed out that in all the devices analyzed, the neotissutal ingrowth appeared to be an accumulation of scar tissue rather than an arrangement of well aligned collagen fibers.

Standardized surgical protocols must be addressed in order to avoid stress concentration and fiber abrasion. Novel structures and resorbable biomaterials must be introduced to mimic the complex physical features of the native tissue and to induce ingrowth of a functional neotissue which contributes to the mechanical stability of the whole implant.

An innovative device has been proposed by Cavallaro et al. (1994). This device is constituted by several hundreds of de novo reconstituted collagen fibers, which are braided together. The prosthesis has been implanted in a canine model. Twelve weeks follow up study showed good integration of the device within bony tunnels and the remodeling of the intrarticular segment. In particular, neosynthesized collagen fibers have been observed in the midsubstance.

It is clear that simple textile structures cannot capture the whole complexity of the viscoelastic and anisotropic mechanical properties of the native tendons and ligaments. The introduction of composite structures seems, therefore, mandatory. Migliaresi and Nicolais (1980) proposed a composite ligament made of polyhyroxyethilmethacrylare (pHEMA) reinforced with PET fibers. The composite has been produced following the filament winding technique. Its structure mimics the one of native ligaments and the composite shows a non linear stress–strain response that is a common feature of all connective tissues. Moreover, coating the rather hydrophobic PET with a hydrophilic matrix would minimize the inflammatory response and prevent fiber-on-fiber abrasion. Later Ambrosio et al. (1998) investigated the viscoelastic behavior of an analogous composite structure. They showed that the dynamic mechanical parameters of the composite possesses the same constitutive dependency upon the kinematic variables as natural tissue ones. As a further attempt towards clinical application of composite structures, Davis et al. (1991) implanted in a rabbit model a resorbable composite tendon constituted by pHEMA/PCL blend matrix reinforced with helically wound poly-L-lactic acid (PLLA) fibers. They carried out a 45 days follow up study which demonstrated negligible inflammatory response and, more importantly, the degradation of the fiber/ matrix composite, and the ingrowth of neosynthesized collagenous tissue.

Tissue engineering could provide a promising therapeutic alternative for the treatment of ACL ruptures. A tissue engineered ligament can obviate most of the drawbacks possessed by the autogenous and allogenous grafts.

Several natural materials have been used for the production of scaffolds. Of course, collagen has been extensively used and the potentials of collagen-based scaffold have been investigated. Dunn et al. (1992) reported preliminary encouraging results of a de novo reconstituted collagen fibers within a collagen matrix. Explanted devices, from rabbit model, showed neotissue ingrowth having morphology similar to native ligament. The resorption of the fibrous reinforcement depended on its crosslinking density. This device, however, did not possess an adequate mechanical stability. Collagen based scaffold are still under investigation. The major concern is related to the immunogenic potential of collagen from animal sources and to mechanical limitations. This has suggested the use of alternative materials. Lin et al. (1999) studied the biocompatibility of a unbraided polycaprolactone coated – polyglycolic acid fibers scaffold. Fibroblasts seeded within this scaffold synthesized neomatrix and responded positively to chemical and mechanical stimuli. Polylactic acid fibers have been used for the construction of scaffold as well (Lu et al. 2005, Freeman et al. 2006). This material is in principle appealing because of its good mechanical properties and slow degradation times. In particular, mechanical conditioning is believed to profoundly affect the biosynthetic activity of cells within a construct. Several theories have been tried to explain this effect. It is generally accepted that mechanical signal are sensed by the cytoskeleton which is linked to the surrounding extracellular environment by transmembrane adhesion proteins. The deformations of the cytoskeleton may trigger up-regulation of ECM synthesis (Petrigliano et al. 2006).

Although much advancement has been made towards a functional tissue engineered ligament replacement, no device has proven to possess appropriate biophysical and biochemical features for and efficient in vivo application.

2.8 Skin Replacements

Closure of skin wounds usually occurs in adults by a repair process, which involves contraction and eventually scar formation. The wound healing process follows a temporary ordered sequence of events. More specifically, it is an event driven process where signals from one cell type trigger a cascade of events of other cell types. The early response of skin to a full thickness wound is hemostasis which is accomplished by the formation of a temporary fibrin clot. This plug prevents bleeding and also fluid and electrolytes loss. Circulating platelets are entrapped within the fibrin network and various inflammatory factors (TGF-α, TGF-β, PDGF, and EGF) are released. Chemotaxis is one of the main aspects which governs the inflammation and several cells, like neutrophils, monocytes, and lymphocytes are recruited to the injured site. Theses cells play an important role in cleaning the wound from debris and microbial pathogens. Moreover, inflammatory cells stimulate migration of keratinocytes, fibroblasts, and endothelial cells which eventually produce a de novo synthesized fibrovascular tissue (Metcalfe and Ferguson 2006). In the meanwhile, a resolution of inflammation is observed.

Severe injuries caused by trauma and burns, deeply damage the cell source which can restore the epidermal barrier, the fibrous dermis, and the vascular plexus. The replacement of epidermis is an absolute requirement in order to protect the wound bed from microbial attack and to prevent fluid loss. Dermal replacement in turn is necessary to guarantee mechanical support, and revascularization is of course essential for the supply of nutrients. Moreover, in chronic wounds the cascade of events that lead to complete healing does not follow the normal pathway and the restoration of the tissue is incomplete. This could be arise from particular pathologies, from the use of certain drugs or repeated trauma. These observations stimulated research activities targeted at developing novel skin replacement by ex vivo and in vitro techniques. As stated in the previous section, skin has a very complex multilayered structure and accomplishes different physiological tasks. Unfortunately, a skin replacement able to capture all the functions and structures of native skin is still not available. Thus far, split thickness autografts represent the gold standard to resurface skin wounds. However, it has to be pointed out that in large burns, the skin sites for autografting could be limited. Moreover, autografting cannot be pursued in those patients having impaired healing functions. Allogenic temporary skin grafts have been proposed in order to overcome this availability problem. It has been shown that allografts can keep the wound bed clean and vascularized. Therefore, they can be involved in the early step of burn treatment but they inevitably undergo to immune rejection in the later stages of healing. Thus allografts must eventually be replaced by a thin split autografts. Moreover, as all cadaveric tissues, allografts bring in a minor chance of infection transmission. This has fostered the attention towards synthetic skin replacements. An ideal skin replacement should possess the structural and functional feature of native tissue, but also it has to promote the incorporation in the wound site without scarring.

Skin substitutes can be produced using natural biopolymers, artificial materials or both of them. According to Jones et al. (2002) skin substitutes can be roughly divided in wound cover and wound closure agents. The former are designed to create a barrier against microbial attack and fluid loss. This provides the wound bed with an adequate environment for epidermis regeneration.

Wound closure devices, in turn, are designed to provide an epidermal barrier and to become incorporated within the defect.

Some of the replacements that have been successfully applied in clinics will be reviewed in the following section and in Chapter 5 (http://www.emedicine.com/plastic/topic477.htm).

2.8.1 Skin Substitutes for Wound Cover

2.8.1.1 Biobrane®

Biobrane® is a bilayered structure. It is composed by a lower nylon mesh covered by a thin silicone membrane (www.burnsurgery.com/Modules/skinsubstitutes/sec5.htm). The nylon filaments are coated with collagen-derived peptides, which are thought to improve adherence of the device to the wound bed. The silicone surface is semipermeable and prevents fluid loss and microbial attack. Biobrane is marketed as a temporary cover and it is removed either when the wound is healed or prior autografting.

2.8.1.2 Transcyte®

Transcyte® is very similar in composition and structure to the Biobrane® device (www.bu.edu/woundbiotech/bioengineered/TransCyteProdPres/index.htm). Its collagen coated nylon substratum is seeded with allogenic fibroblasts extracted from neonatal foreskin. These fibroblasts secrete extracellular matrix macromolecules and growth factors, which are believed to improve healing properties. Before grafting the device is subjected to freezing treatment which destroys cell culture but preserves the de novo synthesized proteins. This treatment is performed in order to reduce the risks of immune reaction. Since nylon is not absorbable, it is not intended to act as a dermal substitute. It is rather used in partial thickness burns treatment as wound cover before autografting.

2.8.1.3 Apligraf®

Apligraf® is a bilayered biohybrid which is composed of a collagen gel seeded with neonatal allogenic fibroblasts. The gel is covered by a layer of allogenic neonatal keratinocytes. Apligraf® is used in the treatment of chronic ulcers (www.apligraf.com).

2.8.1.4 Dermagraft®

Dermagraft® is a poly(glicolide co lactide) scaffold seeded with allogenic neonatal fibroblasts (www.dermagraft.com). The cells secrete dermal matrix components and subsequently are cryopreserved to maintain viability after storage. Allogenic fibroblasts show increased longevity in an immunocompetent host than epithelial cells. However, the matrix they produce is eventually replaced. Dermagraft® is marketed for stimulating ingrowth of fibrovascular tissue from the wound bed. It improves the epithelialization process from the wound edges but it does not close the wound by itself. Dermagraft® has been used for the treatment of chronic lesions rather than burn wounds.

2.8.2 Skin Substitutes for Wound Closure

2.8.2.1 Alloderm®

Alloderm® is human cadaveric skin (www.lifecell.com/products/95/). This device is screened to reduce the risk of transmission of infections. The epidermal layer is removed as well as all the cells within the dermis. The acellular scaffold is finally freeze dried. These treatments aim at reducing the risks of immune response yet preserving the native structure. It is rehydrated prior grafting and it is handled like an autograft. Alloderm® becomes repopulated by host fibroblasts and eventually incorporated into the wound site. It is used as template for dermal regeneration in full thickness wounds and burns. After incorporation into the host tissue, the device is covered by an ultrathin split autograft.

2.8.2.2 Integra®

Integra is one of the most accepted artificial skin substitute (www.integra-ls.com/products). It has been originally described by Yannas and coworkers (Yannas and Burke 1980, Yannas et al. 1980, Dagalakis et al. 1980). This device has a bilaminar structure: a collagen-GAG sponge, which is covered by a silicone membrane. Theporous sponge allows migration of fibroblasts and other cell form the wound bed. After repopulation cells start to remodel the scaffold and produce de novo synthesized vascularized dermis. The silicone membrane simulates the barrier of the epidermal layer. The membrane is removed after the formation of neodermis (a process which usually takes 3–6 weeks) and substituted with a thin split skin autograft. Integra® is usually indicated as wound closure device for partial and full thickness burns. As stated earlier, epithelialization relies on epidermal cells which reside deep within the wound bed (i.e., debris of hair follicules) and ingrowth from the wound edges. However, very large wounds, like giant nevi or in the case of extended burns, would require extremely long healing period for a complete spontaneous epithelialization. Therefore, an epidermal biopsy is usually required.

2.8.2.3 Cultured Epidermal Autograft (CEA)

Autologous epidermal substitutes also referred to as cultured epidermal auto-grafts are in vitro cultured sheets of autologous keratinocytes. CEA are extre-mely delicate and require secondary devices for adequate healing. CEA have proven to be useful in patients with very large burns (>60% total body surface area). A major drawback of CEA device is their poor mechanical stability. The epidermal layer they form frequently gives rise to blistering many months after implantation even with mild shearing stresses (Horch et al. 2005). This phenom-enon has been attributed to the abnormal structure of the anchoring fibrils at the dermal epidermal junction (DEJ). Probably the maturation of DEJ is altered by the enzymatic treatment that the CEA sheets are subjected prior detachment from the culture flask (Horch et al. 1998, Compton et al. 1989).

2.8.2.4 Cultured Skin Substitutes (CSS)

Cultured skin substitutes have been developed in order to overcome the problem related to the mechanical instability of the CEA grafts. CSS are designed to be permanent replacements of both dermis and epidermis. They are constituted by a collagen-GAG scaffold which is seeded with autologous fibroblasts and keratinocytes at high cell densities. The scaffold is then kept in culture for about two weeks. During culturing period, fibroblasts degrade the scaffold and synthesize new ECM. The keratinocytes produce stratified and cornified layers. Immonological analyses demonstrated the presence of a basement membrane at the dermal epidermal junction (Boyce et al. 2002). The drawback of epithelial blistering is thus avoided since the formation and maturation of a DEJ occurs prior grafting.

Several devices have been designed in order to handle and deliver cellular layers of keratinocytes onto the wound bed. Laserskin displays some noteworthy features. It is a modified HA (benzilic esther) porous membrane. Keratinocytes are seeded on the perforated surface. During the in vitro culturing period epithelial cells colonize both sides of the membrane in a multi-ply fashion. The HA core provides the seeded scaffold with mechanical strength and the whole biohybrid can be easily peeled off the culturing flask without any enzymatic treatment. Moreover, the products of degradation of the HA are believed to stimulate cell migration, proliferation and angiogenesis, thus accelerating the healing of the wound (Slevin et al. 2002, Dechert et al. 2006).

2.9 Vascular Grafts

Cardiovascular diseases are a primary cause of mortality in the western world. Vascular graft transplant is an obliged solution for ischemic hearth diseases and peripheral vascular diseases (see Chapter 7). The gold standard for medium-small blood vessel replacements is the use of autologous grafts (Back and White 1996). However, development of alternative solution is required for large vessels whose diameter is greater than 6–7 mm and whenever the autologous graft option cannot be pursued, for example for those patients affected by arteriopathy. Newer grafts made of synthetic polymers allowed the reconstruction of large diameter, high flow, vessels such aorta. The patency of synthetic prostheses in small arteries, whether bypass or reconstruction, is still unsatisfactory (Bordenave et al. 1999, van der Zijpp et al. 2003). An ideal vascular graft should mimic the mechanical properties of the tissue to which it is anastomosed. Moreover, it should prevent the formation of thrombi and should allow the generation of an endothelial cell layer in the inner lumen. Above all, the synthetic graft must remain patent for long period of time. Interesting data sheet of several commercial devices for vascular surgery can be found at www.bostonscientific.com/med_specialty/specialtyMain.jsp?sectionId = 4&relId = 6&task = tskSpecialtyMain.jsp.

The most used polymers are Dacron$^{©}$ (polyethylene terephthalate, PET) and polytetrafluoroethylene (PTFE).

Dacron$^{©}$ grafts have been successfully used in large vessel reconstruction. The devices are produced in the form of woven and knitted fabrics. The arrangement of the fibers in the textiles determines mechanical properties of the construct. PET fabrics are usually characterized by high stiffness and low extensibility. Woven fabrications have lower porosity and do not bleed excessively at implantation. Knitted prostheses, in turn, require preclotting before implantation. The tubes have usually crimped walls which provide them with increased flexibility and prevents kinking when bending occurs. Newer knitted Dacron grafts possess velour filaments which form loops on the outer and inner surfaces. This particular feature confers the prostheses increased integration with surrounding tissues, reduced clotting time during preclotting procedure, and it is believed to enhance endothelialization of the inner lumen (Sato et al. 1989).

Expanded PTFE is the material of choice for the production of small diameter grafts. It is produced in the form of small tubes whose walls display a microporous structure. The hydrophobicity of the PTFE makes the inner surface inert toward the adsorption of those proteins and growth factors which lead to the formations of occlusions. The efficiency of PTFE grafts in above-knee by-pass is good. However, contradictory results have been published so far concerning its superiority on autogenous grafts (Wilson et al. 1995, Allen et al. 1996). However, it displays unsatisfactory results in below-knee applications (Charlesworth et al. 1985).

Both Dacron$^{©}$ and PTFE grafts have porous structure. Porosity is an important design parameter since it is believed that pores can host neotissue ingrowth. The formation of a neoendothelial cell layer is a key feature to avoid thrombus formation and occlusions. The presence of a fibrous lining is considered pseudointima. Fibrous components penetrate porous surfaces in a few weeks after implantation. This penetration is not essential for the formation of a pseudointimal layer, but the support of surrounding connective tissue is essential for its long term stability (Back and White 1996).

Synthetic prostheses retrieved from human patients display inner surface coated with fibrin and practically devoid of endothelial cells (Berger et al. 1972). Fibrin coating is the result of a cascade of biochemical events triggered by the prosthetic surface. These events eventually lead to the formation of thrombi. Therefore, the adsorption of the factors which regulate the coagulation events must be minimized in order to improve long term stability of the synthetic graft.

Polyurethanes (PU) are another class of polymeric materials which have found application in vascular device production. Polyurethanes have very interesting features like mild thrombogenic potential, mechanical properties similar to those of native arteries (Zdrahala 1996). PUs indeed have been extensively used in cardiovascular applications different from graft production like intra-aortic balloons and as elements of artificial hearts. PU grafts can be produced as porous tubes using salt leaching technique. Although positive preliminary results obtained in animal models, a major concern of PU grafts resides in their limited patency caused by early degradation. It has been shown that PU in vivo is susceptible to hydrolysis, enzymatic degradation and oxidation of the ester linkages (Back and White 1996). Some grafts have been coated with silicone rubber in order to improve long term stability. However, this approach hinders the benefits derived from using PU.

Attempts to improve long-term stability of prosthetic devices include the incorporation of "passivating" agents. As an example albumin has been used to coat the inner lumen of synthetic grafts (Kang et al. 1997). Albumin is believed to impair platelet aggregation (de Jonge and Levi 2001). Heparin is a well characterized anticoagulant factor. Heparin coated surface displayed better result than non-coated prostheses. The beneficial effect of the heparin is, however, limited in time since it rapidly diffuses to the blood stream.

The patency of small diameter artificial blood vessels is impaired by lacking in preventing blood coagulation and platelet deposition. These functions are normally performed by endothelial cells. Therefore, the use of these cells seems mandatory in order to reduce the formation of occlusions. The adhesion of endothelial cells on PET, PTFE, and PU is poor prior implantation (Bhat et al. 1998). Endothelialization only occurs 1–2 cm near the anastomotic site. In particular the cell that form the lining are believed to come from the adjacent native arterial segment and do not arise from a genuine endothelialization process of a penetrating tissue. To overcome this limit, the graft surface has been functionalized with bioactive molecules like RGD, fibronectin, and growth factors. Unfortunately, these solutions do not improve significantly graft patency in humans. A two stage process has been proposed which includes cell harvesting and subsequent in vitro culturing, and finally implantation of a biohybrid graft characterized by a confluent lining of endothelial cell on the inner surface.

The production of tissue-engineered grafts has also been proposed, by using biodegradable scaffolds. The neotissue is expected to grow and remodel within the scaffold as it degrades. Of course, in order to avoid biohybrid mechanical disruption, scaffold resorption must be balanced by neotissue ingrowth. Polyglycolic acid, lactide and glicolyde copolymers, polydioxanone have shown better results than non resorbable synthetic grafts (PTFE and Dacron©), in terms of mechanical properties and endothelialization potential. A modified form of HA has been proposed as well for the production of a bioresorbable scaffold. HA has the advantage that its degraded products stimulate the synthesis of ECM components and angiogenetic events. However, HA based scaffolds display unsatisfactory starting mechanical properties.

Even though tissue engineering is a very promising technique, several problems still need to be addressed. It should be pointed out that the in vitro cell expansion requires the optimization of numerous processing parameters. Culturing medium composition, type of culture, mechanical stimuli, all affect the final outcome up to a certain extent, but their relative influence on the biophysical properties is still not clear.

2.10 Cartilage Replacement

Cartilage replacement is particular required for the treatment of articular cartilage defects and reconstruction of maxillofacial zones. The causes of cartilage damaging and defects are usually ascribed to osteoarthritis and traumatic events (see Chapters 12 and 13).

In general, two kinds of defect are present in cartilage. Superficial defect usually trigger a metabolic response, which, however, does not lead to an efficient repair of the defect. In the case of deeper defects (ostechondral defect), the wound site is flooded by marrow. The simultaneous release of cytokine and growth factor and promotes the accumulation of mesenchimal progenitor cells. These cells differentiate and can produce fibrous cartilage with low mechanical properties or bone. Only in the case of small defect this process is able to generate hyaline cartilage (Laurencin et al. 1999).

2.10.1 Joint Resurfacing

This technique poses the principle on the capability of subchondral bone to release bone marrow and initiate to generate fibrocartilage. In general, subchondral bone is shaved or drilled, or damaged cartilage is excised in order to promote bone marrow release (see Chapter 13). Although these procedures have had some success, the formation of hyaline cartilage is rare and the high failure incidence makes this procedure questionable and not suitable for long term viability (Goldberg and Caplan 1994).

2.10.2 Biological Autograft

Healthy cartilaginous tissue is removed from the patient, machined and reimplanted into the patient defect site. Ostechondral defect, perichondrum, periosteum, meniscus, and mandibular cartilage have been repaired with this procedure. The drawbacks of this procedure are constituted by the limited source of autologous tissue to harvest, by inflammatory response of the host caused by mechanical overload and the mechanical mismatch between the donor and host tissue (Pap and Krompecher 1961).

2.10.3 Total Joint Replacement

The most widely treatment used is the total joint replacement in case of severe damage of articular cartilage. The majority of joint replacement implants are made of metals (such as titanium) and polymeric component such as polyethylene. Corrosion and wear of the implant and shielding are problems that more often arise by using this technique.

2.10.4 Tissue Engineered Constructs

Interest of tissue engineering in cartilage regeneration is motivated by the inability of adult cartilage to repair itself after damage by disease or injury, the high frequency of osteoarthritis and cartilage trauma in the population, and limitations of current treatment methods. Cartilage regeneration is obtained in vitro by culturing a cell seeded porous scaffold into a bioreactor and allowing the cell to synthesize neotissue. Cellular component consists in animal chondrocytes such as bovine, horse, or rabbit. Very few studies are available concerning the use of human chondrocytes, however, there is increasing interest in using stem cells and progenitor cells for cartilage engineering.

Several issues have demonstrated that in vitro culture of chondrocytes should be performed by seeding the cells within a three dimensional matrix (e.g., agar or a polymeric support) in order to promote accumulation of synthesized matrix molecules and to prevent the loss of phenotype. Cells grown in a monolayer dedifferentiate and produce smaller amounts of type II collagen, which is found in the cartilage of articulating joints, and greater amounts of type I collagen, characteristic of fibrous tissue. Production of GAG is also diminished in monolayer culture. Chondrocytes immobilized in a three-dimensional support have been cultured for days to months in Petri dishes, well plates, spinner flasks, custom-designed bioreactors, perfusion, and rotating wall bioreactors which simulate microgravity conditions (Heath and Magarit 1996).

The production of tissue engineered cartilage constructs has been so far performed following one of this approach: by dynamic cultivation of a cell seeded porous polymeric scaffolds; by cultivating a gelling solution of chondrocytes and polymer (agarose, pluronic, alginate, peg); by cultivating seeded microbeads of alginate.

However, it has to be pointed out that the production of tissue engineered fibrous cartilage constructs could be different form the production of the articular cartilage one.

Pluronic. Fibrous cartilage has been successfully reconstructed in the shape of ear by Saim et al. (2000) by using chondrocytes harvested by Yorkshire swine and cultured in Pluronic F-127. Pluronic F-127 consists by weight of approximately 70% ethylene oxide and 30% propylene oxide. This material is slowly dissolved and cleared by renal and biliary excretion. Pluronic F-127 is soluble in water and becomes a hydrogel at room temperature (25°C) in a concentration of 20% or greater. The chondrocytes / Pluronic F-127 suspension was injected in a human ear shaped mould created within the dermis of the swine. The vascular plexus of the swine dermis provides the injected suspension with nutrients thus allowing a correct maturation and development of the constructs. The final construct displays glistening, flexible tissue that can be compressed, and that instantly recoils. This study demonstrates that the hydrogel copolymer scaffold Pluronic F-127 enables the engineering of autologous chondrocytres in a pre-determined shape and produce elastic cartilage of useful quality.

Replacement of articular cartilage is actually a challenging task because the requirement is to obtain a material highly specialized to withstand the critical physiologic function in the articular joint. So far numerous investigations have been performed producing information on the influence of material properties and culture condition on tissue growth, assembly, and physical properties evolution, but the replacement of articular defect has not been obtained yet. During chondrocyte cultivation in polymeric scaffold, the resulting cartilage is often fibrous, with physical properties very far from the actual articular carti-lage. The resulting construct is prevalently formed by type I collagen instead of type II. Other problems involved in tissue engineering cartilage is the deposition of tissue confined in a very thin layer of the scaffold in contact with culture media, the prevalence of the tissue in the inner part is necrotic or fibrous. To make and idea on the task that would reach in articular cartilage replacement, a 5 mm cubic construct should be performed. So the crucial aspect in tissue engineering articular cartilage is the knowledge of physical parameters, tissue composition, and assembly evolution with culture conditions.

Agarose. The first work in this direction was performed by Buschmann et al. (1992). Calf bovine condrocytes was seeded in agarose gel (1 mm thick) at high cellular density (10^7 cell/ml). Physical properties of agarose/chondrocyte con-structs were monitored up to 35 days of culture. In parallel biosynthetic activity was assessed by GAG and collagen deposition. During culture time, chondro-cytes deliver a functional extracellular matrix. Hydraulic permeability, dynamic stiffness, and streaming potential change as GAG deposition occurs. By one month, the dynamic stiffness and streaming potential were about one-fourth that of native articular cartilage, consistent with a GAG density and cell density of approximately one-fourth that of the native tissue. Furthermore, the response of the oscillatory stiffness and streaming potential upon the frequency was also suggestive of a mechanical functional cartilage-like matrix. Neotissue ingrowth alters the structure of the agarose matrix. These structural changes in agarose matrix cause modifications in stress–strain response (Fig. 2.10a), non linear strain dependent hydraulic permeability (Fig. 2.10b), free swelling trans-port properties (Fig. 2.10c, d). Functional relationships which predict the response of permeability, diffusivity, and stiffness on the GAG content have also been obtained (Buschmann et al. 1995, De Rosa et al. 2006).

The effect of culture conditions, (static or dynamic compression) on biosyn-thetic activity of chondrocytes in agarose gel was also investigated. Static

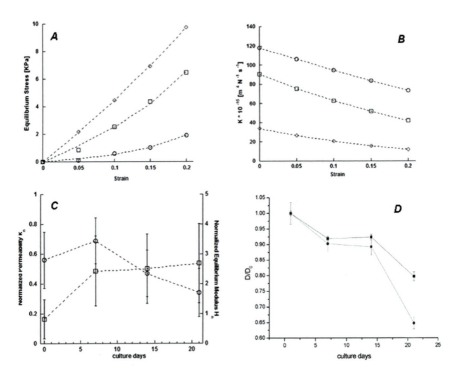

Fig. 2.10 Mechanical and transport properties of agarose-chondrocytes constructs (**a**) Experimental stress strain curve of the construct at 0 day (*circle*), 14 days (*square*) and 21 days (*diamond*). (**b**) hydraulic conductivity – strain curve at 0 day (*circle*), 14 days (*square*) and 21 days (*diamond*). (**c**) Normalized modulus H_n (*square*) and permeability k_n (*circle*) as ratio between the properties of the construct and control at same culture time. (**d**) Normalized diffusion coefficients variation with culture time of dextran (*circle*) and BSA (*square*) in cellular scaffold respect the control gels changes in culture time. From De Rosa et al. 2006 with permission

compression results in an inhibition of tissue development in terms of matrix deposition and change in physical properties. Small oscillatory amplitude was found to stimulate biosynthetic activity, suggesting the importance of reproducing physiological condition during culture process. Optimization of dynamic culture condition was suggested by Chowdhury et al. (2003). Intermittent and continuous compression may differently modulate chondrocyte metabolism. Proteoglycan synthesis was found to increase with number of cycles during continuous solicitations and regular bursts of intermittent compression enhanced proteoglycan synthesis further. Cellular proliferation was found to be inhibited by regular bursts in the intermittent culture. Tissue growth in agarose gel can be modulated and tailored by choosing optimal load history.

PGA. The effect of external stimulation on cartilage deposition and its physical and biochemical evolution has been extensively studied by using a well known polymeric material approved by FDA, PGA. Chondrocyte metabolism and tissue deposition has been shown to be modulated by hydrodynamic condition and mechanical stimulation as well. A hydrodynamic condition stimulates the deposition of a more homogeneous tissue within the PGA scaffold compared to static condition (Obradovic et al. 2000). This is probably caused by an enhanced mass transfer and transmission of shear force, which stimulates the cellular transduction of mechanical stimuli. Samples cultured under hydrodynamic conditions show increased elastic modulus, streaming potential, and hydraulic permeability, producing a more functional tissue. Chondrocytes are also sensitive to the kind of hydrodynamic regimes: turbulent regime produces construct with fibrous zone localized near to the turbulent eddies (spinner flask), whilst laminar flow produces construct with histological organization very close to the native cartilage (Freed et al. 1997). Together with hydrodynamic solicitations, mechanical stimulation has been shown to modulate tissue growth in cartilage explants and in agarose gels. Because agarose in not suitable for tissue replacement due to its low possibility in modulating the degradation kinetics mainly, study on chondrocytes seeded in polymeric matrices such as PGA has been performed. Compressive static offset was found to have inhibitory effect on the biosynthesis of total collagen and sulfated GAG. The inhibition effect is more evident at 50% of offset rather than to 10% compared to a free swelling culture control. Furthermore it has been evaluated the effect of a superimposed sinusoidal compression on a precompressed seeded construct. The effect of this dynamic conditioning seems to depend on both the frequency of the applied waveform and on the extent of precompression. In particular, it has been demonstrated that biosynthesis is maximum in static condition at 10% of precompression. In the case of dynamic condition, the optima set is 0.001 Hz and 50% precompression (Davisonn et al. 2002).

Porous Structure. Other approaches suggest the possibility to use Rapid Prototyping and CAD/CAM design to produce porous scaffold in order to obtain polymeric support with optimized mechanical and morphological properties. By using fiber deposition technique, it was possible to produce PEGT/PBT block co-polymer with 100% interconnecting pore with different architecture and mechanical properties very close to the elastic modulus of native cartilage (0.2 MPa) (Woodfield et al. 2003). Bovine chondrocyte seeded in this structure was able to colonize the full thickness of the scaffold (4 mm) and elaborate cartilaginous tissue both in vitro and in vivo under dynamic culture condition. The authors also investigated the possibility to culture human chondrocyte obtaining good results in vitro but no information is available about in vivo culture.

Questions/Exercises

1. Discuss the mechanical properties of a typical soft tissue in the light of its histological composition.
2. Indicate common features of the main commercially available skin replacements.

3. List the main biomaterials used for the manufacturing of artificial vascular graft and discuss their clinical performances.
4. Describe the main rationale underpinning the design of artificial vascular grafts for large diameter vessels.
5. Provide examples of biomaterials for artificial ligament manufacturing and highlight their biomimicking properties.
6. Explain cartilage functionality in the light of its mechanical properties.
7. Present the features of tissue engineering constructs for 3 distinct clinical applications in regenerative medicine
8. Highlight the main types of biomaterials for tissue engineering applications and discuss their physico-chemical properties and biocompatibility.
9. Critically discuss the main features necessary to 3D scaffolds for tissue engineering
10. Provide a critical discussion of the performances of tissue engineering products already available in clinics. In the light of the information of this chapter and of the following chapters, take into consideration both physico-chemical features and biocompatibility issues.

References

Allen BT, Reilly JM, Rubin BG, Thompson RW, Anderson CB, Flye MW, Sicard GA (1996) Femoropopliteal bypass for claudication vein vs PTFE. Ann Vasc Surg 10: 178–185

Ambrosio L, Apicella A, Mensitieri M, Nicolais L, Huang SJ, Marcacci M, Peluso G (1994) Physical and chemical decay of prosthetic ACL after in vivo Implantation. Clin Mater 15:29–36

Ambrosio L, De Santis R, Iannace S, Netti PA, Nicolais L (1998) Viscoelastic behavior of composite ligament prostheses. J Biomed Mater Res 42:6–12

Ateshian GA, Warden WH, Kim JJ, Grelsamer RP, Mow VC (1997) Finite deformation biphasic material properties of bovine articular cartilage from confined compression experiment. J Biomech 30:1157–1164

Back MR, White RA (1996) Biomaterials in vascular surgery. In: Wise DL et al. (eds) Human biomaterials applications. Humana Press, Totowa, NJ, pp 257–298

Bailey AJ, Paul RG, Knott L (1998) Mechanisms of maturation and ageing of collagen. Mech Ageing Dev 106:1–56

Berger K, Sauvage LR, Rao AM, Wood SJ (1972) Healing of arterial prostheses in man its incompleteness. Ann Surg 175:118–127

Bhat VD, Klitzman B, Koger K, Truskey GA, Reichert WM (1998) Improving endothelial cell adhesion to vascular graft surfaces: clinical need and strategies. J Biomat Sci. Polymer Edition 9:1117–11135

Birk DE (2001) Type V collagen: heterotypic type I/V collagen interactions in the regulation of fibril assembly. Micron 32:223–237

Birk DE, Mayne R (1997) Localization of collagen types I, III and V during tendon development. Changes in collagen types I and III are correlated with changes in fibril diameter. Eur J Cell Biol 72:352–361

Bordenave L, Remy-Zolghadri M, Fernandez P, Bareille R, Midy D (1999) Clinical performance of vascular grafts lined with endothelial cells. Endothelium 6:267–275

Boyce ST, Supp AP, Swope VB, Warden GD (2002) Vitamin C regulates keratinocyte viability, epidermal barrier, and basement membrane in vitro, and reduces wound contraction after grafting of cultured skin substitutes. J Invest Dermatol 118:565–572

Buschmann MD, Gluzband YA, Grodzinsky AJ, Kimura JH, Hunziker EB (1992) Chondrocytes in agarose culture synthesize a mechanically functional extracellular matrix. J Orthop Res 10:745–758

Buschmann MD, Gluzband YA, Grodzinsky AJ, Hunziker EB (1995) Mechanical compression modulates matrix biosynthesis in chondrocyte/agarose culture. J Cell Sci 108:1497–1508

Cavallaro JF, Kemp PD, Kraus KH (1994) Collagen fabrics as biomaterials. Biotechnology and Bioengineering 43:781–791

Charlesworth PM, Brewster DC, Darling RC, Robison JG, Hallet JW (1985) The fate of polytetrafluoroethylene grafts in lower limb bypass surgery: a six year follow-up. Br J Surg 72:896–899

Chien S (2006) Molecular basis of rheological modulation of endothelial functions: importance of stress direction. Biorheology 43:95–116

Chowdhury TT, Bader DL, Shelton JC, Lee DA (2003) Temporal regulation of chondrocyte metabolism in agarose constructs subjected to dynamic compression. Arch Biochem Biophys 417:105–111

Cohen B (1992) Anisotropic Hydrated Soft Tissues in Finite Deformation. Ph.D. Thesis, Columbia University, New York

Comper D, Laurent TC (1978) Physiological function of connective tissue polysaccharides. Physiol Rev 58:255–315

Compton CC, Gill JM, Bradford DA, Regauer S, Gallico GG, O'Connor NE (1989) Skin regenerated from cultured epithelial autografts on full-thickness burn wounds from 6 days to 5 years after grafting. A light, electron microscopic and immunohistochemical study. Lab Invest 60:600–612

Cox RH (1978) Regional variation of series elasticity in canine arterial smooth muscles. Am Physiol 234:542–551

Dagalakis N, Flink J, Stasikelis P, Burke JF, Yannas IV (1980) Design of an artificial skin Part III: Control of pore structure. J Biomed Mater Res 14:511–528

Davis PA, Haung SJ, Ambrosio L, Ronca D, Nicolais L (1991) A Biodegradable composite artificial tendon. J Mater Sci Mater Med 3:359–364

Davisonn T, Kuing S, Chen A, Sah R, Ractliffe A (2002) Static and dynamic compression modulate matrix metabolism in tissue engineering cartilage. J Orthop Res 20:842–848

de Jonge E, Levi M (2001) Effects of different plasma substitutes on blood coagulation a comparative review. Criti Care Med 29:1261–1267

De Rosa E, Urciuolo F, Borselli C, Gerbasio D, Imparato G, Netti P.A. (2006) Time and space evolution of transport properties in agarose-chondrocyte constructs. Tissue Eng 12:2193–2201

Dechert TA, Ducale AE, Ward SI, Yager DR (2006) Hyaluronan in human acute and chronic dermal wounds. Wound Repair Regen 14:252–258

Driessen NJ, Bouten CV, Baaijens FP (2005) A structural constitutive model for collagenous cardiovascular tissues incorporating the angular fiber distribution. J Biomech Eng 127:494–503

Dunn MG, Tria AJ, Kato YP, Bechler JR, Ochner RS, Zawadsky JP, Silver FH (1992) Anterior cruciate ligament reconstruction using a composite collagenous prosthesis. A biomechanical and histologic study in rabbits. Am J Sports Med 20:507–515

Duval N, Chaput C (1997) A classification of prosthetic ligament failures. In: L'Hocine Y (ed) Ligaments and ligamentoplasties. Springer-Verlag, Berlin, pp 168–191

Eshel H, Lanir Y (2001) Effects of strain level and proteoglycan depletion on preconditioning and viscoelastic responses of rat dorsal skin. Ann Biomed Eng 29:164–172

Font B, Eichenberger D. Rosenberg LM, van der Rest M (1996) Characterization of the interactions of type XII collagen with two small proteoglycans from fetal bovine tendon, decorin and fibromodulin. Matrix Biol 15:341–348

Fowler PJ (1993) Synthetic agumentation. In: Jackson DW et al. (eds) The anterior cruciate ligament: Current and future concepts. Raven Press, New York, NY, pp, 339–341

Frank EH, Grodzinsky AJ (1987) Cartilage electromechanics – I. Electrokinetic transduction and the effect of electrolyte pH and ionic strength. J Biomech 20:615–627

Freed LE, Langer R, Martin I, Pellis NR, Vunjak-Novakovic G (1997) Tissue engineering of cartilage in space. Proc Natl Acad Sci U S A 94:13885–13890

Freeman JW, Woods MD, Laurencin CT (2006) Tissue engineering of the anterior cruciate ligament using a braid-twist scaffold design. J Biomech e-pub. Nov 2006.

Goldberg VM, Caplan AI (1994) Biological resurfacing: an alternative to total joint arthroplasty. Orthopedics 17:819–821

Gregersen H. Emery JL, McCulloch AD (1998) History-dependent mechanical behavior of guinea-pig small intestine. Ann Biomed Eng 26:850–858

Grodzinsky AJ (1983) Electromechanical and physicochemical properties of connective tissue. Crit Rev Biomed Eng 9:133–199

Grodzinsky AJ, Levenston ME, Jin M, Frank EH (2000) Cartilage tissue remodelling in response to mechanical forces. Ann Rev Biomed Eng 2:691–713

Guidoin MF, Marois Y, Bejui J, Poddevin N, King MW, Guidoin R (2000) Analysis of retrieved polymer fiber based replacements for the ACL. Biomaterials 21:2461–2474

Harris RI, Wallace AL, Harper GD, Goldberg JA, Sonnabend DH, Walsh WR (2000) Structural properties of the intact and the reconstructed coracoclavicular ligament complex. AmJ Sports Med 28:103–108

Heath CA, Magarit SR (1996) Mechanical Factors Affecting Cartilage Regeneration In Vitro. Biotechnol Bioeng 50:430–437

Hierner R, Degreef H, Vranckx JJ, Garmyn M, Massage P, van Brussel M (2005) Skin grafting and wound healing – the "dermato-plastic team approach". Clin Dermatol 23:343–352

Holmes MH, Mow VC (1990) The nonlinear characteristics of soft gels and hydrated connective tissues in ultrafiltration. J Biomech 23:1145–1156

Holzapfel GA, Gasser TC, Ogden RW (2000) A new constitutive framework for arterial wall mechanics and a comparative study of material models. J Elast 61:1–48

Holzapfel GA, Sommer G, Gasser CT, Regitnig P (2005) Determination of layer-specific mechanical properties of human coronary arteries with nonatherosclerotic intimal thickening and related constitutive modeling. Am J Physiol. Heart and Circ Physiol 289:2048–2058

Horch RE, Corbei O, Formanek-Corbei B, Brand-Saberi B, Vanscheidt W, Stark GB (1998) Reconstitution of basement membrane after 'sandwich-technique' skin grafting for severe burns demonstrated by immunohistochemistry. J Burn Care Rehabil 19:189–202

Horch RE, Kopp J, Kneser U, Beier J, Bach AD (2005) Tissue engineering of cultured skin substitutes. J Cell Mol Med 9:592–608

Huyghe JM, Janssen JD (1997) Quadriphasic mechanics of swelling incompressible porous media. Internationa J Eng Sci 35:793–802

Jenkins RB, Little RW (1974) A constitutive equation for parallel fibered elastic tissues. J Biomech 7:397–402

Jones I, Currie L, Martin R (2002) A guide to biological skin substitutes. Br J Plast Surg 55:185–193

Kang SS, Petsikas D, Murchan P, Cziperle DJ, Ren D, Kim DU, Greisler HP (1997) Effects of albumin coating of knitted Dacron grafts on transinterstitial blood loss and tissue ingrowth and incorporation. Cardiovasc Surg 5:184–189

Keene DR, Marinkovich MP, Sakai LY (1997) Immunodissection of the connective tissue matrix in human skin. Microsc Res Techniq 38:394–406

Khan SN, Cammisa FP Jr, Sandhu HS, Diwan AD, Girardi FP, Lane JM (2005) The biology of bone grafting. J Am Acad Orthop Surg 13:77–86

Koob TJ, Vogel KG (1987) Site-related variations in glycosaminoglycan content and swelling properties of bovine flexor tendon. J Orthop Res 5:414–424

Kronick PL, Sacks MS (1994) Matrix macromolecules that affect the viscoelasticity of calf-skin. J Biomech Eng 116:140–145

Lanir Y (1978) Structure-strength relations in mammalian tendon. Biophys J 24:541–554

Laurencin CT, Ambrosio AMA, Borden MD, Cooper JA Jr (1999) Tissue engineering othopeadic applications. Ann Rev Biomed Eng 1:19–46

Levenston ME, Frank EH, Grodzinsky AJ (1998) Variationally derived 3-field finite element formulations for quasi static poroelastic analysis of hydrated biological tissues. Comput Methods Appl Mech Eng 156:231–246

Lin VS, Lee MC, O'Neal S, McKean J, Sung KL (1999) Ligament tissue engineering using synthetic biodegradable fiber scaffolds. Tissue Eng 5:443–452

Lu HH, Cooper JA Jr, Manuel S, Freeman JW, Attawia MA, Ko FK, Laurencin CT (2005) Anterior cruciate ligament regeneration using braided biodegradable scaffolds: in vitro optimization studies. Biomaterials 26:4805–4816

Mauck RL, Clarck TH, Athesian GA (2003) Modelling of neutral solute transport in a dynamically loaded porous permeable gel: implication for articular cartilage engineering. J Biomech Eng 125:602–614

Metcalfe AD, Ferguson WJ (2006) Tissue engineering of replacement skin: the crossroads of biomaterials, wound healing, embryonic development, stem cells and regeneration. J R Soc Interface 3:1–25

Migliaresi C, Nicolais L (1980) Composite materials for biomedical applications. Int J Artif Organs 3:114–118

Nagaraj A, Kim H, Hamilton AJ, Mun JH, Smulevitz B, Kane BJ, Yan LL, Roth SI, McPherson DD, Chandran KB (2005) Porcine carotid arterial material property alterations with induced atheroma: an in vivo study. Med Eng Phys 27:147–156

Netti PA, D'Amore A, Ronca D, Ambrosio L, Nicolais L (1996) Structure-mechanical properties relationship of natural tendons and ligaments. J Mater Sci Mater Med 7:525–530

Netti PA, Travascio F, Jain RK (2002) Coupled macromolecular transport and gel mechanics: Poroviscoelastic approach. AIChE J 49:1580–1596

Noyes FR, Matthews DS, Mooar PA, Grood ES (1983) The symptomatic anterior cruciate-deficient knee. Part II: the results of rehabilitation, activity modification, and counseling on functional disability. J Bone Joint Surg. American Volume 65:163–174

Obradovic B, Meldon JH, Freed LE, Vunjak-Novakovic G (2000) Glycosaminoglycan deposition in engineered cartilage: Experiments and mathematical model. Am Inst of Chem Eng J 46:1860–1871

Oxlund H, Manschot J, Viidik A (1988) The role of elastin in the mechanical properties of skin. J Biomech 21:213–218

Pap K, Krompecher S (1961) Arthroplasty of the knee, experimental and clinical experiences. J Bone Joint Surg [Am] 43:523–527

Petrigliano FA, McAllister DR, Wu BM (2006) Tissue engineering for anterior cruciate ligament reconstruction: a review of current strategies. Arthroscopy 22:441–451

Pins GD, Christiansen DL, Patel R and Silver FH (1997) Self-assembly of collagen fibers. Influence of fibrillar alignment and decorin on mechanical properties. Biophys J 73:2164–2172

Poole AR, Kojima T, Yasuda T, Mwale F, Kobayashi M, Laverty S (2001) Composition and structure of articular cartilage: a template for tissue repair. Clin Orthop Relat Res 391S:S26-S33

Rachev A, Greenwald SE (2003) Residual strains in conduit arteries. J Biomech 36:661–670

Romanic AM, Adachi E, Kadler KE, Hojima Y, Prockop DJ (1991) Copolymerization of pNcollagen III and collagen I. pNcollagen III decreases the rate of incorporation of collagen I into fibrils, the amount of collagen I incorporated, and the diameter of the fibrils formed. J Biolog Chem 266:12703–12709

Roolker W, Patt TW, van Dijk CN, Vegter M, Marti RK (2000) The Gore-Tex prosthetic ligament as a salvage procedure in deficient knees. Knee Surgery, Sports Traumatology, Arthroscopy 8:20–25

Saim AB, Cao Y, Weng Y, Chang CN, Vacanti MA, Vacanti CA, Eavey RE (2000) Engineering autogenous cartilage in the shape of a helix using an injectable hydrogel scaffold. Laryngoscope 110:1694–1697

Sato O, Tada Y, Takagi A (1989) The biologic fate of Dacron double velour vascular prostheses - a clinicopathological study. Jpn J Surg 19:301–311

Shadwick RE (1999) Mechanical design in arteries. J Experimen Biol 202:3305–3313

Slevin M, Kumar S, Gaffney J (2002) Angiogenic oligosaccharides of hyaluronan induce multiple signaling pathways affecting vascular endothelial cell mitogenic and wound healing responses. J Biolog Chem 277:41046–41059

Thacker JG, Stalnecker MC, Allaire PE, Edgerton MT, Rodeheaver GT, Edlich RF (1977) Practical applications of skin biomechanics. Clin Plast Surg 4:167–171

van Andel CJ, Pistecky PV, Borst C (2003) Mechanical properties of porcine and human arteries: implications for coronary anastomotic connectors. Ann Thorac Surg 76:58–64

van der Zijpp YJ, Poot AA, Feijen J (2003) Endothelialization of small-diameter vascular prostheses. Arch Physiol Biochem 111:415–427

Ventre M (2006) Supramolecular assembly and mechanical properties of dermis. Doctoral Thesis. University of Naples Federico II

Vogel KG, Koob TJ (1989) Structural specialization in tendons under compression. Int Rev Cytol 115:267–293

Watanabe T, Hosaka Y, Yamamoto E, Ueda H, Sugawara K, Takahashi H, Takehana K (2005) Control of the collagen fibril diameter in the equine superficial digital flexor tendon in horses by decorin. J Vet Med Sci 67:855–860

Williamson AK, Chen CA, Sah RL (2001) Compressive properties and function–composition relationships of developing bovine articular cartilage. J Orthop Res 19:1113–1121

Wilson YG, Wyatt MG, Currie IC, Baird RN, Lamont PM (1995) Preferential use of vein for above-knee femoropopliteal grafts. Eur Vasc Endovasc Surg 10:220–225

Wolinsky H, Glagov S (1964) Structural basis for the static mechanical properties of the aortic media. Circ Res 14:400–413

Wolinsky H, Glagov S (1967) A lamellar unit of aortic medial structure and function in mammals. Circ Res 20:99–111

Woo SL, Abramowitch SD, Kilger R, Liang R (2006) Biomechanics of knee ligaments injury, healing, and repair. J Biomech 39:1–20

Woodfield TBF, Malda J, de Wijn J, Peters F, Riesle J, van Blitterswijk CA (2003) Design of porous scaffolds for cartilage tissue engineering using a three-dimensional fiber-deposition technique. Biomaterials 25:4149–4161

Yannas IV, Burke JF (1980) Design of an artificial skin. I. Basic design principles. J Biomed Mater Res 14:65–81

Yannas IV, Burke JF, Gordon PL, Huang C, Rubenstein RH (1980) Design of an artificial skin. II. Control of chemical composition. J Biomed Mater Res 14:107–132

Young BB, Zhang G, Koch M, Birk DE (2002) The roles of types XII and XIV collagen in fibrillogenesis and matrix assembly in the developing cornea. J Cell Biochem 87:208–220

Zdrahala RJ (1996) Small caliber vascular grafts. Part II: Polyurethanes revisited. J Biomater Appl 11:37–61

Zimny ML, Schutte M, Dabezies E (1986) Mechanoreceptors in the human anterior cruciate ligament. Anatomical Record 214:204–209

Chapter 3
Biomaterials for Tissue Engineering of Hard Tissues

Elisabeth Engel, Oscar Castaño, Emiliano Salvagni, Maria Pau Ginebra, and Josep A. Planell

Contents

3.1   Function and Structure of Bone.................................... 56
3.2   Bone Engineering 56
     3.2.1   Cells... 57
     3.2.2   Growth Factors.. 58
     3.2.3   Scaffolds 59
3.3   Synthetic Material for Bone Repair Biodegradable Scaffolds 60
     3.3.1   Bioactive Glass Ceramics 64
3.4   Natural Biodegradable Scaffolds for Bone Repair 66
3.5   Bio-Stable Materials.. 69
     3.5.1   Natural: Hydròxyapatite 69
     3.5.2   Synthetic Materials 72
3.6   Composite Materials 77
     3.6.1   Why Porosity is Needed? 80
     3.6.2   PLA-Glasses 81
     3.6.3   Collagen-Cement Composites............................... 81
     3.6.4   Hyaluronic Acid-collagen 81
     3.6.5   Coated Scaffolds 82
     3.6.6   Self Assembly Nanofibers for Biomineralization 82
3.7   Characterization Methods 83
     3.7.1   Porosity.. 83
     3.7.2   Microstructure 83
     3.7.3   Surface Properties 84
     3.7.4   Dynamic and Static Contact Angles Measurement 85
     3.7.5   Mechanical Properties..................................... 85
     3.7.6   Z Potential 85
     3.7.7   Solubility of the Different Compounds 86
     3.7.8   Adsorbed Protein Amount 86
     3.7.9   Statistical Study 86

J.A. Planell (✉)
Biomaterials, Biomechanics and Tissue Engineering Research Group, Institut de Bioenginyeria de Catalunya, IBEC. Universitat Politècnica de Catalunya. Baldiri Reixac, 13 08028 Barcelona, Spain
e-mail: josep.a.planell@upc.edu

M. Santin (ed.), *Strategies in Regenerative Medicine,*
DOI 10.1007/978-0-387-74660-9_3, © Springer Science+Business Media LLC 2009

3.8   Bone Disease Society Impact and Market 87
      3.8.1   Incidence of Bone Fractures 87
      3.8.2   Business Market on Bone Tissue Engineering 87
Questions/Exercises.. 89
References... 90

3.1 Function and Structure of Bone

Bone is a specialized connective tissue that, as a part of the skeletal system, has three functions: (1) mechanical: support and site of muscle attachment for locomotion; (2) protective: for vital organs and bone marrow; and (3) metabolic: as a reserve of ions, specially calcium and phosphate, for the maintenance of serum homeostasis, which is essential to life (see Chapters 14 and 15).

Fundamental constituents of bone are cells and the extracellular matrix (ECM). This ECM is formed by an organic and a mineral phase. The organic phase is mainly formed by collagen fibers (Collagen type I) and correspond to 90% of the total protein. The 10% left corresponds to other non-collagenous proteins form the organic matrix, as proteoglycans and glycoproteins (as osteoclacin, osteonectin, and osteopontin), and several growth factors. These proteins are incorporated on the collagen matrix during its formation or afterwards. Even though their function is not completely understood, they play a role on cell adhesion, regulation of the mineral deposition, and the control of osteoblasts and osteoclasts (bone matrix producing and resorbing cells) activity. The mineral phase is mainly formed by calcium and phosphorous which form spindle- or plate-shaped crystals of hydroxyapatite $[3Ca_3(PO_4)_2 \cdot (OH)_2]$. (A more detailed explanation of bone biology can be found in Chapter 14.)

Bone tissue is capable of maintaining an optimal shape and structure throughout life via a continual process of renewal through which it is able to respond to changes in its mechanical environment by "remodeling" to meet different loading demands, so maintaining an optimal balance between form and function. However, as a living tissue, bone requires a constant supply of oxygen and nutrients, and is limited in the size of fracture or defect to be able to restore healthy tissue. Furthermore, bone can suffer from pathological conditions, (e.g., cancer) and is subject to degeneration as a result of age and disease (i.e., osteoporosis). In these cases, patient, comfort and bone function can only be restored by surgical reconstruction.

3.2 Bone Engineering

Bone engineering is a good alternative to conventional treatments for fracture, nonunion, spinal fusion, joint replacement, and pathological loss of bone (Fig. 3.1).

Fig. 3.1 Scaffolds for bone regeneration. (**a**) PLA/bioactive glass (**b**) Calcium phosphate cement

A typical approach in bone tissue engineering is the use of cells entrapped into scaffolds (e.g., artificial ECM) which exhibit the ability to support cells adhesion, growth, and differentiation, and eventually induce formation of mineralized bone tissues. Many researchers have proven that all these components or a combination of some are fundamental for successful applications of bone tissue regeneration (Arnaud et al. 1999, Boden 1999, Solheim 1998, Niederwanger and Urist 1996). However, a "simpler" approach is to design intrinsic "osteoinductivity" into the scaffold, that is, the capability to recruit and stimulate the patient's own growth factors and stem cells. Through investigation of the mechanisms controlling bone repair in bone graft substitutes (BGSs), linking interactions between the local chemical and physical environment, scientists are currently developing osteoinductive materials (promoters of bone formation) that can stimulate bone regeneration through control of the scaffold chemistry and structure.

3.2.1 Cells

Cell sourcing is the first issue to address in the development of bioengineered bone. These cells can be immature osteoblasts or mesenchymal stem cells (MSCs). MSCs are non-hematopoietic, stromal cells that exhibit multilineage differentiation capacity being capable to give rise to diverse tissues, including bone, cartilage, adipose tissue, tendon, and muscle.

MSC for tissue engineering applications are harvested from circulating blood, and most commonly from the bone marrow. In vitro cultivation of cells from bone marrow was first explored by Alexander Friedestein and colleagues. They isolated the cells from long bones from guinea pig and later Maureen Owen isolated the cells from rabbit and also obtained osteocytes when implanting these cells into recipient animals.

The first isolation, in vitro cultivation and characterization of human mesenchymal cells (hMSCs), was carried out by Steve Haynesworth and Arnold Caplan (Haynesworth et al. 1992).

With the aim of tissue regeneration, successful engraftment of a population of stem progenitor cells is established by showing the long-lasting viability of physically and functionally integrated donor cells. Bone marrow stem cells (BMSCs) have been shown to be capable of functional grafting. As assessed in laboratory animals, BMSCs of donor origin support host hematopoiesis in the ectopic ossicle formed at the transplant site. Locally delivered together with proper biomaterials, BMSCs efficiently allow for healing bone defects of critical size (Petite et al. 2000, Kon et al. 2000).

Thus, BMSCs represent a clinically useful source of progenitors for bone and possibly other connective tissues (Prockop 1997). However, the increasing number of studies that provide evidence of unorthodox plasticity, together with attempts to cure more generalized mesenchymal diseases, has prompted investigation of the feasibility of systemic injection of BMSC.

Evidences that hMSCs should prove safe to administer to patients was shown by the reinfusion of autologous hMSCs into volunteers, with no deleterious effects after more than five years (Lazarus et al. 1995). Additionally, children with the debilitating disease osteogenesis imperfecta have been infused with bone marrow that contains hMSCs, and the early results suggested a promising therapeutic effect (Horwitz et al. 2002).

Another application for MSC for the repair of musculoskeletal tissues is the use of autologous MSCs that have been genetically modified to secrete BMPs. Although the significant clinical potential, their use may be financially impractical, because it would require the harvesting, isolation, and expansion of cells from every patient requiring treatment. The development of allogeneic or xenogeneic MSC based gene delivery systems may, therefore, prove to be more feasible for widespread clinical use, yet the complex immune responses leading to allogeneic or xenogeneic MSC rejection need to be characterized and attenuated before these approaches can be applied clinically. The transplantation of allogeneic organs typically results in graft rejection unless the host immune responses are decreased using long-term immunosuppressive therapies. The immunological issues related to long-term MSC survival are similar to those for routine organ transplants. (Helm and Gazit 2005).

3.2.2 Growth Factors

Growth factors (GFs) are peptides that regulate cell growth, function, and motility, resulting in the formation of new tissue. Bone GFs influence the synthesis of new bone by acting on the local cell population present in bone marrow and on bone surfaces. They either act directly on specific osteoblasts as local regulators of cell growth and function or by inducing angiogenesis (vascularization) (basic fibroblast growth factors 1 and 2, bFGF-1/2; vascular endothelial growth factor, VEGF), or osteogenesis by promoting endothelial or osteoprogenitor cell migration and differentiation. Bone matrix contains a great number of growth factors (Solheim 1998) including fibroblast growth factors (FGFs), insulin-like growth factor I and II (IGF-I, IGF-II), platelet-derived growth factors (PDGF), and the transforming growth factor beta (TGF-β) supergene family, which currently has 43 members and includes, among others, TGF-β 1–5 and the bone morphogenic proteins, BMP 2–16. The proteins of the TGF-β superfamily regulate many different biological processes, including cell growth, differentiation, and embryonic pattern formation.

BMPs play a critical role in modulating mesenchymal differentiation inducing the complete sequence of endochondral bone formation where cartilage forms first and is subsequently replaced by bone. Other GFs such as TGF-β, IGF, and FGFs all affect the already differentiated bone-forming cells, causing them to divide or increase secretion of ECM, and proteins.

3.2.3 Scaffolds

There are several requirements that a three-dimensional scaffold should fulfil (1) a scaffold has to have high porosity (such as $\geq$ 90%) (Karageorgiou and David 2005) and proper pore size (i.e., pores greater than 50 µm in diameter) (King 2004); (2) a high surface area; (3) biocompatibility (i.e., no cito-toxic); (4) proper mechanical integrity of the scaffold is needed to retain the predesigned tissue structure; (5) biodegradability is usually required, and the degradation rate should match the rate of neotissue formation; (6) the scaffold should interact with cells, promoting cell adhesion, growth, migration, and differentiation; (7) osteoconductivity (induction of osteoblastic cells adhesion, growth, and differentiation) and osteoinductivity; (8) if possible, the scaffold material should be radiographically distinguishable from the new bone, easy to produce, sterilize, and handle in the surgery room (Logeart-Avramoglou et al. 2005).

These constructs should have mechanical properties matching those of the host bone, although they have also been considered not essential (Logeart-Avramoglou et al. 2005). The ideal scaffold should bind to the host bone with a stable interface, should avoid formation of scar tissue, and then resorption in the body in concert with bone growth and eventually degrade in non-toxic excretable products thus leaving the newly regenerated bone tissue in its original state (Jones et al. 2006).

In addition, cells are sensible to the surface properties, bulk geometry (Magan and Ripamonti 1996), and topography of the biomaterial. Characteristics such as surface charge and wettability (hydrophobicity or hydrophilicity of the surface) can affect protein adsorption and consequently cell interactions, and the process of osteogenesis (i.e., the capacity to generate bone tissue de novo), linked to gene expression, and protein adsorption (King 2004).

In the last 10 years, much effort has been made to engineering the ideal matrix material and porous 3D polymeric scaffolds for bone regeneration. The main

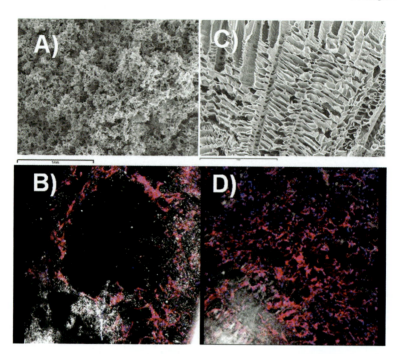

Fig. 3.2 Focus Ion beam scanning electron microscopy images of PLA/bioactive glass scaffolds fabricated by (**a**) solvent casting and (**b**) phase separation. Confocal microscopy images of osteoblast grown on (**c**) Solvent casting, and (**d**) phase separation

techniques, which have been developed to achieve this objective, are solvent casting and particulate leaching, scaffold coating, rapid prototyping, thermally induced phase separation, and microsphere coating, among others. Scientists have been investigating and developing osteoinductive materials that can promote the bone tissue regeneration by controlling the structure and chemistry of the scaffolds. The materials applied for these purposes can be summarized in three main groups: (a) synthetic and natural polymers, (b) inorganic materials (ceramics, glasses, and metals), and (c) a combination of both commonly known as composite materials. This matter will be herein discussed (Fig. 3.2).

3.3 Synthetic Material for Bone Repair Biodegradable Scaffolds

Many advantages make synthetic polymers suitable candidates for bone tissue engineering applications. They can reproducibly and convincingly be prepared in large scale, and compared to natural polymers, reduce the risk of infection, toxicity, and avoid immunologic troubles. A further advantage is that being the synthetic product of monomeric units they present good flexibility in their design. Thus, their properties can be controlled and tuned. For instance, their mechanical, physical, and chemical properties, such as elastic modulus, tensile strength, biodegradability, acidity, hydrophilicity, and hydrophobicity, can be tuned to fit the features of the host bone (Burg et al. 2000). In addition, they can incorporate growth factors, drugs, and other compounds to create multiphase delivery systems (Parikh 2002). Some properties of the discussed polymers are summarized in Table 3.1 (Peter et al. 1998)

The most widely used synthetic polymers for bone tissue engineering are two saturated poly-α-hydroxy esters: poly (lactic acid) (PLA) (Kulkarni et al. 1971, Zhang and Ma 1999) and poly (glycolic acid) (PGA) (Hollinger and Battistone 1986, Ishaug et al. 1997, Karp et al. 2003, Ma and Zhang 2001, Peter et al. 1998, and Yang et al. 2001). Due to their proven biocompatibility and biodegradability, these polymers also have the approval of the Food and Drug Administration (FDA). They undergo non-enzymatic hydrolysis and the products of degradation are natural non-toxic metabolites, being the last product of their metabolic pathway water and carbon dioxide. That means that they can be gradually destroyed, integrated, and eliminated by excretory system which removes metabolic wastes from the body and avoid an acidification of the media. Initially applied for surgical sutures, they successively revealed to be suitable as biodegradable fracture fixation implants. These polymers are synthesized first by condensation of glycolic or lactic acid in to low molecular weight polymers that after thermal treatment are converted in to the cyclic dimeric units, glycolide, and lactide, respectively. The dimeric units are then purified and polymerized by ring opening thus affording the high molecular weight polymers. As lactic acid is a chiral molecule, PLA exists in three different isomeric forms: L-PLA (PLLA), D-PLA (PDLA), and racemic mixture of D,L-PLA (PDLLA) (Fig. 3.3) (Yang et al. 2001).

Although PGA and PLA are very similar in their structures, the pendant methyl group on the alpha carbon of PLA renders these two polymers significantly different in their chemical, physical, and mechanical properties. PGA and the three isomeric forms of PLA vary in their crystallinity, molecular weight, and molecular weight distribution. In addition, PGA is more hydrophilic and acidic and PLA more hydrophobic and less acidic. Consequently,

Table 3.1 Properties of synthetic biodegradable polymers

Polymers type	Melting point (°C)	Glass. trans. temp. (°C)	Degradation time (months)[a]	Tensile strength (MPa)	Elongation (%)	Modulus (GPa)
PGA	225–230	35–40	6–12	>68.9	15–20	6.9
PLLA	173–178	60–65	>24	55.2–82.7	5–10	2.8–4.2
PDLLA	Amorphous	55–60	12–16	27.6–41.4	3–10	1.4–2.8
PLGA	Amorphous	45–55	Adjustable	41.4–55.2	3–10	1.4–2.8
PCL	58–63	65	>24	20.7–34.5	300–500	0.21–0.34

[a]Time to complete mass loss. Rate also depends on part geometry.

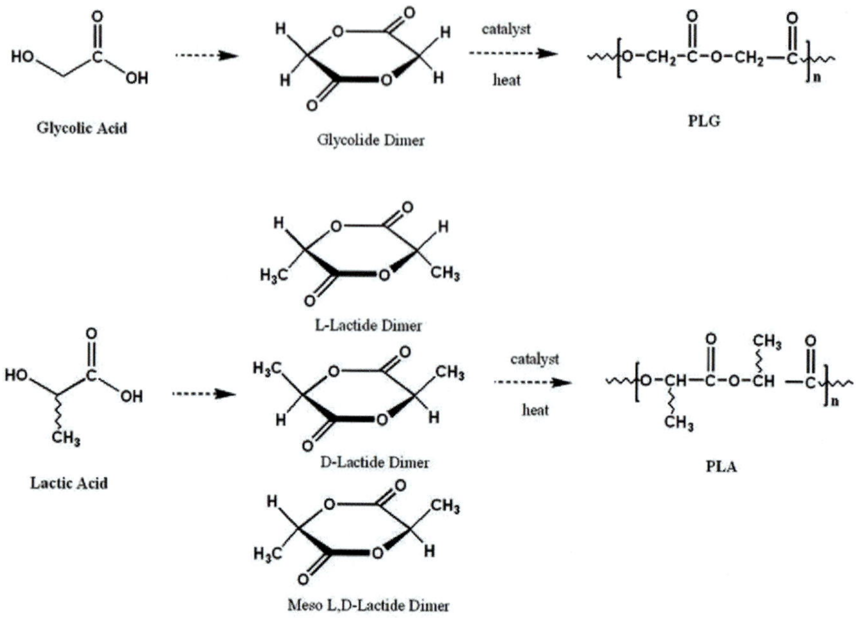

Fig. 3.3 Schematic synthetic pathway for the PGA and PLA synthesis

they differ in their bio-response and degradation rate when in contact with aqueous solution or a body fluid.

There are many factors that accelerate the degradation rate of polymers: more hydrophilic backbone, more hydrophilic end-groups, more reactive hydrolytic groups in the backbone, less crystallinity, and porosity. Therefore, a common strategy to control their biocompatibility and degradation rate is to modify the polymeric structure, for instance by synthesizing a racemic mixtures (D,L-PLA, PDLLA) or copolymers with different LA/GA ratios (PLGA) (Liu and Ma 2004). Indeed, PLLA and PGA exhibit high degree of crystallinity and degrade slowly, whereas the amorphous racemic polymer PDLLA and the copolymer PLGA show an increase on the degradation rate (Table 3.1, Fig. 3.4) (Miller et al. 1997).

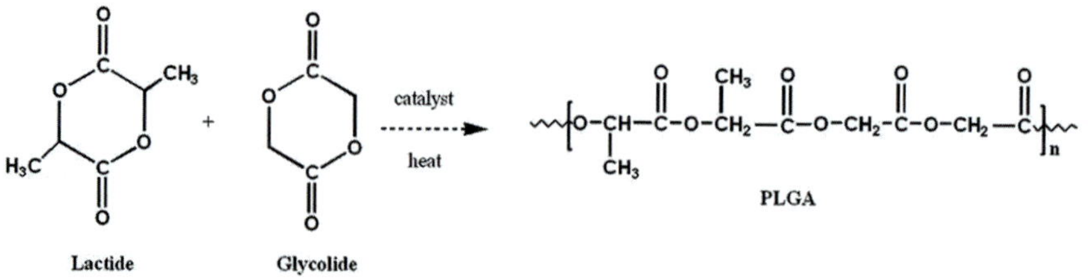

Fig. 3.4 Schematic synthesis of PLGA

Poly (ε-caprolactone) PCL is another aliphatic polyester that has been extensively studied for bone tissue regeneration. The rate of hydrolysis of this polymer is much slower than PGA and PLA being estimated up to several years (Yang et al. 2001, Rich et al. 2002) making it unsuitable for many tissue engineering applications. The rate of hydrolysis of these polymers can be summarized as follows (Rezwana et al. 2006) (Fig. 3.5):

$$PGA > PDLLA > PLLA > PCL$$

The slow degradation rate makes PCL a suitable candidate for long-term tissue regeneration purposes. Exploiting this feature many researchers have synthesized PCL-based copolymers with improved degradation properties (Choi and Park 2002).

Also Polypropylene fumarate (PPF, Fig. 3.6), an unsaturated biocompatible polyester, is appropriate for bone tissue engineering applications. The dissolution rate and mechanical properties can be regulated by changing the molecular weight. The products of degradation (i.e., propylene glycol and fumaric acid) are easily excreted by the body. Some researchers have proposed that PPF can be used as scaffold for guided tissue regeneration or as part of an injectable bone replacement composite (Yaszenski et al. 1995, Peter et al. 2000).

Fig. 3.5 Degradation time (half-life) of PGA and PLA polyemer and copolymers implanted in rat tissue (Miller et al. 1997)

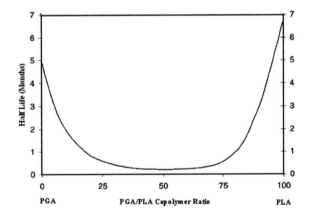

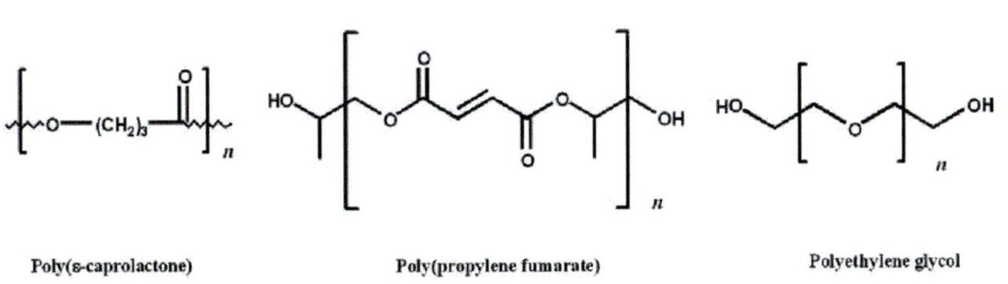

Fig. 3.6 Structure of PLC, PPF, and PEG

However, there are some limitations associated with polyesters used for bone tissue engineering purposes. The polyester acidity auto-catalyzes the rate of hydrolysis and locally lowers the pH thus giving undesirable tissue reactions. In addition, the lack of cell-recognition sites (e.g., pendant chemical functionalities) prevents or limits their bioactive behavior such as cell adhesion at the biomaterial host-tissue interface.

As a way to improve this situation, many researchers have investigated and developed materials that mimic the cell-extracellular matrix interactions, incorporating cell-binding peptides such as the arginine-glycine-aspartic acid (RGD) sequences. For instance, Polyethilene glycol-based hydrogels (PEG, Fig. 3.6) modified with covalently bound RGD peptide sequences showed an improved osteoblast attachment and spreading at high RGD concentrations (Behravesh et al. 2003).

3.3.1 Bioactive Glass Ceramics

Several inorganic compounds have also been applied for bone tissue regeneration. The most commonly used inorganic materials can be classified as bioactive glasses (BGs) and calcium phosphates. The use of these materials as bone substitutes started more than 30 years ago (El Gannham 2005) and over the time they have been extensively used as bone defect fillers (Vogel et al. 2001). The similarities between the bone mineral phase and their structural and surface features are accountable for their high bioactive properties. Indeed, bioactive ceramics have been reported to be biocompatible, osteoconductive, osteoinductive, and bind to bone with no mediation of a fibrous connective tissue interface (Meffert et al. 1985, Schepers et al. 1991) (Table 3.2).

Table 3.2 Biologically relevant calcium phosphate compounds (The symbol represents a vacancy in the crystallographic lattice) and their Ca/P ratios

Hydroxyapatite (HAP)	$Ca_{10}(PO_4)_6(OH)_2$	1.67
Amorphous calcium phosphate (ACP)	$Ca_x(PO_4)_y \, nH_2O$	1.2–2.2
Biological apatite (BA)	$Ca_{8.3}(PO_4)_{4.3}(CO_3-HPO_4)_{1.7}(OH)_{0.3}BA$ = carbonated CDHAP ($x = 1.7$)	1.38–1.93
Calcium-deficienthydroxyapatite (CDHAP)	$Ca_{10-x}(HPO_4)_x(PO_4)_{6-x}(OH)_{2-x}$ ($0 < x < 2$)	1.33–1.67
Dicalcium phosphate (phase α) (α-DCP)	$Ca_2P_2O_7$ (orthorhombic)	1
Dicalcium phosphate (phase β) (calcium pyrophosphate) (β-DCP)	$Ca_2P_2O_7$ (tetragonal)	1
Dicalcium phosphate anhydrate (monetite) (DCPA)	CaHPO	1
Dicalcium phosphate dihydrate (brushite) (DCPD)	$CaHPO_4 \cdot 2H_2O$	1

Table 3.2 (continued)

Fluorapatite (FA)	$Ca_{10}(PO_4)_6F_2$	1.67
Monocalcium phosphate anhydrate (MCPA)	$Ca(H_2PO_4)_2$	0.5
Monocalcium phosphate monohydrate (MCPM)	$Ca(H_2PO_4)_2 \cdot H_2O$	0.5
Octacalcium phosphate (OCP)	$Ca_8(HPO_4)_2(PO_4)_4 \cdot 5H_2O$	1.33
Oxyapatite (OA)	$Ca_{10}O(PO_4)_6$	1.67
Oxyhydroxyapatite (OHAP)	$Ca_{10}(PO_4)_6(OH)_{2-2x}O_{xx}$ $(0 < x < 1)$	1.67
Tetracalcium phosphate (TTCP)	$Ca_4O(PO_4)_2$	2
Tricalcium phosphate (α-TCP)	$Ca_3(PO_4)_2$ (monoclinic)	1.5
Tricalcium phosphate (whitlockite) (β-TCP)	$Ca_3(PO_4)_2$ (rhombohedral)	1.5

The most regularly used calcium phosphates as cell scaffolds for bone tissue engineering are hydroxyapatite [HAP, $Ca_{10}(PO_4)_6(OH)_2$] and β-tricalcium phosphate [(β-TCP), $Ca_3(PO_4)_2$], their derivatives, and their combinations. Due to their synthetic process, these two materials present different physical and chemical properties (El Gannham 2005). HAP shows good bioactive properties, however, under physiological conditions is chemically stable and remains almost insoluble. This occurrence reduces HAP bioactivity and resorbability hence leaving the material integrated into the regenerated bone tissue (Takahashi et al. 2005). Conversely, compared to HAP, TCP presents scarce porosity, smaller size grains, and high dissolution rate. TCP is, therefore, rapidly reabsorbed (Ginebra et al. 2006).

The bioactive glasses have been reported to induce bone growth three times faster than HAP, when implanted in bones (Fujishiro et al. 1997). BGs' network is made of silica (SiO_2), as the former component, and alkaline earth metals (e.g., magnesium and calcium), or alkali metals (e.g., sodium and potassium) as the network modifiers. When in contact with physiological solution as part of bone defect fillers, the bioactive glasses release ions that precipitate into a bone-like apatite on the host-bone surface, thus inducing bioactivity reactions (Ohtsuki et al. 1991, Neo et al. 1993, Anderson and Kangansniemi 1991). Due to their mechanical properties (tensile strength 40–60 MPa, elastic modulus 30–35 GPa), BGs are not suitable for load bearing applications. However, BGs revealed to be successful when applied as nonload-bearing material for bone repair in dental and orthopedic surgery (Ogino et al. 1980, Schepers et al. 1993, Stanley et al. 1997).

Although these materials show very high bioactive properties, they are mechanically brittle and, therefore, difficult to process in highly porous structures. A way to overcome this drawback is to combine them with synthetic or natural polymeric matrix. These composite materials can be elaborated in to high porous scaffolds with improved mechanical properties, which have been reported to promote the regeneration of three dimensional (3D) bone tissues.

3.4 Natural Biodegradable Scaffolds for Bone Repair

Proteins and polysaccharides herein discussed are natural polymers that have been used for tissue-engineered repair of bone.

The fibrous protein Collagen is the main non-mineralized component (70–90%) in the extracellular matrix of mammalian tissues, such as bone, cartilage, tendon, ligament (Thomson et al. 1995, Kadler et al. 1996, Badylak 2002, Girton et al. 2002, and Karp et al. 2003). Type I collagen is the most abundant among the 25 different types of collagen, and it has been reported to induce osteoblastic differentiation in vitro and osteogenesis in vivo (Mizuno et al. 1997). Collagen degradation is promoted by lysosomal enzymes. Enzymatic attack to collagen and, therefore, its resorption rate can be regulated by changing the extent of intermolecular cross linking and the density of its implants. Also mechanical properties, such as tensile strength and flexibility, can be adjusted by varying intermolecular cross-linking (Yarlagadda et al. 2005).

In addition to collagen, also Fibrin sealants are biological matrices totally reabsorbable and biocompatible. Due to their hemostatic and adhesive properties the fibrin sealants have been extensively used for bone repair. However, their capacity to stimulate bone tissue response and bone healing is still an open debate (Le Guéhennec et al. 2004) (Fig. 3.7).

Alginate and their derivatives are natural polysaccharides that have also been exploited for bone tissue engineering. These biocompatible hydrogels are

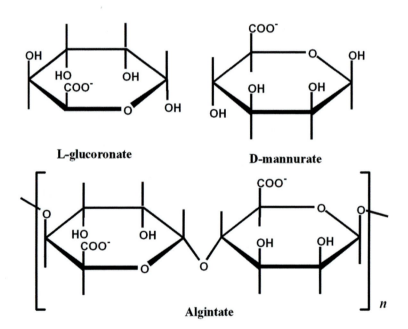

Fig. 3.7 Chemical structures of D-mannurate, L-glucoronate, and Alginate

derived from seaweed and can be injected into the body for cell and protein delivery (Rowley et al. 1999). The monomeric units of these copolymers are D-mannuronate and L-glucuronate (Fig. 3.5), and the concentration of the poly-glucuronate blocks in the alginate chains seems to affect their mechanical properties, which, however, can vary only in a limited range (Rowley et al. 1999). Some researchers have recently investigated paths to obtain biodegradable alginate with an improved range of the mechanical properties (Chen et al. 1999). A further drawback of alginate is the lack of a biological recognition domain (Fig. 3.8).

Other natural polymers used for tissue engineering are chitosan and hyaluronate (Fig. 3.6). Chitosan is a biodegradable and biocompatible polysaccharide obtained from the deacetylation of chitin, which is derived from exoskeletons of crustaceans. Chitosan has been recently used as osteogenic bone substitute showing a remarkable potential for bone tissue engineering applications (Shi et al. 2006). Hyaluronate or hyaluronic acid is a glycosaminoglycan and is one of the components of the extra cellular matrix found in connective, epithelial, and neural tissues. This material has lately gained reputation in tissue engineering. Recent studies have reported that hyaluronate, compared to the porous calcium phosphate ceramics, showed the capability to attach a greater number of cells per unit volume of implant (Rowley et al.1999) (Table 3.3).

Polyhydroxyalkanoates (PHA) are natural polyesters extracted by microorganisms that have also been explored for bone tissue engineering applications. Particularly, poly-3-hydroxybutyrate (PHB, Fig. 3.6), has been

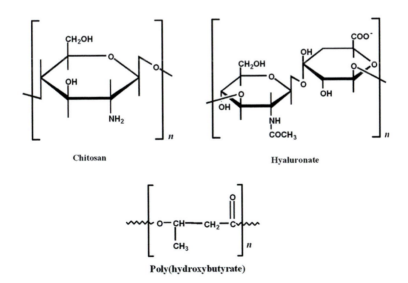

Fig. 3.8 Structure of Chitosan, Hyaluronate, and Poly(hydroxybutirate)

Table 3.3 Typical 3D scaffold composites

Scaffold composite		Percentage of ceramic (%)	Porosity (%)	Pore size (µm)	Compressive (C), tensile (T), flexural (F) strength (MPa)	Modulus (MPa)	Reference
Ceramic	Polymer						
1. Dense composites							
HAP fiber	PDLLA	2–10.5 (vol.)	—	—	45 (F)	$1.75–2.47 \times 10^3$	(Deng et al. 2001)
	PLLA	10–70 (wt.)	—	—	50–60 (F)	$6.4–12.8 \times 10^3$	(Kasuga et al. 2001)
HAP	PLGA	40–85 (vol.)	—	—	22 (F)	1.1×10^3	(Xu et al. 2004)
β-TCP	PLLA-co-PEH	75 (wt.)	—	—	51 (F)	5.18×10^3	(Kikuchi et al. 1999)
	PPF	25 (wt.)	—	—	7.5–7.7 (C)	191–134	(Meter et al. 2000)
A/W	PE	10–50 (vol.)	—	—	18–28 (B)	$0.9–5.7 \times 10^3$	(Juhasz et al. 2004)
Cortical bone					50–150(T)	$12–18 \times 10^3$	(Seal et al. 2001)
					130–180 (C)		
2. Porous composites							
Amorphous CaP	PLGA	28–75 (wt.)	75	>100		65	(Ambrosio et al. 2001)
HAP	PLLA	50 (wt.)	85–96	100×300	0.39 (C)	10–14	(Zhang 1999)
	PLGA	60–75 (wt.)	81–91	800–1800	0.07–0.22 (C)	2–7.5	(Guan et al. 2004)
	PLGA		30–40	110–150		337–1459	(Devin et al. 1996)
Bioglass®	PLGA	75 (wt.)	43	89	0.42 (C)	51	(Stamboulis et al. 2002)
	PLLA	20–50 (wt.)	77–80	~100 (macro) ~10 (micro)	1.5–3.9 (T)	137–260	(Zhang et al. 2004)
	PLGA	0.1–1 (wt.)		50–300			(Boccaccini et al. 2005)
	PDLLA	5–29 (wt.)	94	~100 (macro) 10–50 (micro)	0.07–0.08	0.65–1.2	(Terrier et al. 2004)
Phosphate glass A/W	PLA-PDLLA	40 (wt.)	93–97		0.017–0.020 (C)	0.075–0.12	(Li and Chang 2004)
Cancellous Bone	PDLLA	20–40 (wt.)	85.5–95.2	98–154	4–12 (C)	100–500	(Giesen et al. 2001)

investigated and showing promising results not giving any inflammatory response after 12 months. The disadvantage of these polymers is that their extraction process is very expensive.

There are, however, several disadvantages associated with the use of natural polymers. For instance, they may cause pathogen transmission and immuno rejection, and their biodegradability and mechanical properties are difficult to regulate. A further drawback is that their mechanical strength is too weak to provide adequate protection and structural support for the seeded osteoblasts. In addition, many of them are not readily available and expensive (Badylak 2002).

3.5 Bio-Stable Materials

3.5.1 Natural: Hydroxyapatite

Hydroxylapatite (hydroxyapatite or HAP) is the most chemically similar compound to natural bone mineral phase, which has as a formula $Ca_5(PO_4)_3(OH)$, but usually it is written $Ca_{10}(PO_4)_6(OH)_2$. The crystal lattice cell comprises two molecules. It crystallizes in the hexagonal crystal system, space group $P63_{/m}$ where $a = b = 9.422$ Å and $c = 6.880$ Å as a lattice cell parameters. The hydroxyl groups (OH^-) are aligned parallel to C-axis. Ca^{2+} cations are located in two different positions: around OH^- groups and in the center of each triangular prism which form the hexagon (Fig. 3.9).

Small carbonated-calcium deficient hydroxyapatite crystals are the mainly mineral component of bones, which are also formed by other phosphates such as Whitlockite [ß-Ca3(PO4)2] and octacalcium phosphate (OCP [Ca_8H_2 $(PO_4)6.6H_2O$]), etc. Calcium phosphates compounds can be found in the bone between 30 and 40%, joint with an organic matrix (mostly collagen 34–42%) and water (16–27%) (Table 3.4).

Pure HAP powder is white, however, natural apatites can also have brown, yellow or green colorations, because OH^- ion can be replaced by fluoride,

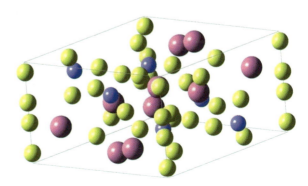

Fig. 3.9 Schematic drawing of the HAP cell where yellow atoms are OH^-, pink are calciums, and blue are phosphorus atoms

Table 3.4 Physical properties of hydroxyapatite

Properties	Typical values
Density (g cm^{-3})	3.15
Young's modulus (GPa)	85–90
Knoop hardness (MPa)	3450
Tensile strength (MPa m^{-2})	120
Poisson coefficient	0.3
thermal expansion ($\times\ 10^{-6}\ K^{-1}$)	11
Melting point (°C)	1660
Specific heat (cal g$^{-1}\ K^{-1}$)	0.15
Thermal conductivity (W cm$^{-1}\ K^{-1}$)	0.01

chloride or carbonate, showing different colors. OH^- can be substituted by carbonate groups (CO_3^{2-}), however, this is not usual. Instead, planar CO_3^{2-} groups can mainly replace phosphate groups (PO_4^{3-}) (Legeros 1991), generating a charge polarization that can be neutralized by a Ca^{2+} deficiency or by sodium (Na^+) substitutions.

Nowadays, therapies for any bone traumatism are mainly focused in surgeries which mainly require the substitution of damaged bone tissue in order to a correct regeneration. Therefore, there are two strategies to repair a damaged tissue: replace it (bioceramic coatings) or, on the other hand, substitute it, and regenerating it (bioactive ceramics).

3.5.1.1 Bioceramic Coatings

Hydroxyapatite coatings are mainly applied on metallic implants surfaces (most commonly titanium/titanium alloys and stainless steels). As a result, it is possible to obtain a hydroxyapatite-type material and avoid immunologic difficulties. It has showed a promising potential for use as a coating on metallic orthopedic and dental prostheses.

There are several methods for preparing hydroxyapatite coatings on metallic surface implant materials. Physical techniques used for processing hydroxyapatite coatings comprise vacuum plasma spraying (Chang et al. 1999), atmospheric plasma spraying (Heimann and Wirth 2006), detonation-gun spraying (Gledhill et al. 2001), RF magnetron sputtering (van Dijk et al. 1996, Nelea et al. 2005), electron beam evaporation (Lee et al. 2000, Lee et al. 2002), pulsed laser deposition (PLD)(Lee et al. 2002), ion beam sputtering (Wang et al. 2001), ion-beam assisted deposition (IBAD) and simultaneous vapor deposition (Hamdi et al. 2000). Chemical methods, including sol–gel, dip coating, hot-isostatic pressing (Onoki and Hashida 2006), electrophoretic deposition (Wang et al. 2002), electrochemical deposition (Zhang et al. 2006), μ-emulsion routes (Lim et al. 1997), and frit enameling (US patent 5916498), have also been used.

3.5.1.2 Bioactive Ceramics

A bioactive material means that not only it is able to be incorporated to the body bone structures without collapsing, dissolving or reacting, but also it can link itself to body tissues. HAP may be employed in forms such as cement-foams or as porous 3D scaffolds. These may take place when large part of bone has been damaged or when bone augmentations are required. The bone cement will provide injectable foam or a 3D scaffold promoting the lack of bone by naturally activating the formation of new tissue. In most cases, healing times and pain are reduced.

HAP or its precursors made in laboratory as a bulk do not have any osteogenic or oesteoinductive properties for its own alone, and show a low mechanical resistance and minimal immediate support. However, spontaneously osteoid is formed once it is attached to the bone surface due to the lack of soft interfacial material. Therefore, osteoid mineralizes and the resulting neobone is submitted to a remodified. In this way, osteoblastic and osteoclastic activities are promoted within the reabsorbed graft. HAP has generally a low resorption degree due to its high crystalline grade, which derivates in a mechanical stress focus. As a result, it is frequently mixed with other materials to improve their resorption characteristics.

Particularly, 3D high porous scaffolds and high surface area is a very good candidate to be vascularized and bony ingrowth. It is a perfect environment for bone cells and bone morphogenetic proteins (BMPs). After implantation, new bone gradually acquires a bone-likely mechanical strength.

On the other hand, apatitic or calcium phosphates cements (CPCs) have the properties to be injectable, hardening inside the damaged bone tissue, and generating a low heat transfer to avoid the premature death of cells. Thus, an inorganic-water mix is prepared joint a surfactant to obtain a suitable foam that can be easily injected promoting hardening and HAP phase transformation within the body (Traykova et al. 2006) (Fig. 3.10).

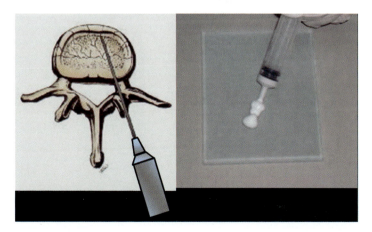

Fig. 3.10 An example of injectable calcium phosphate cement for bone regeneration

3.5.2 Synthetic Materials

3.5.2.1 Metals

Metallic materials are habitually used as biomaterials to replace structural components of the human body. Compared to polymeric and ceramic materials, they possess more superior tensile strength, fatigue strength, and fracture toughness. Hence, metallic biomaterials are used in medical devices such as artificial joints, bone plates, screws, intramedullar nails, spinal fixations, spinal spacers, external fixations, pace maker cases, artificial heart valves, wires, stents, and dental implants (Fig. 3.11).

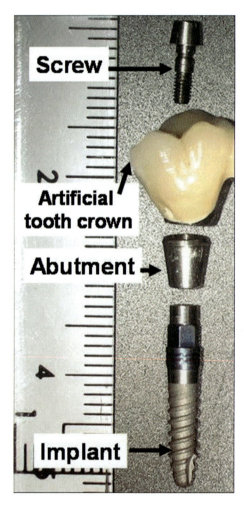

Fig. 3.11 Typical titanium dental implant. (Klockner SA Spain)

A big diversity of porous coatings and metals surfaces have been used as fix surfaces for bone ingrowth proposed to integrate implanted components to bone. Among them there are sintered cobalt-chrome beads, titanium alloy fiber metals, and plasma-sprayed surfaces. Recently, new techniques allow fabricate much more porous materials such as tantalum, which can be fabricated as metallic foam-like structure with interconnecting pores. In this case, it allows exceptionally rapid and complete bone ingrowth. Tantalum foam surface can be coated with a more biocompatible material and allowing a direct bonding of bone to porous surface (Soballe et al. 1993, Tisdel et al. 1994) (Fig. 3.12).

Titanium and its alloys are considered as biocompatible materials and are widely used in biomedical devices and components, especially as hard tissue replacements as well as in cardiac and cardiovascular applications, due to their advantageous mechanical properties, biocompatibility, corrosion resistance, and compatibility with magnetic resonance imaging procedures. Nevertheless, with their introduction came the risk of unintentional mixing of compatible systems with stainless steel components. In addition, surgeons may wish to mix components of different metals in order to use the best properties from them (Serhan et al. 2004).

Because of the above mentioned desirable properties, titanium and titanium alloys are widely used as hard tissue replacements in artificial bones, joints, and dental implants. As a hard tissue replacement, the low elastic modulus of titanium and its alloys is generally viewed as a biomechanical advantage because the smaller elastic modulus can result in smaller stress shielding (Fig. 3.13).

Type 316L stainless steels, cobalt–chromium alloys, commercially pure titanium, and Ti–6Al–4 V alloys are typical metallic biomaterials used for implants devices (Sumita et al. 2000, Sumita et al. 2004, ASTM Standards 2000, Williams 1998, Black and Hastings 1998). In spite of these materials were designed for industrial purposes, they show a relatively high corrosion resistance and excellent mechanical properties. However, when used as biomaterials, these materials pose several problems. These problems include toxicity of corrosion products and fretting debris to the human body, fracture due to corrosion fatigue and

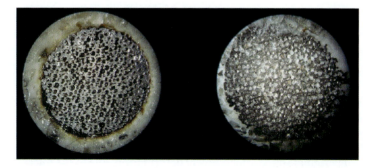

Fig. 3.12 Osteointegration in porous tantalum scaffold (Zimmer Dental Inc. USA)

Fig. 3.13 Titanium Hip prosthesis with ceramic cup and roughed metallics cup for improving osteointegration

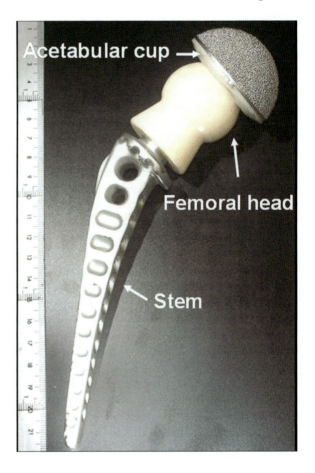

fretting corrosion fatigue, lack of biocompatibility, and inadequate affinity for cells and tissues. In particular, the toxicity problem brought about other problems such as allergy reaction, tumor formation, teratogenicity, and inflammation (IARC 1999, Teoh 2000, Nakayama et al. 1989).

In this sense, a functionalization of the metallic surface is required to avoid external corrosion, avoid toxicity, and mechanical problems (Liu et al. 2004). Therefore, some research groups have been doing a big effort to make them suitable as biomaterial. One option is hydroxyapatite, which shows very good corrosion resistance and biocompatibility. Although, in this case a good crystallinity microstructure is needed, similar to teeth enamel in contrast with the high amorphous degree of bone mineral phase, in contrast an amorphous structure is also needed to promote the bone tissue linkage to the own body.

Shape memory alloys (SMA) have a very promising future inside biomaterials (also memory metals, smart wires or smart alloys), which are alloys that return to its original shape after a severe deformation, due to the microstructure can switch easily from bcc phase to martensitic. This process can be promoted by heating (one-way effect) or unloading (superelasticity). These properties are acquired due to a temperature-dependent phase transformation from a low-symmetry to a highly symmetric crystallographic structure.

However, different works on nickel allergic factors on human body have shown a 7% present allergic reaction in different patients. Therefore, to improve mechanical alloys biocompatibility used as implants, it is necessary to study and develop Ni free shape memory alloys with low elastic modulus.

Materials having the memory effect are (Duerig et al. 1990, Shimizu and Tadaki 1987):

- Ag-Cd 44/49 at.% Cd
- Au-Cd 46.5/50 at.% Cd
- Cu-Al-Ni 14/14.5 wt.% Al and 3/4.5 wt.% Ni
- Cu-Sn approx. 15 at.% Sn
- Cu-Zn 38.5/41.5 wt.% Zn
- Cu-Zn-X (X = Si,Sn,Al) a few wt.% of XXX
- In-Ti 18/23 at.% Ti
- Ni-Al 36/38 at.% Al
- Ni-Ti 49/51 at.% Ni
- Fe-Pt approx. 25 at.% Pt
- Mn-Cu 5/35 at.% Cu
- Fe-Mn-Si
- Pt alloys
- Co-Ni-Al
- Co-Ni-Ga

Most Important Applications of SMA Devices in Biomedicine are the

Dental and Orthodontic Archwires – These work similar to a spring. They apply corrective and supportive appliances, braces, to realign crooked teeth, contrasting to the periodic, and painful tightening required by stainless steels.

3.5.2.2 Polymers

Prostheses since the 1960s have been produced by Poly(methyl methacrylate) (PMMA) bone cement (Charnley 1960) and nowadays, it rest as the most commonly utilized compound for fixation of orthopedic joint substitutions. In dentistry, zinc polycarboxylate and glass polyalkenoate cements received a major research interest from the 1970s to the present day (Kenny and Buggy 2003).

Polymeric biomaterials have several important uses in addition to tissue reconstruction. Examples include PMMA bone cement, PGA degradable sutures, PGA/PLA bone screws, and poly(vinyl siloxane) dental impression materials. Polymers such as PEG often are used to extend the circulation half-life of some drugs. Poly(hydroxyethyl methacrylate) is used to create soft contact lenses. Therefore, they have drastically improved the quality of life of many people (Seal et al. 2001)

Adsorbed, adhesive proteins, however, also mediate interactions of eukaryotic cells on biomaterials. In fact, the clinical success of bone implants is critically dependent on strong bonding of the surrounding bone tissue on implant-material surfaces. At the cellular level, such outcome requires specific adsorbed proteins mediating select cell (e.g., osteoblast but not fibroblast) adhesion and subsequent functions pertinent to neotissue formation. Today, these aspects, which are of critical importance in orthopedics and dentistry, have not been investigated (Anagnostou et al. 2006).

Several proteins (fibronectin, fibrinogen, vitronectin, and collagen) which are adsorbed on biomaterial surfaces may encourage bacteria adhesion/colonization that results in infections (Vaudaux et al. 1990). In particular, fibronectin is considered and described as the model host protein which promotes S. aureus attachment to biomaterial surfaces (Delmi et al. 1994, Peacock et al. 2000). To date, material-modification strategies, for example, coating biomaterials surfaces either with heparin (Lundberg et al. 1998) or with hydrophobic polymers (Jansen 1990) to reduce both protein adsorption and subsequent bacterial adhesion have been studied. Moreover, ionic groups such as sulfonate and carboxylate randomly distributed along the macromolecular of select polymers such as polystyrene (Berlot-Moirez et al. 2002) and PMMA (El Khadali et al. 2002 and Latz et al. 2003) also promote specific interactions with adhesive proteins leading to subsequent inhibition of S. aureus adhesion (Berlot et al. 2002). Specifically, silicone prostheses coated with a carboxylate- and sulfonate-bearing copolymer-induced inhibition of bacteria adhesion in the range 40–90% (Cremieux et al. 2003). Among the polymers used in biomedical applications, PMMA or bone cement is widely used (either moulded, in bulk or as coatings on implant materials) to secure bone implants in situ and, thus, meet various clinical needs.

Several exothermic reaction in hardening problems have been also observed in PMMA which leads to the prosthesis to fail by aseptic loosening (Stauffer 1982), through necrosis of the surrounding tissue. Therefore, in an effort to prolong prostheses lifetime, several research groups have focussed their activities in ionic polymer cements such as zinc and glass polyalkenoates in the late 1960s and early 1970s (Smith and Dent 1968, Wilson and Kent 1971) which eliminated the problems related to the exothermic reaction of PMMA cements. However, zinc promotes the formation of a fibrous collagen capsule around the zinc polycarboxylate cement in vivo (Peters et al. 1972, Nicholson et al. 1988) Demineralization of environment bone was also detected (Blades et al. 1998).

Resin modified glass ionomer were the result of gradually incremental changes to improve cement systems. In this material, the original acid-base reactions have been complemented via monomers and initiators which are able to be photoactive in order to polymerize (Mathis and Ferrancane 1989).

Several products are currently in the market such as hydroxyethyl methacrylate (HEMA) and bisphenol A-glycidylmethacrylate (Bis-GMA), etc. Their easy handling, aesthetics, and water resistance convert them in a very promising option against polyalkenoate cements. However, due to their polymerization shrinkage, they can cause with marginal sealing and retention (Cook et al. 1999). Then they need an external bonding agent to avoid this phenomenon, in dentistry.

3.6 Composite Materials

Polymer matrix reinforced with fibers or particles produce a family of polymer materials widely known as polymer matrix composites. Fibers are typically made of glass or carbon, polymers, and can be either long (continuous) or short (discontinuous) (Table 3.5).

The advance in composite scaffold materials is promising as gainful properties of two or more types of materials that can be associated to better fulfil the mechanical and physiological demands of the host tissue. Two special properties from both materials are combined and synergically promoted such as the polymer formability and the bioceramic mechanical properties (Boccaccini

Table 3.5 Several typical 3D scaffolds preparation methods

Preparation method	Advantages	Drawbacks
Solvent casting/particle leaching (SCPL)	• Controlled porosity • Controlled interconnectivity (particles sinteritzation)	• Isotropic structures • Use of cell-toxic organic solvents • Thick scaffolds can contain residual salt particles inside
Thermally induced phase separation (TIPS)	• Tunable pore size and microstructure • High porosities (~95%) • Anisotropic pores • Interconnected pores micro-structures • No scaffold size limitations	• Shrinkage problems • Non scalable production • Use of cell-toxic organic solvents • Long sublimation times (2 days)
Microsphere sintering	• Grade and controlled porosity microstructures. • Adapted to complex shapes	• Use of cell-toxic organic solvents • Pores interconnection problems
Scaffold coating	• Fast and simple	• Pores can hinder the coating procedures • Use of cell-toxic organic solvents • Delamination
Rapid prototyping (solid free form)	• Porous miscrostructures can suit into host • Protein and cell encapsulation possible • Good interface	• Micro-scale resolution need to be improved • Sometimes use of cell-toxic organic solvents

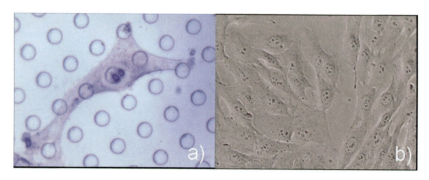

Fig. 3.14 Optical microscopy images of osteoblasts grown on (**a**) micro printed PMMA. (10 μm holes) and (**b**) flat PMMA

et al. 2003, Boccaccini and Maquet 2003, Ramakrishna 2001). In addition, poor bioactivity of some polymers can be avoided (Fig. 3.14).

Polymer/glass composite scaffolds achieve bioactive behavior by inclusions of coatings, the degree of bioactivity is tuneable by several parameters which range from the macro to the nanoscale, such as the ceramic volume fraction, particle size, and shape (Cao et al. 2006), type of particles inclusions, surface area, hydrophilicity, porous interconnectivity, topography, even using fibers instead of particles (Jaakkola et al. 2004) (Table 3.6).

In this sense, synthesis of 3D scaffolds has become in a very useful tool in order to combine both materials controlling their 3D structure, which should be biodegradable. Parallel to this process, the organism has to link the scaffold and renovated it into new bony tissue. Preferably, the degradation and resorption kinetics of composite scaffolds are planned to smooth the progress of cells to proliferate and growing blood vessels, while the scaffold is substituted and they are replaced by the body's own cells. In this sense, they finally leave place for new cell and bony tissue growth (Fig. 3.15).

Consequently, the physical support provided by the 3D scaffold should be maintained until the regenerated tissue has appropriate mechanical resistance. In related research, it has been studied that calcium phosphate derived compounds particles/polymer composites hydrolyzes homogeneously leading to degradation of all the volume at the same time (Shikinami and Okuno 2001).

Surface chemistry, topography, and surface energy become a key issue to have success in proteins attachment and adhesion, through which cell adhesion take place. This process can influence the kind or proteins adsorbed their conformation, orientation, and kinetics adsorption and, consequently, can modify the affect different aspects of cell behavior such adhesion, proliferation, differentiation, morphology, orientation, contact guidance, tissue organization, mechanical interlocking, production of local factors and microenvironments, and also cell selection (Navarro et al.2006, Zinger et al. 2004) through integrins linkage (Fig. 3.16).

Table 3.6 Overview of several surface modification methods

Surface fuctionalization method	Layer characteristics	Goal
Chemical methods		
Chemical treatments		
Acid etching	~10 nm of surface oxide layer	Remove oxide scales and contamination
Alkaline etching	~1 μm of sodium titanate gel	Improve biocompatibility, bioactivity, or bone conductivity
Peroxide etching	~ 5 nm of surface oxide layer. Dense inner and porous outer	Improve biocompatibility, bioactivity, or bone conductivity
Sol-Gel	~10 μm thick film: calcium phosphate, TiO_2, SiO_2, HAP.	Improve biocompatibility, bioactivity, or bone conductivity
Anodic oxydation	~10 nm to 400 μm of TiO_2 layer. Possibility of adsorption and incorporation of electrolyte anions.	Particular surface topographies; improve corrosion resistance; improve biocompatibility, bioactivity, or bone conductivity
MOCVD	~1 μm of TiN, TiC, TiCN, Diamond, and diamond-like carbon thick films	Improve wear and corrosion resistance, and blood compatibility
Biochemical methods	Surface functionalization by silanized TiO_2, photochemistry, self-assembled monolayers, protein-resistance, etc.	Induce specific cell and tissue response by means of surface-immobilized peptides, proteins, or growth factors.
Physical methods		
Thermal spray Flame spray Plasma spray HVOF DGUN	~10 to ~200 μm coating such as TiO_2, ZrO_2, Al_2O_3, calcium silicate, HAP	Improve wear and corrosion resistance, and biological properties.
Contact printing Micro Nano	~10 nm to ~10 mm surface topography modification ~1 nm to ~1 μm surface topography modification	Modulate cell behavior
PVD Evaporation Ion plating Sputtering	~1 μm of TiN, TiC, TiCN, Diamond, and diamond-like carbon thick films.	Improve wear and corrosion resistance and blood compatibility

Particularly, human osteoblasts respond to controlled nanometer topography. They also respond to surface topography with altered morphology, proliferation, and adhesion. The behavior of the cells is influenced differently by the nanotopography, the micro topography or their combination (Firkowska et al. 2006). The suitable pore diameter to regenerate mineralized bone is usually compressed between 50 and 100 μm for non-load-bearing conditions.

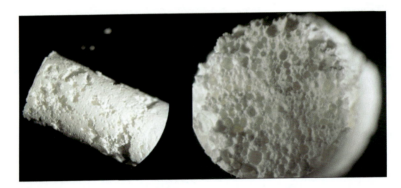

Fig. 3.15 High porous cement 3D scaffold

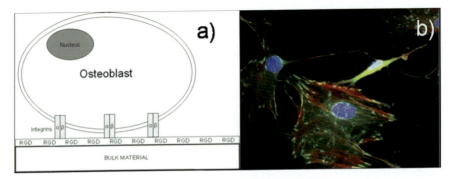

Fig. 3.16 **(a)** Eschematic picture of interaction of osteoblast on RGD grafted material surface by means of integrin receptors. **(b)** Confocal microscopy image of an osteoblast onto a RGD grafted surface where integrins anchorages appears in *red* color

3.6.1 Why Porosity is Needed?

Porosity is defined as the percentage of void space in a solid and it is a morphological property independent of the material. Macroporosity (pore size >50 μm) has a critical impact on, bone formation while microporosity (pore size ~10 μm) and pore wall roughness also play an important role because they contribute to increase the surface area and to higher protein adsorption, to ion exchange, and to bone-like apatite formation by dissolution and re-precipitation (Yuan et al. 1999). Surface roughness improves the integrin-dependent attachment of bone cells.

Currently, is commonly accepted that 3D scaffolds have to contain nano-porosity to allow diffusion of molecules for nutrition and signalling, micropores to ensure cell migration, and capillary formation as well as macropores for arteries and veins.

3.6.2 PLA-Glasses

Calcium phosphate glasses (CPG) are very good candidates for bone remodelling due to their mineral bone-like composition and their solubility easily tuned by adjusting chemical composition. Combination, of a bioabsorbable material such as PLA and soluble CPG leads to a highly degradable composite material and, therefore, to a vary promising and competitive alternative for tissue engineering.

The most widely glass system used is $CaO-TiO_2-Na2O-P_2O_5$ where CaO and TiO_2 enhance glass stability (Navarro et al. 2006).

3.6.3 Collagen-Cement Composites

The mineralization of collagen in vitro is of great interest for the understanding of the mechanisms underlying the mineralization in vivo as well as for the synthesis of improved bone grafts. The substitution of bone is a frequent issue in medicine. Currently autogenous bone, that is, bone which is taken from another place of the same patient, is the best available material for this application. Obviously, it is desirable to find an artificial material for this purpose. The scientists' aim is to develop a synthetic material which is as similar as possible to natural bone. Since collagen and calcium phosphate are the major constituents of human bone, bone implant materials consisting of these substances have been developed.

To achieve properties that are similar to the properties of bone, it is necessary to mimic not only the composition of the natural bone but also its structure. The structure of bone has been revealed down to the structure of the mineralized fibril, though some questions remain on the organization on the molecular level. Up to now, it has not been possible to produce a material in vitro with a structure similar to the structure of bone. This goal is to approach such a structure by biomimetic synthesis of the collagen-calcium phosphate composite (Bradt et al. 1999). In agreement with the biomimetic concept, the general feature of the processes occurring in the living organism during bone growth has been used as guidance. Learning from nature, it is intended to study the essential mechanisms of the formation of a bonelike hydroxyapatite-collagen composite.

3.6.4 Hyaluronic Acid-collagen

The fabrication of scaffolds from natural materials, such as hyaluronic acid and collagen, can impart intrinsic signals within the structure that can enhance tissue formation (Segura et al. 2005). Hyaluronic acid is a glycosaminoglycan copolymer of glucuronic acid and acetyl-glucosamine, and is a major intracellular component of connective tissues such as the synovial fluid of joints, vitreous fluid of the eye, and the scaffolding within cartilage and the umbilical cord. Hyaluronic acid has been shown to play an important role in lubrication, cell differentiation, and cell growth (Lee et al. 1994). Cellular interactions with hyaluronic acid occur through cell surface receptors (CD44, RHAMM, ICAM-l) and influence processes such as morphogenesis, wound repair, inflammation, and metastasis. Soluble hyaluronic acid has been used in clinical applications including ocular surgery, visco-supplementation for arthritis, and wound healing (Solchaga et al. 1999). However, the poor mechanical properties, rapid degradation, and clearance in vivo of uncrosslinked soluble hyaluronic acid limit many direct clinical applications. Soluble hyaluronic acid has been blended within collagen gels to provide the biological benefits of hyaluronic acid without the mechanical limitations (Park et al. 2002).

Applications of Polymer Matrix Composites:

Prostheses for bone fractures;
Hip prostheses;
Ligament and tendon prostheses;
Intervertebral disc prostheses;
Coating of intraosseus/intramedullar prostheses;
Bone repair;
Scaffolds for tissue engineering;
Gel-like structures.

3.6.5 Coated Scaffolds

Foams (Roether et al. 2002), fibrous bodies (Boccacini et al. 2003) or meshes (Day et al. 2004, Stamboulis 2002) can be synthesized by slurry dipping or electrophoretic deposition (EPD). Roether et al. (Roether et al. 2002) developed by slurry dipping macroporous PDLLA foams coated with Bioglass®. On the other hand, EPD was studied as an alternative being less suitable for injecting foams. Otherwise, EPD is highly recommended to introduce nanoparticles in a macroporous structures such as 3D scaffolds.

Coating HAP scaffolds with a HAP/poly(ε-caprolactone) composite generally enhances mechanical properties: higher amounts of the composite coating (more polymer) gradually increase compressive strength and elastic modulus.

3.6.6 Self Assembly Nanofibers for Biomineralization

Samuel Stupp and his group (Hartgerink et al. 2001) have proposed another approach for controlling crystal nucleation to fabricate a material that can resemble bone. The initiative is to prepare a nanostructured composite material that recreates the structural orientation between collagen and hydroxyapatite observed in bone. The composite is prepared by self-assembly, covalent

Fig. 3.17 Schematic draw showing self-assembled PA molecules into a cylindrical micelle

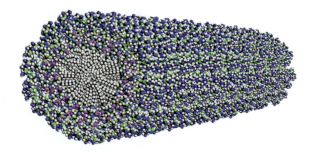

capture, and mineralization of a peptide-amphiphile (PA). This self-assembly is pH controlled and reversible. The mineralization experiments showed that the PA fibers were able to nucleate hydroxyapatite on their surfaces. This highly dynamic system of self-assembly and covalent capture may be easily modified through the selection of different amino acids for other applications in tissue engineering and mineralization (Fig. 3.17).

3.7 Characterization Methods

3.7.1 Porosity

To determine something so crucial such as porosity, porous distribution and connection, and specific surface we will use mercury porosimetry or N_2 B.E.T. porosimetry, which also allows measure specific surface and particle size. Specific surface is useful, for example, to evaluate the useful surface available for the protein and cellules adhesion. On the other hand, porosimetry is a useful tool to have a better acknowledgement of the fluid ability to pass through the material in order to predict the in vivo permeability.

3.7.2 Microstructure

The microstructure is usually evaluated by conventional scanning electron microscopy (SEM) and environmental scanning electron microscopy (ESEM). Both also allow observe surfaces characteristics. Focused Ion Beam-SEM (FIB-SEM) has currently emerged as a powerful technique to observe surface due to its high resolution. A modern FIB can deliver tens of nanoamps $(1 \times 10^{-9}$ A) of current to a sample, or can image the sample with a spot size on the order of a few nanometers. The basic different with SEM is that whereas the SEM uses a focused beam of electrons to image the sample in the chamber, a FIB instead uses a focused beam of gallium ions.

Fig. 3.18 Microtomography model of a hyaluronic acid/ bioactive glass 3D scaffolds where pores appears in red color

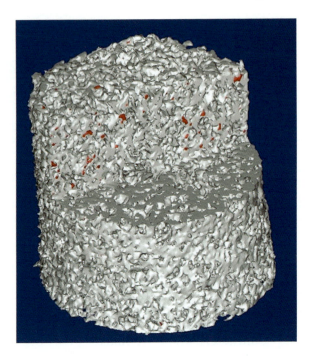

Microtomography, is a novel, powerful, and promising characterization method that allows to characterize 3D microstructual. It uses x-rays to create cross-sections of a 3D-object that later can be used to recreate a virtual model without destroying the original model (Fig. 3.18).

3.7.3 Surface Properties

It is shown that the contact between the biological material and the synthetic one occurs on the surface. A combination of several techniques to characterize different 2D and 3D structures from the developed materials is required to have the main comprehensive understanding of the mechanism which leads formation and properties of the different surfaces. Optic interferometer and SEM are commonly and widely used to for the surface characterization. First one is used to have a 2D and 3D view of the surface roughness. On the other hand, SEM is more focused to the qualitative side of the surface characterization. In addition, SEM allows having a semi-quantitative chemical composition analysis via energy dispersive spectroscopy couple (EDS).

TOF-SIMS is used to obtain compositional maps from the surface and atomic force microscopy (AFM) allows an accurate evaluation of topography in a nanometric scale.

3.7.4 Dynamic and Static Contact Angles Measurement

Onto disk and plates of the developed materials, the contact angle is measured by the *sessil drop* method in order to obtain the surface energy parameters of the surface. Angle measurement of different protein solvents with a different polarity allows obtaining different parameters related to the surface energy such as the dispersive one and the polar one. Within the polar forces are included the basic and the acid groups. Consequently, the measures of dynamic contact angles in biological fluids (water included), the culture medium and the mono-protein solutions allow to elucidate the behavior changes of the physico-chemical interaction which take place in the early phases of contact between substrate and biological medium. Finally, the contact angle hysteresis (difference between advanced and receding contact angle) allows searching the configuration changes related to the protein adsorption. These changes influence dramatically the hysteretic behavior of the contact angle due to its sensitivity to chemical and topographical heterogeneities. Therefore, the dynamic study of the contact angle also is very useful in the substrate characterization with a certain roughness, which, as it is known, influences the biological response *in vitro* as well as *in vivo*.

3.7.5 Mechanical Properties

The biological response to the mechanical stimulus is associated to the mechanical properties of a synthetic scaffold. A compression stress assays allow knowing the modulus of elasticity and the compression mechanical resistance for different materials (Fig. 3.19).

3.7.6 Z Potential

It consists in the net surface electric charge (sign and magnitude) of the implant materials in contact with biological fluids. To date it has been a neglected property. However, taking into account the especial influence that protein adsorption has over the biological comprehensive behavior of materials, the Z potential characterization can be crucial. Chemical and topography surfaces modifications induce a change in the Z potential. On the other hand, each protein in a particular medium has an isoelectric point (IP); in other words, the net charge of the protein. The electrostatic interaction between proteins and the implant material is described by the relation between the Z potential and the IP. This can allow scientists to predict protein affinity for material surfaces and conformational changes upon adsorption.

Fig. 3.19 Typical
compression assay of a
animal vertebra

3.7.7 Solubility of the Different Compounds

Usually it is determined by weight loss measurements in simulated body
fluid (SBF). Ions liberation to the medium is analyzed by Induced Coupled
Plasma-Mass Spectroscopy (ICP-MS) analysis studying the different degradation steps.

3.7.8 Adsorbed Protein Amount

It is studied by Fourier transformed infra-red spectroscopy (FTIR), conformational analysis. The protein distribution is also studied by confocal microscopy
and AFM.

3.7.9 Statistical Study

Data are usually analyzed through suitable statistical methods comparing
mean values from different materials (ANOVA with multiple comparison
procedures such as Fisher's Least Significant Differences or Turkey's Honestly
Significant Differences).

3.8 Bone Disease Society Impact and Market

Bone and joint diseases affect hundreds of millions of patients throughout the world, and are the leading cause of pain and disability, having a huge impact on individuals, families, societies, and economies. Bone and joint diseases account for half of all chronic conditions in people over 50 years of age in developed countries, which are set to sharply increase due to a predicted doubling in the number of people in the world over 50 years of age by 2020.

In developing countries, 10–15 million people are injured or disabled every year from road accidents, many of whom are young people.

3.8.1 Incidence of Bone Fractures

Osteoporosis is a systemic skeletal disease characterized by low bone mass and microarchitectural deterioration of bone tissue, with a consequent increase in bone fragility and susceptibility to fracture. (Anonymous. Consensus Development Conference: diagnosis, prophylaxis, and treatment of osteoporosis. Am J Med. 1993; 94:646-650.)

Osteoporosis affects an estimated 75 million people in Europe, USA, and Japan. 30–50% of women and 15–30% of men will suffer a fracture related to osteoporosis in their lifetime. By 2050, the worldwide incidence of hip fracture in men is projected to increase by 310 and 240% in women. (http://www.iofbo nehealth.org/facts-and-statistics.html#factsheet-category-16, http://www.bone andjointdecade.org/)

Osteoporotic fractures are generally assumed to be a major public health problem in the western world. Cost calculations are based primarily on hip fractures, which are easy to count since most require hospital admission and surgical intervention. The estimates of the cost, which are $10–20 billion annually in the United States to $0.3 billion in England and Wales are, therefore, likely to be underestimates of the total cost of the disease. As populations age throughout the world, the costs worldwide can only increase, and estimates are staggering. Within 50 years, the cost of hip fracture alone in the United States may exceed $240 billion.

3.8.2 Business Market on Bone Tissue Engineering

The potential of the tissue engineering market is enormous, despite the difficulty transitioning from a development stage industry to one with a successful product portfolio. A table trying to summarize the potential business on bone engineering due to the burden of skeletal diseases worldwide can be found below (Tables 3.7 and 3.8).

Table 3.7 Market on bone devices and biomaterials

Clinical procedure	Estimated incidence worldwide	Estimated number of procedures
Vertebral augmentation caused by osteoporotic bone or bone cancer	700,000	260,000
Screw augmentation		150,0000
Pedicle screws implanted/Long bone screw augmentation		280,000/35,000
Compression Screw Augmentation	750,000 hip fractures	250,000
Bone grafts trauma related	100,000	
Grafts for spinal fusion		200,000
Harvested material in non-spinal fusion		100,000
Spinal grafting		500,000
Bone defect grafting		700,000

Table 3.8 Overview of important companies in the bone implant and tissue engineering markets

Company	Products	Valued market (2002)	Applications
IsoTis (USA)	OsSatura SynPlug	646,000 €	Acellular scaffold for TE Cement for hip replacement
BioTissue Technologies (Germany)	BioSeed®-Oral Bone	250,000 €	Jawbone graft
Co.don (Germany)	Co.don osteotransplant	250.000 €	Tumors, pseudoarthroses, sarcomata, etc.
Osiris Therapeutics, Inc (USA)	Osteocel®	No data	Cellular product for TE
Cell factors (K)	Skeletex™		Growth factors and collagen for TE
ApaTech (UK)	ApaPore Actifuse	No data	Synthetic and porous Hydroxiapatite Silicate substituted calcium phosphate
Stryker Orthopedics (USA)	BoneSave	$3 billion (Stryker Corporation) (2005)	Granules 80% Tricalcium phosphate/20% Hidroxyapatite
Synthes (USA)	Norian	$2 billions (2005)	Injectable carbonated apatite cement
Orthovita	Vitoss	$ 27,672,221 (2005)	β Tricalcium phosphate
Interpore Cross	Pro Osteon	$ 8,037,326,823	Many products for hip, knee, elbow replacement

Other companies are commercializing growth factors. **Curis Inc** (USA, www.curis.com), focused on regulatory pathways that control repair and regeneration, the Hedgehog (Hh) pathway and the Bone Morphogenetic Proteins (BMPs). Several therapeutic products are on early to late preclinical stages. **Wyeth** (USA, www.wyeth.com), is carrying out studies on hBMP-2 and rhBMP-2. **Medtronic Sofamor Danek** (USA, www.sofamordanek.com), besides different allografts for spine treatment, cervical trauma, orthopedics applications, also have developed a rhBMP-2 (recombinant human bone morphogenetic protein- 2) formulation combined with a bovine-derived absorbable collagen sponge (ACS) carrier. These are only some examples from a quite long list of companies developing new products on this direction.

The field of bone engineering has an enormous potential, and this can be seen from the increase on number of biomedical companies that produce materials and scaffolds for bone grafting. However, there is still difficult the transition from discovery level science to clinical markets. Thus, the development of a system that fulfils completely the clinical and industrial needs is still far to be achieved.

Acknowledgments The authors would like to thank to The *Spanish Ministerio de Educación y Ciencia*, to the *Departament d'Innovació, Universitats i Empresa de la Generalitat de Catalunya*, and to the *Marie Curie Fellowship Association* for the research grants.
Some of the work described in this manuscript has been supported by Commission European Union: European project SMART-CaP ,Injectable Macroporous Biomaterials Based on Calcium Phosphate Cement Bone Regeneration, contract number: NMP3-CT-500465 ; European Project STEPS, A System Approach to Tissue Engineering Processes and Products, contract number FP6-500465.
The authors also would like to thank Dr. Samuel Stupp for kindly providing figure 17, and the companies Klockner SA, Spain, Zimmer Dental Inc. USA, as well as all the personnel of the Grup de Biomaterials, Biomecànica i Enginyeria de Teixits (BIBITE) for their contributions.

Questions/Exercises

1. Define the difference between inertness and bioactivity.
2. Provide a definition of polymers and give examples of biomaterial polymers for bone repair.
3. What are the differences between synthetic and natural polymers?
4. Why PMMA has been gradually substituted by alternative materials in orthopedic applications?
5. Why surface coating technology is important in biomaterials for bone applications?
6. Difference between hyaluronic acid and alginic acid in terms of structure and function.
7. Which are the advantages and drawbacks of using metals as biomaterials.
8. What does alloy mean?
9. Differentiate between ceramic and glass.

10. What is the main inorganic component of bones in teeth?
11. Define a bioactive 3D scaffold.

References

Ambrosio AM, Sahota JS, Khan Y, Laurencin CT (2001) A novel amorphous calcium phosphate polymer ceramic for bone repair: I. synthesis and characterization. J Biomed Mater Res 58:295–301

Anagnostoua F, Debeta A, Pavon-Djavidb G, Goudabyb Z, Hélaryb G and Migonneyb V (2006) Biomaterials 27:3912–3919

Anderson OH and Kangansniemi L (1991) Calcium phosphate formation at the surface of bioactive glass in vivo. J Biomed Mater Res 25:1019–1030

Annual Book of ASTM Standards 2000, Section 13 Medical Devices.

Arnaud, E, De Pollak C, Meunier A, Sedel L, Damien C, Petite H (1999) Osteogenesis with coral is increased by BMP and BMC in a rat cranioplasty. Biomaterials 20:1909–1918

Badylak SF (2002) Modification of natural polymers: Collagen. In: Atala A, Lanza RP (eds) Methods of tissue engineering. San Diego, CA: Academic Presspp. pp 505–514

Behravesh E, Zygourakis K, Mikos AG (2003) Adhesion and migration of marrow-derived osteoblasts on injectable in situ crosslinkable poly(propylene fumarate-co-ethylene glycol)- based hydrogels with a covalently linked RGDS peptide. J Biomed Mater Res Part A 65:260–270

Berlot, S, Aissaoui Z, Pavon-Djavid G, Belleney J, Jozefowicz M, Helary G, Migonney V (2002) Biomimetic poly(methyl methacrylate)-based terpolymers: modulation of bacterial adhesion effect. Biomacromolecules 3:63–68

Berlot-Moirez S, Pavon-Djavid G, Montdargent B, Jozefowicz M, Migonney V (2002), Modulation of *Staphylococcus aureus* adhesion by biofunctional copolymers derived from polystyrene. ITBM 23:102–108

Black J., Hastings G. (1998) Handbook of biomaterials properties, Chapman & Hall, London.

Blades MC. Moore DP. Revelli PA, Hill RG (1998) J Mat Sci Mat Med 9:701

Boccaccini AR, Blakera JJ, Maquet V, Day RM, Jérôme R (2005), Preparation and characterisation of poly(lactide-co-glycolide) (PLGA) and PLGA/Bioglass(R) composite tubular foam scaffolds for tissue engineering applications. Mater Sci Eng C 25:23–31

Boccaccini AR, Stamboulis AG, Rashid A, Roether JA (2003) Composite surgical sutures with bioactive glass coating. J Biomed Mater Res B: Appl Biomater 67B:618–626

Boccaccini AR, Maquet V (2003) Bioresorbable and bioactive polymer/Bioglass(R) composites with tailored pore structure for tissue engineering applications. Compos Sci Technol 63:2417–2429

Boden SD (1999) Bioactive factors for bone tissue engineering. Clin Orthop Relat Res 367:S84–94

Bradt J, Mertig M, Teresiak A, Pompe, W (1999) Biomimetic mineralization of collagen by combined fibril assembly and calcium phosphate formation. Chem Mater 11: 2694–2701

Burg KJL, Porter S, Kellam JF (2000) Biomaterial developments for bone tissue engineering. Biomaterials 21:2347–2359

Cao Y, Mitchell G, Messina A, Price L, Thompson E, Penington A, Morrison W, O'Connor A, Stevens G, Cooper-White JJ (2006) The influence of architecture on degradation and tissue ingrowth into three-dimensional poly(lactic-co-glycolic acid) scaffolds in vitro and in vivo Biomaterials 27:2854–2864

Chang C, Huang J, Xia J, Ding C (1999) Study on crystallization kinetics of plasma sprayed hydroxyapatite coating. Ceramics Int 25:479–483

Charnley J (1960) Anchorage of the femoral head prosthesis to the shaft of the femur. J Bone Joint Surg Br 42-B:28–30

Chen F, Yoo JJ, Atala A (1999) Acellular collagen matrix as a possible "off the shelf" biomaterial for urethral repair. Urology 54:407–410

Choi SH, Park TGJ (2002) Biomater Sci Polym 13:1163–1174

Cook WD. Forrest M, Goodwin AA. (1999) A simple method for the measurement of polymerization shrinkage in dental composites. Dent Mater 15:447–449

Cremieux AC, Pavon-Djavid G, Saleh Mghir A, Helary G, Migonney V (2003) Bioactive polymers grafted on silicone to prevent Staphylococcus aureus prosthesis adherence: in vitro and in vivo studies. JABBS 1:178–185

Day RM, Boccaccini AR, Shurey S, Roether JA, Forbes A, Hench LL, Gabe SM (2004) Assessment of polyglycolic acid mesh and bioactive glass for soft-tissue engineering scaffolds, Biomaterials 25:5857–5866

Delmi M, Vaudaux P, Lew DP, Vasey H (1994) Role of fibronectin in staphylococcal adhesion to metallic surfaces used as models of orthopaedic devices. J Orthop Res 12:432–438

Deng X, Hao J, Wang C (2001) Preparation and mechanical properties of nanocomposites of poly(d,l-lactide) with Ca-deficient hydroxyapatite nanocrystals. Biomaterials 22: 2867–2873

Devin JE, Attawia MA, Laurencin CT (1996) Three-dimensional degradable porous polymer-ceramic matrices for use in bone repair. J Biomater Sci Polym Ed 7:661–669

Duerig, TW. Melton KN. Stöckel D, Wayman CM (1990) "Engineering aspects of shape memory alloys". ISBN 0-750-61009-3. London: Butterworth Heinemann.

El Gannham A (2005) Bone reconstruction: from biocermamics to tissue engineering. Exper Rev Med Devices 2:87–101

El Khadali F, Hélary G, Pavon-Djavid G, Migonney V (2002) Modulating fibroblast cell proliferation with functionalized poly(methyl methacrylate) based copolymers: chemical composition and monomer distribution effect. Biomacromolecules 3:51–56

Firkowska I. Giannona S, Rojas-Chapana R. and Giersig M (2006) Qualitative evaluation of the response of human osteoblast cells to nanotopography surfaces based on carbon nanotubes. Technical Proceedings of the 2006 NSTI Nanotechnology Conference and Trade Show, Volume 2, Nanotech 2006 Vol. 2

Fujishiro Y, Oonishi H, Hench LL (1997) Quantitative comparison of in vivo bone generation with particulate Bioglass®. In: Sedel L, Rwy C (Eds) Bioceramics 10. Elsevier, NY, USA, pp 283–286

Giesen EB, Ding M, Dalstra M, van Eijden TM (2001) Mechanical properties of cancellous bone in the human mandibular condyle are anisotropic. J Biomech 34:799–803

Ginebra MP, Traykova T, Planell JA (2006) Calcium phosphate cements as bone drug delivery systems: A review. J Controlled Release 113:102–110

Girton TS, Barocas VH, Tranquillo RT (2002) Confined compression of a tissue-equivalent: collagen fibril and cell alignment in response to anisotropic strain. J Biomech Eng 124:568–575

Gledhill HC, Turner IG, Doyle C (2001) In vitro fatigue behaviour of vacuum plasma and detonation gun sprayed hydroxyapatite coatings. Biomaterials 22:1233–1240

Guan L, Davies JE (2004) Preparation and characterisation of a highly macroporous biodegradable composite tissue engineering scaffold. J Biomed Mater Res 71A:480–487

Hamdi M, Hakamata S, Ektessabi AM (2000) Thin Solid Films 377/378:484–489

Hartgerink JD, Beniash E, Stupp SI (2001) Self-assembly and mineralization of peptide-amphiphile nanofibers. Science 294:1684

Haynesworth SE, Goshima J, Goldberg VM, Caplan AI (1992) Characterization of cells with osteogenic potential from human bone marrow. Bone 13:81–88

Heimann RB, Wirth R (2006) Biomaterials 27:823–831

Helm GA, Gazit Z (2005) Future uses of mesenchymal stem cells in spine surgery. Neurosurg Focus 19:E13

Hollinger JO, Battistone GC (1986) Biodegradable bone repair materials—Synthetic-polymers and ceramics. Clin Orthop Rel Res. 290–305

Horwitz EM, Gordon PL, Koo WKK, Marx JC, Neel MD, McNall RY, Muul L, Hofmann T (2002) Isolated allogenic bone marrow-derived mesenchymal cells engraft and stimulate growth in children with osteogenesis imperfecta: Implications for cell therapy of bone PNAS 25:8932–8937 http://www.mindat.org – the mineral and locality database

IARC Monographs (1999) on the Evaluation of Carcinogenic Risks to Humans: Surgical Implants and Other Foreign Bodies, Lyon 74:65

Ishaug SL, Crane GM, Miller MJ, Yasko AW, Yaszemski MJ, Mikos AG (1997) Bone formation by three-dimensional stromal osteoblast culture in biodegradable polymer scaffolds. J Biomed Mater Res 36:17–28

Jaakkola T, Rich J, Tirri T, Narhi T, Jokinen M, Seppala J, Yli-Urpo A (2004) In vitro Ca-P precipitation on biodegradable thermoplastic composite of poly([epsilon]-caprolactone-co-lactide) and bioactive glass (S53P4). Biomaterials 25:575–581

Jansen B (1990) Bacterial adhesion to medical polymers—use of radiation techniques for the prevention of materials-associated infections. Clin Mater 6:65–74

Jones JR, Ehrenfried LM, Hencha LL (2006) Optimising bioactive glass scaffolds for bone tissue engineering. Biomaterials 27:964–973

Juhasz JA, Best SM, Brooks R, Kawashita M, Miyata N, Kokubo T, Nakamura T, Bonfield W (2004) Mechanical properties of glass-ceramic A-W-polyethylene composites: effect of filler content and particle size. Biomaterials 25:949–955

Kadler KE, Holmes DF, Trotter JA, Chapman JA (1996) Collagen fibril formation. Biochem J 316:1–11

Karageorgiou V, David K (2005) Porosity of 3D biomaterial scaffolds and osteogenesis. Biomaterials 26:5474–5491

Karp JM, Rzeszutek K, Shoichet MS, Davies JE (2003) Fabrication of precise cylindrical three-dimensional tissue engineering scaffolds for in vitro and in vivo bone engineering applications. J Craniofac Surg 14:317–323

Kasuga T, Ota Y, Nogami M, Abe Y (2001) Preparation and mechanical properties of polylactide acid composites containing hydroxyapatite fibres. Biomaterials 22:9–23

Kenny SM, Buggy M (2003) Bone cements and fillers: a review. J Mater Sci Mater Med 14:923–938

Kikuchi M, Tanaka J, Koyama Y, Takakuda K (1999) Cell culture tests of TCP/CPLA composite, J Biomed Mater Res 48:108–110

King KA (2004) Bone repair in the twenty-first century: biology, chemistry or engineering? Phil Trans R Soc Lond A 362:2821–2850

Kon E, Muraglia A, Corsi A, Bianco P, Marcacci M, Martin I, Boyde A, Ruspantini I, Chistolini P, Rocca M, Giardino R, Cancedda R, Quarto R (2000) Autologous bone marrow stromal cells loaded onto porous hydroxyapatite ceramic accelerate bone repair in critical-size defects of sheep long bones. J Biomed Mater Res 49:328–37

Kulkarni RK, Moore EG, Hegyeli AF, Leonard F (1971) Biodegradable poly(lactic acid) polymers, J Biomed Mater Res 5:169–181

Latz C, Pavon-Djavid G, Helary G, Evans MD, Migonney V (2003) Alternative intracellular signaling mechanism involved in the inhibitory biological response of functionalized PMMA-based polymers. Biomacromolecules 4:766–771

Lazarus HM, Haneysworth SE, Gerson SL, Rosenthal NS, Caplan AI (1995) Ex vivo expansion and subsequent infusion of human bone marrow derived stromal progenitor cells (mesenchymal progenitors cells): Implications for therapeutic use. Bone Marrow Transplant 15:935–942

Le Guéhennec L, Layrolle P, Daculsi G (2004) Review of bioceramics and fibrin sealant. Eur Cells Mater 8:1–11

Lee B, Litt M, Buchsbaum G (1994) Rheology of the vitreous body: part 3. Concentration of electrolytes, collagen and hyaluronic acid. Biorheology 31:339–351

Lee IS, Whang CN, Kim HE, Park JC, Song JH, Kim SR (2002) Mater Sci Eng: C 22:15–20

Lee SH, Kim HW, Lee EJ, Li LH, Kim HE (2000) Hydroxyapatite-TiO$_2$ hybrid coating on Ti implants. Biomaterials 21:469–473

Legeros RZ (1991) Calcium phosphates in Oral biology and Medicine, Karger, New York

Li H. and Chang J (2004) Preparation and characterisation of bioactive and biodegradable wollastonite/poly(d,l-lactic acid) composite scaffolds. J Mater Sci. Mater Med 15: 1089–1095

Lim GK, Wang J, Ng SC, Chew CH, Gan LM (1997) Processing of hydroxyapatite via microemulsion and emulsion routes. Biomaterials 18:1433–1439

Liu X, Ma PX (2004) Polymeric Scaffolds for Bone Tissue Engineering. Ann Biomed Eng 32:477–486

Liu X, Chu PK, Ding C (2004) Surface modification of titanium, titanium alloys, and related materials for biomedical applications. Mater Sci Eng: R: Rep 47:49–121

Logeart-Avramoglou D, Anagnostou F, Bizios R, Petite H (2005) Engineering bone: challenges and obstacles. J Cell Mol Med 9:72–84

Lundberg F, Gouda I, Larm O, Galin MA, Ljungh A (1998) A new model to assess staphylococcal adhesion to intraocular lenses under in vitro flow conditions. Biomaterials 19:1727–1733

Ma PX, Zhang RY (2001) Microtubular architecture of biodegradable polymer scaffolds. J Biomed Mater Res 56:469–477

Magan A, Ripamonti U (1996) Geometry of porous hydroxyapatite implants influences osteogenesis in baboons (Papio ursinus). J Craniofac Surg 7:71–78

Mathis RL, Ferrancane JL (1989) Dent Mat 5:355

Meffert R, Thomas J, Hamilton K, Brownstein C (1985) Hydroxylapatite as an alloplastic graft in the treatment of periodontal osseous defect. J Periodontol 56:63–73

Miller RA, Brady JM, Cutright DE (1997) Degradation rates of oral resorbable implants and (polylactates and polyglycolates): Rate modification with changes in PLA/POA ratios. J Biomed Mater Res 11:711

Mizuno M, Shindo M, Kobayashi D, Tsuruga E, Amemiya A, Kuboki Y (1997) Osteogenesis by bone marrow stromal cells maintained on type I collagen matrix gels in vivo. Bone 20:101–107

Nakayama Y, Yamamuro T, Kotoura Y, Oka M (1989) In vivo measurement of anodic polarization of orthopaedic implant alloys: comparative study of in vivo and in vitro experiments. Biomaterials 10:420–424

Navarro M, Aparicio C, Charles-Harris M, Ginebra MP, Ángel E, Planell JA (2006) Development of biodegradable composite scaffolds for bone tissue engineering: physicochemical, topographical, mechanical degradation, and biological properties. Adv Polym Sci 200:209–231

Nelea V, Morosanu C, Iliescu M, Mihailescu IN (2005) Hydroxyapatite thin films grown by pulsed laser deposition and radio-frequency magnetron sputtering: comparative study. Appl Surf Sci 228:346–356

Neo M, Nakamura T, Ohtsuki C, Kokubo T, Yamamuro T (1993) Apatite formation on three kinds of bioactive material at an early stage in vivo: a comparative study by transmission electron microscopy. J Biomed Mater Res 27:999–1006

Nicholson JW, Brookman PJ, Lacy OM, Savers GS, Wilson AD (1988) J Biomed Mater Res 22:623

Niederwanger M, Urist MR (1996) Demineralized bone matrix supplied by bone banks for a carrier of recombinant human bone morphogenetic protein (rhBMP-2), a substitute for autogeneic bone grafts. J Oral Implantol 22:210–215

Ogino M, Ohuchi F, Hench LL (1980) Compositional dependence on the formation of calcium phosphate films on bioglass. J Biomed Mater Res 14:55–64

Ohtsuki C, Kushitani H, Kokubo T, Kotani S, Yamamuro T (1991) Apatite formation on the surface of Ceravital-type glass-ceramic in the body. J Biomed Mater Res 25: 1363–1370

Onoki T, Hashida T (2006) New method for hydroxyapatite coating of titanium by the hydrothermal hot isostatic pressing technique. Surf Coat Technol 200:6801–6807

Parikh SN (2002) Bone graft substitutes: past, present, future. J Postgrad Med 48:142–148

Park SN, Park JC, Kim HO, Song MJ, Suh H (2002) Characterization of porous collagen/ hyaluronic acid scaffold modified by 1-ethyl-3-(3-dimethylaminopropyl)carbodiimide cross-linking. Biomaterials 23:1205–1212

Peacock SJ, Day NP, Thomas MG, Berendt AR, Foster TJ (2000) Clinical isolates of Staphylococcus aureus exhibit diversity in fnb genes and adhesion to human fibronectin, J Infect 41:23–31

Peter SJ, Lu L, Kim DJ and Mikos AG (2000) Marrow stromal osteoblast function on a poly(propylene fumarate)/β-tricalcium phosphate biodegradable orthopaedic composite. Biomaterials 21:1207–1213

Peter SJ, Miller MJ, Yasko AW, Yaszemski MJ, Mikos AG (1998) Polymer concepts in tissue engineering. J Biomed Mater Res 43:422–427

Peters WJ, Jackson RW, Iwano K, Smith DC (1972) The. biological responses to zinc polyacrylate cement. Clin Orthop 88:228

Petite H, Viateau V, Bensaid W, Meunier A, de Pollak C, Bourguignon M, Oudina K, Sedel L, Guillemin G (2000) Tissue-engineered bone regeneration. Nat Biotechnol 18:959–63

Prockop DJ (1997) Marrow stromal cells as stem cells for nonhaematopoietic tissues. Science 276:71–4

Ramakrishna S, Mayer J, Wintermantel E, Leong KW (2001) Biomedical applications of polymer-composite materials: a review. Compos Sci Technol 61:1189–1224

Rezwana K, Chena QZ, Blakera JJ, Boccaccini AR (2006) AR, Biodegradable and bioactive porous polymer/inorganic composite scaffolds for bone tissue engineering. Biomaterials 27:3413–3431

Rich J, Jaakkola T, Tirri T, Narhi T, Yli-Urpo A, Seppala J (2002) In vitro evaluation of poly([var epsilon]-caprolactone-co-DL-lactide)/bioactive glass composites. Biomaterials 23:2143–2150

Roether JA, Boccaccini AR, Hench LL, Maquet V, Gautier S, Jerjme R (2002) Development and in vitro characterisation of novel bioresorbable and bioactive composite materials based on polylactide foams and Bioglass$^{(R)}$ for tissue engineering applications. Biomaterials 23:3871–3878

Rowley JA, Madlambayan G, Mooney DJ (1999) Alginate hydrogels as synthetic extracellular matrix materials. Biomaterials 20:45–53

Schepers E, de Clercq M, Ducheyne P, Kempeneers R (1991) Bioactive glass particulate material as a filler for bone lesions. J Oral Rehabil 18:439–452

Schepers EJ, Ducheyne P, Barbier L, Schepers S (1993) Bioactive glass particles of narrow size range: a new material for the repair of bone defects. Implant Dent 2:151–156

Seal BL, Otero TC, Panitch A (2001) Polymeric biomaterials for tissue and organ regeneration , Mater Sci Eng: R: Rep 34:147–230

Segura T, Show, Anderson BC, Chung PH, Webber RE, Shull KR, Shea LD (2005) Crosslinked hyaluronic acid hydrogels: A strategy to functionalize and pattern. Biomaterials 26(4):359–371

Serhan H, Slivka M, Albert T, Kwak SD (2004) Is galvanic corrosion between titanium alloy and stainless steel spinal implants a clinical concern?, Spine J 4:379–387

Shi C, Zhu Y, Ran X, Wang M, Su Y, Cheng T (2006) Therapeutic potential of chitosan and its derivatives in regenerative medicine. J Surg Res 133:185–192

Shikinami Y, Okuno M (2001) Bioresorbable devices made of forged composites of hydroxyapatite (HA) particles and poly -lactide (PLLA). Part II: practical properties of miniscrews and miniplates. Biomaterials 22:3197–3211

Shimizu K, Tadaki T (1987) Shape Memory Alloys, In: Funakubo H (ed) Gordon and Breach Science Publishers, New York.

Smith DC (1968) Br Dent J 125:381

Soballe K, Hansen ES, Brockstedt-Rasmussen H, Bunger C (1993) Hydroxyapatite coating converts fibrous tissue to bone around loaded implants. J Bone Joint Surg Br 75:270–8

Solchaga LA, Dennis JE, Goldberg VM, Caplan AI (1999) Hyaluronic acid-based polymers as cell carriers for tissue-engineered repair of bone and cartilage. J Orthop Res 17:205–213

Solheim E (1998) Growth factors in bone. Int Orthopaedics 22:410–416

Stamboulis AG (2002) Novel biodegradable polymer/bioactive glass composites for tissue engineering applications. Adv Eng Mater 4:105–109

Stanley HR, Hall MB, Clark AE, King C, Hench LL, Berte JJ (1997) Using 45S5 bioglass cones as endosseous ridge maintenance implant to prevent alveolar ridge resorption: a 5 year evolution. Int J Maxillofac Implants 12:95–105

Stauffer RN (1982) Ten-year follow-up study of total hip replacement. J Bone Joint Surg Am 64:983–990

Sumita M, Hanawa T, Teoh SH (2004) Development of nitrogen-containing nickel-free austenitic stainless steels for metallic biomaterials-review. Mater Sci Eng C, 24:753–760

Sumita M, Ikada Y, Tateishi T (2000) Metallic Biomaterials—Fundamentals and Applications, ICP, Tokyo p. 629

Takahashi Y, Yamamoto M, Tabata Y (2005) Osteogenic differentiation of mesenchymal stem cells in biodegradable sponges composed of gelatin and β-tricalcium phosphate. Biomaterials 26:3587–3596

Teoh SH (2000) Fatigue of biomaterials: a review. Int J Fatigue 22:825–837

Thomson RC, Wake MC, Yaszemski MJ, Mikos AG (1995) Biodegradable polymer scaffolds to regenerate organs. Adv Polym Sci 122:245–274

Tisdel CL, Goldberg VM, Parr JA, Bensusan JS, Staikoff LS, Stevenson S (1994) The influence of a hydroxyapatite and tricalcium phosphatae coating on bone growth into titanium fiber-metal implants. J Bone Joint Surg Am 76:159–71

Traykova T, Aparicio C, Ginebra MP, Planell JA (2006) Bioceramics as nanomaterials, Nanomedicine 1:91–106

United States Patent 5916498.Method of manufacturing a dental prosthesis.

van Dijk K, Schaeken HG, Wolke JGC, Jansen JA (1996) Biomaterials 17:405–410

Vaudaux P, Yasuda H, Velazco MI, Huggler E, Ratti I, Waldvogel FA, Lew DP, Proctor RA (1990) Role of host and bacterial factors in modulating staphylococcal adhesion to implanted polymer surfaces. J Biomater Appl 5:134–153

Verrier S, Blaker JJ, Maquet V, Hench LL, Boccaccini AR (2004) PDLLA/Bioglass® composites for soft-tissue and hard-tissue engineering: an in vitro cell biology assessment. Biomaterials 25:3013–3021

Vogel M, Voigt C, Gross U, Müller-Mai C (2001) In vivo comparison of bioactive glass particles in rabbits. Biomaterials 22:357–362, 26:359–371

Wang C, Ma J, Cheng W, Zhang R (2002) Thick hydroxyapatite coatings by electrophoretic deposition. Mater Letters 57: 99–105

Wang CX, Chen ZQ, Guan LM, Wang M, Liu ZY, Wang PL (2001) Beam interactions with materials and atoms, fabrication and characterization of graded calcium phosphate coatings produced by ion beam sputtering/mixing deposition. Nucl Instrum Methods Phys Res B 179:364–372

Williams DF (1998) Medical and dental materials, materials science and technology vol. 14, VCH, Weinheim

Wilson AD, Kent BE (1971) J Appl Chem Biotechnol 21:313

Xu HHK, Simon JCG (2004) Self-hardening calcium phosphate composite scaffold for bone tissue engineering. J Orthop Res 22:535–543

Yang S, Leong KF, Du Z, Chua CK (2001) The design of scaffolds for use in tissue engineering. Tissue Eng. 7:679–689

Yarlagadda PKDV, Chandrasekharan M, Shyan JYM (2005) Recent advances and current developments in tissue scaffolding. Bio-Med Mater Eng 15:159–177

Yaszenski MJ, Payne RG, Hayes WC, Langer R, Aufdemorte TB, Mikos AG (1995) The ingrowth of new bone tissue and initial mechanical properties of a degrading polymeric composite scaffold. Tissue Eng 1:41–52

Yuan H, Kurashina K, de Bruijn JD, Li Y, de Groot K, Zhang X (1999) A preliminary study on osteoinduction of two kinds of calcium phosphate ceramics. Biomaterials 20:1799–1806

Zhang K, Wang Y, Hillmyer MA, Francis LF (2004) Processing and properties of porous poly(l-lactide)/bioactive glass composites. Biomaterials 25:2489–2500

Zhang RY, Ma PX (1999) Poly(alpha-hydroxyl acids)/hydroxyapatite porous composites for bone-tissue engineering. I. Preparation and morphology. J Biomed Mater Res 44:446–455

Zhang Y, Tao J, Pang Y, Wang W, Wang T (2006) Transactions of nonferrous metals society of China, vol. Elsevier, New York, pp 633–637

Zinger O, Anselme K, Denzer A, Habersetzer P, Wieland M, Jeanfils J, Hardouin P, Landolt D, (2004) Time-dependent morphology and adhesion of osteoblastic cells on titanium model surfaces featuring scale-resolved topography. Biomaterials 25:2695–2711

Chapter 4
Biomimetic and Bio-responsive Materials in Regenerative Medicine

Intelligent Materials for Healing Living Tissues

Jacob F. Pollock and Kevin E. Healy

Contents

4.1	Introduction to Biomimetic Materials	98
4.2	Biomimetic Rationale and Design Principles	99
4.3	Mammalian Tissue and Natural ECM as Design Guides	102
	4.3.1 ECM Composition and Structure	102
4.4	Tissue Dynamics	106
	4.4.1 ECM Metabolism	106
	4.4.2 Enzymatic Catabolism	108
4.5	Strategies for Biomimetic Tissue Regeneration	109
	4.5.1 Matrices and Tissue Conduction	110
	4.5.2 Growth Factors and Cellular Induction	111
	4.5.3 Cells and Tissue Neogenesis	113
4.6	Building Biomimetic Materials	115
	4.6.1 Starting Materials	115
	4.6.2 Natural Biopolymers	115
	4.6.3 Synthetic Polymers	117
	4.6.4 Hybrid and Composite Materials	118
4.7	Biomimetic Material Synthesis	118
	4.7.1 Macromers	118
	4.7.2 Conjugation Methods	119
	4.7.3 In Situ Material Synthesis	119
	4.7.4 Free Radical Polymerization	120
	4.7.5 Step-Growth Polymerization	120
	4.7.6 Physical Association	121
	4.7.7 Molecular Self-Assembly	122
4.8	Biomimetic Elements	122
	4.8.1 Cell Adhesion Domains	122
	4.8.2 Substrate Mechanics	131
	4.8.3 Enzymatic Degradability	133
	4.8.4 Growth Factor Activity	137
4.9	Neglected Topics in Biomimetic Materials	144
	Questions/Exercises	145
	Top Ten Original Publications from the Last Decade	146
	References	146

J.F. Pollock (✉)
Department of Bioengineering, University of California, Berkeley

M. Santin (ed.), *Strategies in Regenerative Medicine*,
DOI 10.1007/978-0-387-74660-9_4, © Springer Science+Business Media, LLC 2009

4.1 Introduction to Biomimetic Materials

Modern biomaterials are a key component of successful tissue regeneration strategies. The importance of biomaterials has been demonstrated in both clinical devices and experimental biomaterials as vehicles for the delivery of drugs and biologics, structural reinforcements, as well as templates for tissue growth (Ratner and Bryant 2004, http://www.tesinternational.org/ – Tissue Engineering and Regenerative Medicine International Society). Consequently, many advances in regenerative medicine have been correlated with improvements in biomaterial performance. Recent developments in the field of advanced biomaterials, such as *cell-mediated morphogen release* and *in situ material formation* (such as for cell encapsulation), indicate that biomaterials are critical for integrating the other components of tissue engineering into effective systems for properly controlling tissue production. Advances in polymer chemistry, chemical biology, and bioconjugate techniques have allowed for the improvement and creation of novel biomaterials with unique potential to control cell behavior and generate new tissue (Chaikof et al. 2002). The power of biomaterials science has become clear, not only in developing new materials for tissue regeneration, but also in understanding the performance of medical implants and devices.

Despite the clear beneficial results gained through the use of early-generation biomaterials, these devices serve mainly structural functions in reconstructive surgery and have limited integration with surrounding tissue. An initial strategy to improve the biological response of these materials was to modify them in order to minimize the default biological response to implanted materials, characterized by non-specific protein adsorption and inflammatory cell activation, which result in chronic inflammation and fibrosis. However, there has been a transition in biomaterials development away from modifying industrially available (i.e., "off-the-shelf") materials towards bio-inspired materials designed to specifically control cell behavior in order to promote tissue regeneration. The interfacial and degradation characteristics of these materials are engineered to interact at the cellular and molecular level via biochemical mechanisms, such that they regenerate specific tissue types and structures.

This chapter provides an overview of biomimetic materials that exploit natural mechanisms of cell response and tissue growth with an emphasis on in situ forming hydrogels. The rationale and principles of biomimetic design for tissue regeneration are covered along with the natural extracellular matrix (ECM) composition, structure, and dynamics that guide them. The fundamental strategies and properties of biomimetic materials are introduced, followed by a discussion of recent materials and methods as well as the key results that they have revealed. The chapter concludes with a critical perspective on neglected topics and potential future developments in the field of advanced biomaterials for tissue regeneration.

Information about the basic principles of material sciences, biomaterials, and polymers can be explored at the valuable educational websites listed below:

Material science
http://www.mrs.org/ – Materials Research Society
http://www.matweb.com/ – MatWeb Material Property Data
http://www.azom.com/ – The A to Z of Materials

Biomaterials
http://www.uweb.engr.washington.edu/research/tutorial.html – University of Washington Engineered Biomaterials Tutorial
http://www.btec.cmu.edu/tutorial/biomaterials/biomaterials.htm – Carnegie Mellon Biomaterial Tutorial
http://www.biomat.net/ – Biomaterials Network
http://www.biomaterials.ca/ius-bse.html – International Union of Societies for Biomaterials Science and Engineering

Polymers
http://www.cem.msu.edu/~reusch/VirtualText/polymers.htm – Michigan State University
http://www.pslc.ws/macrog.htm – Macrogalleria
http://plc.cwru.edu/tutorial/enhanced/files/polymers/Intro.htm – Case Western University Polymer Tutorial
http://www.chemistryland.com/PolymerPlanet/Polymers/PolymerTutorial.htm – ChemistryLand Polymer Tutorial
http://www.ausetute.com.au/polymers.html – AUS-e-TUTE Polymer Tutorial

4.2 Biomimetic Rationale and Design Principles

Natural processes of tissue physiology involve precise spatial and temporal coordination of physical and chemical cues. The micro-environments surrounding cells participating in these processes provide the critical signals necessary for properly regulating cellular activities related to tissue development, wound healing, and remodeling (http://www.uoguelph.ca/zoology/devobio/210labs/histo1.html – University of Guelph Developmental Biology Online). Biomimetic materials for regenerative medicine aim generally to recapitulate the natural processes of proper tissue dynamics and morphogenesis by providing guidance through cell-instructive and biologically responsive materials. This biomimetic rationale incorporates entire strategies for tissue regeneration into novel materials, diverging from previous paradigms regarding biomaterials in tissue engineering as simply depots for controlled release of molecules and scaffolds for the conduction of tissue growth.

Natural biological systems demonstrate profound diversity and functionality with integration of structures ranging from macromolecules to organisms to ecosystems. Biological macromolecules exhibit selective recognition and specific function that are critical to physiological processes (http://web.mit.edu/esgbio/www/lm/lmdir.html – Massachusetts Institute of Technology Large

Molecules Hypertextbook). For obvious reasons, engineers and material scientists are often inspired and guided by natural designs and mechanisms. The natural ECM serves as a model for the development of novel biomimetic materials and tissue constructs. Natural ECMs perform the functions desired for tissue formation, providing a powerful means of controlling the biological performance of regenerative materials. Therefore, the design principles of biomimetic engineering for regenerative medicine are guided by mammalian ECM composition, structure, and function, particularly biochemical reactions and interactions (http://www.visualhistology.com/Visual_ Histology_Atlas/ – Visual Histology, http://www.siumed.edu/~dking2/index.htm – Southern Illinois University School of Medicine Histology).

In natural developmental and healing processes, loose macromolecular networks serve as initial provisional matrices that are replaced by functional tissue. Matrix components are produced and deposited by resident cells, depending on their phenotype and surroundings. In this sense, biomimetic materials may only need to set in motion the natural process of tissue regeneration, as the ECM, signals, and cellular connections produced by cells within the construct may stimulate and guide later stages of tissue growth.

The principles of biomimetic materials design are derived from the natural processes which they intend to imitate. These mechanisms require precise coordination through macromolecular recognition, a defining characteristic of biomimetic biomaterials. Native tissue structures, receptor-ligand dynamics, and enzymatic activity involving such molecular coordination can be mimicked in biomaterials in order to the direct growth of the desired tissues. Native ECM structures have been mimicked through hydrogel and nano-fiber materials as well as through engineered macromolecular self-assembly.

Extrinsic cell response to the surrounding micro-environment include interactions with other cells, the surrounding matrix, and soluble signals through receptor-ligand binding, specific interactions at the plasma membrane that are amplified and translated into cellular response and activity by intracellular signaling, gene regulation, and metabolic pathways. Receptor-ligand interactions have been utilized in materials for regenerative medicine by incorporation and controlled release of ligands.

Enzymes are responsible for many cellular activities including degradation and remodeling of macromolecules of the ECM. Blood coagulation, clot resorption, and ECM metabolism involve enzymatic activity and have important functions in the natural wound-healing response. Enzymatic substrates in biomimetic materials have several advantages that go beyond those of traditional, non-specifically degradable biomaterials (Table 4.1). Enzyme action also enables novel methods of material formation and modification. Information about the specific role of enzymes relevant to regenerative medicine can be found at the educational website shown below:

http://www.fibrinolysis.org/NewFiles/Proteins.html   –   Fibrinolysis.org
    Proteases

Table 4.1 Comparison of the benefits of biomaterial degradation

Degradability	Enzymatic Biodegradability
	Permits cell-mediated migration and remodeling of the material
Allows the material to be completely replaced by regenerated tissue	Allows feedback between the rates of material degradation and tissue generation
Diminishes issues of chronic inflammation caused by long-term implants	Promotes formation of cell and material morphologies amenable to functional tissue formation
Eliminates the need for removal ,and, possibly, revision surgeries	Improves infiltration of native tissue through creeping substitution to maintain material integrity
Allows engineered release and degradation modes with pre-defined temporal profiles	Allows cell-demanded release of soluble components

http://wiz2.pharm.wayne.edu/biochem/enz.html – ACS Division of Medicinal Chemistry Enzyme Chemistry

http://web.mit.edu/esgbio/www/eb/ebdir.html – Massachusetts Institute of Technology Enzyme Biochemistry

http://users.rcn.com/jkimball.ma.ultranet/BiologyPages/E/Enzymes.html – Enzyme Information

http://users.rcn.com/jkimball.ma.ultranet/BiologyPages/E/EnzymeKinetics.html – Enzyme Kinetics Information

Integration of these design parameters allows the creation of artificial extracellular matrices capable of a bidirectional interaction with cells that approximates the reciprocal dynamics of natural tissue regulation.

A successful methodology of biomimetic biomaterials development has been to identify critical biological mechanisms, minimize the involved components, and integrate them into functional biomaterials. Biomimetic building blocks have been created to engineer materials with specific bioactivity and degradation behavior. Interchangeable functional components can be integrated in order to produce materials for tissue- and application-specific regenerative therapies. Cell fate and tissue morphogenesis can be controlled through independent modulation of biomimetic parameters (Table 4.2) (Saha et al. 2007). Precise control over biomaterial characteristics allows for the comprehensive study of biological response as well as powerful materials design and performance optimization platforms. It allows for parametric engineering design

Table 4.2 Biomimetic biomaterial design parameters

Structural architecture

Insoluble attachment and inductive ligands

Soluble effector molecules

Enzymatically degradable linkages

Mechanical properties

guided by biological rationales. The creation of precursor libraries additionally allows for high throughput screening to analyze biological performance and optimize composition for cell phenotype and tissue regeneration (Hubbell 2004). Furthermore, artificial micro-environments can be exploited as in vitro models for in vivo simulations of tissues to deconvolute the complex dynamics of cell-ECM interactions, provide valuable information about cell and tissue biology, and better understand and improve regenerative therapies (Abbott 2003, Griffith and Swartz 2006). Recent common themes incorporated into advanced biomimetic materials include receptor binding and signaling, cell-mediated degradation, and incorporation and controlled release of effector molecules.

4.3 Mammalian Tissue and Natural ECM as Design Guides

4.3.1 ECM Composition and Structure

4.3.1.1 ECM Types and Function

Tissues are composed of various cell populations residing within or on the surface of a particular ECM (http://web.indstate.edu/thcme/mwking/extracel lularmatrix.html – Indiana State University Medical Biochemistry ECM, http://www.ncbi.nlm.nih.gov/books/bv.fcgi?rid = mboc4.section.3532 – The National Institutes of Health Molecular Biology of the Cell ECM). The ECM anchors cells, provides a structural support, and performs critical biological functions in tissue development, homeostasis, and repair. The ECM provides environmental cues about tissue state which dictate gene expression and result-ing phenotype and stage of the cell cycle. It further regulates tissue by binding and limiting the diffusion of macromolecules. The ECM contains an abundance of biochemical information relevant to tissue physiology in the form of macro-molecular structures and interactions (Fig. 4.1).

Connective tissue has been the prototypical model for tissue structure and the target for many regenerative therapies due to its relatively low complex-ity, low cellularity, and potential for regeneration. Connective tissue is a heterogeneous and often anisotropic composite of ECM and cells that pro-duce and remodel the ECM, such as fibroblasts, chondroblasts, or osteo-blasts, as well as immune cells. Another common tissue structure is the epithelia found at luminal surfaces involved in protection, secretion, and absorption. Basement membranes are continuous sheets of ECM underly-ing epithelia of the luminal tracts, including blood vessels and glands. Base-ment membranes are very thin and acellular but critical to organ and tissue organization and dynamics. The basal lamina provides signals that pro-mote either self-renewal or differentiation and plays an important role in tissue development, maintenance, and regeneration. The basal lamina also

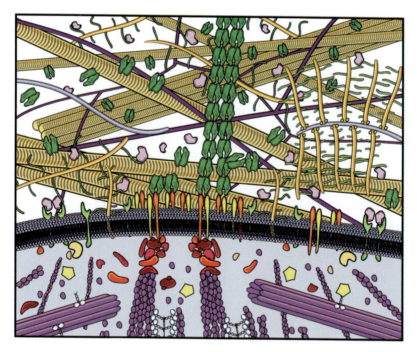

Fig. 4.1 Cells detect and engage their local physical and chemical environment through membrane receptors. Shown here is a generalized interaction between cells and the extra-cellular matrix (ECM). The ECM is composed of insoluble structural proteins as well as soluble polysaccharides and glycoproteins. Structural proteins include bundled collagen fibrils and a cross-linked elastin network. Glycosaminoglycans, some of which have an affinity for particular growth factors, are typically found bound to a protein core in the form of proteoglycans, which are further structured into aggregates on chains of hyaluronic acid. Glycoproteins, such as fibronectin, have multi-valent interactions with other ECM components, such as collagen, and integrin receptors. Membrane receptors, including integrins, trans-membrane proteoglycans, and growth factor receptors, bind to ECM components, serving as a sensory and mechanical link between the cell and its environment. Receptor binding induces intracellular signaling through membrane-bound and cytosolic signaling molecules and secondary messengers. Integrin engagement and clustering results in cytoskeletal organization through focal adhesion complexes and associated bridging proteins

segments and seals tissues, such as muscle and nerve bundles, into compartments. It serves as a barrier to cell migration and a selective filter for macromolecule diffusion. The importance of the basal lamina to both cell self-renewal and differentiation is demonstrated by the fact that epithelial cells, such as those of the skin and gut, are continually lost and replaced by proliferating stem cells, while those lining glands differentiate during development to form specialized terminal cell types (Giancotti and Ruoslahti 1999).

4.3.1.2 ECM Structure

The ECM of soft tissues is essentially a fiber-reinforced hydrogel composite of structural proteins, polysaccharides, and multi-adhesive glycoproteins. The aqueous portion of the ECM is considered to be highly organized by surrounding hydrophilic macromolecules, influencing molecular interactions and diffusion. Macromolecules of the ECM are characterized by complex, multivalent interactions. The combination of many weak intermolecular interactions guides self-assembly and provides integrity to the gel. ECM components form associative gels, the properties of which are naturally modified through biochemical cross-linking. ECM composition depends on tissue type (e.g., tendon, cartilage, bone, skin amongst connective tissues) and determines their mechanical properties and physiological role.

4.3.1.3 ECM Components

Structural proteins are the main source of mechanical integrity in the ECM. Specific intermolecular interactions result in the assembly of these proteins into fibrillar, rope-like, and sheet-like network structures, which resist tensional stresses experienced by tissues. The major structural components of mammalian ECM are collagens, a family of heterotrimeric proteins that share similar primary sequence repeat patterns of modified amino acids, but vary widely in ultra-structure and functionality. Collagen fibers, which can be several microns in diameter and longer than cells, are composed of inter- and intra-molecularly cross-linked protein strands assembled via enzymatic cleavage of precursors within the golgi apparatus prior to secretion. Certain collagens are known to greatly impact cell attachment, growth, and differentiation and have been used extensively in cell culture and developed as clinical medical treatments. Elastin is another structural protein that forms a cross-linked network of fibers which provide elastic deformability and resilience to certain connective tissues, such as those in the skin and bronchioles.

Polysaccharides are another important biopolymer component of ECMs. Glycosaminoglycans (GAGs) are complex, amino-substituted carbohydrates that tightly bind water, impart viscoelastic hydrogel characteristics to tissue, and resist compression (Table 4.3). They are typically composed of repeating disaccharide units and may be biochemically modified by sulfonation resulting in highly charged macromolecules that further resist hydrostatic stresses by electrostatic repulsion and osmotic pressure. GAGs are present commonly in ECM as proteoglycans, in which one or more GAGs are covalently tethered to a

Table 4.3 GAGs of the mammalian ECM

Dermatin (Sulfate)	Heparin (Sulfate)
Chondroitin (Sulfate)	Hyaluronan
Keratin (Sulfate)	

polymorphic protein core. Some GAGs and proteoglycans bind to growth factors, increasing local concentrations as a reservoir for delayed release, protecting them from degradation, influencing activity, and coordinating multivalent interactions with receptors. For example, basic fibroblast growth factor (bFGF) has an affinity for heparin sulfate, which effectively increases its activity through localized concentration, while transforming growth factor β (TGF-β) binds to the protein core of decorin, reducing its ability to bind receptors. Another major GAG of the ECM is hyaluranon, a very high molecular weight, linear chain which is non-sulfated and does not form proteoglycans. Hyaluronic acid plays an important role in tissue development and healing, forming a loose, porous network that provides an open space for further tissue growth. It is known to resist attachment of certain cell types but also plays a role in some initial and long-range cell adhesion events. GAGs interact with cells through non-integrin receptors, such as CD44, a cell surface glycoprotein, for hyaluronan, and syndecan, a trans-membrane proteoglycan, for other GAGs. Cell membrane-bound proteoglycans can also bind and localize growth factors to the cell-surface, enhancing their activity. Large, hierarchical, brush-like molecules of the ECM are composed of proteoglycans, such as aggrecan, terminally connected via linker proteins along a hyaluronic acid core and are responsible for some of the unique mechanical properties to cartilage.

Glycoproteins, also referred to as matricellular or multi-adhesive proteins, mediate interactions between matrix macromolecules and cell surface receptors. Glycoproteins are proteins modified by glycosylation with short, branched oligosaccharides. They contain multiple functional domains that bind to various other matrix components and to cell surface receptors. Glycoproteins are important for coordinating the assembly and structure of other ECM components and cell-matrix adhesions. Fibronectin is a high molecular weight, homodimeric glycoprotein that is largely responsible for cell adhesion, and influences cytoskeletal organization, cell cycle progression, and cell survival through interactions with integrins. Fibronectin has affinity for collagens, heparin, fibrin, and integrin receptors. It is assembled into fibrils at the cell surface which align with filaments of the cytoskeleton. Laminin is another large glycoprotein which is found predominantly in the basement membrane. It is a heterotrimeric protein that can form networks and bind to collagen IV, heparin, heparin sulfate, and integrin receptors. Laminin is known to influence cell attachment, differentiation, phenotype maintenance, and promote cell survival. Variation amongst similar glycoproteins and proteins of the ECM can arise from closely related gene families or alternative gene splicing. Many other glycoproteins are specific to particular ECMs, such as bone sialoprotein

4.3.1.4 ECM Receptors

Cell adhesion to the ECM and to other cells is mediated by cell-surface receptors. Integrins are a large family of divalent cation-binding, trans-membrane glycoprotein ECM receptors that bind to specific sequences of ECM molecules and form some cell-cell adhesions. They are composed of non-covalently linked heterodimers of α and β sub-units and form the structural link between the cytoskeleton of the cell and the ECM. High numbers of low affinity interactions with ECM macromolecules at localized concentrations result in adhesive forces responsible for cell attachment and migration. Various cell types express different amounts of each integrin receptor, which results in phenotype-specific response to the micro-environment, important for multi-cellular processes of tissue formation. Integrins and their associated ECM domains have specific binding affinities for ECM ligands but can bind more than one ligand type with varying strength; similarly, the ECM ligands often have varying affinity for multiple integrin combinations (Humphries et al. 2006). Integrins complement soluble ligand receptors in transmitting information about the micro-environment and allow substrate interrogation by cells. Differential binding of various integrin receptor types results in differing adhesion and interaction with tissue-specific ECMs and is a mechanism by which cells can interpret their biochemical and physical environment. Exposure and binding of certain ECM components upon injury, disease, or remodeling can initiate cellular mechanisms of wound healing and tissue generation. Futher information about integrins can be found at the educational website listed below:

http://walz.med.harvard.edu/Research/Cell_Adhesion/Integrins.php – Harvard Medical School Integrins
http://en.wikipedia.org/wiki/Integrin – Wikipedia Integrin Information

4.4 Tissue Dynamics

4.4.1 ECM Metabolism

Natural tissues and ECMs are amazing materials, not only because of their compositional complexity and important biological function, but also due to the fact that they are not static structures, but instead are capable of changing to maintain integrity, repair damage, or respond to applied stresses. Proper tissue homeostasis requires dynamic reciprocity between cells and their environment, in which cells maintain the ECM which, in turn, regulates cell behavior. Due to the inherent instability of biological macromolecules, most tissues, like cells, have defined but adjustable rates of turn-over, or replacement. Cell replacement is accomplished by mitosis and differentiation of stem and progenitor cells to continuously replace cells lost to damage, disease, or healthy tissue processes. Matrix replacement, on the other hand, is governed by macromolecular metabolism involving enzymatic breakdown and construction of ECM components. The feedback between cells and the ECM involves detection of adjacent cells and matrix, as well as local soluble signals, through cell surface receptors, integration of these inputs via intracellular signaling pathways, and alteration in gene expression pathways. Changes in cell phenotype are exhibited through

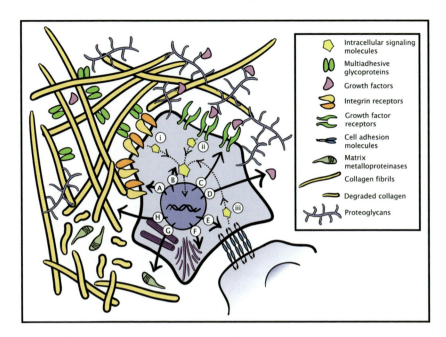

	Intracellular signaling molecules
	Multiadhesive glycoproteins
	Growth factors
	Integrin receptors
	Growth factor receptors
	Cell adhesion molecules
	Matrix metalloproteinases
	Collagen fibrils
	Degraded collagen
	Proteoglycans

Fig. 4.2 Bidirectional feedback between the cell and its environment results in reciprocal dynamics. Binding events of (i) integrin receptors, (ii) growth factor receptors, and (iii) cell-cell adhesions induce intracellular signaling that cross-talk through common transduction molecules. These signals are also integrated through transcription factor and gene expression networks to alter protein translation. Production of receptors (A,C, E) and intracellular signaling molecules (B) affect how cells sense their surroundings. Secretion of ECM components (H) and ECM degrading enzymes modify the surrounding matrix. Changes in the cytoskeleton (F) alter cell morphology, mechanics, and migration while secretion of growth factors (D) affects autocrine and paracrine signaling

altered production of membrane receptors, inductive signals, matrix molecules and precursors, enzymes, and intracellular signaling molecules and accessory proteins (Fig. 4.2). Impressively, tissues are able to maintain their mechanical integrity throughout the process of replacement during remodeling, as is particularly clear in the case of bone. This natural quality is an inspiration and guide to achieve the same property in biomimetic materials.

The processes of proper tissue formation involve multiple cell types that express various phenotypes depending on their relative position and the stage of tissue formation. These cells, of course, play an integral role in mediating the processes of tissue dynamics. Recruitment of particular progenitor cells and directed movement of cells during ECM construction and modification are achieved by migration in response to gradients in insoluble ECM-bound ligands and soluble signaling ligands, referred to as haptotaxis and chemotaxis. Gradients also play critical roles in embryogenesis and development; sharp local gradients result in cell polarization or migration and smoother gradients may result in segmentation due to differential threshold response. Cells cooperatively coordinate tissue dynamics through intercellular communication via endocrine (i.e., distant, global) and paracrine (i.e., proximate, local) signaling. Autocrine signaling, in which cells detect the same signal that they produce, is another a method by which cells can interpret their surroundings and establish feedback systems.

4.4.2 Enzymatic Catabolism

ECM degradation is an important process in tissue dynamics and is governed by enzymes that cleave structural macromolecules. Matrix metalloproteinases (MMPs) are a family of proteases characterized by a zinc-binding catalytic domain that are important to cell migration and tissue healing and remodeling (Lee and Murphy 2004). MMPs cleave ECM proteins and have been historically classified as collagenases, gelatinases, stromelysins, and membrane-type MMPs (MT-MMPs) (Steffensen et al. 2001). MMPs must be produced as precursors, known as pro-enzymes or zymogens, which are activated by other soluble MMPs or by MT-MMPs which are themselves activated in the golgi. Initiation of an auto-catalytic proteolytic cascade at the cell surface (Fig. 4.3) allows for local degradation, cell migration and extension, and cell-mediated release of growth factors (Stamenkovic 2003). MMPs can also activate or deactivate growth factors and cell-surface receptors by enzymatic cleavage or allosteric binding. Tissue inhibitors of matrix proteases (TIMPs) inhibit MMP activity with differing inhibitory specificity for the various MMPs. The activity of proteases is also diminished by more general protease inhibitors, which are present in plasma, and by membrane-anchored glycoproteins. Another important example of inhibition of enzymatic activity in tissue dynamics is the regulation of thrombin clotting activation by binding of anti-thrombins. Plasmin, a serine protease activated by tissue plasminogen activator, not only degrades fibrin clots, but can also activate MMPs.

MMPs are up-regulated during tissue morphogenesis and healing. Angiogenesis is a good example of MMP action in complex tissue development and remodeling (Rundhaug 2005). During angiogenesis, MMPs cleave cell-cell adhesions and detach cells from the ECM. They also release growth factors, expose cryptic integrin adhesion sites, and free ECM fragments which may promote or inhibit cell migration. MMPs play a role in ECM assembly as well by cleavage of precursor macromolecules (Kleiner and Stetler-Stevenson 1999). Cell migration in dense tissue in which the pore size is much smaller than that of the cell, such as the basement membrane, requires proteolytic degradation. In cancer, MMPs are responsible for metastasis and have been a recent target for inhibitors in clinical therapies (Forget et al. 1999). Cancer cells have also been shown to change motility modes, adopting an amoeboid pattern of migration achieved by cell and matrix deformation.

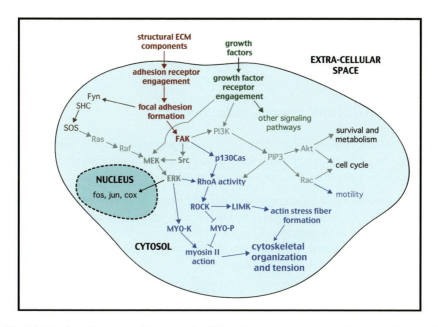

Fig. 4.3 Tensional homeostasis is maintained through trans-membrane and intracellular signal transduction and gene transcription. Cells engage ligands through receptors at the cell surface. Adhesion receptor engagement and focal contact formation are required for attachment, (bridging the mechanics of the cell and substrate), spreading, and subsequent cellular contractility. Intracellular signaling molecules transmit the activation of focal adhesions, by focal adhesion kinase (FAK), and growth factor receptors. These cross-talking signaling networks activate RhoA, which activates ROCK, leading to actin stress fiber development and myosin II action and ultimately cytoskeletal organization and tension

Enzymes other than proteases are also responsible for ECM component degradation and important to ECM catabolism. These include polysaccharide-cleaving enzymes such as hyaluronidase, chondroitinase, and heparinase, which degrade hyaluronan, chondroitin, and heparin, respectively. The expression of these enzymes, like proteases, is strongly influenced by growth factor signaling. Hyaluronan has relatively high rate of turnover in the ECM. The molecular weight of hyaluronan chains or fragments determines their effect on cell behavior, likely due to competitive binding of surface receptors and multivalent effects.

4.5 Strategies for Biomimetic Tissue Regeneration

Biomimetic tissue engineering aims to promoting regeneration of tissues by mimicking aspects of their structure and functionalities. Further information about the tissue engineering discipline can be found at

http://www.ptei.org – Pittsburgh Tissue Engineering Initiative
http://www.nidr.nih.gov/spectrum/NIDCR4/4menu.htm – The National
   Institutes of Health Tissue Engineering

4.5.1 Matrices and Tissue Conduction

The initial stages of wound healing and tissue regeneration involve formation of
a temporary matrix, such as a fibrin blood clot or a loose reticular network,
which allows cells to invade, replace, and remodel the tissue space

http://www.emedicine.com/plastic/topic411.htm – WebMD Wound Healing
http://www.uweb.engr.washington.edu/research/tutorials/woundhealing.html –
   University of Washington Engineered Biomaterials Wound Healing
http://www.emedicine.com/plastic/topic457.htm#section~types_of_ound_heal
   ing – eMedicine Wound Healing.

Therefore, many roles of biodegradable scaffolds in tissue regeneration
are biomimetic (Table 4.4). Materials and methods for forming artificial
ECMs, and biological responses to them, have been the focus of biomaterials
development for over a decade. Biomaterials for tissue engineering have
conventionally been considered scaffolds for the conduction of tissue growth.
Tissue conduction involves a material serving as a bridge to allow cell
migration and as a substrate upon which to build new ECM and ultimately
functional or scar tissue. Tissues with high natural regenerative capacity,
such as bone, often require only an interconnected scaffold placed within a
defect to promote new tissue formation. And, while many materials may act
as substrates for the growth of certain tissues, they may not control it very
well, resulting in fibrous scar tissue formed within or around them. Tissue
response indeed depends not only on the scaffold, but also on the site of
implantation; certain tissues will not regenerate properly when simply provided a physical template.

Table 4.4 Roles of biomaterials as artificial cellular matrices

Provide a physical form and space for cells to form tissue
Resist implantation and in vivo mechanical forces until reinforced and replaced by native tissue
Provide a background upon which to present cell adhesion ligands
Store, maintain the activity of, and control the release of soluble components
Act to deliver and spatially organize cells and inductive signals
Morphologically guide tissue growth
Allow transport of oxygen, nutrients, metabolites, and degradation products betweens cells and surrounding tissue
Act as surrogate micro-environments to direct cell behavior

Synthetic matrices are typically produced in the form of porous solids, hydrogels, or nano-fibrous materials. Hydrogels are composed of hydrophilic polymers, natural or synthetic, that are rendered insoluble through intermolecular covalent (chemical) or associative (physical) bonding. Hydrogels are predominantly composed of water, which results in very low surface energy in aqueous environments which limits non-specific protein adsorption and cell interaction.

An important chemical characteristic of biomaterial scaffolds is hydrophilicity. The hydrophilic quality of the base material influences its water content and, consequently, mechanical, permeability, and degradation characteristics. While the interconnected nature of *solid porous scaffolds* promotes mass transport, facilitates cell seeding and migration, and improves integration with existing tissue, they appear as two-dimensional surfaces at the scale of cells. *Hydrogels* and *nano-fibers* resemble the structure of the natural ECM and provide cells with a true three-dimensional environment with which to interact. These materials have high diffusion rates due to their low mass and volume fraction. Nano-fibers have a very high surface area that allows for growth factor adsorption and release while hydrogels control macromolecule release through molecular encapsulation and entanglement.

4.5.2 Growth Factors and Cellular Induction

Inductive signaling molecules produced by cells govern patterns of tissue development and regulate interactions in tissue physiology. In biomaterials, these serve as a means of controlling cell behavior. They provide the proper biochemical cues to complement the chemical and mechanical cues from matrix. Inductive signals can be classified as mitogens, which induce proliferation, morphogens, which induce differentiation, and cytokines, which promote cell migration. The same signal may produce multiple effects, depending on the cell type and other incidental signals. These inductive signals are generally referred to as "growth factors." Basic principles of biochemistry, growth factors, and cell metabolism and signaling can be found at the educational websites listed below:

Biochemistry
http://web.indstate.edu/thcme/mwking/home.html – Indiana State University Medical Biochemistry Page
http://employees.csbsju.edu/hjakubowski/classes/ch331/bcintro/default.html – St. John's University Biochemistry Online

Molecular and Cell Biology
http://www.cytochemistry.net/cell-biology/ – University of Arkansas Cell Biology Reference

http://web.mit.edu/esgbio/www/cb/cbdir.html – Massachusetts Institute of Technology Cell Biology Chapter

Growth factors

http://web.indstate.edu/thcme/mwking/growth-factors.html – Indiana State University Medical Biochemistry Growth Factors

http://www.copewithcytokines.de/cope.cgi – Cytokines & Cells Online Pathfinder Encyclopaedia

Signal transduction

http://www.signaling-gateway.org/ – University of California San Diego Signaling Gateway

http://web.indstate.edu/thcme/mwking/signal-transduction.html – Indiana State University Medical Biochemistry Signal Transduction

http://www.rpi.edu/dept/bcbp/molbiochem/MBWeb/mb1/part2/signals.htm – Rensselaer Polytechnic Institute Signal Transduction Cascades

The associated cell receptors have high affinity for these ligands, which, when combined with intracellular signal amplification, result in sensitivity to even very low concentrations of growth factors. Various families of growth and differentiation factors induce cells to progress through the cell cycle or express various phenotypes. Gradients in such signals organize cells during embryogenesis to produce correct tissue structures and anatomic forms. Growth factors are responsible for many types of autocrine, endocrine, and paracrine signaling. Their expression is up-regulated during development and wound healing.

Growth factors were actually harnessed prior to their discovery through tissue transplantation, as in the case of bone grafts which contain and present growth factors in a natural setting to promote regeneration. While originally derived in small amounts from tissues and cell-lines, genetic engineering now permits the production of human growth factors on an industrial scale. These inductive signals are promising therapies for tissue regeneration, but limited success has been achieved with free, soluble growth factors in a clinical practice. For example, growth factors are used clinically at much higher doses than present naturally in tissues (micromolar vs. nanomolar), which can induce undesired effects.

Growth factors are detected by cells through membrane receptors and interpreted by intracellular signaling pathways. Altered gene expression, and consequently cell behavior and fate, result from regulation of transcription factor activity at the end of the signaling cascade. Cell response is typically dose-dependent for many growth factors; different concentrations may result in different phenotype expression. Due to complex interconnectivity of signaling pathways, growth factors can alter the effects of one another, resulting in synergistic, additive, and antagonistic effects on cell behavior. Alternatively, simultaneous combinations of growth factors may have completely different effects than exposure to them individually. Likewise, the temporal order of exposure to various growth factors and concentrations can affect differentiation and cell fate.

In natural tissue dynamics, growth factors interact with the ECM, a feature that has been incorporated as a design handle in biomimetic materials. Growth factor binding to heparin-containing proteoglycans of the ECM and the cell surface results in efficient activity and produces cooperative, multivalent interactions (Pantoliano et al. 1994, Cai et al. 2005). Cell-mediated release is achieved by network degradation of the ECM and activation by proteolytic precursor cleavage.

In biomimetic materials, inductive factors induce cells within the scaffold to express a particular phenotype and produce a specific type of tissue. Alternatively, cytokine signals may attract surrounding cells and entice them to produce tissue. An inductive material will cause the production of the desired tissue, even when implanted within different type of tissue. For example, scaffolds containing bone morphogenic proteins (BMPs) will induce bone formation in soft connective tissue while the same scaffolds without those factors would form fibrous soft tissue (Whang et al. 1998).

4.5.3   Cells and Tissue Neogenesis

Proper regulation of cell behavior is central to biomimetic tissue regeneration. Platelets and inflammatory cells are involved in the initial stages of wound healing and biological response to materials by forming an initial matrix and signaling to other cells. Many other cells, such as fibroblasts and osteoblasts, primarily produce and maintain the ECM while others, such as neurons and muscle cells, form multi-cellular organizations to create new tissue. However, like direct growth factor application, directly injected cells have had little success in tissue regeneration. Without the correct micro-environment, cells do not survive for extended periods or may migrate away from the site of implantation. Cells that remain at the site often fail to correctly produce the desired tissue, particularly if the degenerative or diseased phenotype is a result of disease state or age. In the process of tissue neogenesis, cells infiltrate the site and degrade the existing matrix, proliferate to populate the area with new cells, signal to other cells for recruitment and concerted tissue formation, and deposit and remodel the ECM that comprises new tissue.

The native progenitor cells of the host may be sufficient to regenerate tissue given induction by the correct material environment. However, many patients that require particular assistance in tissue repair may have diminished capacity for cellular renewal and differentiation. A common approach in tissue engineering is the seeding of cells in scaffolds prior to implantation. Due to significant issues involved in cell sourcing, it is desirable to utilize a minimal number of stem or progenitor cells implanted in regenerative therapies or, alternatively, to harness the regenerative capacity of native host cells. Biomimetic materials can be designed to entice resident or surrounding cells to migrate, proliferate, differentiate, and conduct tissue generation. Autologous cells from the patient, which present the same cell-surface antigens as native cells, may be harvested, expanded, and re-implanted without little risk of immune rejection.

Stem cells from various anatomical and physiological niches supply the cells that produce and occupy new tissue. They are capable of differentiating into a variety of cell lineages depending on direction from ECM components and growth factors. Embryonic stem cells from which all cells and tissues are derived, mesenchymal stem cells found in adults, and further differentiated tissue-specific cell line progenitors can be used in tissue regeneration therapies. Genetic modification of cells may be used to improve their immune response, induce cells to express a particular phenotype, or overexpress a particular protein or gene product.

The sourcing and preservation of cells is a major challenge to the success of tissue-engineering strategies. Efforts to expand stem cell lines without differentiation are currently being pursued as a continuous source of multi-potent cells for tissue regeneration therapies. Like others, such cells behave differently in traditional tissue culture than in vivo due to altered gene expression and biomimetic materials can help them retain, or even regain, their phenotype. Cell function in homeostasis is indicated by the fact that cells produce more functional proteins in the presence of ECM ligands but produce more adhesive and structural proteins in the absence of these signals (Itle et al. 2005).

Regenerative biomaterials function to control cell behavior and direct tissue neogenesis. They are also used to deliver cells or signaling molecules and retain them at the site of implantation. Proposed strategies in regenerative medicine utilize incorporated cells, such as progenitor cells from established cell lines, those derived from autologous or allogenic sources, or recruited native cells to promote tissue regeneration. Precise control over cell fate is particularly important when manipulating multi-potent cells, as differentiation may lead to an irreversible, undesirable phenotype. Regenerative biomimetic materials must also control cell behavior in a type-specific manner. They often suppress immune cell and fibroblast activity and differentially direct other cell types, such as endothelial and smooth muscle cells in vascular applications.

Basic principles about cells can be found at the educational websites such as:

Cells
http://www.ncbi.nlm.nih.gov/About/primer/genetics_cell.html – The National Institutes of Health What is a Cell?

Stem Cells
http://www.explorestemcells.org/ – ExploreStemCells.com

http://www.isscr.org/science/faq.htm – International Society for Stem Cell Research FAQ

http://stemcells.nih.gov/info/faqs.asp – The National Institutes of Health Stem Cell FAQ

Cell migration
http://www.cellmigration.org/index.shtml – Nature Cell Migration Gateway

4.6 Building Biomimetic Materials

4.6.1 Starting Materials

The starting components of artificial ECMs are critical to their performance because material composition dictates potential chemical, mechanical, and degradation properties and the resulting biological response. Hydrogels, which have found particular usefulness as biomimetric materials may be based on a range of hydrophilic natural or synthetic polymers and are formed by chemical cross-linking, non-covalent association, or self-assembly into ultrastructures such as fibrils. Polymers are selected from *natural biopolymers, synthetic polymers*, or *hybrids* of these materials, which may be produced by a variety of methods (Table 4.5).

4.6.2 Natural Biopolymers

Regenerative biomaterials have historically been based on naturally derived biological materials due to their broad availability and inherent bioactivity. Many early materials for tissue repair were created from purified and modified ECM components with the concept that: *cells will function properly when presented with an environment that resembles their native one.* The chemical composition and mechanical properties of these materials often match biological tissues, presenting a complex set of biochemical signals which is difficult to reproduce synthetically. In addition, natural products typically degrade to byproducts that can be benignly metabolized. Biopolymers usually contain chemical sites amenable to conjugation or cross-linking, which allows physical and chemical modification to improve the mechanical properties, degradation rate, and biological performance.

However, the use of natural materials has several major drawbacks. Due to variations in source as well as methods of isolation and purification, the composition of biological materials is often ill-defined. Also, animal-derived products may illicit a strong immunological and inflammatory response due to their antigenicity and may also serve as vectors for viral transmission.

Table 4.5 Biomimetic material building block synthesis methods

Purification and modification (chemical, thermal, physical) of natural biopolymers
Synthetic chemical synthesis
Solid-phase peptide synthesis
Genetic engineering
Chemical coupling of hybrid building blocks

Natural biopolymers used in biomimetic materials include proteins, such as collagen and fibrin, and polysaccharides, including hyaluronic acid, alginate, cellulose, and chitin. Animal ECM components are typically isolated from raw tissues in which they are abundant and from cell-lines derived either from animals or from genetically modified organisms that produce matrix components. Collagen has been a popular and successful biopolymer for tissue regeneration and the creation of biomimetic tissue scaffolds. Collagen can be chemically cross-linked, enzymatically modified, thermally processed, or physically formed into specific structures to alter its characteristics. It has been used for wound dressings, space-fillers, haemostatic agents, and artificial tissue grafts and can be implanted in solid form or injected as a solution that gels at physiological temperature.

Fibrin glues have also found extensive usage in clinical surgical procedures as sealants, haemostats, drug delivery vehicles, and tissue scaffolds. These materials mimic or stimulate the final stages of the natural blood clotting process. Fibrinogen can be isolated by coagulation of or cryoprecipitation from plasma. Fibrin glues are typically applied as two-component solutions of thrombin and fibrinogen with factor XIII. When mixed, thrombin cleaves factor XIII to factor XIIIa and fibrinogen to fibrin which polymerizes to form strands that are cross-linked by factor XIIIa, forming a solid protein network. Alternatively, thrombin can be directly applied to stimulate clotting of native fibrinogen and activate platelets.

Matrigel^{TM}, a solubilized preparation resembling the basement membrane, is extracted from a mouse sarcoma cell-line and has been used to induce in vivo cell behavior in vitro, to control stem cell self-renewal, and, experimentally, as a thermo-responsive, injectable tissue scaffold. These proteinaceous materials contain adhesion ligands that allow for cellular interaction and migration, and contain sites for cleavage by proteases that promote cellular infiltration and replacement by tissue.

Polysaccharides from animals, plants, algae, and bacteria represent a variety of biopolymers that can be used for the creation of biomaterials. Hyaluronan has been used in ophthalmic surgery, joint disease treatment, post-operative adhesion prevention, soft tissue augmentation, and tissue engineering. Hyaluronan is known to have little non-specific cell interaction but can influence the behavior of some cells through receptor binding and is natively enzymatically degradable. It was historically derived from rooster comb, but is now produced mainly by microbial fermentation. Other glycosaminoglycans, such as chondroitin sulfate and heparin sulfate, have also been used as tissue engineering scaffolds and in therapeutic applications. Cellulose, the most abundant biopolymer on earth, forms the structural matrix of plant cell walls and can be chemically modified to create polymers useful for biomaterials. Chitosan, a linear glucosamine created by de-acetylation of chitin, the structural protein of crustacean shells and second most abundant biopolymer, has been used for biomedical applications, such as blood-clotting wound dressings and as tissue engineering scaffolds. It is biocompatible and degradable and its cationic nature makes it naturally antibacterial and a potential material for gene delivery. Polysaccharides can also be derived from the cell walls of algae or fungus. Agar and alginate, which are widely used in cell culture, are obtained from red and brown seaweed, respectively. Alginate solutions can be gelled by the addition of divalent cations that cross-link the polymer, providing a gentle method of cell encapsulation. Alginate has also been used as a delivery vehicle and modified to form tissue engineering scaffolds.

Mammalian cells have no natural receptors for, or specific native mechanisms for degradation of, exogenous biopolymers. However, they provide a blank slate for presentation of desired biological signals and incorporation of growth factors, cells, and degradable linkages (Augst et al. 2006). Endogenous natural biopolymers can be functionalized for incorporation as biomimetic material building blocks as they may contain biological elements, such as cell-binding domains, specifically enzymatically cleavable chemistries, or growth factor binding ability.

4.6.3 Synthetic Polymers

Synthetic polymers provide the opportunity for enormous chemical diversity and range of properties due to the huge library of monomers and wide variety of potential co-polymer structures. Biomimetic hydrogels have been created from polyacrylamides, polyacrylates, polyethers, polyesters, polyhydroxy acids, polyfumarates, polyphosphazenes, and polypeptides produced by solid phase synthesis and recombinant methods. Hydrogel architectures include homopolymers and random, block, graft, and dendritic copolymers cross-linked by covalent reaction or physical association. Synthetic polymers have many features that have led to attempts to substitute them for natural systems. They may be utilized as the background to present synthetic biomimetic elements or coupled with naturally produced biopolymers to form hybrid materials.

Co-polymers incorporating oligo- or polyethylene glycol (PEG, also known as polyethyleneoxide, PEO) have been particularly successful in creating biomimetic materials due to their low protein adsorption and non-specific biological interaction (http://en.wikipedia.org/wiki/Polyethylene_glycol, http://www.chemicalland21.com/arokorhi/industrialchem/organic/POLYETHYLENE%20GLYCOL.htm). Functionalized PEG polymers and derivatives have been used as the base material and as the conjugation spacer for peptides in biomimetic materials. The flexibility of PEG chains in aqueous solution, due to freely rotating ether linkages in the polymer backbone, makes them useful as linkages to peptide sequences to allow for the proper orientation of ligands to potentially engage with receptors. Multifunctional PEG diacrylate, PEG divinyl sulfone, PEG diacrylamide, and multi-arm PEG macromers, can be used to form hydrogels. For example, bifunctional PEG can be used to cross-link collagen matrices or to couple proteins to them via amino groups (Chen et al. 2002).

Non-specific amino acid sequences, such as repetitive glycine, have also been used as spacer arms for coupling peptides or proteins to biomaterials (Marler et al. 2000, Sakiyama-Elbert et al. 2001). Peptides produced by solid phase synthesis allow for biologically inspired interaction without issues typical of naturally derived materials. Puramatrix™ is a synthetic oligopeptide containing repeating amino acids which self-assembles into nano-fibers upon exposure to physiological ionic concentrations. It has been used for creating in vitro cellular microenvironments and for enhancing tissue regeneration. Synthetic polymer backbones must be modified with bioactive or responsive elements in order to produce materials that mimic the natural biochemical composition of tissue.

Further information about natural hydrogels currently available in clinics can be found at the websites listed below:

Fibrin Glue
http://www.surgeryencyclopedia.com/Ce-Fi/Fibrin-Sealants.html – Surgery
   Encyclopedia Fibrin Glue

Puramatrix™
http://www.puramatrix.com/

Matrigel™
http://www.bdbiosciences.com/discovery_labware/products/display_product.
   php?keyID = 230

4.6.4 Hybrid and Composite Materials

Biohybrid materials are composed of natural or synthetically produced bio-mimetic elements integrated with synthetic biomaterials (Rosso et al. 2005). Composites consist of a reinforcing material, such as inorganic or organic fibers or particles, supported by a different matrix material, such as a ceramic or polymer. Composites offer significant mechanical performance advantages, reflect natural tissue structure, and can benefit tissue in-growth and regeneration. Composite and biohybrid materials have the potential to combine the best properties of man-made and natural materials.

4.7 Biomimetic Material Synthesis

4.7.1 Macromers

One objective in biomimetic material synthesis is to construct modular components containing biomimetic elements and molecular handles for incorporating them into a material. Libraries of biomimetic building blocks containing one or multiple biomimetic elements can then be independently selected and composed into a wide range of biomaterials. This method can also allow functional materials to be constructed in situ around cells without requiring further processing or modification. Macromers, or macromolecular mono-mers, are high molecular weight precursors that are incorporated to form polymers and hydrogels and are often utilized to construct biomimetic mate-rials. Macromers can be used as multi-functional building blocks to form materials in situ with lower reactivity and decreased toxicity from than low molecular weight monomers.

4.7.2 Conjugation Methods

Biomimetic elements inherent to the starting materials or incorporated as building blocks include *cell-binding motifs, enzymatically degradable linkages, biological binding* domains, and *inductive factors*. These elements can be incorporated into materials or macromers using chemical reaction, including bioconjugate techniques (http://chem.ch.huji.ac.il/~eugeniik/edc.htm – Hebrew University of Jerusalem Institute of Chemistry). Specific chemical reaction schemes, developed in various fields of the life sciences and chemical biology, are used to covalently bind biomolecules while preserving their activity (Drotleff et al. 2004). These include reaction of maleimides with thiols, hydrazides with carboxylic acids, succinimides with primary amines, or carbodiimide/succinimide chemistry to couple carboxylic acids with pri-mary amines. Covalent bonding between more reactive moieties, such as aldehydes or epoxides, can also be used for coupling, but may require harsh chemical conditions or lead to undesired side reactions that affect peptide presentation and diminish activity. Conjugation can be used to prepare pre-cursors which are then incorporated by further reaction schemes or, alterna-tively, can be utilized as the final step of material formation. Phase-transition to form a hydrogel may be accomplished by *step-growth conjugate addition, free-radical addition polymerization, associative cross-linking,* or *self-assembly*. Michael addition and photo-initiated free-radical polymerization have found particular recent success as synthetic routes to forming biomimetic materials in situ (Elbert et al. 2001, Anseth et al. 2002).

4.7.3 In Situ Material Synthesis

The importance of delivery vehicles and three-dimensional micro-environments surrounding cells has led to the development of materials capable of forming in situ around cells and macromolecules as well as within the body. Materials that transform from liquid to solid can be implanted as a large form through a minimally invasive injection. Such materials can take complex shapes and form an intimate interface with existing tissue which promotes implant integration.

Various methods of in situ formation have been invented or developed for regenerative materials: *thermal and thermo-reversible gelation via macromolecular association* (e.g., some ECM components and other thermo-responsive polymers), *ionic association* (e.g., repeating ionic peptides, alginate gels, polyelectrolytes), *precipitation of a water-insoluble material* by exchange of a water-miscible solvent with water, *chemical reaction* between reactive components (e.g., Michael-type addition of thiol groups to unsaturated carbon double bonds), *enzymatic activity* (e.g., fibrin-mimetic polymerization), and *polymerization or cross-linking of precursors* initiated by chemical or photo-induced decomposition (e.g., free-radical polymerization and photo-polymerization of oligomers). Maintaining cell viability and the activity of biologics is a challenge for in situ biomaterial development due to the chemically reactive nature of such systems and potential toxicity of initiators and catalysts, and has resulted in a search for benign, bio-orthogonal (i.e., does not react with cells or biologics) reaction schemes with practical kinetic efficiency to sufficiently stabilize within a defect or around cells.

4.7.4 Free Radical Polymerization

Addition polymerization methods that are non-toxic are useful for in situ material formation. Methacrylic anhydride has been reacted with the hydroxyl groups of alginate and hyaluronan to create acrylate functional polysaccharides. Hydrogels were then formed by photo-initiated polymerization of these macromers (Smeds et al. 2001). Similarly, reaction of chitosan with azidobenzoic acid rendered it photo-polymerizable (Yeo et al. 2006). Condensation reaction between vinylbenzoic acid and the amino groups of collagen produces a photo-polymerizable collagen hybrid material (Hoshikawa et al. 2006). Methacrylation of hyaluronic acid via carbodiimide activated reaction between carboxylic acid groups of hyaluronic acid and the amine group of a methacrylating agent allow photo-initiated co-polymerization with PEG-diacrylate (Park et al. 2003). Oligo(PEG fumarate) is a degradable, biocompatible synthetic polymer containing a double bond in the backbone which allows it to be cross-linked via free radical polymerization initiated by redox reaction of ammonium persulfate and ascorbic acid (Jo et al. 2001).

4.7.5 Step-Growth Polymerization

Step-growth conjugation reactions of macromers to form hydrogels have also been utilized to form biomimetic materials. A variety of methods are available for enzymatic modification and chemical cross-linking of natural proteins, such as collagen, by reaction of amines with multifunctional aldehydes, and polysaccharides, such as alginate and hyaluronic acid, by reaction of their carboxylic acid groups with hydrazides or carbodiimide coupling with amines, to form materials for tissue regeneration with controlled degradability, mechanical properties, and antigenicity (Lee and Mooney 2001). Hyaluronic acid and chondroitin sulfate can be hydrazide functionalized by carbodiimide activated condensation of carboxylic acids with dihydrazide and then cross-linked with homo-bifunctional cross-linkers such as PEG dialdehyde (Pouyani and Prestwich 1994, Luo et al. 2000, Kirker et al. 2002, Peattie et al. 2004). Hydrazide functional heparin was formed into a hydrogel by reaction with succinimide-bifunctional PEG (Tae et al. 2006).

Recently, conjugate reactions have been developed that allow for in situ formation and safe cellular encapsulation to enhance tissue regeneration. Michael-type conjugate addition of free thiol groups, which are typically present only in non-disulfide bonded cysteine amino acids of peptides and proteins, with reactive double bonds, such as those of acrylates, acrylamides, and vinyl sulfones, has been extensively used to create functionalized bioactive and bio-responsive macromers. Thiols were conjugated to gelatin, hyaluronan, and chondroitin sulfate via hydrazide chemistry. The protein and GAG building blocks were then cross-linked by disulfide bond formation (Shu et al. 2003) or by Michael-type addition reaction with PEG-diacrylate both individually and in various ratios to give gelatin-hyaluronan and gelatin-chondroitin sulfate hybrid gels (Shu et al. 2006). Michael addition has also been utilized to functionalize PEG precursors with bioactive components and cross-link them with enzymatically degradable peptides (Lutolf et al. 2003, Zisch et al. 2003, Pratt et al. 2004, Park et al. 2004, Seliktar et al. 2004, Raeber et al. 2005).

4.7.6 Physical Association

Some natural polymers are environmentally responsive, which can be utilized for in situ formation, such as the thermal gelation of collagen and Matrigel^{TM} or the ionic cross-linking of alginate and Puramatrix^{TM}. Natural polymers can also be modified to form in situ by physical association. Substituted cellulose ethers exhibit a lower critical solution (LCST) temperature, above which the polymer precipitates from solution forming a physically cross-linked hydrogel. Injectable, thermo-responsive PEG-grafted chitosan formulations have been created for effective incorporation of bioactive molecules. Schiff base formation of imide bonds between aldehyde-functional PEG and the amino groups of chitosan followed by reductive amination was used for conjugate PEG grafting (Bhattarai et al. 2005). Collagen and hyaluronan were rendered thermoresponsive by photo-initiated graft co-polymerization of N-isopropyl acrylamide (Ohya et al. 2001, Morikawa and Matsuda 2002). Additionally, laminin has been tethered to methyl cellulose to form thermoresponsive hydrogels for neural tissue engineering. This was accomplished via Schiff base reaction between aldehyde groups of oxidized methyl cellulose and free amine groups of laminin (Stabenfeldt et al. 2006).

Many synthetic hydrogels have been developed that respond to environmental stimuli, such as temperature, pH, ionic strength, light, electric fields, and specific macromolecules (Kopecek 2003, Ruel-Gariepy and Leroux 2004, de Las Heras Alarcon et al. 2005). These materials can be modified to be biomimetic and used to form materials in situ by exposure to physiological temperature, pH, and ionic strength (Fisher et al. 2004, Kim et al. 2005). Synthetic polymers that exhibit LCST behavior are typically amphiphilic polymers and can be used to form solid gels from liquid solutions after exposure to physiological temperature. These include homo- or co-polymers of N-isopropyl acrylamide (NIPAAm), block co-polymers of PEG and poly(propylene glycol) (PPG), block or graft copolymers of PEG and poly(lactic-co-glycolic acid) (PLGA), and polyorganophosphazene derivatives.

LCST behavior and Michael-type addition reaction have been used to create polymer systems that gel by tandem physical associative gelling and covalent chemical reaction. Thermoresponsive four-arm poly(propylene glycol-block-ethylene glycol) (PPG-b-PEG) co-polymers were terminally functionalized with either acrylic esters or thiols as conjugate reaction partners (Cellesi and Tirelli 2005). Similarly, acrylate functionalized, thermo-responsive p(NIPAAm-co-HEMA) was combined with a tetra-functional thiol to form a system that gels by tandem association and reaction (Lee et al. 2006). These systems possess properties of quick gelation useful for encapsulation and injection as well as the stability and mechanical properties of chemically cross-linked hydrogels.

4.7.7 Molecular Self-Assembly

Self-assembly is an ordered form of association in which precursors form morphological ultra-structures. Natural biopolymers such as collagen form fibrillar architectures and networks. Synthetic repeating peptides with alternating positively and negatively charged ionic hydrophilic and hydrophobic amino acids that form nano-fiber hydrogels have been extensively developed as in situ forming regenerative matrices (Zhang et al. 2003, Zhang 2003). Peptide amphiphiles with a hydrophilic protein head and hydrophobic alkyl tail are also capable of self-assembly into nano-fiber hydrogels (Hartgerink et al. 2002, Guler et al. 2006, Hosseinkhani et al. 2006, Mardilovich et al. 2006).

4.8 Biomimetic Elements

4.8.1 Cell Adhesion Domains

4.8.1.1 Cellular Adhesion

Cell attachment, spreading, migration, differentiation, and other cellular activities necessary for tissue regeneration depend on engagement of integrin and other cell-surface receptors with complementary cell-adhesion ligands of

Table 4.6 Advantages of minimal peptides over complete proteins

Increased efficiency of ligand presentation with lower density required for adhesion and spreading
Incorporation is facilitated by relatively small size and fewer conjugation sites
Decreased probability of altering structure and binding affinity upon conjugation
Fewer steric restrictions for effective presentation
Do not display exogenous epitopes due to spreading or denaturing
Controlled interaction with specific cell types due to fewer functional domains
Decreased cost of production and purification

the material or tissue interface. This interaction is specifically controlled in biomimetic materials by presenting cell-binding domains, typically short peptide sequences (Table 4.6), upon or within a material that otherwise inhibits protein adsorption and cell attachment. Because different cell types express different amounts of each integrin, this serves as a means of achieving cell-specific adhesion and response.

The affinity between a specific integrin type and a cell-binding domain depends largely on the primary amino acid sequence, and contributes largely to the adhesion force between a cell and a substrate (Gallant and García 2007). Hundreds of peptide sequences from ECM molecules have been identified that are recognized by integrins and other membrane receptors (Table 4.7). Many of

Table 4.7 Adhesion peptide sequences used in biomimetic materials

Sequence	Origin	Receptor	Source
RGD(X) (and di-cysteine cyclic forms)	Collagen (X = T), Fibronectin (X = S), Vitronectin (X = V), Laminin (X = N), Thrombospondin, Osteopontin, Bone sialoprotein	$\alpha v\beta 3$, $\alpha 5\beta 1$, $\alpha 3\beta 1$, and other integrins	(Hern and Hubbell 1998; Schense and Hubbell 2000; Shin et al. 2002; Harbers and Healy 2005) (Brandley and Schnaar 1988; Hirano et al. 1991)
DVDVPDGRGDLAYG	Osteopontin	$\alpha 5\beta 1$ integrin	(Shin et al. 2004)
CGGNGEPRGDTYRAY	Bone sialoprotein	αv and $\beta 1$ integrins	(Harbers and Healy 2005)
FHRRIKA	Bone sialoprotein	transmembrane proteoglycans	(Rezania and Healy 1999)
TMKIIPFNRLTIGG	Fibrinogen-γ	$\alpha M\beta 2$ integrin	(Gonzalez et al. 2006)
KQAGDV	Fibrinogen-γ	$\alpha 2b\beta 3$ integrin	(Mann et al. 2001; Mann and West 2002)
PHSRN	Fibronectin FNIII-9	$\alpha 5\beta 1$ integrin synergy	(Mardilovich et al. 2006)

Table 4.7 (continued)

Sequence	Origin	Receptor	Source
REDV	Fibronectin FN-IIICS-5	$\alpha4\beta1$ integrin	(Massia and Hubbell 1992; Drumheller and Hubbell 1994; Heilshorn et al. 2003; Girotti et al. 2004)
LDV	Fibronectin FN-IIICS-1	$\alpha4\beta1$ integrin	(Jiang et al. 2006)
DGEA	Collagen-I	$\alpha2\beta1$ integrin	(Schense and Hubbell 1999; Gilbert et al. 2003)
GFOGER	Collagen-I-$\alpha1$	$\alpha2\beta1$ integrin	(Reyes and Garcia 2004)
IKVAV	Laminin-α	$\alpha 1\beta1$ and $\alpha3\beta1$ integrins	(Ranieri et al. 1994; Silva, Czeisler et al. 2004)
YIGSR	Laminin-β-1	110 kDa and 67 kDa laminin binding protein	(Massia and Hubbell 1990; Schense et al. 2000)
RNIAEIIKDI	Laminin-γ		(Schense et al. 2000)
VAPG	Elastin	non-integrin glycoside binding protein	(Mann and West 2002; Gobin and West 2003)
KHIFSDDSSE	Neural cell adhesion molecule (N-CAM)	N-CAM	(Ranieri, Bellamkonda et al. 1994)
HAV	N-cadherin	N-cadherin	(Schense et al. 2000)

these sequences maintain significant activity when reduced to the minimal peptide sequences necessary for receptor binding (Shin et al. 2003). The prototypical cell-binding motif found in several molecules of the ECM, including collagen, fibronectin, vitronectin, and laminin, is the tri-peptide sequence arginine-glycine-aspartic acid ("RGD").

While particular sequences are responsible for receptor engagement, binding is affected and can be modulated by the amino acids flanking the central domain (Harbers and Healy 2005). Cell binding motifs are typically derived from the sequence of native ECM proteins, but may be altered through replacement of specific amino acids or selected through the display, binding, elution, and repeat stages of bio-panning (Li et al. 2003) or by directed evolution. Many sequences for integrin binding identified from the native ECM and by synthetic means share similar patterns, illustrating common mechanisms of receptor binding and specificity.

4.8.1.2 Conjugation of Cell-Binding Motifs to Polymers

Cell-adhesive peptides can be conjugated to the functional groups of natural biopolymers and synthetic polymers, which can then be formed into hydrogels via covalent reactions of native or imparted reactive groups, or by associative gel formation. Photo-crosslinkable hyaluronic acid has been functionalized

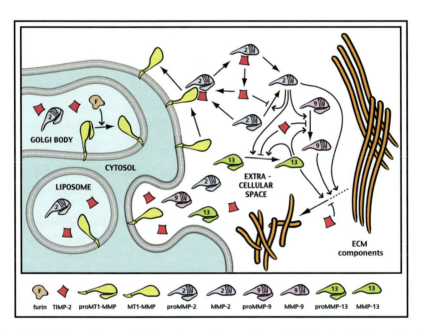

Fig. 4.4 Initiation of a cascade of matrix metalloproteinase (MMP) activation from the cell surface results in local, cell-mediated ECM degradation. A few steps in the complex network of MMP activation are illustrated. Membrane-type MMPs (MT-MMPs) are activated in the golgi apparatus through cleavage by furin. MT-MMPs, MMP precursors (proMMPs), and tissue inhibitors of metalloproteases (TIMPs) are transported to the cell surface and extra-cellular space in liposomes. Complex formation between MT1-MMP, TIMP-2, and MMP-2 results in surface induced MMP-2 activation. Autocatalytic activation of proMMP-2 can then occur. MMP-2 also activates MMP-9 and MMP-13. MMP-13 cleaves proMMP-13 as well as proMMP-9. MMPs-2, 9, and 13, as well as other MMPs and proteases involved in the complex activation network, specifically degrade extra-cellular matrix components. Local concentrations of TIMPs inhibit MMP activation and prevent ECM degradation

with cell-binding peptides by carbodiimide activated reaction between its carboxylic acid groups and the amino terminus of the peptide (Leach et al. 2004). Cell-binding domains were conjugated to type I collagen prior to fibrillar gel formation using a heterobifunctional linker that creates amide bonds with amino groups of lysine residues of collagen and disulfide links with the free sulfhydryl, i.e., thiol, group of the cysteine residue of the peptide (Myles et al. 2000). A recombinant peptide sequence from the FNIII-7-10 domain of fibronectin, containing integrin binding and synergy sites, was coupled to bovine serum albumin, which was adsorbed onto a poly(styrene) surface, using homo- and heterobifunctional cross-linkers to react with free amines and free sulfhydryl groups, respectively, at various sites of the protein and peptide (Cutler and Garcia 2003).

Synthetic polymers also offer methods of direct peptide conjugation to form biomimetic materials. Hydroxyl groups of graft co-polymers were carboxylated using succinic anhydride. The carboxyl group was then succinimide functionalized by carbodiimide activated reaction with N-hydroxysuccinimide (NHS) which was further reacted with the N-terminus of the peptide (Irvine et al. 2001, Koo et al. 2002). Oligo(PEG fumarate) has been terminally functionalized with, alternatively, NHS or 4-nitrophenylchloroformate and reacted with the N-terminus of a peptide as another method of RGD conjugation (Behravesh et al. 2003). Integrin-binding peptides have been terminally conjugated to PEG polymers that interpenetrate another polymer network. Diamino PEG was reacted with the carboxylic acid groups of the IPN surface via EDC (1-ethyl-3-[3-dimethylaminopropyl]-carbodiimide) activation and succinimide (sulfo-NHS) stabilization and subsequently maleimide functionalized by reaction of the amino groups with the succinimide group of a heterobifunctional SMCC linker (succinimidyl-4-[N-maleimidomethyl]cyclohexane-1-carboxylate). Peptide conjugation was accomplished by thiol addition to the maleimide group of SMCC (Barber et al. 2003).

Thermoresponsive polymers and networks can be peptide-functionalized in order to form biomimetic gels at physiological temperature. Integrin-binding RGD peptide sequences have been grafted along the backbone of linear poly (acrylic acid) (pAAc) chains that are incorporated into thermo-responsive semi-interpenetrating poly(NIPAAm-co-AAc) (poly(N-isopropylacrylamide-co-acrylic acid). Conjugation was accomplished by carbodiimide (EDC) and succinimide (sulfo-NHS) activated reaction between carboxylic acid groups of pAAc and the hydrazide groups of a heterobifunctional linker (EMCH – N'-[E-Maleimidocaproic acid) hydrazide]) followed by conjugation of the thiol groups of a cysteine containing peptide to the maleimide functionality of EMCH (Kim et al. 2005). These interpenetrating hydrogels allow for independent modulation of chemical, mechanical, and biological properties. Cell-binding peptides were also tethered directly to thermoresponsive co-polymers containing succinyl PEG grafts by EDC activated conjugation to the N-terminus of an RGD peptide (Park et al. 2004).

4.8.1.3 Self-Assembling Nanofibers Featuring Cell-Binding Motifs

The high surface area to volume ratio of nano-fiber materials offers the ability to present a remarkably high density of ligand. Self-assembling peptide-amphiphiles have been designed that present the cell-binding motifs of fibronectin and laminin at the surface of the nano-fibers formed at physiological pH and ionic strength (Silva et al. 2004, Guler et al. 2006). Peptide-amphiphiles have also been created that incorporate the cell-binding RGD FNIII-10 domain of fibronectin and the synergistic PHSRN FNIII-9 domain with a peptide spacer arm between the sequences (Mardilovich et al. 2006).

4.8.1.4 Reactive Macromers Containing Cell-Binding Motifs

Peptides can be vinyl functionalized for incorporation in materials via subsequent free radical polymerization or conjugate hydrogel formation. This has been achieved, for example, by Michael-type addition of vinyl groups of a multifunctional linker, such as terminally functional PEG macromers, with the free thiol groups of a cell-binding peptide. Conjugate reaction between bifunctional PEG acrylates and thiol-containing cysteine groups of a peptide or protein is referred to as *PEGacrylation* to distinguish it from more traditional *PEGylation* which does not impart acrylate functionality. Reaction of an excess of homobifunctional PEG with cysteine terminated cell-binding peptides has been utilized to create reactive peptide-PEG macromers that were incorporated upon photo-initiated polymerization with the remaining PEG cross-linker (Elbert and Hubbell 2001). Hydrogels containing cell-binding domains have been created from peptide-PEG-acrylate by Michael-type addition reaction with thiolated hyaluronic acid, functionalized by hydrazide reaction, followed by cross-linking with PEG-diacrylate (Shu et al. 2004).

Michael-type addition of thiols to vinyl groups have been utilized to bind cysteine-terminated cell-binding peptide sequences to vinyl sulfone terminated PEG macromers. The same peptide-PEG-vinylsulfone macromers were then conjgated to thiol terminated protease substrate peptides to form regenerative hydrogels (Lutolf et al. 2003, Zisch et al. 2003, Park et al. 2004, Pratt et al. 2004, Seliktar et al. 2004, Raeber et al. 2005). Fibronectin domains responsible for RGD-based $\alpha5\beta1$ and $\alpha v\beta3$ integrin binding (FNIII-8-11), heparin II binding site for CD44 and syndecan receptors (FNIII-12-15), and binding sites for $\alpha4\beta1$ integrin (FNIII-CS) were recombinantly produced with C-terminal cysteines for bioconjugation into hydrogels. The peptides were then reacted with either excess homobifunctional PEG-diacrylate or PEG-divinyl sulfone to produce cell-binding domain precursors. Thiol-functional hyaluronan was added to the mixture of the precursors and unreacted PEG cross-linker to form gels under mild reaction conditions.

The utility of this type of conjugate reaction is indicated by the associated number of characterizations and optimizations. Reaction of thiol containing hyaluronan with PEG divinylsulfone was demonstrated to be more hydrolytically stable than reaction with PEG diacrylate (Ghosh et al. 2006). The amide linkage formed between peptides and PEG diacrylamide has also been shown to be more stable than ester bonds formed by reaction with PEG diacrylate (Elbert and Hubbell 2001). Conjugate reaction between thiols and unsaturated bonds is clearly faster than disulfide bond formation, which has also been used to form biomaterials in situ and likely occurs over time in Michael-type addition systems (Shu et al. 2006). In these systems, some reaction also occurs between the unsaturated double bonds and functional groups present in proteins other than thiols, such as primary amines (Zisch et al. 2003). Conjugation rate can be enhanced by placing positively-charged amino acids in proximity to the cysteine residue of the peptide (Lutolf et al. 2001).

Alternatively, cell-binding peptides can be PEGacrylated by conjugation with heterobifunctional acryloyl-PEG-N-hydroxysuccinimide (Ac-PEG-NHS). The succinimide group reacts with the N-terminus of cell-adhesion peptides for incorporation in free radical photo-polymerized hydrogels (Yeo et al. 2006). Reactive PEG-acrylate tethered cell-binding peptide sequences and PEG-acrylate terminated degradable sequences have been created for incorporation during photo-initiated PEG-diacrylate hydrogel formation (Mann et al. 2001). Acryloyl-PEG-RGDS has been incorporated into functionalized chitosan through photo-polymerization (Yeo et al. 2006). Peptide-PEG-acrylates have been incorporated into gels of methacrylated hyaluronan, itself created via reaction of hyaluronic acid with glycidyl methacrylate, by photo-initiated free-radical co-polymerization with multi-arm PEG-acrylate (Leach et al 2004). Peptide-PEG-acrylates created in this manner have also been used to impart bioactivity to degradable hydrogels created from oligo(PEG fumarate). Free radical polymerization can be alternatively used as a method of forming gels from precursors, including PEGacrylated peptide and PEG-diacrylate (Shin et al. 2002).

4.8.1.5 Parameters and Effects of Cell-Binding Motif Incorporation

The type, density, and presentation of the selected ligand or ligands alter cellular adhesion to biomimetic materials and ensuing effects on cytoskeletal organization, intracellular signaling, and gene expression. Materials with cell-adhesion domains support cell attachment and spreading while the protein-resistant synthetic analogs do not (Drumheller and Hubbell 1994, Bearinger et al. 1998). The specific sequence alters cell adhesion, proliferation, and matrix production (Harbers et al. 2005). Many studies have shown that the RGD, compared to non-sense RGE or RDG peptides or no peptides at all, promotes cell adhesion and spreading. Lower densities of high affinity cell-adhesion ligands, which can be costly to produce and incorporate, are necessary for cell attachment, spreading, and migration (Schense and Hubbell 2000, Harbers and Healy 2005). Cell attachment and spreading typically increase with cell-binding domain density (Shin et al. 2004). Cell adhesion can be quantified by the force required to detach them from a surface (Hertl et al. 1984, Rezania and Healy

1999, Koo et al. 2002, Harbers et al. 2005). Cell detachment force is a linear function of the ligand and receptor density product, which is predicted by simple receptor-ligand binding equilibrium (Garcia and Boettiger 1999). Exposure to specific, competitive soluble cell-binding domains, and integrin blocking antibodies inhibits cell attachment, further supporting the concept of receptor-mediated adhesion (Hern and Hubbell 1998, Schense and Hubbell 2000, Halstenberg et al. 2002, Shin et al. 2004, Zaman et al. 2006).

The effect of the minimal RGD sequence on cell behavior can be strongly modified by selecting appropriate flanking amino acids. Fibroblasts were shown to proliferate faster on hydrogels modified with the RGD sequence from osteopontin compared to those modified with generic RGD peptide, while ostoblast proliferation was the same on both materials. Osteoblasts migrated farther on osteopontin-mimetic hydrogels than on those modified with simple RGD, and migration distance increased with peptide concentration for each. Furthermore, osteoblasts migrated faster than fibroblasts seeded at the same density through the osteopontin-mimetic materials (Shin et al. 2004).

The combined interaction of multiple integrin types can result in different cell behavior than the individual integrins (Gonzalez et al. 2006). For example, a biomimetic hydrogel containing multiple fibronectin functional domains supported much greater fibroblast spreading and migration that simple RGD-functional equivalents (Ghosh et al. 2006). Similarly, combined fibronectin domains spaced with a biomimetic spacer resulted in stronger adhesion than equimolar amounts of combined individual domains or whole fibronectin (Mardilovich et al. 2006).

In addition to ligand type and density, the presentation (i.e., the attachment method, conformation, mobility, clustering or avidity, and gradient) of adhesion ligands influences cell binding and migration (Fig. 4.5). This reflects receptor-ligand binding dynamics and multivalent interactions with cell surface receptors that occur in the natural ECM and result in altered receptor clustering, adhesive complex formation, and cytoskeletal organization. Proteins have three-dimensional tertiary structures that present sequences for adhesion ligands in a specific spatial arrangement that promotes binding to receptors. Ligand conformation and orientation are, therefore, an important aspect of conjugating cell-binding domains to materials. Integrin engagement with the RGD sequence can be enhanced by presenting it in cyclic form that resembles its native configuration in a hairpin loop of several ECM molecules (Xiao and Truskey 1996, Schense and Hubbell 2000).

A spacer arm between the base material and cell-binding peptide sequence enhances specific recognition of adhesion ligands and increases resulting cell attachment, spreading, and growth (Hern and Hubbell 1998, Park et al. 2004). The length of PEG, or other non-fouling, grafts relative to peptide spacer length also alters integrin engagement, and cell attachment (Shin et al. 2002). The non-fouling polymer graft length should be long enough to prevent non-specific adhesion but the relative length of the peptide spacer should allow

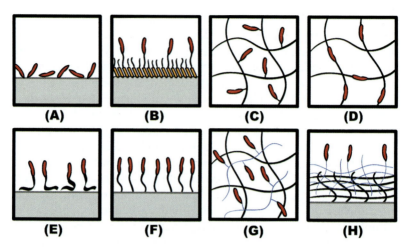

Fig. 4.5 Cell-binding domains can be incorporated into materials by a variety of methods. These include (**A**) simple surface adsorption; (**B**) self-assembling monolayers; (**C**) grafting from polymer hydrogel; (**D**) incorporation as polymer backbone of a hydrogel; (**E**) amphiphilic molecule adsorption; (**F**) surface grafting or modification; (**G**) grafted linear chains of semi-interpenetrating network; and (**H**) grafting to interpenetrating network surface

bioavailability of the ligands (Behravesh et al. 2003). The linkage method, as well as spacer, impact bioavailability and cell-spreading. Terminal linkage, as opposed to pendant-like linkage, with longer spacer length has been shown to result in increased cell spreading (Shu et al. 2004). As with many other studies, the results did not directly translate to encapsulated cells in vivo, indicating the distinction between cell response in 2D and 3D culture as well as in vitro and in vivo (Burdick and Anseth 2002).

Integrin binding site clustering and gradients are important aspects of natural ECM dynamics and can be controlled in biomimetic materials. Peptide sequences can be presented in clustered organization by tethering to dendritic polymers to form star co-polymers (Maheshwari et al. 2000) or to linear polymers to form comb-like graft co-polymers (Irvine et al. 2001, Kim et al. 2005). RGD clustering reduces the ligand density required to support cell migration. Ligand clustering increases cell migration speed and it promotes actin filament organization and stress fiber formation, illustrating the multivalent effects on cell-adhesion, the result of integrin clustering and adhesion complex formation (Maheshwari et al. 2000). Ligand clustering improves adhesive strength and can induce adhesion reinforcement when focal adhesions are stressed by mechanical stimulation, a result of intracellular signaling (Koo et al. 2002). Additionally, clustered ligands allow strengthening of ligand-receptor-cytoskeleton connections through complex formation and cytoskeletal organization.

Gradients in insoluble ligands are responsible for many processes of tissue dynamics. Gradients of cell-binding domains can be created by controlled flow of macromer solutions. Cell alignment and migration have been shown to depend on the slope of concentration gradients of adhesion ligands and occur along the direction of the gradient (DeLong et al. 2005).

4.8.2 Substrate Mechanics

Through mechanisms often under-rated by the biological research community, but fundamental to tissue physiology, cells have the ability to sense and respond to the mechanical properties and stress state of their microenvironment (Table 4.8). Mechanical stresses in tissue are supported by the surrounding matrix and distributed to individual cells. Cells react in a specific physiologically relevant manner to shear, compressive, tensile, hydrostatic, and multi-axial stress states. Tissue cells of interest to regenerative medicine are typically not specialized mechanosensory cells; however, response to the mechanics of the matrix are critical to their roles in tissue physiology, both from a physicochemical standpoint and through cellular responses to physiological environmental stresses. Several mechanisms have been identified in mechanotransduction including stretch-sensitive ion channels and integrin-mediated cytoskeletal and signaling responses (Alenghat and Ingber 2002). Cellular translation of force fields in the micro-environment requires molecular linkage to surroundings via biochemical contact formations that transfer mechanical loads. Therefore, cells and the ECM or tissue engineering material exert forces on one another, with larger cell deformations on stiff matrices and larger matrix deformation on softer matrices (Engler et al. 2006).

Cells maintain a mechanical homeostasis with the surrounding microenvironment through intracellular signaling and gene networks (Fig. 4.6). Tissue mechanics determine how cells sense stress or deformation, which in turn affects how they modify the surrounding matrix by ECM production in order to alter the mechanical properties and maintain adhesion and contractile forces (Paszek et al. 2005). Sensing of the mechanical characteristics of a material feeds back on adhesion complex formation, cytoskeletal organization, and contractile forces from the cell (Discher et al. 2005). In effect, cells modulate their own stiffness in response to the compliance of the material to

Table 4.8 Mechanical properties of instructive biomaterials

Elastic moduli

Dynamic mechanical properties (visco-elasticity)

Failure properties such as yield and ultimate stresses and strains

Temporal changes upon in situ material formation and degradation

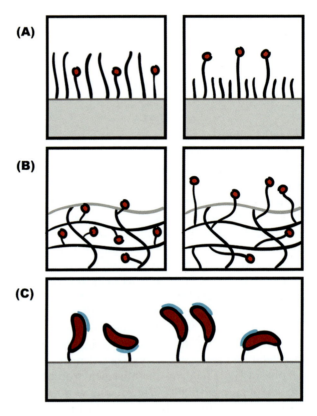

Fig. 4.6 Grafting method alters ligand bioavailability. (**A**) The length of the ligand spacer relative to the non-functional grafts used to reduce non-specific adsorption on 2-D surfaces affects how receptors can interact with them. (**B**) Similarly, ligand spacer length and cross-linking in a 3-D hydrogel alters their availability at interfaces. (**C**) The grafting method can also reduce ligand binding efficiency due to altered presentation of the binding domain relative to freely soluble ligand (highlighted)

which they are adhered. Pre-stresses are developed by actin-myosin molecular motor action on tensed microfilaments; tensional integrity is supported by compressed microtubules in order to maintain cell shape stability, resist deformation, and apply forces to the substrate (Wang et al. 2001). Thus, cells adhere and spread more on cell-adhesive surfaces with higher stiffness (Elbert and Hubbell 2001, Chung et al. 2006). As a result, cells accumulate at the stiff end when seeded on a gel of gradient stiffness (Zaari et al. 2004). In addition, cell response to mechanical properties affects cell morphology, migration, and regulation of tissue dynamics.

Base material selection and modification determines matrix mechanical properties, such as by cross-linking, while attached proteins or peptides are used to control ligand density (Pelham and Wang 1998). Smooth muscle cell spreading on collagen I coated polyacrylamide gels can be modeled as a function of ligand density and substrate compliance. Smooth muscle cell spreading was biphasic with collagen ligand density and increased with higher gel modulus (Engler et al. 2004). In addition to cell behavior, matrix compliance also influences cell fate. Mesenchymal tem cells differentiated along neurogenic, myogenic, or osteogenic cell lineages according to substrate stiffness upon which they were cultured and differentiation markers peaked at the physiologically relevant substrate elasticity. Cells had the similar response to substrate stiffness in the presence of inductive media, indicating synergy between matrix and soluble induction. Extended exposure to inductive matrices resulted in commitment to a particular phenotype, as occurs after prolonged exposure to inductive media. Inhibition of non-muscle myosin II, responsible for intracellular contractile forces, blocked inductive effects of matrix elasticity (Engler et al. 2006).

The interrelated effects of integrin binding and matrix mechanics have been demonstrated by the differing response of cells to ligand density depending on matrix stiffness. Artificial EMCs were created with stiffness modulated by Matrigel^{TM} concentration and receptor-ligand engagement controlled by fibronectin content and integrin-blocking antibodies. Cell migration in gels with low pore size depended on cell adhesion ligand density, proteolysis, and matrix stiffness. Cell migration speed in 3D was biphasic with adhesiveness, characterized by the product of available receptors and presented ligands, as well as with ECM stiffness (Zaman et al. 2006). Studies utilizing independent control of the ligand density and dynamic modulus of biomimetic, thermoresponsive semi-interpenetrating networks have elucidated the response surface for matrix stiffness and peptide concentration on primary osteoblast proliferation (Chung et al. 2006). Utilization of these model systems allows the generation of response surfaces for cell behavior with respect to independent physical and biochemical matrix parameters for controlling and optimizing tissue formation.

4.8.3 Enzymatic Degradability

4.8.3.1 Engineering Degradability into Biomaterials

Biomaterial based tissue regeneration depends on material degradation for replacement by functional tissue. The focus of the previous generation of degradable biomaterials (e.g., polyesters) was to control the non-specific degradation mechanisms of synthetic materials. Degradation in these materials occurs by bulk or surface erosion and is dominated by hydrolytic cleavage and dissolution. The degradation rate of materials was controlled by engineering chemical compositions, as with co-polymer chemistry, crystallinity, and macromolecular structure, as well as with specific morphologies, such as with porous architectures.

The degradation rate and mechanism greatly influences a material's regenerative capacity. Slow degradation results in impaired healing, chronic inflammation, and fibrous encapsulation while a fast degrading material may not persist for long enough to serve its role in tissue regeneration. Developments in early degradable materials faced difficulties in matching the kinetics of tissue ingrowth with that of degradation. Cell-mediated degradation and release of growth factors, which is much more efficient than traditional methods of controlled release and degradation modes with pre-defined temporal profiles.

Hydrogels can be engineered to resist hydrolytic degradation and undergo enzymatic degradation exclusively (Seliktar et al. 2004, West and Hubbell 1999). Although degradation is meant to be cell-mediated, several design handles impact on cell and material response. The sequence, concentration, and incorporation method of degradable linkage and the polymer architecture affect network degradation characteristics. Like receptor-ligand binding, enzymatic activity depends on specific molecular recognition. A variety of peptide sequences of structural ECM molecules that are sites of specific cleavage by particular enzymes have been identified (Table 4.9). These sequences have varying degrees of reactivity and specificity to different proteases and can be reduced to short sequences that maintain specificity as enzyme substrates. Integration of enzymatically degradable linkages can be achieved using similar methods as cell binding domains but are internal, rather than tethered, to building block backbones due to need for them to impact network connectivity. Enzyme degradable sequences can be incorporated as part of natural biopolymers or recombinant ECM proteins or as peptide cross-linkers or network precursors. Enzyme-substrate kinetic parameters can be measured in solution in order to design biomaterials with particular enzyme specificity and degradation rates (Lutolf et al. 2003). The Michaelis-Menten degradation kinetics of enzymatically cleavable peptides are well preserved upon conjugation to synthetic polymers, allowing for engineered proteolytic degradation.

The principles of cleavable sequence selection are also akin to those of cell binding domain selection. They are chosen according to up-regulation during tissue development and healing (Kim et al. 2005). This serves as an additional method of imparting cell-type specific interaction in biomimetic materials.

Table 4.9 Enzymatic substrate peptide sequences

Substrate sequence	Enzyme	Source
GPQGIWGQ, GPQGIAGQ	MMP-1 and MMP-2	(Mann et al. 2001; Seliktar et al. 2004; Rizzi and Hubbell 2005)
QPQGLAK	MMP-13	(Kim et al. 2005)
LGPA, APGL	Collagenase	(West and Hubbell 1999; Gobin and West 2002)
YKNR	Plasmin	(Pratt et al. 2004; Raeber et al. 2005)
NNRDNT, YNRVSED	Plasmin	(Halstenberg et al. 2002)
LIKMKP	Plasmin	(Schmoekel et al. 2004; Ehrbar et al. 2005)
VRN	Plasmin	(West and Hubbell 1999; Gobin and West 2002)
AAAAAAAA	Elastase	(Mann et al. 2001; Gobin and West 2002)
NQEQVSP	Factor XIIIa	(Zisch et al. 2001)
GLVPRG	Thrombin	(Halstenberg et al. 2002; Zisch et al. 2003)

4.8.3.2 Methods of Imparting Enzymatic Degradability

Natural ECM biopolymers, including proteins, such as collagen and fibrinogen, and glycosaminoglycans, such as hyaluronan, chondroitin, and heparin, can be utilized as the degradable elements of artificial ECMs. These biopolymers are naturally degradable by enzymes involved in tissue dynamics, including MMPs, plasmin, hyaluronase, chondroitinase, and heparinase. Degradation can be altered by component selection and chemical or physical modification to form biomimetic materials that can be specifically degraded and remodeled by cells. For example, cross-linked HA hydrogels are readily degraded by hyaluronidase (Park et al. 2003). The relative composition of degradable elements affects the susceptibility of the gels to specific enzymes combinations (Shu et al. 2006).

Proteolytically degradable cross-linkers can be created by acrylation of the primary amine groups of MMP-degradable peptides by acryloyl chloride, which can then be incorporated into free-radical polymerized hydrogels (West and Hubbell 1999, Kim and Healy 2003, Kim et al. 2005). PEG systems based on Michael-type addition of thiols to vinyl sulfones have been created with cell-binding domains and degradable peptide sequences. Cysteine capped adhesion peptides and cysteine terminated MMP- and plasmin-degradable peptide sequences were reacted with multi-arm PEG-vinylsulfone to form specifically degradable, biomimetic hydrogels (Park et al. 2004, Raeber et al. 2005). PEG divinyl sulfone has been cross-linked with a synthetic peptide containing three cysteine residues and a plasmin-degradable sequence between each one (Pratt et al. 2004). Synthetic protein polymers were recombinantly produced to contain a repeat of a fibrinogen-derived sequence with a cell-adhesion RGD domain, a plasmin-cleavable domain, and a collagen-derived MMP-cleavable domain. Thiol groups of cysteine residues between the degradation sites and at the termini of the repeat unit were reacted with PEG divinylsulfone to form a cross-linked hydrogel (Rizzi and Hubbell 2005). Peptide cross-linkers sensitive to degradation by elastase, collagenase, or plasmin have also been created by reaction of hydrazide-functional acryloyl-PEG-NHS with the N-terminus and primary amine of C-terminal lysine to create acrylate terminated peptides. These peptides were then cross-linked via photo-polymerization with acryloyl-PEG-RGDS and PEG-diacrylate to create synthetic extracellular matrix analogues (Mann et al. 2001, Gobin and West 2002, Gobin and West 2003).

4.8.3.3 Effects of Enzymatic Degradability

Proteolytic cell migration in 3D is dependent on cell-binding domains, matrix stiffness, and enzymatically degradable linkages (Zaman et al. 2006). Both proteolytic degradability and cell-adhesion peptides are necessary for cell migration through non-porous hydrogels. Peptide cross-linkers containing scrambled or random sequences do not support proteolytic degradation (Zisch et al. 2003). Furthermore, studies have indicated that enzymatic degradation and integrin-binding are also required for in vitro cellular outgrowth into hydrogels (Lutolf et al. 2003). Cleavage of degradable cross-links results in decreased network connectivity, swelling, and eventual mass loss of proteolytically degradable gels (Gobin and West 2002). Cell migration through collagen-mimetic hydrogels containing cell-binding domains and degradable cross-links is similar to migration through collagen gels. Fibroblasts were able to migrate through MMP-sensitive hydrogels but not through plasmin-sensitive hydrogels. Migration was dependent upon MMP proteolytic activity as up-regulation of MMP with TNF-α increased migration while inhibition of MMP activity effectively suppressed migration. MMP modulation has less effect on other natural, biomimetic materials such as fibrin due to their porous, nano-fibrous nature which allows amoeboid migration (Raeber et al. 2005). Aprotinin, a serine-protease inhibitor, prevented cell migration through plasmin-degradable artificial protein-PEG hybrids (Halstenberg et al. 2002). Fibroblast outgrowth and migration occurred in three-dimensions through a plasmin-degradable gel with cell-adhesive ligands while plasmin-insensitive gels did not allow outgrowth. Bone regeneration in vivo depends on the presence of both plasmin-degradable cross-links and bone morphogenic protein (BMP). The inclusion of covalently bound heparin-binding peptide and heparin, which is expected to bind the growth factor and control its release, further increases bone regeneration (Pratt et al. 2004). In other systems, the combination of both BMP growth factor and enzymatic degradability greatly enhances bone regeneration (Lutolf et al. 2003, Rizzi et al. 2006).

Tissue infiltration is also affected by polymer architecture; a more tightly cross-linked network resulted in less bone in-growth in vivo (Lutolf et al. 2003). Illustrating this principle, the structural properties of hybrid fibrino-gen-PEG hydrogels influenced their swelling and degradation as well as cellular morphology and migration (Dikovsky et al. 2006). Hydrogels with cross-links that are not degradable resist tissue in-growth, especially compared to enzymatically degradable gels (Chung et al. 2006). Increased proteolytically-cleavable cross-link density and hydrogel concentration in recombinant protein-PEG systems result in lower rates of cell invasion, density of cellular outgrowths, and cell interconnectivity (Rizzi et al. 2006). MMP degradability has been shown to permit chondrocyte migration, resulting in the formation of cell clusters in some hydrogels. Gels with lower elastic modulus due to lower cross-link density resulted in larger clusters and more diffuse matrix production. Analysis of gene expression of entrapped cells indicated that ECM component expression is increased in proteolytically degradable gels while MMPs are up-regulated in non-degradable gels (Park et al. 2004). Other biomimetic hydrogel systems have also indicated that native matrix synthesis is increased in degradable gels compared to non-degradable gels (Mann et al. 2001). The expression profiles for specific ECM components and MMPs for particular cells, both in response to different biomaterials and during natural tissue physiology, likely reflects their adaptive mechanisms in vivo, and can be used to design tissue-specific regenerative materials.

4.8.4 Growth Factor Activity

4.8.4.1 Considerations for Growth Factor Incorporation

Administration of growth factors for tissue regeneration was originally considered an extension of drug delivery to higher molecular weight macromolecules and focus was placed on sustained release at appropriate doses by diffusion from delivery vehicles. However, it is now clear that many biologics require further control over delivery than has been traditionally applied for small molecule drugs. Macromolecules can be quickly cleared, degraded, and deactivated in physiological and intracellular environments and local delivery to specific cells enhances effectiveness.

In the natural ECM, growth factors interact with surrounding macromolecules. Certain inductive molecules bind to glycosaminoglycans, such as heparin, and proteins, such as collagen, which alters their solubility, increases their local concentration, prevents degradation, and promotes their cell-mediated release. Interactions of growth factors with the surrounding matrix may also increase their effective activity, due to multi-valency effects or inhibited internalization and degradation, or, alternatively, decrease it, due to competitive or allosteric binding.

A new generation of delivery vehicles has been designed to present and deliver growth factors in a biomimetic fashion. Inductive biologic macromolecules can be incorporated by *encapsulation*, the release of which is dictated by *diffusion* or controlled by *degradability*, by *tethering* to natural or synthetic materials, which may be released by *enzymatic degradation*, or by *affinity binding* to heparin or heparin-binding components (Fig. 4.7).

4.8.4.2 Growth Factor Encapsulation

Simple encapsulation of growth factors can be achieved during liquid-to-solid transition under physiological conditions, such as addition of BMP-7 to thermo-reversible p(PF-co-EG). Increasing concentrations of BMP in the hydrogel improved chondrocyte viability but decreased proteoglycan synthesis (Fisher et al. 2004). Transforming growth factor β-3 (TGF-β3) was embedded in thermo-responsive poly(NIPAAm) co-polymers blended with hyaluronan. Inclusion of the growth factor increased initial chondrocyte proliferation and cartilage ECM production in gels injected subcutaneously (Na et al. 2006). Vascular endothelial growth factor (VEGF) encapsulated in photo-polymerized chitosan-RGD gels was retained over long periods under non-degrading conditions (Yeo et al. 2006). Photo-crosslinkable chitosan has also been used to incorporate fibroblast growth factor 2 (FGF-2), which remained active and was released by degradation of the gel (Obara et al. 2005). Delivery of growth factors from such systems is governed by non-specific degradation, adsorption, partitioning into the matrix, and diffusion within and from the matrix, as well as external convection in some cases. However, the effect of simply encapsulated

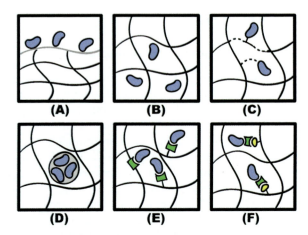

Fig. 4.7 The method of growth factor incorporation and release can greatly impact the regenerative capability of materials. Growth factor release can be accomplished by: (A) surface adsorption; (B) encapsulation, absorption, or bulk adsorption; (C) grafting in a degradable network or by a degradable linker; (D) particle or capsule entrapment; (E) grafting or backbone incorporation of growth factor binding domains; (F) grafting of heparin binding domains to form sandwich structures of bound growth factors

growth factors in hydrogels that are not specifically degraded in vivo has not proven to profoundly improve tissue regeneration.

Synthetic nano-fiber matrices have been used to deliver growth factors due to their high surface area to volume ratio which promotes adsorption and sustained delivery. Peptide-amphiphiles that form 3D nano-fiber scaffolds allow encapsulation of growth factors. Such nano-fibers supported sustained release of basic fibroblast growth factor (bFGF) in vitro and in vivo. Delivery of bFGF with peptide-amphiphile nano-fiber matrices improved neovascularization surrounding the injected gel, compared to administration of either nano-fibers or freely soluble growth factor alone (Hosseinkhani et al. 2006). Controlled delivery of growth factors can also be accomplished by incorporation in self-assembling repeating peptides prior to induced gelation. Platelet-derived growth factor (PDGF) was shown to bind peptide nano-fibers in vitro and demonstrated sustained release in vivo. Growth factor delivery promoted cardiomyocyte survival after myocardial infarction via receptor-induced signaling (Hsieh et al. 2006). PDGF delivered by a nano-fiber matrix reduced infarct volume and improved ventricular function without pulmonary toxicity in a myocardial infarct model (Hsieh et al. 2006). Insulin-like growth factor 1 (IGF-1) was bound to nano-fiber matrices via a biotin sandwich method. Biotinylated IGF-1 was complexed to streptavidin and subsequently bound to biotinylated self-assembling peptides. Bound IGF-1 was retained in the nano-fiber scaffold in vivo, induced specific intracellular signaling, and protected implanted cardiomyocytes from apoptosis following myocardial infarction (Davis et al. 2006).

The effects of inductive molecules encapsulated in biomaterials can be enhanced by imparting through enzymatic degradability to the material. Cytokines have been incorporated by entrapment in cross-linked hyaluronan gels which are degradable by hyaluronase. Incorporation of growth factors resulted in substantially more blood vessel formation than hyaluronan hydrogels with no cytokine. There was a synergistic effect on vessel density of VEGF administered with hyaluronan films compared to the growth factor or hyaluronan alone which was not seen with hyaluronan and bFGF (Peattie et al. 2004). Incorporation of VEGF or keratin growth factor (KGF) in crosslinked hyaluronan resulted in intact microvessel formation with well-defined borders. An additive effect was found upon co-administration of VEGF and KGF (Peattie et al. 2006). Cell-mediated BMP release has been achieved by entrapment in enzymatically degradable hydrogels. Growth factor release was stimulated by addition of exogenous MMP that degraded the hydrogel (Pratt et al. 2004, Lutolf et al. 2003).

4.8.4.3 Growth Factor Conjugation

Growth factors can be conjugated to polymers such as PEG for covalent incorporation into hydrogels. They are able to maintain significant morphogenic and mitogenic activity when conjugated to polymers (Mann et al. 2001, Zisch et al. 2001, Gobin and West 2003, Koch et al. 2006). VEGF was necessary for endothelial cell migration through MMP-degradable gels containing cell-adhesion peptides. In vitro, hydrogels containing bound VEGF resulted in a highly localized angiogenic response with the development of well-defined vessels. The presence of VEGF and degradable peptide cross-links was required for cellular infiltration and vascularization of the gels (Zisch et al. 2003). VEGF has also been conjugated to collagen matrices via reaction with excess PEG disuccinimide, the remainder of which cross-linked the collagen gel. Conjugation, as opposed to simple mixing of VEGF with cross-linked collagen, resulted in increased endothelial cell proliferation in vitro and capillary formation in vivo. VEGF release could be induced in vitro by addition of collagenase, again indicating cell-demanded degradation as the release mechanism (Koch et al. 2006).

Growth factors, including basic fibroblast growth factor (bFGF) (DeLong et al. 2005), TGF-β (Mann et al. 2001), and epidermal growth factor (EGF) (Gobin and West 2003), have been functionalized with acryloyl-PEG-NHS to produce acrylate-functional growth factors with spacer arms for photo-initiated co-polymerization with PEG-diacrylate. Tethered EGF enhanced cell migration through RGD-functional, collagenase-degradable PEG hydrogels (Gobin and West 2003). A concentration gradient of immobilized bFGF resulted in smooth muscle cell alignment with and migration up the gradient (DeLong et al. 2005).

Cell adhesion elements in biomimetic materials, while increasing cell attachment and spreading, can reduce ECM production (Mann et al. 1999). Certain growth factors, such as TGF-β, are known to increase ECM protein production. Covalent incorporation of TGF-β in hydrogels containing cell-binding domains was able to increase matrix production of entrapped smooth muscle cells to levels greater than those observed in the absence of adhesive peptides. Although the polymer-tethered growth factor is not as active as unmodified growth factor when presented in solution (Sakiyama-Elbert et al. 2001), immobilized TGF-β increased ECM production more than soluble TGF-β in the same matrix (Mann et al. 2001), possibly due to cooperative binding or reduced internalization.

A BMP-2 derived peptide sequence was produced by solid-phase synthesis and conjugated to cross-linked alginate by carbodiimide activation and succinimide stabilized coupling. Implantation of the peptide-linked alginate resulted in ectopic bone formation, as is expected of BMP-loaded inductive scaffolds (Suzuki et al. 2000). Immobilized growth factors also allow cell adhesion and spreading by direct binding to integrins (Hutchings et al. 2003).

Variants of growth factors can be genetically engineered for incorporation into biomaterials. VEGF has been engineered to have an additional, unpaired cysteine residue at the C-terminus for covalent conjugation to functionalized PEG (Seliktar et al. 2004). Incorporation of tethered VEGF increased MMP-2 expression in endothelial cells. Immobilized and locally released growth factor was more effective than soluble growth factor. However, activation of the zymogen to the active isoform required stimulation of cell-mediated regulation by addition of soluble TGF-β (Seliktar et al. 2004). Similarly, tumor necrosis factor α (TNF-α) has also been shown in induce pro-MMP-9 production in fibroblasts (Raeber et al. 2005).

Growth factor retention can be affected by altering their chemistry for extended delivery from biomaterial matrices. A non-glycosylated version of BMP-2 with low solubility in fibrin was recombinantly produced and incorporated into fibrin matrices by precipitation. Insoluble BMP-2 in fibrin induced more bone growth than more soluble BMPs, demonstrating a correlation between growth factor solubility, retention, and tissue healing. This BMP-2 variant delivered in fibrin enhanced healing in the first weeks compared to autograft, and performed comparably thereafter (Schmoekel et al. 2004).

VEGF variants have also been recombinantly engineered to contain a transglutaminase substrate for incorporation by factor XIIIa into fibrin during coagulation. Fibrin-bound VEGF induced a stronger mitogenic response than soluble VEGF and, like soluble VEGF, had a biphasic dose-dependence on endothelial cell proliferation (Zisch et al. 2001). One chimeric version included a sequence cleavable by plasmin, which degrades fibrin clots, coupled between the sequence for attachment to fibrin and the growth factor for specific cell-mediated release of intact VEGF. Native VEGF variants resulted in burst release of soluble VEGF, fibrin-bound VEGF resulted in slow cell-demanded release via fibrin degradation, and engineered plasmin-cleavable fibrin-bound VEGF demonstrated accelerated cell-demanded release. Endothelial cell proliferation was increased by plasmin-releasable VEGF. The VEGF variants were also capable of positively modulating endothelial progenitor cell maturation (Ehrbar et al. 2005).

The same approach has been used to enzymatically incorporate BMP-2 (Schmoekel et al. 2005) and nerve growth factor (NGF) (Sakiyama-Elbert et al. 2001) into fibrin and allow enzymatic release via a plasmin degradable linkage. BMP-2 engineered for incorporation into fibrin matrices provided enhanced healing of critical-size defects compared to wild-type BMP-2 in fibrin. The same regenerative fibrin matrix performed as well as cancellous bone autograft in a pre-clinical bone fusion study. Similarly, fibrin-bound NGF enhanced neurite extension compared to soluble NGF in fibrin. Susceptibility to plasmin-mediated release improved the recombinant NGF activity, as demonstrated by enhanced neurite extension. An optimal concentration for neurite outgrowth was found for this fibrin-immobilized, plasmin-cleavable NGF. The transmembrane cell-surface protein ephrin B has also been genetic engineering to couple it to the transglutaminase substrate in order to impart fibrin gels with growth factor activity (Zisch et al. 2004).

Fibrin degradation alone results in cell-demanded growth factor release from these matrices. Fibrinolytic inhibitors which prevent plasmin from degrading fibrin and releasing the bound growth factor have been used to demonstrate this mechanism. Using a different method, the amino groups of recombinant keratinocyte growth factor (KGF) were conjugated to a synthetic peptide containing the factor XIIIa substrate via a heterobifunctional cross-linker containing succinimide and maleimide groups. Cell-mediated activation of plasminogen to plasmin resulted in release of active KGF, enhancing epithelial cell proliferation and migration. The fibrin-bound KGF resulted in enhanced wound closure compared to fibrin or fibrin with topical KGF (Geer et al. 2005).

4.8.4.4 Heparin and Heparin-Binding Incorporation

Utilizing a matrix that is capable of interacting with growth factors, even regardless of their exogenous addition, is another approach to improve tissue regeneration. It has been proposed that chondroitin sulfate hydrogels can serve as a repository of growth factors produced by regenerating mucosa, enhancing tissue growth (Gilbert et al. 2004). Acellular tissue containing glycosaminoglycans bind FGF and enhance new tissue formation and angiogenesis (Lai et al. 2006). Sustained release of growth factors can also be accomplished by covalent incorporation of heparin or heparin-binding domains into biomimetic hydrogels (Table 4.10). Heparin has been cross-linked with alginate by carbodiimide activated reaction of carboxylic acids with ethylene diamine for sustained release of absorbed bFGF. Inclusion of the heparin component suppressed burst release of the growth factor and greatly enhanced angiogenesis

Table 4.10 Heparin binding peptide sequences

Sequence	Origin	Source
BBXB, XBBXBX	Consensus (X = hydrophilic, B = basic amino acid)	(Massia and Hubbell 1992)
KRSR	Proteoglycan-binding in bone	(Dee et al. 1998)
PRRARV	Fibronectin	(Kao et al. 2001)
FAKLAARLYRKA	Anti-thrombin-III	(Sakiyama-Elbert and Hubbell 2000; Seal and Panitch 2003)
FHRRIKA	Bone sialoprotein	(Rezania and Healy 1999)
RHRNRK	Vitronectin	(Dettin et al. 2005)
LRKKLGKA		(Rajangam et al. 2006)
KHKGRDVILKKDV	N-CAM	(Sakiyama et al. 1999)
YKKIIKKL	Platelet factor 4	(Sakiyama et al. 1999)

in vivo (Tanihara et al. 2001). Heparin has been modified with thiol groups for co-crosslinking with either chemically modified hyaluronan or gelatin. VEGF or bFGF were incorporated prior to cross-linking with PEG diacrylate. Growth factor release decreased monotonically with increasing heparin concentration. Effective growth factor release was demonstrated in vivo by enhanced vascularization of gels containing both cross-linked heparin and growth factor (Pike et al. 2006). A modified chitosan gel blended with oxidized heparin has been used to control release of FGF-2 (Fujita et al. 2005). Additionally, heparin was conjugated to cross-linked collagen matrices using carbodiimide/succinimide (EDC/NHS) coupling. Binding of bFGF to the heparinized collagen matrices increased with increasing heparin content up to saturation (Wissink et al. 2001). Hydrazide functional heparin gels covalently cross-linked with PEG discuccinimide demonstrated sustained release of VEGF. Endothelial cell proliferation increased with increasing VEGF concentration. Immunostaining of subcutaneous implants indicated a strong angiogenic response surrounding the gel (Tae et al. 2006).

Heparin-binding peptide sequences enhanced cell spreading and mineralization when combined with surfaces containing cell-adhesion domains (Healy et al. 1999, Rezania and Healy 1999). A heparin-binding peptide sequence has been conjugated to IPN hydrogel surfaces to enhance osteoblast proliferation and matrix-production (Harbers and Healy 2005). Multiarm PEG macromers have also been functionalized with heparin-binding peptides and physically cross-linked with heparin. The gels were able to sequester and release exogenous heparin-binding peptides at a rate dependent on their relative affinity to heparin (Seal and Panitch 2003). Futhermore, self-assembling peptide-amphiphiles have been designed with a peptide head sequence that binds heparin. The presence of heparin chains induced nanofiber formation and sustained fibroblast growth factor 2 (FGF-2) release. Peptide-amphiphile heparin gels loaded with FGF-2 and VEGF promoted vastly improved neovascularization compared to other delivery methods after implantation in the cornea (Rajangam et al. 2006).

Fibrin gels do not naturally bind growth factors and the enzymatic activity of factor XIIIa can be used to incorporate bioactive components during coagulation. Bi-domain peptides have been synthetically produced that contain the factor XIIIa substrate from α-2-plasmin inhibitor at one end and a heparin-binding domain from anti-thrombin III at the other. The fibrin system containing this peptide non-covalently binds heparin, which, in turn, binds heparin-binding growth factors such as bFGF and demonstrated enhanced neurite outgrowth into the gel (Sakiyama et al. 1999). Covalent incorporation of heparin-binding domains led to sustained, effective delivery of growth factor (Ehrbar et al. 2004). Growth factor release decreased with increasing ratio of heparin to growth factor. An optimal dose of bound bFGF was found for enhanced neurite extension. Inclusion of free bFGF in unmodified fibrin or heparin-bound VEGF did not enhance neurite outgrowth. All components of the delivery system, including heparin-binding peptide, heparin, and heparin-binding growth factor, were necessary for slow growth factor release and increased neurite extension (Sakiyama-Elbert and Hubbell 2000). These systems have also been used to deliver nerve growth factor β (NGF-β), a non-heparin-binding growth factor, by slow diffusion-based release. Fibrin matrices containing a large molar excess of heparin interacted with neurotrophins to enhance neurite outgrowth by up to 100% compared to unmodified fibrin. Free neutrophins within fibrin did not enhance neurite extension (Sakiyama-Elbert and Hubbell 2000).

4.8.4.5 Multiple Growth Factor Release Profiles

Temporal presentation of multiple growth factors is critical to organogenesis and can be achieved by biomimetic material design. The method of growth factor incorporation controls its release rate and different release mechanisms can be utilized to engineer the delivery profile of multiple growth factors. This has been accomplished with VEGF and platelet-derived growth factor (PDGF) delivered from porous poly(lactide-co-glycolide). Lyophilized VEGF was simply mixed with polymer particles prior to material formation while PDGF was preencapsulated in micro-spheres prior to incorporation (Richardson et al. 2001). Covalently bound, engineered VEGF has been co-delivered with encapsulated soluble TGF-β (Seliktar et al. 2004). Also, composite oligo(poly(ethylene glycol) fumarate) (OPF) hydrogels have been created by co-encapsulated TGF-β loaded gelatin micro-particles and chondrocytes. Cells proliferated to a greater extent in the presence of loaded microspheres while apparently maintaining their phenotype (Park et al. 2005). This composite system allows different growth factors to be entrapped in each phase and the release rate can be modified by parameters such as cross-linking, allowing controlled dual growth factor delivery (Holland et al. 2005).

4.9 Neglected Topics in Biomimetic Materials

Biomaterials are currently experiencing a renaissance and biomimetic materials lie at the forefront of a more general materials revolution. While many advances have been made recently in the development of biomimetic materials for tissue regeneration, there is plenty of room for the field to grow and more progress must be made before wide clinical implementation of many promising novel technologies can be realized.

There are many opportunities for new composite and biohybrid materials that combine synthetic biomimetic materials and natural or more traditional biomaterials. New material synthesis and manufacturing techniques could lead to the production of more complex, hierarchical structures that more closely approximate native tissue properties. Composites allow for temporal control through differential degradation and spatial control by evolution of microstructural architecture. Using current methods and components, a more diverse set of tissue-specific artificial ECMs can be designed beyond those for connective tissues. Improved methods of integrating vascularization and nerve growth will improve the prospects of engineered and regenerated tissues.

Cell-cell adhesion and signaling are as important as cell-matrix interactions during development and in many types of tissues. Materials that mimic the mechanisms involved in such interactions and induce cellular self-assembly into relevant tissue structures could be better utilized for tissue regeneration. Deconvolution of the effects of growth factors, ECM receptor engagement, and cell-cell contacts and subsequent transduction into cell phenotype and behavior will help to realize the potential of biomimetic biomaterials. Furthermore, biomaterial studies using co-culture of multiple cell types will lead to better understanding of in vivo material performance and improved regenerative therapies.

In the near future, new methods of bio-orthogonal in situ formation will likely be developed that allow for safe and fast formation of tissue with controlled biomimetic characteristics and robust mechanical properties. On the other hand, effective methods of tissue adhesion with biomaterials that promote healing would be of great benefit to surgical medicine.

Gene expression analysis throughout healing and regeneration processes as well as the effect of biomaterials on such profiles would aid in developing and improving biomimetic biomaterials. High throughput screening of biomimetic material formulations using micro-array analysis has yet to be fully implemented in biomaterials development. Further collaboration with cell biologists and new methods of identification and selection could lead to improved utilization of existing and new biomimetic components. For example, characterizing more receptors specific to certain cell types and their associated ligands as well as natural mechanisms of ECM remodeling would enhance directed tissue formation.

The use of biomimetic materials in tissue model systems will aid in their development through mapping of cell response to multiple input parameters. Complementary mathematical and computational modeling will help to understand multivalent receptor-ligand dynamics and ECM macromolecular dynamics. This knowledge can be integrated in biomimetic biomaterials to better control the bidirectional feedback between cells and their surrounding matrix. Work will likely focus on identifying and understanding cell lineage progression and associated checkpoints as well as cell-signaling and transcription factor networks.

Recent progress has focused on cell-mediated degradation and growth factor release which will continue to improve the effectiveness of biomaterials. The clearly apparent differences in performance in vitro and in vivo as well as between cell behaviors in two- and three-dimensions must be reconciled to truly understand and engineer the desired biological response. Better knowledge of the degradation products of biomaterials, their effect on cell behavior, and their physiological clearance mechanisms would also improve biological performance.

Finally, a focus must be placed on commercialization of these new products with an emphasis on clinical demand and regulatory approval. Cell and biologic molecule sourcing and large-scale production are critical issues that must be addressed. Practical and ethical solutions to the problems of stem cell sourcing and scaling must be adopted. Many biomimetic regenerative therapies combine materials, drugs or biologics, and cells and will face new regulatory hurdles on the path to widespread clinical implementation.

Questions/Exercises

1. Explain the differences between traditional, early generation biomaterials and modern, biomimetic materials for regenerative medicine. How has knowledge of tissue physiology altered biomaterial design philosophy? What are the differences in host response between these types of materials?
2. Which aspects of natural tissues and ECMs have been mimicked in biomaterials? How do these influence cell function and tissue formation?
3. Describe the mechanisms of biomimetic materials that can be utilized to achieve cell- and tissue-specific response. Explain with examples.
4. Describe the general roles of receptor-ligand dynamics in cell behavior. What are the similarities, differences, and interactions of the receptor-ligand dynamics of growth factors and cell-adhesion ligands? Name at least three roles of enzymes in tissue dynamics.
5. Give examples of tissue dynamic processes. How are the feedback mechanisms between cells and the ECM bidirectional? What is the term for these interactions?
6. What are three fundamental strategies of biomaterials for directed tissue growth? How are these complementary? Give specific examples of their use in advanced biomimetic materials design.
7. Describe several methods can be used to form biomaterials in situ? What are the advantages of in situ biomaterial formation for tissue regeneration?

8. List aspects of cell-adhesion domains which can be controlled in biomimetic materials? How do these aspects influence cell behavior? What methods are used to incorporate cell-adhesion domains?
9. What are the advantages of enzymatic degradability in biomimetic materials? Describe specific ways that this property can be imparted?
10. List at least three methods of incorporating growth factor activity into biomimetic materials. Predict the release profiles of growth factors from these materials.

Top Ten Original Publications from the Last Decade

Ehrbar M, Metters A, Zammaretti P, et al. (2005) Endothelial cell proliferation and progenitor maturation by fibrin-bound VEGF variants with differential susceptibilities to local cellular activity. J Control Release 101(1–3):93–109.
Elisseeff J, Anseth K, Sims D, et al. (1999) Transdermal photopolymerization for minimally invasive implantation. Proc Natl Acad Sci U S A 96(6):3104–3107.
Engler AJ, Sen S, Sweeney HL, et al. (2006) Matrix elasticity directs stem cell lineage specification. Cell 126(4):677–689.
Ghosh K, Ren XD, Shu XZ, et al. (2006) Fibronectin functional domains coupled to hyaluronan stimulate adult human dermal fibroblast responses critical for wound healing. Tissue Eng 12(3):601–613.
Kim S, Chung EH, Gilbert M, et al. (2005) Synthetic MMP-13 degradable ECMs based on poly(N-isopropylacrylamide-co-acrylic acid) semi-interpenetrating polymer networks. I. Degradation and cell migration. J Biomed Mater Res A 75(1):73–88.
Kuhl PR, Griffith-Cima LG, (1996) Tethered epidermal growth factor as a paradigm for growth factor-induced stimulation from the solid phase. Nat Med 2(9):1022–1027.
Lutolf MP, Weber FE, Schmoekel HG, et al. (2003) Repair of bone defects using synthetic mimetics of collagenous extracellular matrices. Nat Biotechnol21(5):513–518.
Rajangam K, Behanna HA, Hui MJ, et al. (2006) Heparin binding nanostructures to promote growth of blood vessels. Nano Lett 6(9):2086–2090.
West JL, Hubbell JA. (1999) Polymeric biomaterials with degradation sites for proteases involved in cell migration. Macromolecules 32(1):241–244.
Zaman MH, Trapani LM, Sieminski AL, et al. (2006) Migration of tumor cells in 3D matrices is governed by matrix stiffness along with cell-matrix adhesion and proteolysis. Proc Natl Acad Sci U S A 103(29):10889–10894.

References

Abbott A (2003) Cell culture: biology's new dimension. Nature 424:870–872
Alenghat FJ, Ingber DE (2002) Mechanotransduction: all signals point to cytoskeleton, matrix, and integrins. Sci STKE 2002:PE6

Anseth KS, Metters AT, Bryant SJ, et al. (2002) In situ forming degradable networks and their application in tissue engineering and drug delivery. J Control Release 78:199–209

Augst AD, Kong HJ, Mooney DJ (2006) Alginate hydrogels as biomaterials. Macromol Biosci. 6:623–633

Barber TA, Golledge SL, Castner DG et al. (2003) Peptide-modified p(AAm-co-EG/AAc) IPNs grafted to bulk titanium modulate osteoblast behavior in vitro. J Biomed Mater Res A 64:38–47

Bearinger JP, Castner DG, Healy KE (1998) Biomolecular modification of p(AAm-co-EG/ AA) IPNs supports osteoblast adhesion and phenotypic expression. J Biomater Sci Polym Ed 9:629–652

Behravesh E, Zygourakis K, Mikos AG (2003) Adhesion and migration of marrow-derived osteoblasts on injectable in situ crosslinkable poly(propylene fumarate-co-ethylene glycol)-based hydrogels with a covalently linked RGDS peptide. J Biomed Mater Res A 65:260–270

Bhattarai N, Matsen FA, Zhang M (2005) PEG-grafted chitosan as an injectable thermo-reversible hydrogel. Macromol Biosci 5:107–111

Brandley BK, Schnaar RL (1988) Covalent attachment of an Arg-Gly-Asp sequence peptide to derivatizable polyacrylamide surfaces: support of fibroblast adhesion and long-term growth. Anal Biochem 172:270–278

Burdick JA, Anseth KS (2002) Photoencapsulation of osteoblasts in injectable RGD-modified PEG hydrogels for bone tissue engineering. Biomaterials 22:4315–4323

Cai S, Liu Y, Zheng Shu X, et al. (2005) Injectable glycosaminoglycan hydrogels for controlled release of human basic fibroblast growth factor. Biomaterials 26:6054–6067

Cellesi F, Tirelli N (2005) A new process for cell microencapsulation and other biomaterial applications: Thermal gelation and chemical cross-linking in "tandem". J Mater Sci Mater Med 16:559–565

Chaikof EL, Matthew H, Kohn J, et al.(2002) Biomaterials and scaffolds in reparative medicine. Ann N Y Acad Sci 961:96–105

Chen JS, Noah EM, Pallua N, et al. (2002) The use of bifunctional polyethyleneglycol derivatives for coupling of proteins to and cross-linking of collagen matrices. J Mater Sci Mater Med 13:1029–1035

Chung EH, Gilbert M, Virdi AS, et al. (2006) Biomimetic artificial ECMs stimulate bone regeneration. J Biomed Mater Res A 79A:815–826

Cutler SM, Garcia AJ (2003) Engineering cell adhesive surfaces that direct integrin $\alpha5\beta1$ binding using a recombinant fragment of fibronectin. Biomaterials May 24:1759–1770

Davis ME, Hsieh PC, Takahashi T, et al. (2006) Local myocardial insulin-like growth factor 1 (IGF-1) delivery with biotinylated peptide nanofibers improves cell therapy for myocardial infarction. Proc Natl Acad Sci U S A 103:8155–8160

de Las Heras Alarcon C, Pennadam S, Alexander C (2005) Stimuli responsive polymers for biomedical applications. Chem Soc Rev 34:276–285

Dee KC, Andersen TT, Bizios R (1998) Design and function of novel osteoblast-adhesive peptides for chemical modification of biomaterials. J Biomed Mater Res 40:371–377

DeLong SA, Gobin AS, West JL (2005) Covalent immobilization of RGDS on hydrogel surfaces to direct cell alignment and migration. J Control Release 109:139–148

DeLong SA, Moon JJ, West JL (2005) Covalently immobilized gradients of bFGF on hydrogel scaffolds for directed cell migration. Biomaterials 26:3227–3234

Dettin M, Conconi MT, Gambaretto R, et al. (2005) Effect of synthetic peptides on osteoblast adhesion. Biomaterials 26:4507–4515

Dikovsky D, Bianco-Peled H, Seliktar D (2006) The effect of structural alterations of PEG-fibrinogen hydrogel scaffolds on 3-D cellular morphology and cellular migration. Biomaterials 27:1496–1506

Discher DE, Janmey P, Wang YL (2005) Tissue cells feel and respond to the stiffness of their substrate. Science 310:1139–1143

Drotleff S, Lungwitz U, Breunig M, et al. (2004) Biomimetic polymers in pharmaceutical and biomedical sciences. Eur J Pharm Biopharm 58:385–407

Drumheller PD, Hubbell JA. (1994) Polymer networks with grafted cell adhesion peptides for highly biospecific cell adhesive substrates. Anal Biochem 222:380–388

Ehrbar M, Djonov VG, Schnell C, et al. (2004) Cell-demanded liberation of VEGF121 from fibrin implants induces local and controlled blood vessel growth. Circ Res 94:1124–1132

Ehrbar M, Metters A, Zammaretti P, et al. (2005) Endothelial cell proliferation and progenitor maturation by fibrin-bound VEGF variants with differential susceptibilities to local cellular activity. J Control Release 101:93–109

Elbert DL, Hubbell JA (2001) Conjugate addition reactions combined with free-radical cross-linking for the design of materials for tissue engineering. Biomacromolecules 2:430–441

Elbert DL, Pratt AB, Lutolf MP, et al. (2001) Protein delivery from materials formed by self-selective conjugate addition reactions. J Control Release 76:11–25

Engler A, Bacakova L, Newman C, et al. (2004) Substrate compliance versus ligand density in cell on gel responses. Biophys J 86:617–628

Engler AJ, Sen S, Sweeney HL, et al. (2006) Matrix elasticity directs stem cell lineage specification. Cell 126:677–689

Fisher JP, Jo S, Mikos AG, et al. (2004) Thermoreversible hydrogel scaffolds for articular cartilage engineering. J Biomed Mater Res A 71:268–274

Forget MA, Desrosiers RR, Beliveau R (1999) Physiological roles of matrix metalloproteinases: implications for tumor growth and metastasis. Can J Physiol Pharmacol 77:465–480

Fujita M, Ishihara M, Morimoto Y, et al. (2005) Efficacy of photocrosslinkable chitosan hydrogel containing fibroblast growth factor-2 in a rabbit model of chronic myocardial infarction. J Surg Res 126:27–33

Gallant ND, Garcia AJ (2007) Quantitative analysis of cell adhesion strength. Methods Mol Biol 370:83–96

Garcia AJ, Boettiger D (1999) Integrin-fibronectin interactions at the cell-material interface: initial integrin binding and signaling. Biomaterials 20:2427–243

Geer DJ, Swartz DD, Andreadis ST (2005) Biomimetic delivery of keratinocyte growth factor upon cellular demand for accelerated wound healing in vitro and in vivo. Am J Pathol 167:1575–1586

Ghosh K, Ren XD, Shu XZ, et al. (2006) Fibronectin functional domains coupled to hyaluronan stimulate adult human dermal fibroblast responses critical for wound healing. Tissue Eng 12:601–613

Giancotti FG, Ruoslahti E (1999) Integrin signaling. Science 285:1028–1032

Gilbert M, Giachelli CM, Stayton PS (2003) Biomimetic peptides that engage specific integrin-dependent signaling pathways and bind to calcium phosphate surfaces. J Biomed Mater Res A 67:69–77

Gilbert ME, Kirker KR, Gray SD, et al. (2004) Chondroitin sulfate hydrogel and wound healing in rabbit maxillary sinus mucosa. Laryngoscope 114:1406–1409

Girotti A, Reguera J, Rodriguez-Cabello JC, et al. (2004) Design and bioproduction of a recombinant multi(bio)functional elastin-like protein polymer containing cell adhesion sequences for tissue engineering purposes. J Mater Sci Mater Med 15:479–484

Gobin AS, West JL (2002) Cell migration through defined, synthetic ECM analogs. Faseb J 16:751–753

Gobin AS, West JL (2003) Effects of epidermal growth factor on fibroblast migration through biomimetic hydrogels. Biotechnol Prog. 19:1781–1785

Gobin AS, West JL (2003) Val-ala-pro-gly, an elastin-derived non-integrin ligand: smooth muscle cell adhesion and specificity. J Biomed Mater Res A 67:255–259

Gonzalez AL, Gobin AS, West JL, et al. (2006) Integrin interactions with immobilized peptides in polyethylene glycol diacrylate hydrogels. Tissue Eng 10:1775–1786

Griffith LG, Swartz MA (2006) Capturing complex 3D tissue physiology in vitro. Nat Rev Mol Cell Biol 7:211–224

Guler MO, Hsu L, Soukasene S, et al. (2006) Presentation of RGDS epitopes on self-assembled nanofibers of branched peptide amphiphiles. Biomacromolecules. 7:1855–1863

Halstenberg S, Panitch A, Rizzi S, et al. (2002) Biologically engineered protein-graft-poly (ethylene glycol) hydrogels: a cell adhesive and plasmin-degradable biosynthetic material for tissue repair. Biomacromolecules 3:710–723

Harbers GM, Gamble LJ, Irwin EF, et al. (2005) Development and characterization of a high-throughput system for assessing cell-surface receptor-ligand engagement. Langmuir 21:8374–8384

Harbers GM, Healy KE (2005) The effect of ligand type and density on osteoblast adhesion, proliferation, and matrix mineralization. J Biomed Mater Res A 75:855–869

Hartgerink JD, Beniash E, Stupp SI (2002) Peptide-amphiphile nanofibers: a versatile scaffold for the preparation of self-assembling materials. Proc Natl Acad Sci U S A 99:5133–5138

Healy KE, Rezania A, Stile RA (1999) Designing biomaterials to direct biological responses. Bioartificial Organs Ii: Technology, Medicine, and Materials 875:24–35

Heilshorn SC, DiZio KA, Welsh ER, et al. (2003) Endothelial cell adhesion to the fibronectin CS5 domain in artificial extracellular matrix proteins. Biomaterials 24:4245–4252

Hern DL, Hubbell JA (1998) Incorporation of adhesion peptides into nonadhesive hydrogels useful for tissue resurfacing. J Biomed Mater Res 39:266–276

Hertl W, Ramsey WS, Nowlan ED (1984) Assessment of cell-substrate adhesion by a centrifugal method. In Vitro 20:796–801

Hirano Y, Kando Y, Hayashi T, et al. (1991) Synthesis and cell attachment activity of bioactive oligopeptides: RGD, RGDS, RGDV, and RGDT. J Biomed Mater Res 25:1523–1534

Holland TA, Tabata Y, Mikos AG (2005) Dual growth factor delivery from degradable oligo(poly(ethylene glycol) fumarate) hydrogel scaffolds for cartilage tissue engineering. J Control Release 101:111–125

Hoshikawa A, Nakayama Y, Matsuda T, et al. (2006) Encapsulation of chondrocytes in photopolymerizable styrenated gelatin for cartilage tissue engineering. Tissue Eng. 12:2333–2341

Hosseinkhani H, Hosseinkhani M, Khademhosseini A, et al. (2006) Enhanced angiogenesis through controlled release of basic fibroblast growth factor from peptide amphiphile for tissue regeneration. Biomaterials 27:5836–5844

Hsieh PC, Davis ME, Gannon J, et al. (2006) Controlled delivery of PDGF-BB for myocardial protection using injectable self-assembling peptide nanofibers. J Clin Invest 116:237–248

Hsieh PC, MacGillivray C, Gannon J, et al. (2006) Local controlled intramyocardial delivery of platelet-derived growth factor improves postinfarction ventricular function without pulmonary toxicity. Circulation 114:637–644

Hubbell JA (2004) Biomaterials science and high-throughput screening. Nat Biotechnol 22:828–829

Humphries JD, Byron A, Humphries MJ (2006) Integrin ligands at a glance. J Cell Sci 119:3901–3903

Hutchings H, Ortega N, Plouet J (2003) Extracellular matrix-bound vascular endothelial growth factor promotes endothelial cell adhesion, migration, and survival through integrin ligation. Faseb J 17:1520–1522

Irvine DJ, Mayes AM, Griffith LG (2001) Nanoscale clustering of RGD peptides at surfaces using Comb polymers. 1. Synthesis and characterization of Comb thin films. Biomacromolecules 2:85–94

Itle LJ, Koh WG, Pishko MV (2005) Hepatocyte viability and protein expression within hydrogel microstructures. Biotechnol Prog 21:926–932

Jiang XS, Chai C, Zhang Y, et al. (2006) Surface-immobilization of adhesion peptides on substrate for *ex vivo* expansion of cryopreserved umbilical cord blood CD34+ cells. Biomaterials 27:2723–2732

Jo S, Shin H, Mikos AG (2001) Modification of oligo(poly(ethylene glycol) fumarate) macromer with a GRGD peptide for the preparation of functionalized polymer networks. Biomacromolecules 2:255–261

Kao WJ, Lee D, Schense JC, et al. (2001) Fibronectin modulates macrophage adhesion and FBGC formation: the role of RGD, PHSRN, and PRRARV domains. J Biomed Mater Res 55:79–88

Kim S, Chung EH, Gilbert M, et al. (2005) Synthetic MMP-13 degradable ECMs based on poly(N-isopropylacrylamide-co-acrylic acid) semi-interpenetrating polymer networks. I. Degradation and cell migration. J Biomed Mater Res A 75:73–88

Kim S, Healy KE (2003) Synthesis and characterization of injectable poly(N-isopropylacrylamide-co-acrylic acid) hydrogels with proteolytically degradable cross-links. Biomacromolecules. 4:1214–1223

Kirker KR, Luo Y, Nielson JH, et al. (2002) Glycosaminoglycan hydrogel films as bio-interactive dressings for wound healing. Biomaterials 23:3661–3671

Kleiner DE, Stetler-Stevenson WG (1999) Matrix metalloproteinases and metastasis. Cancer Chemother Pharmacol 43:S42–51

Koch S, Yao C, Grieb G, et al. (2006) Enhancing angiogenesis in collagen matrices by covalent incorporation of VEGF. J Mater Sci Mater Med 17:735–741

Koo LY, Irvine DJ, Mayes AM, et al. (2002) Co-regulation of cell adhesion by nanoscale RGD organization and mechanical stimulus. J Cell Sci 115:1423–1433

Kopecek J (2003) Smart and genetically engineered biomaterials and drug delivery systems. Eur J Pharm Sci 20:1–16

Lai PH, Chang Y, Chen SC, et al. (2006) Acellular biological tissues containing inherent glycosaminoglycans for loading basic fibroblast growth factor promote angiogenesis and tissue regeneration. Tissue Eng 12:2499–2508

Leach JB, Bivens KA, Collins CN, et al. (2004) Development of photocrosslinkable hyaluronic acid-polyethylene glycol-peptide composite hydrogels for soft tissue engineering. J Biomed Mater Res A 70:74–82

Lee BH, West B, McLemore R, et al. (2006) In-situ injectable physically and chemically gelling NIPAAm-based copolymer system for embolization. Biomacromolecules. 7:2059–2064

Lee KY, Mooney DJ (2001) Hydrogels for tissue engineering. Chem Rev 101:1869–1879

Lee MH, Murphy G (2004) Matrix metalloproteinases at a glance. J Cell Sci 117:4015–4016

Li R, Hoess RH, Bennett JS, et al. (2003) Use of phage display to probe the evolution of binding specificity and affinity in integrins. Protein Eng 16:65–72

Luo Y, Kirker KR, Prestwich GD (2000) Cross-linked hyaluronic acid hydrogel films: new biomaterials for drug delivery. J Control Release 69:169–184

Lutolf MP, Lauer-Fields JL, Schmoekel HG, et al. (2003) Synthetic matrix metalloproteinase-sensitive hydrogels for the conduction of tissue regeneration: engineering cell-invasion characteristics. Proc Natl Acad Sci U S A 100:5413–5418

Lutolf MP, Tirelli N, Cerritelli S, et al. (2001) Systematic modulation of Michael-type reactivity of thiols through the use of charged amino acids. Bioconjug Chem 12:1051–1056

Lutolf MP, Weber FE, Schmoekel HG, et al. (2003) Repair of bone defects using synthetic mimetics of collagenous extracellular matrices. Nat Biotechnol 21:513–518

Maheshwari G, Brown G, Lauffenburger DA, et al. (2000) Cell adhesion and motility depend on nanoscale RGD clustering. J Cell Sci 113:1677–1686

Mann BK, Gobin AS, Tsai AT, et al. (2001) Smooth muscle cell growth in photopolymerized hydrogels with cell adhesive and proteolytically degradable domains: synthetic ECM analogs for tissue engineering. Biomaterials 22:3045–3051

Mann BK, Schmedlen RH, West JL (2001) Tethered-TGF-beta increases extracellular matrix production of vascular smooth muscle cells. Biomaterials 22:439–444

Mann BK, Tsai AT, Scott-Burden T, et al. (1999) Modification of surfaces with cell adhesion peptides alters extracellular matrix deposition. Biomaterials 20:2281–2286

Mann BK, West JL (2002) Cell adhesion peptides alter smooth muscle cell adhesion, proliferation, migration, and matrix protein synthesis on modified surfaces and in polymer scaffolds. J Biomed Mater Res 60:86–93

Mardilovich A, Craig JA, McCammon MQ, et al. (2006) Design of a novel fibronectin-mimetic peptide-amphiphile for functionalized biomaterials. Langmuir 22:3259–3264

Marler JJ, Guha A, Rowley J, et al. (2000) Soft-tissue augmentation with injectable alginate and syngeneic fibroblasts. Plast Reconstr Surg 105:2049–2058

Massia SP, Hubbell JA (1990) Covalent surface immobilization of Arg-Gly-Asp- and Tyr-Ile-Gly-Ser-Arg-containing peptides to obtain well-defined cell-adhesive substrates. Anal Biochem 187:292–301

Massia SP, Hubbell JA (1992) Immobilized amines and basic amino acids as mimetic heparin-binding domains for cell surface proteoglycan-mediated adhesion. J Biol Chem 267:10133–10141

Massia SP, Hubbell JA (1992) Vascular endothelial cell adhesion and spreading promoted by the peptide REDV of the IIICS region of plasma fibronectin is mediated by integrin alpha 4 beta 1. J Biol Chem 267:14019–14026

Morikawa N, Matsuda T (2002) Thermoresponsive artificial extracellular matrix: N-isopropylacrylamide-graft-copolymerized gelatin. J Biomater Sci Polym Ed 13:167–183

Myles JL, Burgess BT, Dickinson RB (2000) Modification of the adhesive properties of collagen by covalent grafting with RGD peptides. J Biomater Sci Polym Ed 11:69–86

Na K, Park JH, Kim SW, et al. (2006) Delivery of dexamethasone, ascorbate, and growth factor (TGF beta-3) in thermo-reversible hydrogel constructs embedded with rabbit chondrocytes. Biomaterials 27:5951–5957

Obara K, Ishihara M, Fujita M, et al. (2005) Acceleration of wound healing in healing-impaired db/db mice with a photocrosslinkable chitosan hydrogel containing fibroblast growth factor-2. Wound Repair Regen 13:390–397

Ohya S, Nakayama Y, Matsuda T (2001) Thermoresponsive artificial extracellular matrix for tissue engineering: hyaluronic acid bioconjugated with poly(N-isopropylacrylamide) grafts. Biomacromolecules 2:856–863

Pantoliano MW, Horlick RA, Springer BA, et al. (1994) Multivalent ligand-receptor binding interactions in the fibroblast growth factor system produce a cooperative growth factor and heparin mechanism for receptor dimerization. Biochemistry 33:10229–10248

Park H, Temenoff JS, Holland TA, et al. (2005) Delivery of TGF-beta1 and chondrocytes via injectable, biodegradable hydrogels for cartilage tissue engineering applications. Biomaterials 26:7095–7103

Park KH, Kim MH, Park SH, et al. (2004) Synthesis of Arg-Gly-Asp (RGD) sequence conjugated thermo-reversible gel via the PEG spacer arm as an extracellular matrix for a pheochromocytoma cell (PC12) culture. Biosci Biotechnol Biochem 68:2224–2229

Park Y, Lutolf MP, Hubbell JA, et al. (2004) Bovine primary chondrocyte culture in synthetic matrix metalloproteinase-sensitive poly(ethylene glycol)-based hydrogels as a scaffold for cartilage repair. Tissue Eng 10:515–522

Park YD, Tirelli N, Hubbell JA (2003) Photopolymerized hyaluronic acid-based hydrogels and interpenetrating networks. Biomaterials 24:893–900

Paszek MJ, Zahir N, Johnson KR, et al. (2005) Tensional homeostasis and the malignant phenotype. Cancer Cell 8:241–254

Peattie RA, Nayate AP, Firpo MA, et al. (2004) Stimulation of in vivo angiogenesis by cytokine-loaded hyaluronic acid hydrogel implants. Biomaterials 25:2789–2798

Peattie RA, Rieke ER, Hewett EM, et al. (2006) Dual growth factor-induced angiogenesis in vivo using hyaluronan hydrogel implants. Biomaterials 27:1868–1875

Pelham RJ, Jr., Wang YL (1998) Cell locomotion and focal adhesions are regulated by the mechanical properties of the substrate. Biol Bull 194:348–349; discussion 349–350

Pike DB, Cai S, Pomraning KR, et al. (2006) Heparin-regulated release of growth factors in vitro and angiogenic response in vivo to implanted hyaluronan hydrogels containing VEGF and bFGF. Biomaterials 27:5242–5251

Pouyani T, Prestwich GD (1994) Functionalized derivatives of hyaluronic acid oligosaccharides: drug carriers and novel biomaterials. Bioconjug Chem 5:339–347

Pratt AB, Weber FE, Schmoekel HG, et al. (2004) Synthetic extracellular matrices for in situ tissue engineering. Biotechnol Bioeng 86:27–36

Raeber GP, Lutolf MP, Hubbell JA (2005) Molecularly engineered PEG hydrogels: a novel model system for proteolytically mediated cell migration. Biophys J 89:1374–1388

Rajangam K, Behanna HA, Hui MJ, et al. (2006) Heparin binding nanostructures to promote growth of blood vessels. Nano Lett 6:2086–2090

Ranieri JP, Bellamkonda R, Bekos EJ, et al. (1994) Spatial control of neuronal cell attachment and differentiation on covalently patterned laminin oligopeptide substrates. Int J Dev Neurosci 12:725–735

Ratner BD, Bryant SJ (2004) Biomaterials: where we have been and where we are going. Annu Rev Biomed Eng 6:41–75

Reyes CD, Garcia AJ (2004) Alpha2beta1 integrin-specific collagen-mimetic surfaces supporting osteoblastic differentiation. J Biomed Mater Res A 69:591–600

Rezania A, Healy KE (1999) Biomimetic peptide surfaces that regulate adhesion, spreading, cytoskeletal organization, and mineralization of the matrix deposited by osteoblast-like cells. Biotechnol Prog 15:19–32

Richardson TP, Peters MC, Ennett AB, et al. (2001) Polymeric system for dual growth factor delivery. Nat Biotechnol 19:1029–1034

Rizzi SC, Ehrbar M, Halstenberg S, et al. (2006) Recombinant protein-co-PEG networks as cell-adhesive and proteolytically degradable hydrogel matrixes. Part II: biofunctional characteristics. Biomacromolecules. 7:3019–3029

Rizzi SC, Hubbell JA (2005) Recombinant protein-co-PEG networks as cell-adhesive and proteolytically degradable hydrogel matrixes. Part I: Development and physicochemical characteristics. Biomacromolecules 6:1226–1238

Rosso F, Marino G, Giordano A, et al. (2005) Smart materials as scaffolds for tissue engineering. J Cell Physiol 203:465–470

Ruel-Gariepy E, Leroux JC (2004) In situ-forming hydrogels–review of temperature-sensitive systems. Eur J Pharm Biopharm 58:409–426

Rundhaug JE. Matrix metalloproteinases and angiogenesis (2005) J Cell Mol Med 9:267–285

Sakiyama SE, Schense JC, Hubbell JA (1999) Incorporation of heparin-binding peptides into fibrin gels enhances neurite extension: an example of designer matrices in tissue engineering. Faseb J 13:2214–2224

Saha K, Pollock JF, Schaffer DV, et al. (2007) Designing synthetic materials to control stem cell phenotype. Curr Opin Chem Biol 11:381–387

Sakiyama-Elbert SE, Hubbell JA (2000) Controlled release of nerve growth factor from a heparin-containing fibrin-based cell ingrowth matrix. J Control Release 69:149–158

Sakiyama-Elbert SE, Hubbell JA (2000) Development of fibrin derivatives for controlled release of heparin-binding growth factors. J Control Release 65:389–402

Sakiyama-Elbert SE, Panitch A, Hubbell JA (2001) Development of growth factor fusion proteins for cell-triggered drug delivery. Faseb J 15:1300–1302

Schense JC, Bloch J, Aebischer P, et al. (2000) Enzymatic incorporation of bioactive peptides into fibrin matrices enhances neurite extension. Nat Biotechnol 18:415–419

Schense JC, Hubbell JA (1999) Cross-linking exogenous bifunctional peptides into fibrin gels with factor XIIIa. Bioconjug Chem 10:75–81

Schense JC, Hubbell JA (2000) Three-dimensional migration of neurites is mediated by adhesion site density and affinity. J Biol Chem 275:6813–6818

Schmoekel H, Schense JC, Weber FE, et al. (2004) Bone healing in the rat and dog with nonglycosylated BMP-2 demonstrating low solubility in fibrin matrices. J Orthop Res 22:376–381

Schmoekel HG, Weber FE, Schense JC, et al. (2005) Bone repair with a form of BMP-2 engineered for incorporation into fibrin cell ingrowth matrices. Biotechnol Bioeng 89:253–262

Seal BL, Panitch A (2003) Physical polymer matrices based on affinity interactions between peptides and polysaccharides. Biomacromolecules 4:1572–1582

Seliktar D, Zisch AH, Lutolf MP, et al. (2004) MMP-2 sensitive, VEGF-bearing bioactive hydrogels for promotion of vascular healing. J Biomed Mater Res A 68:704–716

Shin H, Jo S, Mikos AG (2003) Biomimetic materials for tissue engineering. Biomaterials 24:4353–4364

Shin H, Jo S, Mikos AG (2002) Modulation of marrow stromal osteoblast adhesion on biomimetic oligo[poly(ethylene glycol) fumarate] hydrogels modified with Arg-Gly-Asp peptides and a poly(ethyleneglycol) spacer. J Biomed Mater Res 61:169–179

Shin H, Zygourakis K, Farach-Carson MC, et al. (2004) Attachment, proliferation, and migration of marrow stromal osteoblasts cultured on biomimetic hydrogels modified with an osteopontin-derived peptide. Biomaterials 25:895–906

Shin H, Zygourakis K, Farach-Carson MC, et al. (2004) Modulation of differentiation and mineralization of marrow stromal cells cultured on biomimetic hydrogels modified with Arg-Gly-Asp containing peptides. J Biomed Mater Res A 69:535–543

Shu XZ, Ahmad S, Liu Y, et al. (2006) Synthesis and evaluation of injectable, in situ cross-linkable synthetic extracellular matrices for tissue engineering. J Biomed Mater Res A 79:902–912

Shu XZ, Ghosh K, Liu Y, et al. (2004) Attachment and spreading of fibroblasts on an RGD peptide-modified injectable hyaluronan hydrogel. J Biomed Mater Res A 68:365–375

Shu XZ, Liu Y, Palumbo F, et al. (2003) Disulfide-crosslinked hyaluronan-gelatin hydrogel films: a covalent mimic of the extracellular matrix for in vitro cell growth. Biomaterials 24:3825–3834

Silva GA, Czeisler C, Niece KL, et al. (2004) Selective differentiation of neural progenitor cells by high-epitope density nanofibers. Science 303:1352–1355

Smeds KA, Pfister-Serres A, Miki D, et al. (2001) Photocrosslinkable polysaccharides for in situ hydrogel formation. J Biomed Mater Res 54:115–121

Stabenfeldt SE, Garcia AJ, LaPlaca MC (2006) Thermoreversible laminin-functionalized hydrogel for neural tissue engineering. J Biomed Mater Res A 77:718–725

Stamenkovic I (2003) Extracellular matrix remodelling: the role of matrix metalloproteinases. J Pathol 200:448–464

Steffensen B, Hakkinen L, Larjava H (2001) Proteolytic events of wound-healing–coordinated interactions among matrix metalloproteinases (MMPs), integrins, and extra-cellular matrix molecules. Crit Rev Oral Biol Med 12:373–398

Suzuki Y, Tanihara M, Suzuki K, et al. (2000) Alginate hydrogel linked with synthetic oligopeptide derived from BMP-2 allows ectopic osteoinduction in vivo. J Biomed Mater Res 50:405–409

Tae G, Scatena M, Stayton PS, et al. (2006) PEG-cross-linked heparin is an affinity hydrogel for sustained release of vascular endothelial growth factor. J Biomater Sci Polym Ed 17:187–197

Tanihara M, Suzuki Y, Yamamoto E, et al. (2001) Sustained release of basic fibroblast growth factor and angiogenesis in a novel covalently crosslinked gel of heparin and alginate. J Biomed Mater Res 56:216–221

Wang N, Naruse K, Stamenovic D, et al. (2001) Mechanical behavior in living cells consistent with the tensegrity model. Proc Natl Acad Sci U S A 98:7765–7770

West JL, Hubbell JA (1999) Polymeric biomaterials with degradation sites for proteases involved in cell migration. Macromolecules 32:241–244

Whang K, Tsai DC, Nam EK, et al. (1998) Ectopic bone formation via rhBMP-2 delivery from porous bioabsorbable polymer scaffolds. J Biomed Mater Res 42:491–499

Wissink MJ, Beernink R, Pieper JS, et al. (2001) Binding and release of basic fibroblast growth factor from heparinized collagen matrices. Biomaterials 22:2291–2299

Xiao Y, Truskey GA (1996) Effect of receptor-ligand affinity on the strength of endothelial cell adhesion. Biophys J 71:2869–2884

Yeo Y, Geng W, Ito T, et al. (2006) Photocrosslinkable hydrogel for myocyte cell culture and injection. J Biomed Mater Res B Appl Biomater. Sep 12 2006.

Zaari N, Rajagopalan P, Kim SK, et al. (2004) Photopolymerization in microfluidic gradient generators: Microscale control of substrate compliance to manipulate cell response. Advanced Materials 16:2133

Zaman MH, Trapani LM, Sieminski AL, et al. (2006) Migration of tumor cells in 3D matrices is governed by matrix stiffness along with cell-matrix adhesion and proteolysis. Proc Natl Acad Sci U S A 103:10889–10894

Zhang S, Holmes T, Lockshin C, et al. (2003) Spontaneous assembly of a self-complementary oligopeptide to form a stable macroscopic membrane. Proc Natl Acad Sci U S A 90:3334–3338

Zhang S (2003) Fabrication of novel biomaterials through molecular self-assembly. Nat Biotechnol 21:1171–1178

Zisch AH, Lutolf MP, Ehrbar M, et al. (2003) Cell-demanded release of VEGF from synthetic, biointeractive cell ingrowth matrices for vascularized tissue growth. Faseb J 17:2260–2262

Zisch AH, Schenk U, Schense JC, et al. (2001) Covalently conjugated VEGF–fibrin matrices for endothelialization. J Control Release 72:101–113

Zisch AH, Zeisberger SM, Ehrbar M, et al. (2004) Engineered fibrin matrices for functional display of cell membrane-bound growth factor-like activities: study of angiogenic signaling by ephrin-B2. Biomaterials 25:3245–3257

Chapter 5
Clinical Approaches to Skin Regeneration

S.E. James, S. Booth, P. Gilbert, I. Jones, and R. Shevchenko

Contents

5.1   Introduction   155
5.2   Normal Skin Structure and Function   156
        5.2.1   Epidermis   157
        5.2.2   Basement Membrane   157
        5.2.3   Dermis   157
        5.2.4   Cellular Component of the Dermis   158
5.3   Mechanisms of Skin Loss   159
        5.3.1   Pathological Damage   159
        5.3.2   Surgical Damage   160
5.4   Healing of Wounds   161
        5.4.1   Wound Closure Categories   161
        5.4.2   Wound Healing Process   162
5.5   Problems of Natural Wound Healing   165
5.6   Current Clinical Approaches to Wound Healing   166
5.7   Development of Novel Approaches to Skin Regeneration   169
        5.7.1   Surgical Use of Cultured Keratinocytes in Burns Patients   169
        5.7.2   Commercially Available Skin Substitutes   174
        5.7.3   Clinical Application of Skin Substitutes   175
Questions/Exercises   185
References   186

5.1 Introduction

Intact human skin is essential for normal body function. It is the largest organ of the body and its primary function is to act as a barrier, keeping environmental insults and invading organisms out, and body fluids in. It can be damaged by many diseases and traumatic processes, resulting in a break or discontinuity of its architecture and in particular, of the surface epithelial layer. Such wounds are then easily colonized by bacteria which can result in further skin damage and loss. The most common causes of skin loss are the result of

S.E. James (✉)
School of Pharmacy and Biomolecular Sciences, University of Brighton, Sussex, UK
e-mail: s.e.james@brighton.ac.uk

M. Santin (ed.), *Strategies in Regenerative Medicine*,
DOI 10.1007/978-0-387-74660-9_5, © Springer Science+Business Media, LLC 2009

burn injuries, trauma or surgical insult. The aim of the plastic surgeon is to restore the damaged skin to as near normal structure and function as possible. This chapter describes the structure and function of normal skin; the mechanisms of skin loss; the normal processes of wound healing, and the clinical approaches to skin regeneration available to today's surgeons.

5.2 Normal Skin Structure and Function

Anatomically and functionally, the skin has two layers, the epidermis and the dermis (Fig. 5.1, www.siumed.edu/~dking2/intro/skin.htm). Appendages such as hair follicles or sweat glands breach both the epidermal and the dermal layers. Except in very thin skin, the junction between the epidermal and dermal layer is irregular and appears undulating, with numerous connective tissue finger-like projections pushing up into the epidermal under-surface and protruding down into the dermis. These are dermal papillae and Rete ridges respectively. In an area where the skin is subjected to greater mechanical stress, these projections are much more closely spaced so there is a much greater surface area interface between the two skin layers.

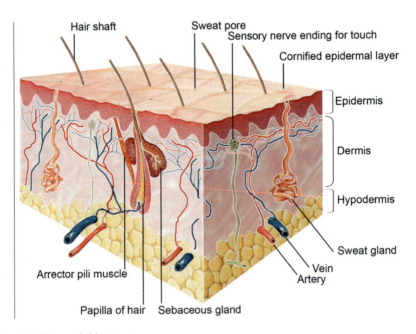

Fig. 5.1 Scheme of skin structure

5.2.1 Epidermis

The epidermis provides a barrier to infection and moisture loss. It varies in thickness with the body location, but is generally about 0.2 mm thick. It consists of several layers of cells with the basal cells forming a single layer of cuboidal nucleated epithelial cells referred to as the skin stem cells. These stem cells are slow cycling cells with a high capacity for self-renewal. They not only exist as the basal epithelial layer, but also line the hair follicles and sweat glands, thus providing a source of epithelial stem cells deep into the dermal layer. They give rise to rapidly proliferating daughter cells, keratinocytes, which shift upwards into the suprabasal layers in a process of upward migration. Keratinocytes differentiate as they move upwards through the epidermal layers, with changes in the quantity and class of expressed cytokeratins. During this migration the keratinocytes change shape from cuboid, to flatter, then completely flat desiccated, anucleated cells in the uppermost layer. This upward migration is a consequence of terminal differentiation resulting in a top layer of cornified, dead squamous cells that are routinely shed from the skin surface.

5.2.2 Basement Membrane

At the basal surface of the epithelium is the basement membrane (BM), responsible for the successful attachment of the epithelial layer to the underlying dermal layer. A failure in epithelial resurfacing of a wound is often a result of a failure to generate an adequate BM. Epithelial cell attachment to the BM is via hemidesmosome integrins binding to members of the laminin family and via protein-glycosaminoglycan chain interactions. The major proteins of the BM are collagen IV, which is organized in a 2 dimensional reticulum rather than in the fibers characteristic of other types of collagen; heparan sulfate proteoglycans and the glycoproteins entactin, and laminin. The basement membrane is tightly bound to the underlying dermis by anchoring fibrils, largely made up of collagen VII possibly connected to fibronectin and or fibrillin. A successful dermo-epidermal junction functions as a key factor in skin repair, strength, and durability.

5.2.3 Dermis

The extra cellular matrix (ECM) of the deeper dermal connective tissue layer provides the elasticity and mechanical integrity of skin. It is made up of collagen, glycoproteins, and proteoglycans.

The components and structure of the ECM include fibrils longer than the connective fibrils of the BM, made up mainly from chondroitin sulfate proteoglycans, elastin, and collagens I, III, and V. Non-fibrillar collagens VI, IX, and

XIV are associated with these fibrils and are responsible for the spatial organization of collagen within the ECM. The ECM in the dermal layer of skin is a
dense connective tissue, where the collagen fibres are closely packed together.

This matrix supports the cells of the dermal layer both physically and by
providing the biochemical signals necessary for cell function and differentiation. It is continually being degraded, re-synthesized, and remodeled locally.

The glycoproteins of the ECM, such as fibronectin, have a role in the
regulation of cellular interactions with the matrix including attachment, migration, and morphology. At least 20 different fibronectin chains are generated by
alternative splicing of the RNA transcript of a single fibronectin gene. Fibronectin is important in wound healing as it facilitates macrophage migration and
initiation of blood clots. It possesses specific high affinity binding sites for cell
surface receptors, collagen, fibrin, and sulfated proteoglycans. These sites contain a specific amino acid sequence: arginine-glycine-aspartic acid (RGD).

Proteoglycans are protein/polysaccharide complexes found in all connective
tissues. They consist of a core protein to which 1 or more glycosamino-glycan
(GAG) polysaccharides attach. GAGs are repeating polymers of specific dis-
accharides, each bearing many negative charges. Proteoglycans are diverse and
named according to the structure of the principle repeating disaccharide, such
as chondroitin sulfate. Proteoglycans are located on the surface of many cell
types, particularly epithelial cells, binding to collagen and to the glycoprotein
fibronectin.

Hyaluronic acid (HA) is a major structural component of complex proteoglycans in the skin. Each HA molecule forms a long rigid rod due to the strong
repulsion from negatively charged carboxylated groups that protrude at regular
intervals. The large number of hydrophilic residues on its surface ensures
binding to a large amount of water even at relatively low concentration. HA
binds to the surface of migrating cells. The loose, hydrated, porous nature of its
coat appears to keep cells apart allowing them the freedom to move and
proliferate. Cessation of cell movement and initiation of cell-cell attachments
frequently correlates with a fall in HA and an increase in hyaluronidase, the
enzyme responsible for HA degradation.

In between the ECM components of the dermis is the interstitial fluid which
mainly consists of HA and blood plasma proteins.

5.2.4 Cellular Component of the Dermis

The cells within the dermis, apart from the cells of the sweat glands and hair
follicles are part of either a constant or a transient population. The constant
population includes fibroblasts, nerve cells, vascular endothelial cells, macrophages, and mast cells. The transient population usually migrates into the
dermis from the blood as part of an inflammatory response to a specific
stimulus and is made up from mainly white blood cells.

The primary cell in the dermis is the fibroblast. They are usually non-dividing and responsible for the normal turnover of the ECM. They produce growth factors as well as the extracellular matrix proteins. Those in the upper region of the dermis are also responsible for the turnover of the basement membrane proteins. They are a major contributor to the wound healing process, with the differentiated fibroblasts (myofibroblasts) promoting neo-dermis formation and the normal fibroblasts promoting re-epithelialization to close the wounds.

Endothelial cells form the specialized lining of blood vessels within the dermis and following injury, these cells are responsible for angiogenesis and the revascularization of the newly formed dermis.

Macrophages are phagocytic cells derived from monocytes within the blood stream. Following migration into connective tissue the monocytes mature into macrophages with the ability to ingest unwanted materials from the damaged tissue and act as antigen presenting cells during an immune response.

5.3 Mechanisms of Skin Loss

5.3.1 Pathological Damage

The depth of the damage to skin is an important factor in the ability to heal the wound. Descriptions of wound depth (Fig. 5.2) previously included terms such as first degree, second degree, and third degree depths of burn or injury. Alternative terms such as partial, intermediate, and full thickness wounds are the terms most commonly used today (Fig. 5.2).

5.3.1.1 Wound Depth

Partial thickness injuries, sometimes referred to as superficial, usually involve just the surface epithelium and superficial dermal layers. A graze injury or a mild scald/burn injury, resulting in visible blistering, is usually very painful due

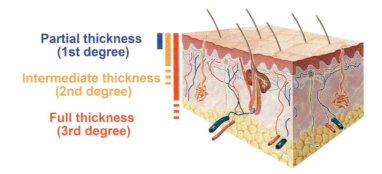

Fig. 5.2 Different depths of wound injury

to the stimulation of nerve ends found in the superficial layers. These wounds will normally heal rapidly providing they do not become infected. Even if the whole of the surface epithelium is removed, it can be readily regenerated from the epithelial cell sources lining the numerous skin appendages. This is the type of wound created surgically when split-thickness skin samples are cropped from healthy areas of skin to provide donor skin for the reconstruction of deeper injuries elsewhere on the body.

Intermediate thickness injuries are where the whole epithelial layer is removed as well as varying depths of the dermal layer. As skin appendages penetrate the dermis to different depths, the deeper the wound the greater the reduction in the number of surviving appendages. Thus those wounds described as "shallow" intermediate will have a large number of epithelial sources (skin appendages) from which to generate new epithelial cover. Those described as "deep dermal" intermediate will have fewer and more widely spaced surviving appendages and, therefore, epithelial regeneration will take longer. These are often the most difficult wounds to treat as it is hard for the surgeon to accurately assess the depth of dermal damage. The use of Laser Doppler imaging of surviving blood flow to a tissue has helped to overcome, but not eliminate, this problem.

Full thickness injuries involve damage to all of the epithelium and all of the dermis. They can extend even deeper to include fatty tissue, muscle and even bone. There are no surviving skin appendages and, therefore, no potential source of epithelial regeneration. They can be deceptive as they are not painful and appear paler than the surrounding inflamed, deep dermal injuries. This is as a result of total destruction of nerves and blood vessels within the tissue. Figure 5.3 shows a burn injury to the arm, chest and abdomen which include the very red, intermediate depth injury to the elbow region; the deep dermal injury across the top of the chest and the paler, almost white, full thickness injury to the rest of the chest and abdomen.

If left untreated, the full thickness damaged tissue will start to break down over the next 24–48 h providing a perfect nutrient source for invading micro-organisms. These serious depths of burn injuries can often be misinterpreted by the inexperienced eyes in a local casualty hospital, resulting in a delay in treatment by the experienced burns surgeon. The use of Telemedicine facilities linked to specialist centers has gone a long way to addressing this issue.

Full thickness injuries must be surgically removed and the skin replaced with either skin grafts or dermal/skin substitutes.

5.3.2 Surgical Damage

Skin loss can be created as a result of surgical intervention, such as the removal of cancerous tissues or simply the deliberate removal of healthy tissue to replace damaged tissue elsewhere on the body (donor skin).

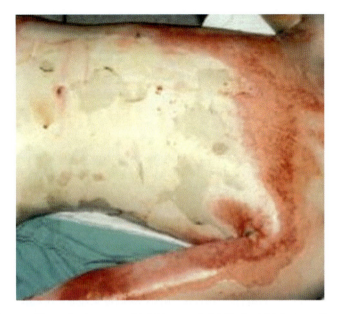

Fig. 5.3 Range of burn depths: normal (pink); superficial (dark pink); intermediate (red) and full thickness (white)

The type of surgical intervention will have a profound effect on the type of wound healing that occurs (www.surgical-tutor.org.uk/core/trauma/skin_grafts.htm).

5.4 Healing of Wounds

Wound healing has been extensively reviewed by Singer and Clark (1999).

Any damage to the skin will result in a wound. The ultimate goal of the wound healing or repair process is to regenerate or reconstruct both the dermal and the epidermal skin components leading to a restoration of function and looks. This is achieved via neo-dermis formation; re-epithelialization of the wound surface and a certain degree of skin contracture. Skin substitutes used to help repair or reconstruct damaged tissue will be required to either shorten the expected healing time or to improve its success rate.

5.4.1 Wound Closure Categories

There are three basic categories of wound closure, irrespective of the type of injury – primary; secondary, and tertiary.

5.4.1.1 Primary Healing

Primary healing (also known as healing by first intention) does not necessarily require any intervention. The wound spontaneously closes within hours of sustaining a full thickness injury. Here the edges of the wound are aligned accurately with no tension and there is minimal intervention or damage to tissues.

5.4.1.2 Secondary Healing

Secondary healing (also known as healing by secondary intention) does not involve any surgical intervention to close the wound, but relies on the body's ability to close a full thickness wound by contraction and re-epithelialization. It usually results in a greater inflammatory response with a large quantity of granulation tissue being produced as well as extensive contraction of the skin.

5.4.1.3 Tertiary Healing

Tertiary healing (also known as delayed primary healing) will require initial surgical excision of damaged tissues, followed by a delayed surgical wound closure. Here the edges of the wound are not necessarily aligned immediately following debridement. This is sometimes desirable if the wound is infected. By about the fourth day, phagocytosis of contaminated tissue is well underway; collagen deposition and epithelial regeneration has started. The wound can be surgically cleaned and closed at this stage. However, if the infection continues beyond this stage then chronic inflammation occurs with resulting permanent hypertrophic scarring.

A fourth category of wound closure was introduced to describe the healing of partial thickness wounds where the epidermis and superficial dermis can be repaired from epithelial elements in surviving skin appendages. The migration of new epithelial cells across the damaged surface closes the wound without the necessity for skin contraction.

5.4.2 Wound Healing Process

The wound healing process is a combination of biological processes including chemotaxis; cell migration; phagocytosis; collagen synthesis and degradation; angiogenesis; ECM remodeling; contraction, and epithelialization. This healing process is dependent on local and systemic factors affecting the balance between the cells, the ECM and a variety of cell mediators (cytokines) as depicted in Fig. 5.4.

There are several overlapping phases of the healing process, described by Cahill and Carroll (1993), Tanenbaum (1995), Habif (1996), Cho and Lo (1998), reviewed by Singer & Clark (1999) and summarized by Rosenberg and de la Torre (2006). These phases are hereinafter discussed.

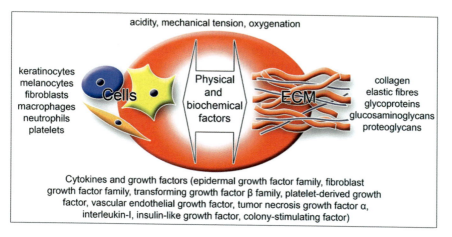

Fig. 5.4 Factors controlling wound healing

5.4.2.1 Hemostasis and Inflammation

Hemostasis starts immediately after tissue injury and may continue for a few days. Inflammation may continue for up to 6 days. Damaged cells release the potent vasoconstrictors thromboxane A2 and prostaglandin 2-alpha. Exposed collagen also activates the clotting cascade and initiates the inflammatory phase. Platelets adhere to the endothelial cells of the damaged blood vessels releasing adenosine di-phosphate which promotes thrombocyte clumping to seal the wound. The platelets release a variety of chemokines including epidermal growth factor (EGF); platelet derived growth factor (PDGF); von Willebrand factor, platelet factor IV, transforming growth factor (TGF)-beta; fibronectin, and fibrinogen. The dense bodies in the thrombocytes release vasoactive amines such as serotonin and histamine which reverse the initial vasoconstriction allowing the infiltration of cells. Fibrinogen is cleaved to fibrin which produces a structural support for the infiltrating cells responding to the released chemotactic factors.

PDGF attracts fibroblasts to the site and together with the TGF-beta, stimulates fibroblast proliferation as well as inflammatory cell recruitment, including polymorphnuclear (PMN) cells (neutrophils), monocytes, and lymphocytes from the surrounding blood vessels. The PMNs are responsible for an initial clean-up of the wound site. Within 6–8 h of the injury the newly formed blood clot is removed by the PMNs. Platelet degranulation also activates complement components C3 and C5a which, together with the PMNs, opsonise contaminating bacteria leading to their removal from the wound. PMN number is maximal at about 24-48 h and has started to reduce by 72 h post injury.

The inflammatory phase continues with the maturation of monocytes into macrophages. The macrophages release further cytokines including fibroblast growth factor (FGF), tumor necrosis factor (TNF) alpha, TGF-beta, and PDGF which continue to attract and stimulate the proliferation of fibroblasts as well as stimulating angiogenesis. Macrophages secrete a variety of interleukins which are able to recruit and stimulate other inflammatory leukocytes as well as initiating the formation of new tissue. Macrophages appear to have an essential role in the transition between inflammation and wound healing as demonstrated by defective wound healing in macrophage depleted animals.

5.4.2.2 Proliferative Phase

This includes the processes of matrix deposition and granulation tissue formation (giving rise to a neo-dermis), angiogenesis, and epithelialization. This phase starts at the end of the inflammation phase and lasts up to 4 weeks. By days 5–7, granulation tissue formation has begun; fibroblasts have infiltrated the wound, started to proliferate and differentiate into myofibroblasts which then lay down new collagens – types I and III. Type III predominates in the early stages but is later replaced by type I. Macrophages secrete enzymes including collagenases which are able to assist in the process.

Tropocollagen is the precursor of all collagen types. It is transformed within the cell's endoplasmic reticulum where proline and lysine are hydroxylated. Disulfide bonds are formed allowing 3 tropocollagen strands to form a triple helix (procollagen). When procollagen is secreted into the extracellular space, peptidases in the cell wall cleave terminal peptide chains giving rise to true collagen fibrils.

Fibroblasts are also responsible for the production of glycosaminoglycans (GAGs) such as heparin sulfate; hyaluronic acid; chondroitin sulfate, and keratin sulfate as well as the production of fibronectin. GAGs covalently linked to a protein core form proteoglycans which contribute to the ECM deposition.

The healing of a gaping wound is achieved by the formation of this granulation tissue which develops from the base of the wound following abundant fibroblast and vascular cell proliferation. This tissue is also high in hyaluronic acid and fibronectin, influencing cell adhesion and migration. Epithelial migration and proliferation occurs at the wound margin following the breakdown of desmosomal junctions between the cells, as well as the hemidesmosomes between epithelial cells and the basement membrane. The integrin receptors on the epithelial cell surface interact with components of the ECM (mainly vitronectin and fibronectin) and together with the production of collagenases by these cells, allow them to migrate across viable tissue separating it from any remaining eschar. This process starts within hours of injury allowing the epithelial cells to advance slowly from the periphery to cover this now chronic wound. Fibroblast contraction at the wound edge also contributes to delayed healing by contraction.

As the wound healing progresses, there is a gradual reduction in the number of myofibroblasts and a return to normal numbers of normal fibroblasts. The presence of normal fibroblasts and collagen type I production is essential for epithelial cell growth over the neo-dermis (Moulin et al. 2000; Suk Wha Kin et al. 2002).

5.4.2.3 Scar Maturation or the Remodeling Phase

This starts at about 4 weeks and can last for years. The wound undergoes constant remodeling. The maximum collagen deposition occurs around the 3-week time point and thereafter an equilibrium between continual collagen degradation and deposition is maintained. The maximum tensile strength of the new tissue is reached by about week 12 but is usually only 80% that of the original normal skin.

5.5 Problems of Natural Wound Healing

Wounds which are left open for prolonged periods of time face the increased possibility of hypertrophic scar formation or secondary infection. There is a good correlation between wound closure time and the development of ugly, raised, hypertrophic scars. Wounds which take longer than 3 weeks to close will invariably result in extensive, pronounced hypertrophic scar formation (Fig. 5.5) probably requiring surgical reconstruction at a later date; whereas

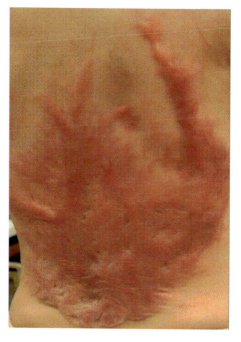

Fig. 5.5 Hypertrophic scarring

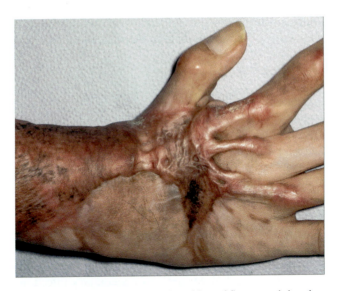

Fig. 5.6 Wound contraction distorting natural position of fingers and thumb

wounds closed by the three week deadline will usually heal without such scar formation (Cubison et al. 2006).

Infected wounds will not normally heal without further intervention. The continued presence of an infection will lead to further tissue breakdown and the increased risk of extensive wound contraction (Fig. 5.6) as well as the development of hypertrophic scarring. All of these will then require more major surgical intervention and longer hospital stay times with an increased risk of requiring reconstructive surgery at a later date.

The issues of delayed wound closure and prolonged wound care treatment will result in a substantial increase in cost to the health care profession. Thus early surgical intervention resulting in early wound closure benefits not only the patients' wellbeing but also results in substantial cost savings. The availability of a suitable skin substitute will contribute to the surgeons' ability to close wounds more quickly, prevent contractures, as well as secondary infections and thereby avoid unnecessary scarring or subsequent reconstructive surgery.

5.6 Current Clinical Approaches to Wound Healing

Wounds can be surgically closed in a variety of ways depending on the nature and depth of the injury. Simple suturing of aligned skin edges, where there is no tissue damage or loss of tissue bulk, will close an incision-type wound with minimal long-term problems. Where there is a loss of tissue, such as in the removal of full thickness skin, perhaps containing a tumor; or in an injury where tissue has been removed or damaged by the trauma, then the skin needs to be replaced.

The gold standard replacement skin is via an autograft where skin is taken, or harvested from a healthy site and moved to the damaged site of the same patient. This avoids all prospects of immune rejection. In order for the moved, or grafted, skin to survive, there is a need for a continuous supply of nutrients via a blood supply. Large areas of damaged skin can be grafted with a piece of full thickness (epidermis + dermis) or a partial thickness (epidermis + small amount of the dermal layer) donor-site skin. For this skin to survive the whole undersurface of the skin graft must be in contact with the wound bed and thereby a supply of nutrients is assured until new blood vessels can establish themselves throughout the tissue. This is referred to as graft "take". If the grafted skin is fed only from the peripheral edges of the wound – as in the case of replacing, for example, an area of facial cheek skin – a blood supply will not be able to re-establish itself before the central area of grafted skin becomes necrotic. In such cases, a skin flap which has been raised from the patient's forearm along with a length of blood vessel (Fig. 5.7), can then be transplanted with its own blood supply system and "plumbed-in" to a local vessel in the neck.

The use of donor-site skin for wound healing raises its own problems. For instance, a burns patient with greater that 50% of their total body surface area (TBSA) damaged, would not have enough undamaged skin to cover the injury. Bearing in mind that donor-sites themselves then become wounds which need to be healed, the ability to make small areas of donor skin cover larger damaged areas is essential to aid the healing process. This can be achieved by meshing the donor skin in such a way as to spread it over a large wound area (Fig. 5.8). The traditional lattice mesh can be expanded up to six times, although at this ratio it is difficult to handle and an expansion ratio of 1:4 or less is more common. The alternative Meek meshing, which produces 3 mm^2 island of skin glued to a pleated dressing via their external surface, can produce an expansion ratio of up

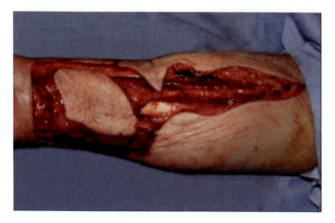

Fig. 5.7 Donor site of a full thickness skin flap being harvested together with a length of blood vessel for grafting elsewhere

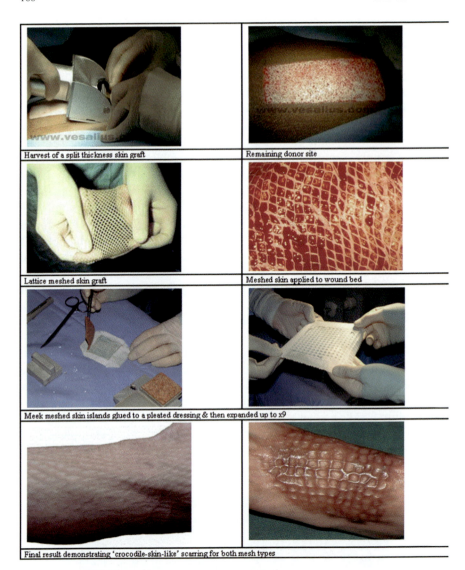

Fig. 5.8 Illustration of the procedures followed to apply a split-thickness skin graft

to 1:9, depending on the depth of pleats in the dressing. These islands of skin are then easy to transfer onto the wound bed while still attached to the dressing.

Wound closure of meshed grafted areas relies on the ability of epithelial cells to grow out from the pieces of skin and to fill the interstitial spaces. Therefore, the wider the expansion ratio, the greater the interstitial space, the longer it takes to close the wound and the greater the risk of scarring.

Not only does creating donor sites increase the wound area of the whole patient, but the donor sites also have their own problems. Partial thickness donor site wounds will normally close with epithelial cell migration from the surviving skin appendages. They usually leave a scar, albeit usually a minor one with altered pigmentation, but if they become infected, then the delayed healing increases the possibility of contraction and deeper scarring. Due to the severing of superficial nerves, donor sites are often very painful and usually more so than the original injury.

Full thickness donor sites are not able to re-epithelialize and close by themselves; they require partial thickness donor skin from yet another site. Skin substitutes or engineered skin replacements have a role to play in cases such as these in order to reduce the number of donor sites cropped.

Donor skin grafts do not always succeed, or "take". If the body fails to re-vascularize the new tissue, perhaps the graft is prevented from keeping in contact with the wound bed due to mechanical sheer forces, then cell infiltration will not take place and the graft will die. Potential skin substitutes or replacements also have to overcome problems of poor "take" in order to succeed.

5.7 Development of Novel Approaches to Skin Regeneration

Alternative live-saving approaches have been sought in the treatment of extensive full-thickness wounds, especially burns, where donor sites for split-skin-graft harvesting are not available. The need to develop skin substitutes was recognized over 30 years ago and since then the field of tissue engineering has advanced rapidly. The development and clinical use of cultured autologous keratinocytes or bioengineered skin substitutes have made great progress in recent years (Horch et al. 2005, Atiyeh and Costagliola 2007, Clark et al. 2007, MacNeil 2007). Off-the-shelf availability of some products or the possibility of producing, in a relatively short period of time, sufficient quantities of epithelial cells to aid in permanent wound closure, make these products sometimes the only mean of help in extensive deep burns treatment.

5.7.1 Surgical Use of Cultured Keratinocytes in Burns Patients

Burns patients are vulnerable to the infection of open wounds or burnt tissue. It is essential to remove dead tissue and close the wound bed as soon as possible after the burn injury. In the case of minor burns this can be achieved by excising the burnt tissue and replacing it with healthy skin from a donor site elsewhere on the patient. This is not always possible with a major burn due to a lack of available donor sites. A solution to this problem has been to take a small sample of the patient's own healthy skin and expand this in the laboratory to give large quantities of cultured skin which can then be grafted back onto the same patient (Fig. 5.9 (a) skin to culture flasks, (b) expansion, (c) sheet, (d) cell suspension).

Although it is possible, within about 4 weeks, to produce enough cultured skin to cover the whole of an adult human body from a postage-stamp size piece of donor skin, this means that there is a necessary delay before being able to

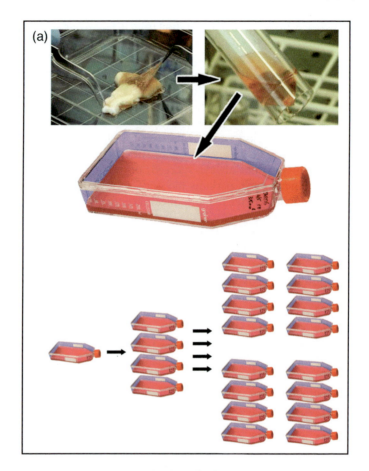

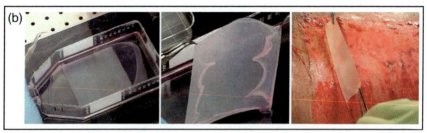

graft such cells back onto the patient. The earliest application of some cells produced in this way is usually about 10–14 days after initiation of the culture. In the mean time, the burn wounds need to be excised and the healing process started. The use of cadaver donor skin, as a temporary dressing on the excised burn, allows the extent of the burn to be more accurately assessed by the surgical team, as well as giving the scientist time to culture the skin cells in the

(c)

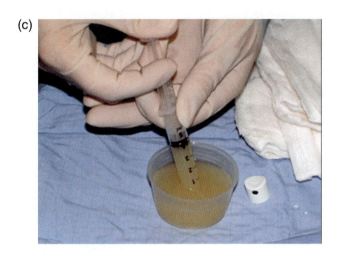

Fig. 5.9 (continued)

laboratory. In some cases, dermal substitutes (see below) have provided an alternative to the cadaver skin.

The technology to grow skin cells in the laboratory has been around since the early 1970s (Rheinwald and Green 1975, Green et al. 1979). The first application of cultured sheets of cells to a burns patient came in the mid 1970s, with some degree of success (reviewed extensively by Wood et al. 2006). However, the level of success was not always acceptable. It is apparent that the sheets of cultured cells contain a mixture of healthy dividing cells at the basal level and older, keratinized dying cells in the upper layers. The longer they are kept in culture (by a matter of days), the greater the proportion of dead and dying cells there are in the basal layer. This is probably due to the basal cells being starved of nutrients by the keratinizing upper layers, thus making the timing of surgery very difficult and success rates unpredictable.

An interest developed in delivering cells to the wound bed while they are all still healthy and dividing. This ruled out the use of cultured cell sheets, which depended upon the presence of the older, dying cells in the upper layers to give stability to the sheet. The cells attached to the plastic of the tissue culture flask proliferate and spread until they cover a maximum of 80% of the total available area – termed "sub-confluent" thus avoiding the differentiation into non-proliferating cells.

Research concentrated on the delivery of these healthy cells either in a suspension or by delivery of the sub-confluent cells on a matrix. The cultured cells needed to be harvested before they started to mature, or needed to be grown on some sort of delivery matrix which could be added directly to the wound bed. The growing, sub-confluent cells are released from the plastic culture flasks and re-suspended in a buffered solution. At this stage, the cells can be drawn up into a standard 5 ml syringe. A commercially available spray nozzle is then attached to the end of this syringe and the cell suspension

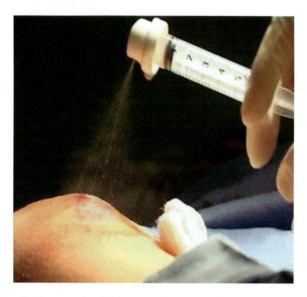

Fig. 5.10 Cultured cell suspension being sprayed onto wound bed

delivered onto the prepared wound bed as a fine spray (Fig. 5.10). The spray delivery system was pioneered in Australia and their clinical experiences have been extensively reviewed (Wood et al. 2006), demonstrating the success of this approach. Such successes have also been achieved by UK teams in Birmingham and East Grinstead.

Scientists from the Blond McIndoe Centre together with the surgeons from the Burns Centre at The Queen Victoria Hospital, in East Grinstead were also

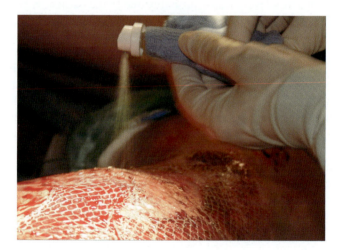

Fig. 5.11 Combination of skin mesh and sprayed cells

able to tackle full thickness burns by combining the use of very widely meshed donor skin with the sprayed cultured cells filling in the gaps between the meshes (Fig. 5.11). This has led to very early wound closure and recovery for patients who would otherwise have died (James et al. 2008).

Figure 5.12 shows the case of an 80-year-old lady with full thickness flame burns to her legs who was failing to heal even after the normal skin grafting procedure. This grafting was repeated with the addition of sprayed cultured cells and 4 weeks later, the patient was discharged from hospital with her legs fully healed (Fig. 5.12).

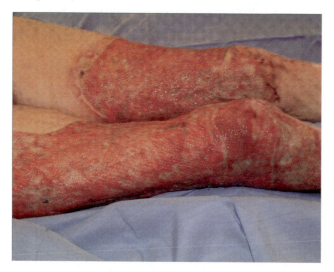

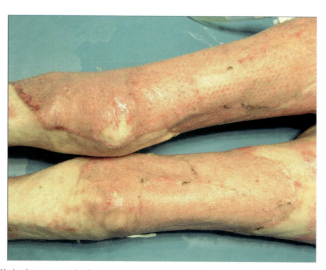

Fig. 5.12 Clinical outcome before and after 4 weeks of treatment with a combination of skin mesh and sprayed cells

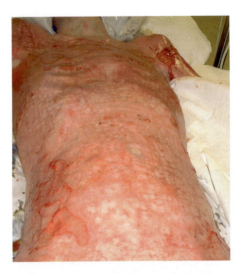

Fig. 5.13 Clinical outcome of a combined Meek meshed skin graft and sprayed cell treatment in a patient who sustained 90% burn

A 27-year-old male doused in petrol and set alight sustained 90% burns, the majority of which were full thickness. With the combination of very widely Meek-meshed donor-site skin (of which there was very little) and sprayed cultured cells, his wounds closed and he recovered from what would otherwise have been fatal injuries (Fig. 5.13).

5.7.2 Commercially Available Skin Substitutes

Due to the great importance and demand for skin substitutes, there is a long history of material development and many research groups from all over the world are focusing on this problem. This has led to some confusion in terminology where skin substitute biomaterials are commonly referred to by many alternative descriptors including bioengineered skin equivalents; tissue-engineered skin; tissue engineered constructs; biological skin substitutes; bioengineered skin substitutes; skin substitute bioconstructs; living skin replacements, and more recently as bioengineered alternative tissue (Kim et al. 2006). Although these terms differ slightly from each other and have imperfections in the true product representation, they are often considered to be equal and interchangeable.

All such products need to comply with three major requirements: (1) to be safe for the patient, (2) to be clinically effective, and (3) to be convenient for handling and the ease of application. Properties of the "ideal" skin substitute for in vivo use were recently reviewed by MacNeil (2007). In general, such biomaterials must not induce a toxic response; must not prompt an immune reaction;

must not result in excessive inflammation; should have no or a low level of transmittable disease risk; be biodegradable; support the reconstruction of normal tissue; have similar physical and mechanical properties to the skin it replaces; should be repairable; should be readily available; should be user-friendly; should possess a long shelf life and should be cost-effective (http:// www.celltran.co.uk/download/macneil%20nature05664.pdf). No currently available tissue-engineered skin replacement biomaterial possesses all of the above mentioned properties, nor can they fully replace the functional and anatomical properties of the native skin. There are, however, some bioengineered skin replacement products which are currently available to clinicians and are used for wound healing purposes. Most of these products target only some aspect of skin replacement, which means that a surgeon may need to select different products to achieve specific functions. Shakespeare (2005) outlines four functions of bioengineered skin replacement products against which the surgeons can assess their requirements – known as the four "Ps": (1) *Protection* – by establishing a mechanical barrier preventing invasion by micro-organism as well as controlling fluid loss from the tissues; (2) *Procrastination* – due to the early wound debridement tactic some sort of wound closure is needed until permanent wound healing can be achieved by other means such as serial skin grafts or cultured autologous cell applications; (3) *Promotion* – delivery of dermal matrix components, cytokines, and growth factors to the wound bed, which can promote and enhance the natural wound healing response; (4) *Provision* – of new structures, such as dermal collagen or cultured cells, which are incorporated into the wound to persist during wound healing and possibly thereafter.

5.7.3 Clinical Application of Skin Substitutes

The surgeon has a number of options for treating wounds of various depths using commercially available natural or tissue engineered epidermal/dermal substitutes or skin equivalents. (Tables 5.1, 5.2, 5.3). In cases of large burn injuries, the option of healing by autograft alone is not always possible. The surgeon requires an alternative. A suitable skin replacement material could substantially improve both the mortality and the morbidity resulting from burns as well as the functional and aesthetic outcome in other reconstructive procedures. Many products are still in development or have recently been introduced onto the market, some of which aim to facilitate early excision of burns whereas others target optimizing the healing of chronic wounds such as diabetic and vascular ulcers. The hard commercial reality is that a product from a pharmaceutical company able to achieve the closure of full thickness major burn wounds will not be economically profitable when compared with the cost of autologous skin grafts. Nevertheless as **mortality** from major burns reduces in developed countries, the desire to produce materials which reduce **morbidity** is increasing interest in designing a skin substitute to repair extensive skin loss.

Table 5.1 Currently commercially available or marketed dermo-epidermal skin constructs

Brand name / manufacturer	Schematic representation	Incorporated human cells	Cell / scaffold source	Scaffold material
Allograft (cadaveric)				
From not for profit skin banks		—	allo /allo	native human skin with dermal and epidermal cells
Apligraf® Organogenesis Inc., Canton, Massachusetts, CA, USA		Cultured keratinocytes and fibroblasts	allo / xeno	bovine collagen (Type I)
PolyActive™ HC Implants B.V., Leiden, Netherlands		Cultured keratinocytes and fibroblasts	auto /—	PEO/PBT
OrCel® Ortec International, Inc., New York, NY, USA		Cultured keratinocytes and fibroblasts	allo/xeno	bovine collagen sponge (Type I collagen)

PEO – polyethylene oxide; PBT – polybutyliterephthalate.

Table 5.2 Currently commercially available or marketed epidermal constructs

Brand name / manufacturer	Schematic representation	Incorporated human cells	Cell / Scaffold source	Scaffold material
Bioseed®-S BioTissue Technologies GmbH, Freiburg, Germany		Cultured keratinocytes	auto / allo	fibrin sealant
CellSpray® Clinical Cell Culture (C3), Perth, Australia		Non-/cultured sprayed keratinocytes	auto / —	—
Epicell® Genzyme Biosurgery, Cambridge, MA, USA		Cultured keratinocytes	auto / —	—
EpiDex™ Modex Therapeutiques, Lausanne, Switzerland		Cultured keratinocytes from outer root sheath of scalp hair follicles	auto / —	—
Laserskin™ or **Vivoderm™** Fidia Advanced Biopolymers, Padua, Italy		Cultured keratinocytes	auto / —	HAM
MySkin™ CellTran Ltd, Sheffield, UK		Cultured sub-confluent keratinocytes	auto / —	silicone support layer with a specially formulated surface coating

Table 5.3 Currently commercially available or marketed dermal constructs

Brand Name / Manufacturer	Schematic Representation	Incorporated Human Cells	Cell/ Scaffold Source	Scaffold Material
AlloDerm® LifeCell Corporation, Branchburg, NJ, USA		—	— / allo	human acellular dermal matrix
Biobrane® / **Biobrane**®-**L** UDL Laboratories, Inc., Rockford,IL, USA		—	— / xeno	silicon film, nylon fabric + collagen
Dermagraft® Advanced BioHealing, Inc., New York, NY and La Jolla, CA, USA		Cultured neonatal fibroblasts from foreskin tissue	allo / —	PGA/PLA + ECM
EZ Derm™ Brennen Medical, Inc., MN, USA.		—	— / xeno	aldehyde crosslinked porcine dermal collagen
GraftJacket® Wright Medical Technology, Inc., Arlington, TN, USA		—	— / allo	human acellular dermal matrix

Table 5.3 (continued)

Brand Name / Manufacturer	Schematic Representation	Incorporated Human Cells	Cell/ Scaffold Source	Scaffold Material
Hyalograft® 3D Fidia Advanced Biopolymers, Abano Terme, Italy		Cultured fibroblasts	auto / —	HAM
Hyalomatrix® PA Fidia Advanced Biopolymers, Abano Terme, Italy		—	— / —	HYAFF layered on silicone membrane
Integra® Dermal Regeneration Template Integra NeuroSciences, Plainsboro, NJ, USA		—	— / xeno	polysiloxane (silicone), cross-linked bovine tendon collagen + GAG
Matriderm® Dr. Suwelack Skin & Health Care AG, Billerbeck, Germany		—	— / xeno	bovine collagen and elastin
OASIS® Wound Matrix Cook Biotech Inc., West Lafayette, IN, USA		—	— / xeno	porcine acellular small intestine submucosa

Table 5.3 (continued)

Brand Name / Manufacturer	Schematic Representation	Incorporated Human Cells	Cell/ Scaffold Source	Scaffold Material
Permacol® Surgical Implant Tissue Science Laboratories plc., Aldershot, UK		—	— / xeno	cross-linked porcine acellular dermal collagen
SureDerm™ HANS BIOMED Corporation, Seoul, Korea		—	— / allo	human acellular dermal matrix
Terudermis® Olympus Terumo Biomaterial Corp., Tokyo, Japan		—	— / xeno	silicone, lyophilized collagen sponge of heat-denatured bovine athelocollagen
TransCyte® Advanced BioHealing, Inc., New York, NY and La Jolla, CA, USA		cultured neonatal fibroblasts	allo / xeno	silicon film, nylon mesh, porcine dermal collagen

PGA – polyglycolic acid (Dexon™); PLA – polylactic acid (Vicryl™); ECM – extracellular matrix, derived from fibroblasts; HAM – Hyaluronic Acid Membrane (microperforated); HYAFF® – a derivative of hyaluronan; GAG – glycosaminoglycan.

Consequently, new skin substitute products are continually being introduced for these applications.

The surgeon assesses the ability of the available skin substitutes to meet their requirements based on several alternative functional considerations (Horch et al. 2005, Clark et al. 2007, Atiyeh et al. 2007, MacNeil 2007) which can be summarized on the basis of:

Anatomical structure: Depending on which component of skin needs to be replaced there is a choice of (a) composite (dermo–epidermal); (b) epidermal; or (c) dermal substitute.

Duration of the cover required as either (a) permanent; (b) semi-permanent; or (c) temporary.

Type of the biomaterial required being (a) biological and either autologous, allogeneic or xenogeneic; or (b) synthetic and either biodegradable or non-biodegradable.

The site of biomaterial reconstitution occurring either (a) in vitro in the laboratory or (b) in vivo on the patient.

Biomaterial composition can be either (a) acellular or (b) cellular

Some of the currently marketed and clinically available tissue-engineered skin substitute products are listed in this chapter (Tables 5.1, 5.2, 5.3), and some are more extensively reviewed in Jones et al. (2002), and Ratner and Bryant (2004). However, many more are still in the process of investigation and are not listed due to the unavailability of the product information or the lack of experimental or clinical results on materials' performances.

These products achieve either wound closure or provide wound cover. Wound closure requires a material to provide an immediate epidermal barrier function (albeit a temporary one). Some component is usually integrated into the healing wound restoring a dermal function. Materials used for wound cover rely on the in-growth of granulation tissue for adhesion and are most suited to more superficial burns. They create an improved environment for epidermal regeneration from surviving dermal appendages such as hair follicles or sweat glands, as well as providing a barrier to infection and controlling water losses.

They can be sub-divided into dermal substitutes (Table 5.1), epithelial replacements (Table 5.2) or composites combining both of the former features (Table 5.3). Tables 5.1, 5.2, 5.3 summarize the features of these materials and further information can be found on the manufacturers' web pages (for details see Chapter 2).

Skin substitutes are applied as a dermal substitute in conjunction with a form of epithelial cover. Their role is to add thickness and elasticity to the graft. The oldest dermal substitutes are, in fact, mainly composites of unprocessed skin products such as living or cadaveric allograft. The pathological immunosuppression present in the first few weeks of a severe burn injury protects the grafts from rejection during this period. Allograft material can be obtained from national skin banks where careful screening of prospective donors and standardized sterilization methodologies are rigorously controlled. This is essential to reduce the risk of potential transmission of infective agents.

Cadaveric allograft is currently the gold standard biological dressing when autografting is not possible. It provides cover for excised deep partial or full thickness burn wounds and decreases the wound microbial count by adhering to and sealing the wound bed. It stimulates hemostasis and simultaneously decreases pain, as well as easing movement.

Cadaveric allograft is available fresh, lyophilized, glycerolized, or cryopreserved. Fresh allograft is highly viable, and superior in terms of rapidity and strength of adherence to the wound, infection control, and rate of angiogenesis but difficult to obtain. Lyophilization destroys all the cells, including the epidermal layer together with its barrier function; hence, its role is reduced to that of a sheet of collagen.

The European skin bank now supplies only glycerolized cadaveric skin, which has a reduced viability without destroying all of the cellular components. They report that the loss of viability resulting from the glycerolization process not only reduces an inflammatory response, but also does not hamper the desired biological effect.

Cryopreservation maximizes the survival of cellular components, but as a consequence, increases the inflammatory response, and subsequent possibility of immunological rejection.

Viable allograft tissue stimulates new blood vessel endothelial cells to migrate from the wound bed and into the dermis. This is a necessary process for permanent graft take, and in many cases results in the bio-integration of the dermal component of the allograft into the healing wound. Rejection of mainly epithelial tissue tends to be naturally delayed, occurring 1–3 weeks post grafting. This is as a consequence of the suppressed immune function characteristic of a patient with massive burns. If rejection is allowed to occur, the underlying bed becomes inflamed and prone to infection, with the risk of systemic shock. In partial thickness burns, the allograft gradually lifts off as the patient's epidermis regenerates.

A major disadvantage of cadaveric allograft is the recognized risk of disease transmission, particularly of viruses, with cases of human immunodeficiency virus transfer, as well as clinical cytomegalovirus infections developing post allogeneic grafting.

More recently, and to try to address the potential problem of transmission of infective agents, processed collagen products have become available, or processed collagen in conjunction with growth factors. There are also bilaminar synthetic products, aiming to simulate a temporary epidermal role until a permanent epidermis either regenerates or is externally applied via a skin graft or cultured keratinocytes to a regenerated dermis.

The most widely used synthetic dermal substitute in burns patients is Integra[TM] described originally by Yannas and Burke (1980). Integra[TM] is bovine Type 1 collagen cross-linked with shark chondroitin-6-sulfate (aGAG) and coated on one side with a silicone rubber membrane to provide a temporary epidermal function. The purpose of the silicone sheet which forms the superficial layer, is to limit granulation tissue formation (Martin and Mehendale 2001) and to protect the wound from infection and excessive loss of moisture. The permeability of the silicone is 5 mg/cm^2/h. The collagen layer has a pore size of 20–120 μm designed to be sufficiently large to allow autologous cell migration necessary for bio-integration, but not so large that there is insufficient area for the endothelial cells and fibroblasts to attach (Tomkins and Burke 1992).

Integra^{TM} is normally applied to a freshly excised wound bed where the collagen layer bio-integrates over a 3–4 week period to give a vascular "neodermis". When the neodermis has formed, the silicone layer peels away easily and an ultra thin split skin graft is applied. This is approximately 0.1 mm thick, whereas an average split-thickness skin graft (SSG) measures 0.2–0.4 mm. Due to the underlying Integra^{TM}, the ultra thin SSG does not contract to the extent that would be anticipated if it were applied directly to the wound bed (Sheridan et al. 1994).

Wound closure with Integra^{TM} requires two operations separated by a minimum time interval of three weeks to allow neodemis formation. If used in combination with cultured epithelial autografts (CEAs) the time delay is an advantage as it allows for expansion of the skin biopsy into CEAs. When used conventionally with ultra thin SSGs it provides a "neodermal" bed improving elasticity and cosmesis with corresponding reduced donor site morbidity compared with standard donor sites. It has become the "gold standard" dermal substitute biomaterial giving a good cosmetic outcome with reduced rates of contraction and scarring (Anthony et al. 2006, Kim et al. 2006). There is a single clinical case reporting the successful graft of CEA sheets onto the Integra^{TM} neodermis (Pandya et al. 1998) (Fig. 5.14).

Attempts have been made to combine this dermal regenerative template with disaggregated epithelial cells in a single stage procedure. Animal studies where the product was seeded with cultured autologous epithelial cells (Jones et al. 2003) (Fig. 5.15) or a non-cultured, freshly isolated cell suspension (Wood et al. 2007) gave promising results showing the production of a surface epithelial layer after approximately 10 days (Jones et al. 2003).

Preliminary clinical studies treating a full thickness donor site (manuscript in preparation) and pre-seeding the Integra^{TM} with cultured autologous cells demonstrated similar results to the animal studies initially. After 3 weeks, the silicone layer became loose revealing the successful production of an epithelial layer (Fig. 5.16). However, this layer subsequently broke down and required a SSG to close the wound. This suggests that the cultured cells were not able to maintain an effective epithelium and that the absence of a niche of epithelial stem cells probably prevented a continuous turnover of the epithelial layer.

It is important to remember that although some of these materials can achieve wound closure, they do not necessarily result in a fully functioning skin. The successful reconstruction of both dermal and epidermal layers still leaves the patient with a lack of sweat glands and hair follicles. Without these skin appendages the patient can neither maintain control of their body temperature, nor can they keep their skin moist without regular applications of creams. So, although many exciting advances have been made over the last 30

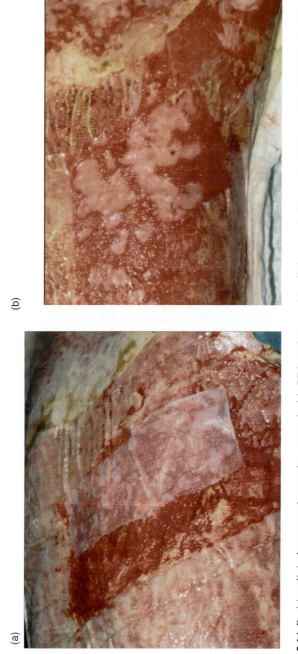

Fig. 5.14 Positive clinical outcome in a patient treated with CEA on Integra. (**a**) cell sheet applied to mature Integra, (**b**) epithelialization by cultured cells on Integra

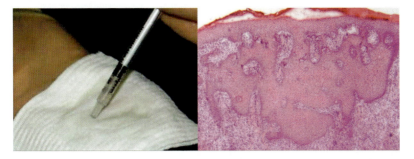

Fig. 5.15 Integra seeded with a suspension of cultured autologous cells and implanted in an animal model giving rise to an intact epithelial surface

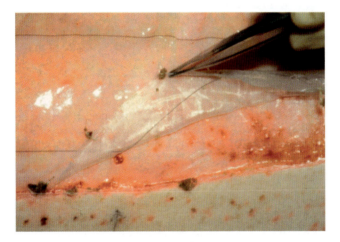

Fig. 5.16 Clinical outcome after 3 weeks in a patient treated with Integra pre-seeded with a suspension of cultured epithelial cells

years, the next milestone in the field of tissue engineered skin will be even more rewarding; when the ultimate goal of an off-the-shelf, fully functional skin substitute will be achieved.

Acknowledgments   SEJ would like to thank the surgeons at the Queen Victoria Hospital, East Grinstead, UK for exposing this scientist to the world of surgery and opening up invaluable communications between scientists and surgeons. Special thanks to Nick Parkhouse; Phil Gilbert; Balj Dheansa; Tania Cubison; John Boorman; Sheraz Daya and Ken Lavery, all of whom have contributed directly or indirectly to this work.

Questions/Exercises

1. What are the main histological features of the dermis and their functions?
2. Describe the histological features characterizing the different depths of burn injury.

3. Present the biochemical and cellular basis leading to hypertrophic scarring.
4. Discuss the main pathological conditions leading to death in burned patients.
5. Critically assess the clinical performance and commercial viability of three typical dermal substitutes and highlight their common and distinct features.
6. Develop a comparative analysis between the clinical use and performance of acellular and cellular dermal substitutes.
7. Illustrate the split skin graft technique and highlight its advantages and limitations
8. Highlight the advances in, and limitations of, the in vitro skin cell expansion technique.
9. Explore the research and clinical outcome of autologous cell spray when applied with dermal substitutes.
10. Provide examples of neglected biomaterials which may lead to the manufacturing of highly performing dermal substitutes.

References

Anthony ET et al. (2006) The development of novel dermal matrices for cutaneous wound repair. Drug Discovery Today: Ther Strateg 3:81–86

Atiyeh BS, Costagliola M (2007) Cultured epithelial autograft (CEA) in burn treatment: three decades later. Burns 33:405–13

Atiyeh BS, Hayek SN, Gunn SW (2005) New technologies for burn wound closure and healing–review of the literature. Burns 31:944–56

Cahill KV, Carroll RP (1993) Principles, techniques, and instruments. In: Stewart WB, (ed) Surgery of the eyelid, orbit, and lacrimal system. USA: Oxford University Press 1:10–11

Cho CY, Lo JS (1998) Dressing the part. Dermatol Clin 16:25–47

Clark RA, Ghosh K, Tonnesen MG (2007) Tissue engineering for cutaneous wounds. J Invest Dermatol 127:1018–29

Cubison TC, Pape SA, Parkhouse N (2006) Evidence for the link between healing time and the development of hypertrophic scars (HTS) in paediatric burns due to scald injury. Burns 32:992–99

Green H, Kehinde O, Thomas J (1979) Growth of cultured human epidermal cells into multiple epithelia suitable for grafting. Proc Natl Acad Sci 76:5665–8

Habif TP (1996) Dermatologic surgical procedures. In: Clinical Dermatology: A Color Guide to Diagnosis and Therapy 3rd ed St Louis: Mosby 809–810

Horch RE, Kopp J, Kneser U, Beier J, Bach AD (2005) Tissue engineering of cultured skin substitutes. J Cell Mol Med 9:592–608

Jones I, Currie L, Martin R (2002) A guide to biological skin substitutes. Br J Plast Surg 55:185–93

James SE, Booth S, Dheansa B, Mann DJ, Reid MJ, Shevchenko RV, Gilbert PM (2008) Sprayed cultured autologous keratinocytes used alone or in combination with meshed autografts to accelerate wound closure in difficult-to-heal burns patients. Accepted for publication in Burns

Jones I, James SE, Rubin P, Martin R (2003) Upward migration of cultured autologous keratinocytes in Integra artificial skin: a preliminary report. Wound Repair Regen 11:132–38

Kim PJ, Dybowski KS, Steinberg JS (2006) Feature: a closer look at bioengineered alternative tissues. Podiatry Today 19:38–55

MacNeil S (2007) Progress and opportunities for tissue-engineered skin. Nature 445:874–80

Martin P, Mehendale F (2001) The cellular and molecular events of wound healing. In: Falanga V, (ed) Cutaneous Wound Healing London, UK: Martin Dunitz pp 15–38

Moulin V, Auger FA, Garrel D, Germain L (2000) Role of wound healing myofibroblasts on re-epithelialization of human skin. Burns 26:3–12

Pandya AN, Woodward B, Parkhouse N (1998) The use of cultured autologous keratinocytes with Integra in the resurfacing of acute burns. Plast Reconstr Surg 102:825–8

Ratner BD, Bryant SJ (2004) Biomaterials: where we have been and where we are going. Ann Rev Biomed Eng 6:41–75

Rheinwald JG, Green H (1975) Serial cultivation of strains of human epidermal keratino-cytes: the formation of keratinizing colonies from single cells. Cell 6:331–43

Rosenberg LZ, de la Torre J (2006) Wound healing growth factors. Emedicine, available at http://www.emedicine.com/plastic/topic457.htm.

Shakespeare PG (2005) The role of skin substitutes in the treatment of burn injuries. Clini Dermatol 23:413–418

Sheridan RL, Hegarty M, Tompkins RG, Burke JF (1994) Artificial skin in massive burns-Results to 10 years. Eur J Plast Surg. 17:91–93

Singer AJ, Clark RA (1999) Cutaneous wound healing. N Engl J Med 341:738–46

Suk Wha Kin, Lee IW, Cho HJ, Cho KH, Han Kim K, Chung JH, Song PI, Chan Park K (2002) Fibroblasts and ascorbate regulate epidermalization in reconstructed human epidermis. J Dermatol Sci 30:215–23

Tanenbaum M (1995) Skin and tissue techniques. In: McCord CD Jr, Tanenbaum M, Nunery WR, eds. Oculoplastic Surgery. 3rd ed. New York: Raven Press 3–4.

Tomkins RG, Burke JF (1992) Burn wound closure using permanent skin replacement materials. World J Surg. 16:47–52

Wood FM, Stoner ML, Fowler BV, Fear MW (2007) The use of a non-cultured autologous cell suspension and Integra(R) dermal regeneration template to repair full-thickness skin wounds in a porcine model: A one-step process. Burns 33:693–700

Wood FM, Kolybaba ML, Allen P (2006) The use of cultured epithelial autograft in the treatment of major burn wounds: eleven years of clinical experience. Burns 32:538–44

Yannas IV, Burke JF (1980) Design of an artificial skin. I. Basic design principles. J Biomed Mater Res 14:65–81

Chapter 6
Angiogenesis in Development, Disease, and Regeneration

Rakesh K. Jain and Dai Fukumura

Contents

6.1   Introduction   190
6.2   Developmental Angiogenesis   190
6.3   Angiogenesis in Adult Life   191
6.4   Endothelial Cell Specialization   193
6.5   Molecular Regulators of Angiogenesis   195
       6.5.1   Vascular Endothelial Growth Factor   197
       6.5.2   Hypoxia-Inducible Factor 1   198
       6.5.3   Angiopoietins and Tie Receptors   199
       6.5.4   Platelet Derived Growth Factor (PDGF) Family   200
6.6   Structure of the Blood Vessel   201
6.7   Angiogenesis and Tissue Development   202
       6.7.1   Pancreas   202
       6.7.2   Liver   202
       6.7.3   Nervous System   203
       6.7.4   Adipose Tissue   204
6.8   Angiogenesis in Pathological Conditions   206
       6.8.1   Angiogenesis in Tumors   207
       6.8.2   Psoriasis   213
       6.8.3   Ocular Neovascularization   213
       6.8.4   Atherosclerosis   213
6.9   Tools to Study Angiogenesis   214
       6.9.1   Matrigel Tube Formation Assay   215
       6.9.2   Cornea Pocket Assay   215
       6.9.3   Chamber Models   215
6.10 Angiogenesis in Regeneration   216
       6.10.1 Growth Factors Based Pro-angiogenic Therapy   219
       6.10.2 Cell-based Pro-angiogenic Therapy   220
6.11 Concluding Remarks   222
6.12 Questions/Exercises   222
References   223

R.K. Jain (✉)
Harvard Medical School and Edwin L. Steele Laboratory for Tumor Biology
Department of Radiation Oncology Massachusetts General Hospital, 100 Blossom St,
Cox 7, Boston, MA 02114
e-mail: jain@steele.mgh.harvard.edu

M. Santin (ed.), *Strategies in Regenerative Medicine*,
DOI 10.1007/978-0-387-74660-9_6, © Springer Science+Business Media, LLC 2009

6.1 Introduction

The cardio-vascular system consists of the heart, arterial networks, capillary plexuses, and venous networks (www.le.ac.uk/pa/teach/va/anatomy/case1/ frmst.html, http://www.angio.org/). Working in parallel with the blood circulation is the lymphatic system that drains interstitial fluid from tissue and returns it back into the venous blood circulation. The circulatory systems arose during evolution to provide transport routes for supplying nutrients, soluble factors and gaseous molecules, and removing waste products and metabolic intermediates from different tissues in the body. These functions are required to maintain homeostasis. When the circulatory system goes awry, it can result in a number of different diseases. For example, when circulation is partially or completely disrupted, or when vessels become structurally or functionally abnormal, the affected tissue rapidly undergoes a number of physiological changes related to the drop in tissue oxygenation (hypoxia). This state is defined as ischemia and can become pathological when hypoxic conditions persist. Ischemia induces cellular dysfunctions that thwart the normal function of the tissue, and depending on the tissue type and severity of ischemia, irreversible damage including cell death may occur. Ischemia can occur in any tissue, but its effects on the heart lead to the highest morbidity and mortality because it directly affects the circulatory system. Once a tissue suffers from injury, it may either undergo healing with scar formation or it may regenerate and regain partial or complete normal function. Therapeutic angiogenesis aims to promote the regeneration of the injured tissue by creating a favorable environment with a rich blood supply for repair.

Pathological states are induced not only by blood vessel disruption or aberrant function, but can also result from blood vessel formation in tissues that are normally avascular (e.g., retina in macula degeneration and cartilage in arthritis) or in newly formed tissues (e.g., tumors or atherosclerotic plaques). Thus, the challenge for therapy in these diseases will be to discover strategies to judiciously promote or halt the formation of new blood vessels, depending on the type of vascular disease. We discuss in this chapter the process of angiogenesis during development and in physiological and pathological settings. We will then cover some of the main molecular pathways involved in angiogenesis and review the tools and techniques that are currently used to study new vessel formation. Finally, we will discuss the concept and application of angiogenic or anti-angiogenic therapy in regenerative medicine.

6.2 Developmental Angiogenesis

The cardiovascular system arises during early embryogenesis, when committed endothelial progenitor cells (known as angioblasts) are fused to form a primitive plexus of blood vessels, in an event known as "vasculogenesis". The development of the vascular system is intricately linked to that of the hematopoietic system. For example, embryonic endothelial cells and hematopoietic cells share a common precursor – the hemangioblast – a cell derived from the mesodermal layer. Blood precursor cells are first formed in the yolk sac as blood islands in a process called primitive hematopoiesis. The outer surface of the blood islands is covered with angioblasts (Ueno and Weissman 2006). These hematopoietic stem cells (HSCs) use blood vessels to migrate to the fetal liver where they self-renew and differentiate into all lineages of cells in the hematopoietic system. After birth, the HSCs migrate again from the fetal liver to the bone marrow. Conversely, the embryonic HSCs migrate in avascular areas and promote angiogenesis (Takakura et al. 2000). The bone marrow HSCs are derived from embryonic precursor cells from the aorta-gonad-mesonephros (AGM) region, and this differentiation process is known as definite hematopoiesis (Medvinsky and Dzierzak 1996).

Hemangioblasts have been generated in vitro by differentiating mouse embryonic stem cells. Several reports have shown that these cells are vascular endothelial growth factor receptor-2 (VEGFR2 or Flk1) positive and that a single cell can produce blast colonies that have the potential to differentiate into hematopoietic and endothelial cells (Choi et al. 1998, Chung et al. 2002). In addition, other reports have demonstrated that VEGFR2$^+$ embryonic stem cells can give rise to both endothelial and smooth muscle cells, the main cellular components of blood vessels (Yamashita et al. 2000). The presence of hemangioblasts or common vascular precursor cells in adults remains controversial. No study has shown definitively that a single adult stem cell can give rise to both hematopoietic and endothelial cells or to both endothelial and smooth muscle cells.

Following the formation of the vascular plexus through vasculogenesis, further growth of the embryonic vascular system occurs predominantly through "angiogenesis". Angiogenesis is defined as the growth of new vessels from pre-existing ones. This can occur through sprouting, bridging, or intussusception (Fig. 6.1). The nascent vessels form networks with a hierarchal organization, with larger vessels sub-dividing into smaller vessels. The endothelial cells become functionally specialized by differentiating into arterial and venous endothelial cells. This sub-specialization occurs before blood flow, thus it is likely determined at least in part by genetic programming and not solely by environmental cues such as hemodynamics.

Angiogenesis accompanies tissue formation and growth throughout development. But as we will review later, the vascular system is also critical in providing inductive signals for the differentiation of different tissues.

6.3 Angiogenesis in Adult Life

In adulthood, most blood vascular endothelial cells are quiescent. The normal turnover rate of endothelial cells varies among different tissues. It has been estimated that it ranges from days in the endometrial vessels to month or years

A   B

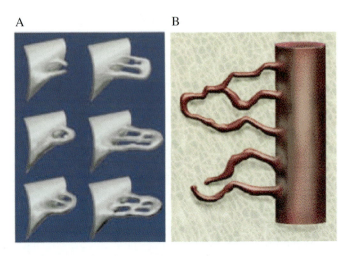

Fig. 6.1 Different modes of angiogenesis. (**A**) Intussusception: enlargement of existing vessels and growth of interstitial tissue or extracellular matrix columns in the enlarged lumen, thereby splitting the enlarged vessel into multiple vessels. (**B**) Sprouting: growth of a new sprout of endothelial cells from an existing vessel, the new sprout then develops a lumen and undergo functional maturation to become a new vessel. Adapted from (Carmeliet and Jain 2000)

in the brain endothelium. Although physiological angiogenesis is less prevalent in adults, it occurs during the menstrual cycle or wound healing, as well as in the rare cases in which adult tissues are growing (e.g., regenerating liver or expanding adipose tissue). Physiological angiogenesis is a tightly regulated process, leading to precise vascular patterns. In the initial phase of angiogenesis, blood vessels revert to a more immature form – vessels become leaky, basement membrane dissolves, and perivascular cells dissociate from the vessels. The "naked" endothelial cell tube is unstable, and in the presence of appropriate pro-angiogenic signals, vascular sprouts and fusions arise to form new vessels. The new vessels are then covered by perivascular cells and extra-cellular matrix, which results in a stable, normal-functioning and mature blood vasculature, whereas excessive vessels are "pruned" by regression (Fig. 6.2). In pathological angiogenesis (e.g., tumors, atherosclerotic plaques), the imbalance between pro- and antiangiogenic factors leads to a chronic stimulation of new vessel formation, which leads to an immature, structurally, and functionally abnormal vasculature.

   Sprouting is the most widely studied mechanism of angiogenesis. The physical process of sprouting angiogenesis is most thoroughly characterized in the retina (Gerhardt et al. 2003), but the same mode of angiogenesis has been reported in other tissues. In the retina, the vascular sprout consists of two functional components: a specialized "tip cell" at the apex of the sprout, and "stalk cells" making up the rest of the sprout (Fig. 6.3). Tip cells are responsible for guiding the migration of the vascular sprout in response to environmental cues, while stalk cells proliferate and lengthen the sprout. Vascular endothelial

Fig. 6.2 Basic structure of a blood vessel. A tube of endothelial cells (*red*) surrounded by perivascular cells (*purple*), and embedded in basement membrane (*blue*)

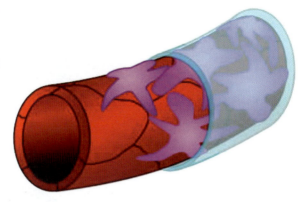

growth factor (VEGF) plays an important role in both guiding tip cell migration and inducing stalk cell proliferation, but in different manners: stalk cell proliferation is regulated by VEGF concentration, whereas tip cell migration responds to VEGF gradient. Furthermore, recent study showed that Delta-like ligand 4 (Dll4)/Notch signaling controls formation of appropriate number of tip cells during angiogenesis (Hellstrom et al. 2007).

6.4 Endothelial Cell Specialization

The vascular system is made up by many different subpopulations of endothelial cells with distinct morphological and molecular features (Chi et al. 2003, Eichmann et al. 2005). Different branches of the vascular tree serve different functions and are exposed to different environments. As a result, the endothelial cells populating these different branches must possess different properties that are suited for their respective functions. For example, the main function of arteries is to transport blood with high efficiency, therefore, arterial endothelial cells must be able to withstand the high fluid pressure and velocity, and thus must be structurally organized to minimize loss of blood from the vessel lumen. On the other hand, a major function of capillaries is to facilitate exchange of nutrients and waste materials between the blood in the vessel lumen and the interstitial fluid from the surrounding tissue. Capillary endothelial cells of different organs are specialized to optimally carry out this function. Similarly, venous cells possess distinctive features adapted to their own unique environment and functions such as facilitating the adherence and migration of immune cells on post-capillary venules when necessary. Different subpopulations of embryonic endothelial cells seem to posses a certain degree of plasticity and take on characteristics of host vessels when transplanted in a new embryo; however, this plasticity is progressively lost with development (Moyon et al. 2001). An example of a consequence of exposing endothelial cells to conditions for which they are not adapted for is demonstrated in the vein graft disease

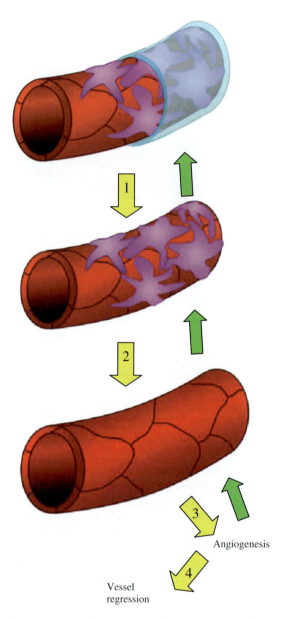

Fig. 6.3 Major steps during angiogenesis. Vessels become leaky, basement membrane dissolves (1), and perivascular cells dissociate from the vessels (2). The "naked" endothelial cell tube is unstable, and in the presence of appropriate pro-angiogenic signals, vascular sprouts and fusions arise to form new vessels (3). New vessels need to be encased by perivascular cells and basement membrane to become stable, normal-functioning, mature blood vessels (*green arrows*), otherwise they will be "pruned" by regression (4)

following arteriovenous grafts. In this case, the injuries to the venous endothelium – that is not designed to handle arterial hemodynamic environment – eventually leads to thrombosis, neointimal hyperplasia, and vein graft atherosclerosis.

In addition to specialization according to vessel type, endothelial cells often adopt tissue-specific features that are critical to the tissue's proper function. Endothelial cells in the kidney glomerulus are flattened and highly fenestrated – features which enable a very high rate of glomerular filtration when necessary. Similarly, liver sinusoidal endothelial cells are highly fenestrated and have no basement membrane. These features facilitate timely processing of the blood contents from the portal vein by enhancing plasma access to hepatic parenchymal cells. In contrast to the leaky glomerular and sinusoidal vasculatures, endothelial cells in the central nervous system (CNS) form a tightly sealed (by intercellular tight junctions) barricade known as the blood brain barrier (BBB). The BBB protects the CNS from harmful chemicals and pathogens by strictly controlling the passage of substances from the blood into the CNS. Recent studies have begun to characterize at the molecular level the specialization of endothelial cells. For example, mapping the distribution of endothelial cell-surface molecules using phage-display has led to the discovery of tissue-specific molecular targets on endothelial cells that may allow targeted delivery of therapeutics (Arap et al. 2002).

The mechanisms responsible for specialization of endothelial cells, both in relation to position in the vascular tree and to different tissue types, are still being actively studied. A thorough understanding of endothelial cell specialization will likely be a key step towards the successful repair or even construction of complex solid tissues.

6.5 Molecular Regulators of Angiogenesis

Since blood vasculature is critical to the development and function of larger organisms, it comes as no surprise that angiogenesis is a highly regulated mechanism involving many different and interconnected molecular pathways. Some of the most studied factors in the regulation of angiogenesis are listed in Table 6.1. Note that the classification of angiogenesis activator vs. inhibitor is somewhat arbitrary, as a number of these molecular pathways could either promote or inhibit angiogenesis, depending on the context and the interaction from other regulators. The next section discusses several of the key molecular regulators in more detail. An exhaustive account of all the mechanisms involved in angiogenesis is beyond the scope of this chapter. The interested reader is encouraged to seek out more comprehensive reviews in the literature (Carmeliet and Jain 2000, Jain 2003, Folkman 2007, Jain et al. 2007a–c).

Table 6.1 Angiogenesis activators and inhibitors. Modified from Jain et al. (2007)

Activators	Function	Inhibitors	Function
VEGF members	Pro-angiogenic, promote leukocyte adhesion, and increase permeability	VEGFR1, soluble VEGFR1, soluble NRP-1	"Sink" for VEGF and family members
VEGFR1/2, NRP-1	Integrate angiogenic and survival signals	Ang2	Antagonist of Ang-1
Ang1, Tie2	Stabilize vessels, inhibits permeability	TSP-1, -2	Decrease endothelial cell migration, adhesion, survival
PDGF receptors	Recruit smooth muscle cells Increase proliferation of endothelial cells	Angiostatin and related plasminogen kringles	Suppress angiogenesis
TGF-β1, endoglin TGF-β receptors	Stimulate extracellular matrix production	Endostatin (Collagen XVIII fragments)	Inhibits endothelial cell survival and migration, binds Integrin $\alpha_5\beta_1$
FGF, HGF, MCP-1	Stimulate angiogenesis	Vasostatin, calreticulin	Inhibit endothelial cell growth
Integrins $\alpha_v\beta_3$, $\alpha_v\beta_5$, $\alpha_5\beta_1$	Receptors for matrix macromolecule and proteinases	Platelet factor 4	Inhibits binding of bFGF and VEGF
VE-cadherin, PECAM1	Endothelial junction molecules	TIMPs, MMP inhibitors	Suppress pathological angiogenesis
Ephrins	Regulate arterial/venous specification	IFN-α, β, γ, IP-10, IL-4, IL-12, IL-18	Inhibit endothelial cell migration, downregulate bFGF
Plasminogen activators MMPs, Cathepsins	Remodel matrix, release VEGF	Prothrombin kringle 2 Antithrombin III fragment	Suppress endothelial cell growth
PAI-1	Stabilize nascent vessels	Prolactin	Inhibits bFGF/VEGF
NOS, COX-2	Stimulate angiogenesis and vasodilatation	VEGI	Modulate cell growth
AC133	Angioblast differentiation	SPARC fragments	Inhibit binding and activity of VEGF
Id1/Id3	Determine endothelial plasticity	Osteopontin fragments	Interfere with integrin signaling
SDF1/ CXCR4	Involved in migration	Delta like 4/ Notch	Contribute to endothelial tip cells specification
Sphingosine 1-phosphate/ S1P1	Involved in migration, vessel maturation	Canstatin, proliferin-related protein	Inhibits endothelial cell survival and migration, binds Integrins $\alpha_v\beta_3$, $\alpha_3\beta_1$
PDGF-A/B/ C/D	Promote angiogenesis, vascular remodeling	Tumstatin (Collagen IV fragment)	Inhibits endothelial cell survival, binds Integrins $\alpha_v\beta_3$, $\alpha_6\beta_1$
		Arresten	Inhibits endothelial cell survival and migration, binds Integrin $\alpha_1\beta_1$

6.5.1 Vascular Endothelial Growth Factor

Vascular endothelial growth factor A (VEGF-A, commonly referred to as VEGF) is a master regulator of angiogenesis. There are three other members in the mammalian VEGF family: VEGF-B, -C, -D, and 4 placenta growth factors (PlGF1-4). The VEGF family members bind to three cognate tyrosine kinase VEGF receptors (VEGFR-1, -2, and -3), as well as to co-receptors such as neuropilins (NRP1 and -2). The detailed functional relationships between the VEGF family of ligands and their receptors, as well as functions of the VEGF signaling pathways in and outside of the vascular system have been comprehensively reviewed in recent literature (see Takahashi and Shibuya 2005, Olsson et al. 2006). Here, we will concentrate on the major functional characteristics of VEGF-A, arguably the most potent and best-characterized angiogenic factor, in relation to new blood vessel formation.

Human VEGF-A has several different isoforms ($VEGF_{204}$, $VEGF_{189}$, $VEGF_{165}$, $VEGF_{145}$, and $VEGF_{121}$) as a result of alternative splicing. The isoforms differ in the length of their C-terminal end of the amino acid sequence. The difference manifests in varying capacity of VEGF isoforms to bind to heparin sulfate on the cell surface and in the extracellular matrix. $VEGF_{121}$ is an acidic protein, does not bind to heparin sulfate, and can freely diffuse in the extracellular matrix. In contrast, $VEGF_{189}$ and $VEGF_{204}$ are highly basic and they tightly bind to heparin sulfate. The high affinity of $VEGF_{189}$ and $VEGF_{204}$ to heparin sulfate results in their sequesteration in the extracellular matrix. The consequence is limited diffusion and biological action of the two isoforms. $VEGF_{165}$ has intermediate properties in that it binds to heparin sulfate less tightly, thus allowing it to diffuse in the extra-cellular matrix. Finally, a subset of MMPs can cleave matrix-bound isoforms of VEGF, releasing biologically active soluble fragments (Lee et al. 2005). The dual actions of binding and diffusing in the extracellular matrix (ECM) generate a concentration gradient of the protein from a source of VEGF production (i.e., cells in hypoxic tissue regions). The VEGF gradient is critical in directing angiogenesis spatially. Mice that have been engineered to express only one VEGF isoform (e.g., $VEGF_{120}$) succumbed to perinatal death as a result of impaired angiogenesis (Carmeliet et al. 1999). Thus, it is likely that for an efficacious use of VEGF for therapeutic angiogenesis, it will be critical to select the appropriate mixture of VEGF isoforms to generate structurally and functionally normal blood vessels.

VEGF has a multitude of effects on blood vessels, including induction of vascular permeability, stimulation of proliferation and inhibition of apoptosis in vascular endothelial cells, and promotion of endothelial cell migration. The regulation of the expression level of VEGF during vessel development is critical since either a half-fold reduction or a two-fold increase in VEGF expression can lead to embryonic lethality (Carmeliet et al. 1996, Ferrara et al. 1996). In adult tissues, VEGF expression level is generally low with a few exceptions (e.g., lung). Recent findings suggest a homeostatic role of VEGF signaling in maintaining normal endothelial functions. Administration of four different VEGF-signaling inhibitors in adult mice resulted in a reduction in vascular density by 20–70% in 11 out of 17 different tissues analyzed, and in particular, in the pancreatic islets and thyroid tissue (Kamba et al. 2006). The extent of vessel regression was dependent on the particular organ, but a common feature was that all the affected endothelia were the fenestrated-type. These data were in line with previous in vitro results, which indicated a role of VEGF in inducing endothelial fenestration (Esser et al. 1998). Upon cessation of VEGF inhibition, the blood vessels were able to re-grow but the vessel density remained close to 20% less than the vehicle control 40 days after the last treatment (Baffert et al. 2006).

Studies of the side effects of anti-VEGF therapy in cancer patients uncovered some previously unknown roles for VEGF in normal vasculature. The adverse effects seen in patients treated with anti-VEGF agents frequently include proteinuria, hypertension, and gastrointestinal perforation. It was recently discovered that VEGF is normally highly expressed in the kidney, specifically by the podocytes (Eremina et al. 2003). Podocytes are situated around the glomerulus where they provide structural support and serve as a barrier between the blood and urinary space. When VEGF signaling is altered in the glomerulus by either deletion or overexpression of VEGF in podocytes, it results in endothelial dysfunction and glomerular disease. Anti-VEGF agents can also have detrimental effects on the CNS (Jain et al. 2007a–c).

6.5.2 Hypoxia-Inducible Factor 1

An interesting question regarding the development of the vascular system is how the density and patterning of blood vessels is controlled in a given tissue. One potential regulatory mechanism is through the sensing of oxygen tension. The cellular machinery for sensing oxygen tension is a transcription factor called hypoxia-inducible factor 1 (HIF-1). HIF-1 consists of two subunits, HIF-1α and HIF-1β. At normal physiological oxygen tension, HIF-1α is bound to HIF-1β in the cytosol and HIF-1α is marked for degradation in the peroxisome by ubiqutination of a hydroxylated proline residue (Bruick and McKnight 2001). Under hypoxic conditions, HIF1-α is not degraded, but instead accumulates and translocates into the nucleus. Once in the nucleus, HIF1-α functions as a transcription factor for driving the transcription of a number of different genes including, most importantly, VEGF (Semenza 1999). As we discussed above, VEGF plays a critical role in providing signals for the proliferation and survival of endothelial cells both in development and in pathology. The general mechanism in HIF-1 induced angiogenesis is presumed to be as follows:

1) Hypoxia activates HIF-1 resulting in an induction of VEGF;
2) Differences in oxygen tension in a given tissue generate a VEGF gradient; and
3) VEGF gradients attract and guide new blood vessels to invade regions that are the most oxygen-deprived

6.5.3 Angiopoietins and Tie Receptors

Angiopoietins are ligands of the Tie family of receptors. Angiopoietins and Tie receptors are important in the regulation of vascular remodeling – particularly in stabilization and maturation of immature vessels as well as destabilizing vessels that need to be remodeled or removed. Of the four angiopoietins (Ang1–4) and two Tie receptors (Tie1 and Tie2), the interactions of Ang1 and Ang2 with Tie2 are by far the best characterized (Shim et al. 2007).

Tie2 (Tek) is selectively express on endothelial cells, but can also be detected on subsets of hematopoietic cells. Tie2 is a tyrosine kinase receptor and undergoes auto-phosphorylation when engaged with its ligand Ang1. Ang1 is expressed at high levels by perivascular cells. In adult tissues, Tie2 is normally expressed and phosphorlyated in endothelial cells of quiescent vasculature including arteries, veins, and capillaries (Wong et al. 1997). One of the most important functions of Ang1-Tie2 signaling is to stabilize blood vessels. This occurs by various mechanisms, including promotion of endothelial cell survival, as well as modulation of interactions between endothelial cells and their supporting cells and/or extracellular matrix. Tie2 signaling has been implicated in the recruitment of perivascular cells to the nascent angiogenic blood vessels in multiple models, but the mechanism of this effect is still unclear (Jain 2003). Genetically engineered mice lacking either Ang1 or Tie2 have very similar phenotypes – both are embryonically lethal due to deficiencies in remodeling of the primary capillary plexus and impaired branching of the myocardial circulation (Sato et al. 1995, Suri et al. 1996).

Ang2 is produced mainly by endothelial cells, and, unlike the widely expressed Ang1, is mainly expressed at sites of active vascular remodeling (Maisonpierre et al. 1997). In many instances, Ang2 functions as an antagonist to the Ang1-Tie2 interaction (mice over-expressing Ang2 have a phenotype that is similar to those of Ang1- or Tie2-deficient mice), leading to destabilization of blood vessels. In the absence of other angiogenic cues (such as VEGF), the unstable vessels will regress and cease to exist, whereas when other angiogenic cues are present, the unstable vessels would begin sprouting and developing in response to those cues, thereby expanding the vascular network.

Some studies have suggested that Ang2 may not always be an antagonist to the Ang1-Tie2 interaction, and that under certain conditions (currently characterized mostly in vitro) Ang2 can act as an agonist and induce Tie2 phosphorylation (Kim et al. 2000). Further characterization of the roles of Ang2 in different contexts, as well as more thorough understandings of the properties of Ang3, Ang4, and Tie1, will likely bring fresh insights to the role of Tie receptor signaling in angiogenesis and vessel maturation.

6.5.4 *Platelet Derived Growth Factor (PDGF) Family*

PDGF is a super family of homodimeric cytokines closely related to VEGF family (Fredriksson et al. 2004). Four members have been characterized to date, PDGF-A through –D, which exist as homo– and heterodimers in vivo. These dimers bind to their cognate PDGF receptors (PDGRα and $-\beta$), which can also homo– or heterodimerize. The isoforms of PDGF have different receptor affinities and thus also differ in their biological activities (Betsholtz 2004). PDGFs were initially discovered and purified from degranulating platelets, but later shown to be expressed by many different cell types including macrophages, endothelial cells, fibroblasts, smooth muscle cells, and malignant cells (Heldin and Westermark 1999, Hoch and Soriano 2003). PDGFs have dual functions. They act as mitogens for cells of mesenchymal (muscle, bone/cartilage, and other connective tissue) and endothelial lineage; but they can also act as chemotactic factors (Hirschi et al. 1998, Heldin and Westermark 1999, Li et al. 2005). The chemotactic function of PDGFs, and of PDGF-BB in particular, is thought to play an important role in the recruitment of perivascular cells to endothelium. In addition, PDGF-BB has been implicated in the process of blood vessel maturation. This has been shown in gene knockout models of PDGF-B and PDGF receptor-β. The two knockout mice share similar phenotypes in that both mice are embryonic lethal due to defects in vascular remodeling. Blood vessels in these mice are hemorrhagic and show a reduction in the number of perivascular cells. When PDGF-B expression is knocked out specifically in endothelial cells, it causes abnormalities in the blood vessels of the heart, kidney, and brain. Collectively, these results suggest that the secretion of PDGF-B from endothelial cells is critical for the formation of mature vessels during development. A recent study has added another layer of complexity – by showing that not only is the secretion of PDGF important, but a gradient of PDGF concentration is required for perivascular cell recruitment. PDGF-B contains, at its carboxyl terminal end, approximately 10 basic amino acid residues, a domain that is known as the retention sequence. It is believed that the retention sequence of PDGF-B interacts with the negatively charged components of ECM such as heparan sulfate on the cell surface. Normally, most of the secreted PDGF-B is retained at the cell surface, thereby limiting its actions on cells in the immediate vicinity. Mice genetically engineered to express PDGF-B deficient in the retention sequence exhibit marked abnormalities in the brain blood vessels due to a reduction in perivascular cells (Lindblom et al. 2003).

We recently attempted to harness the ability of PDGF to recruit perivascular cells, by genetically engineering endothelial cells that overexpress PDGF-BB. We hypothesized that by increasing the production of PDGF-BB by endothelial cells, it may be possible to hasten the recruitment of perivascular cells from the host to stabilize the nascent blood vessels. As expected, PDGF-BB-overexpressing endothelial cells promoted the proliferation and migration of perivascular cells in vitro. However, to our surprise, PDGF-BB overexpression resulted in rapid regression of the endothelial cells after implantation in vivo (Au et al. Submitted). These results suggest that angiogenic growth factors and endothelial cells could interact negatively, and further highlighting the necessity of thorough in vivo testing for any potential regenerative cell/growth factor combinations.

6.6 Structure of the Blood Vessel

Regardless of their tissue location, blood vessels display a fairly consistent basic structure (www.udel.edu/biology/wags/histopage/colorpage/cbv/cbv.htm): a tube formed by endothelial cells and surrounded by perivascular cells (also referred to as mural cells), which include pericytes (around capillaries and veins) and smooth muscle cells (around arteries). The two cell layers are separated by a basement membrane containing various matrix components such as collagen IV and elastin (Fig. 6.4). The structural organization of these components

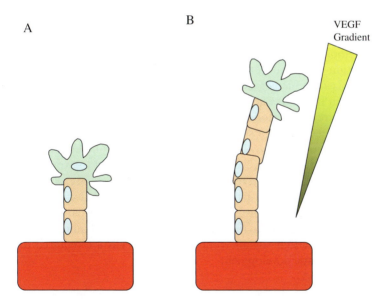

Fig. 6.4 Sprouting angiogenesis. (**A**) Illustration of the cellular structure of an endothelial sprout growing out from an existing vessel (*red*), with the filopodiated tip cell at the apex (*green*), followed by stalk cells (*pink*). (**B**) The tip cell directs the migration of the sprout following the gradient of VEGF in the microenvironment, while the stalk cells lengthen the sprout by proliferating in response to the concentration of VEGF. Sprouts also respond to other factors besides VEGF, but the mechanisms involved are yet to be clarified

varies with the size, location, and function of the blood vessel. For example, the endothelium of the arterioles is always continuous (with no gaps or fenestrations between endothelial cells) and is completely enclosed by tightly packed smooth muscle cell layer. In constrast, capillaries are only sparsely covered by pericytes and capillary endothelial cells may be continuous, fenestrated, or discontinuous, depending on the specific organ. Proper functional and structural association between endothelial cells, perivascular cells, and basement membrane is crucial to the stability and function of blood vessels (Jain 2003).

6.7 Angiogenesis and Tissue Development

Angiogenesis is a critical element of tissue development. In addition to providing oxygen and nutrients to sustain developing tissues, endothelial cells are often the source of critical signals for proper cell differentiation and organogenesis. Several such examples will be discussed below. This interaction between specialized cells in developing organs and their endothelial cells, if properly utilized, may prove beneficial towards regenerative medicine/tissue engineering efforts.

6.7.1 Pancreas

In the developing pancreas, initial pancreatic growth and subsequent endocrine differentiation occurs at contact sites between the endoderm and the major blood vessels (for further details see Chapter 8). Furthermore, pancreatic differentiation from the endoderm requires exposure to endothelial cells: isolated pre-pancreactic endoderm undergoes endocrine pancreatic development (evidenced by insulin expression), and occasionally develops structures resembling pancreatic buds, only when co-cultured with vascular endothelium. Frog embryos with endothelial precursors removed also failed to express endocrine pancreatic genes. Conversely, in another animal model where excess endothelial cells were attracted to non-pancreatic parts of the endoderm by transgenic VEGF expression, ectopic insulin expression was found in parts of the foregut that were directly adjacent to the excess endothelial cells (Lammert et al. 2001). These results show that signals from endothelial cells are required to induce endocrine pancreatic development.

6.7.2 Liver

Interaction between liver and endothelial cells begins very early in life - even before the liver bud emerges in the early embryo. The dependency of liver organogenesis on endothelial cells was demonstrated using VEGFR2-deficient mouse embryos (Matsumoto et al. 2001). Embryos from these mice can form angioblasts, but not mature endothelial cells or blood vessels. Liver organogenesis in these embryos is limited to the formation of a multi-layered liver epithelium – the liver bud is unable to develop further by growing into the surrounding septum transversum. This deficient liver growth was reproducible in vitro in isolated liver buds (thereby removing potential influences outside the liver bud, such as blood-borne factors). When vessel formation in isolated wild-type liver buds was inhibited, the growth of liver epithelium was also impaired. These results show that interaction between liver epithelium and blood vessel endothelium is critical for early liver morphogenesis. Angiogenesis is also important in adult liver regeneration: angiogenic growth factors are expressed in a spatiotemporally regulated manner during liver regeneration (Ross et al. 2001), and inhibition of angiogenesis also inhibits liver regeneration (Drixler et al. 2002). For further information about liver regeneration, see Chapter 9.

6.7.3 Nervous System

The vascular and nervous systems are tightly intertwined (see also Chapter 10). The structural similarities between the nervous and blood vascular networks have been noticed hundreds of years ago. Students in dissection classes are usually taught early on that nerves and blood vessels usually run together in what is known as neurovascular bundles. Recent studies have revealed the coordination between nerves and blood vessels extend beyond their anatomical proximity – the two share common regulatory pathways and are developmentally and functionally dependent on each other (Carmeliet and Tessier-Lavigne 2005, Weinstein 2005).

Neural stem cells reside in concentrated clusters known as vascular niches due to their close proximity to blood vessels (Palmer et al. 2000). Endothelial cells are an important component of these niches – they produce soluble factors that promote neural stem cell proliferation and maintenance of a more primitive phenotype, as well as enhance the ability of neural stem cells to form neurons (Shen et al. 2004). Several prominent neurogenesis and axon guidance pathways, such as the ephrin/Eph pathway, have been shown to also be important for vascular development (Wang et al. 1998, Gerety et al. 1999). VEGF, long known as the master regulator of angiogenesis, also has neurotrophic and neuroprotective properties, and has effects on neurogenesis and cognitive functions (Cao et al. 2004, Storkebaum et al. 2004). There are many other examples of such shared molecular regulators (Park et al. 2003), and many more to be discovered. One example of the interdependent and cooperative relationship between neuron and blood vessel development is in the brains of male adult songbirds, where the higher vocal center (HVC) undergoes seasonal testosterone-driven hypertrophy. Testosterone increases VEGF expression in the HVC (presumably by neurons and/or astrocytes, both of which express androgen receptors) and VEGFR2 expression in endothelial cells, leading to a dramatic expansion of HVC microvasculature. These HVC blood vessels are in turn stimulated by testosterone to produce brain-derived neurotrophic factor (BDNF), which promotes recruitment of new neurons from the HVC ventricular zone (Louissaint et al. 2002).

The inter-dependence between nerves and blood vessels extends beyond the developmental phase. For example, the structural integrity and proper functioning of the blood brain barrier depends on the coordination between brain microvessels, astrocytes, and neurons (Abbott et al. 2006). In addition, dysfunction in endothelial cells can be significant contributors to neurological disorders such as Alzheimer's disease (Kalaria 2002) and amyotrophic lateral sclerosis (Lambrechts et al. 2003).

6.7.4 Adipose Tissue

Fat tissue is highly vascularized (Fig. 6.5). On a per-protoplasm basis fat is more richly perfused than skeletal muscle (Gersh and Still 1945). During embryonic development, fat tissue and its corresponding vasculature develop in close spatial and temporal synchrony; and in some experimental animals, lean and obese fetuses can be distinguished by adipose tissue blood vessels well before major differences in adipocyte size or number are observed (Crandall et al. 1997). It has long been known that adipose tissue is highly angiogenic. Surgeons have been using adipose tissue grafts to promote wound healing and to revascularize ischemic tissue for centuries. This practice dates back in the early 17th century, when, after one of the many battles between the Spaniards and the Dutch at the siege of Ostend, Dutch surgeons 'sallied forth in strength... and brought in great bags filled with human fat, esteemed the sovereignest remedy in the world for wounds and disease' (Motley 1861). To

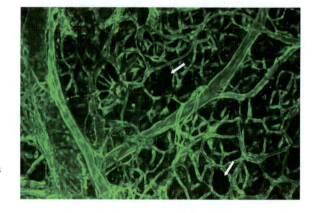

Fig. 6.5 Vascularity of fat tissue. Visualization of blood vessels (*green*) in fat tissue from mice expressing green fluorescent protein in endothelial cells. Notice the loops of small blood vessels (*arrows*) that are pervasive throughout the tissue. Each of these loops circumscribes one fat cell. Adapted from (Duda et al. 2004)

A

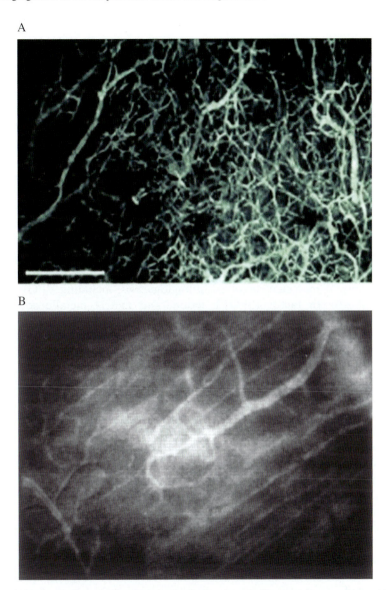

B

Fig. 6.6 Angiogenesis and adipogenesis. Angiogenesis and adipogenesis were observed in the dorsal skin chamber following implantation of pre-adipocytes. (**A**) New blood vessels that were formed as a result of the vigorous angiogenic response that accompanies the development of pre-adipocytes into fat-storing adipocytes. (**B**) When pre-adipocytes did not develop into adipocytes because adipogenesis was inhibited by a dominant negative gene construct, the angiogenic response is almost completely negated (the vessels that are visible in this figure are vessels in striated muscle and subcutaneous tissue beneath the implanted cells in the dorsal skin chamber). (**C**) Conversely, when angiogenesis was inhibited by a pharmaceutical agent (DC101, an anti-VEGFR-2 antibody), adipogenesis was also inhibited, as demonstrated by the low expression level of aP2 (a genetic marker for adipogenesis) in cells in the treated animals compared to the controls (PBS and IgG). Figures adapted from (Fukumura, Ushiyama et al. 2003)

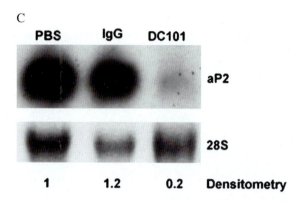

Fig. 6.6 (continued)

this day, surgeons continue to use omentum fat transplants to promote wound healing and revascularization (Maloney et al. 2003). Adipocytes produce a large variety of molecules involved in angiogenesis, including VEGF, basic fibroblast growth factor (bFGF), leptin, and matrix metalloproteinases (MMPs) (Bouloumie et al. 2002). Obesity is associated with elevated serum VEGF (Miyazawa-Hoshimoto et al. 2003). Several recent studies, using anti-angiogenic/anti-vascular agents as well as genetically engineered animal models, have demonstrated the dependency of fat tissue development on angiogenesis (Rupnick et al. 2002, Fukumura et al. 2003, Kolonin et al. 2004, Lijnen et al. 2006).

Work in our lab has revealed the reciprocal regulation between adipogenesis (fat cell differentiation) and angiogenesis, and the role of VEGF signaling in this relationship. After pre-adipocytes were implanted into the dorsal skin chamber model (chamber models will be discussed in more detail later in this chapter), the pre-adipocytes differentiated into fat-storing adipocytes, and this process was accompanied by a vigorous angiogenic response. The differentiation of pre-adipocytes into adipocytes and the growth of new blood vessels are functionally linked together: when adipogenesis was inhibited by a dominant negative gene construct, there was no angiogenesis; on the other hand, when angiogenesis was inhibited by a pharmaceutical anti-VEGF agent, the proliferation, and differentiation of pre-adipocytes was also inhibited (Fig. 6.6). This paracrine reciprocal regulation between endothelial cells and pre-adipocytes is mediated at least in part by a VEGF-induced factor that promotes adipogenesis (Fukumura et al. 2003).

6.8 Angiogenesis in Pathological Conditions

Many diseases are characterized by abnormal angiogenesis, and these abnormalities often contribute to symptoms (Carmeliet and Jain 2000). When angiogenesis is artificially induced, such as in therapeutic angiogenesis, it becomes critical that vascular growth and remodeling proceed in a controlled, regulated manner. Uncontrolled angiogenesis may lead to serious side effects (Dor et al. 2002). Therefore, it is instructive to consider the effects of abnormal angiogenesis seen in pathological conditions prior to attempting the induction of neo-vascularization for regenerative medicine or tissue engineering purposes.

6.8.1 Angiogenesis in Tumors

In normal tissues, blood vessels follow an ordered hierarchy, from arteries to arterioles to capillaries to venules to veins. The different vessel types can be clearly distinguished from each other by distinctive features. They include molecular markers, vessel morphology, cellular constituents, and blood flow profile. Blood flow is normally unidirectional and follows the vascular tree hierarchy, and the flow velocity in normal vessels is dependent on vessel diameter. Branching in the normal vascular tree is mostly dichotomous, where a larger vessel splits into two smaller vessels of similar diameter, and two smaller vessels with similar diameter would join together to form one larger vessel. Blood vessels in healthy tissues are organized such that no cell is more than 100–200 μm away from the nearest capillary vessel. This ensures that adequate oxygen and nutrients diffuse to all the cells. The exchange between blood and interstitial fluids in healthy tissues is well balanced: the amount of fluid exiting blood vessels and entering into tissue interstitium is equal to the amount of fluid removed from the interstitium by lymphatic vessels, so that local fluid build-up is prevented.

Abnormal angiogenesis with marked deficiencies of the vascular network is a hallmark of solid tumors (National Cancer Institute - Understanding Cancer Series: Angiogenesis http://www.cancer.gov/cancertopics/understandingcancer/ angiogenesis, Edwin L. Steele Laboratory for Tumor Biology, http://steele. mgh.harvard.edu/). Although the specific types and extent of abnormalities vary with different tumor types, and even between different regions within the same tumor, several general abnormal features are common to many tumor vessels. Tumors produce excess amount of multiple pro-angiogenic molecules and/or reduced amount of anti-angiogenic molecules. Furthermore, the expression of angiogenic factors is spatially and temporally heterogenous. The imbalance of these pro- and anti-angiogenic molecules causes abnormal structure and function of the tumor vasculature. The endothelial lining (normally a monolayer) of tumor vessels is often uneven – with wide gaps at some sites, and stacked endothelial cell layers at others. Perivascular cells are loosely attached or absent. The basement membrane thickness, normally fairly uniform for each vessel, is highly heterogeneous in tumor vessels. On the vascular structure level, tumor vessels are highly tortuous, uneven in diameter, and chaotically patterned. Instead of the normal dichotomous branching, tumor vessels often have multiple branches of dissimilar diameters (Fig. 6.7). Unlike normal blood vessels, blood flow velocity in tumor vessels is independent of vessel diameter, and the average velocity can be up to an order of magnitude lower in some tumors compared to

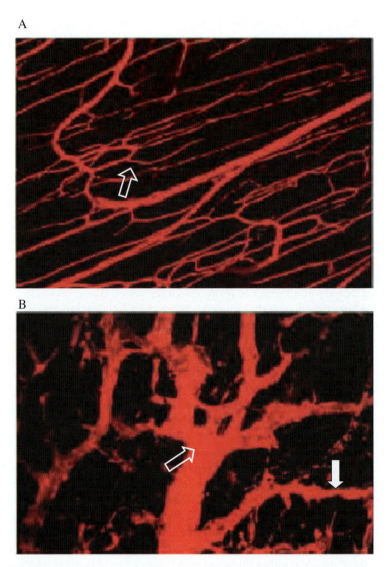

Fig. 6.7 Morphology of normal vs. tumor vessels. Normal blood vessels in mouse skeletal muscle (**A**), contrasted with abnormal vessels in a human tumor xenograft (**B**). While normal vasculature is characterized by dichotomous branching (**A**, *hollow arrow*), tumor vessels often have multiple branches of uneven diameters (**B**, *hollow arrow*). Note also the uneven diameter along a tumor vessel (**B**, *solid arrow*). From (Jain 2005)

normal tissue (Fig. 6.8) (Leunig et al. 1992, Yuan et al. 1994, Fukumura et al. 1997). Within a single tumor blood vessel, blood flow velocity fluctuates widely with time and can even reverse in direction (Endrich et al. 1979, Leunig et al. 1992, Brizel et al. 1993).

A

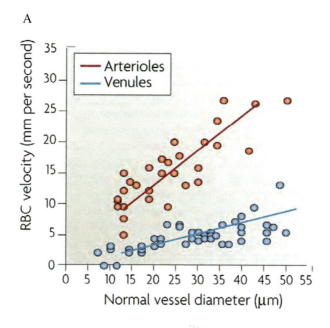

B

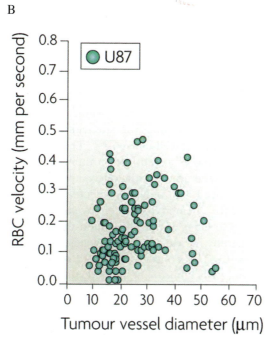

Fig. 6.8 (continued)

The vascular abnormalities in tumors lead to significant consequences. Tumor vessels are often excessively leaky (Fig. 6.9), which, combined with altered vascular resistance due to disordered tumor vessel architecture, and inefficient fluid removal by lymphatics, causes fluid build-up, and highly elevated interstitial fluid pressures in many tumors (Jain 2004). The chaotic vascular pattern and blood flow velocities lead to inefficient transport of oxygen, nutrients, and waste products. The chaotic patterning of tumor vessels results in many regions in a tumor being located farther than 100–200 μm away from the nearest functioning blood vessel. Because they are beyond the diffusion limit of oxygen, these regions are chronically hypoxic, and often necrotic (Thomlinson and Gray 1955, Torres Filho et al. 1994). On the other hand, disorderly blood flow causes oxygen availability to fluctuate. Regions of a tumor subject to these fluctuations may experience intermittent periods of hypoxia when blood flow decreases. These regions are susceptible to reperfusion injury due to the production of free radicals as blood flow (and thus oxygen availability) increases again (Brown and Giaccia 1998, Dewhirst 1998). Inefficient removal of lactic acid and carbonic acid could also cause acidic pH in parts of a tumor (Helmlinger et al. 1997, 2002). There is evidence that these harsh environments may be a source of selective pressure that could push tumor cells into more aggressive and malignant phenotypes (Smalley et al. 2005). Thus the hostile microenvironment caused by abnormal tumor vasculature may directly contribute to the neoplastic progression (Jain et al. 2007a–c).

A second consequence is that the same forces that hinder oxygen delivery and waste removal in solid tumors also impede delivery of therapeutic agents from the systemic circulation into tumors (Jain 2001). Many of the most defective tumor vessels have features that are reminiscent of immature blood vessels – with unstable endothelial tubes that are leaky and poorly supported by perivascular cells and the extracellular matrix. These features may render the same defective vessels more susceptible to damage by anti-angiogenic agents. If anti-angiogenic therapy is judiciously applied so that the most defective vessels are destroyed, and only less defective vessels remain, it may be possible to improve the hemodynamic environment in a tumor (e.g., removing the leakiest vessels may reduce fluid build-up and interstitial hypertension in a tumor), thereby increasing the efficiency of therapeutic delivery into the tumor (Fig. 6.10). This concept is referred to as "vascular normalization" (Jain 2005), and early clinical and animal studies have given support

◄──

Fig. 6.8 (continued) Blood velocity in normal vs. tumor vessels. (**A**) Blood flow velocity in normal arterioles and venules vary linearly with vessel diameter within each vessel type. (**B**) Blood flow velocity in a glioblastoma (U87), note the dissociation between velocity and vessel diameter. Tumor blood flow velocity is also several orders of magnitude lower compared to normal vessels. Adapted from (Jain, di Tomaso et al. 2007)

A

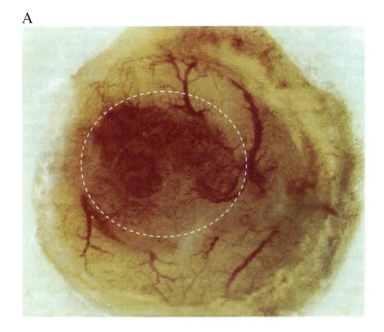

B

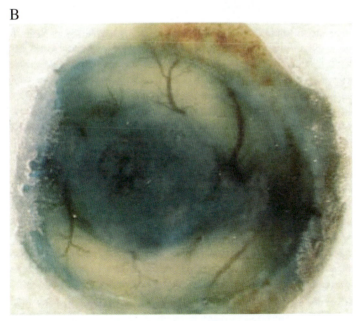

Fig. 6.9 Permeability of normal vs. tumor vessels. Photographs of a mammary adenocarcinoma transplant in a mouse before (**A**, location of tumor high-lighted by *doted line*) and 1 min after (**B**) injection of the Lissamine green dye (relative molecular mass 577), showing significant leakage of the green dye from the tumor. Note also the lack of detectable leakage in the host tissue next to the tumor (the ring of green dye on the periphery came from bone tissue that was disrupted during the preparation of the tumor transplant). Adapted from (Yuan et al. 1994)

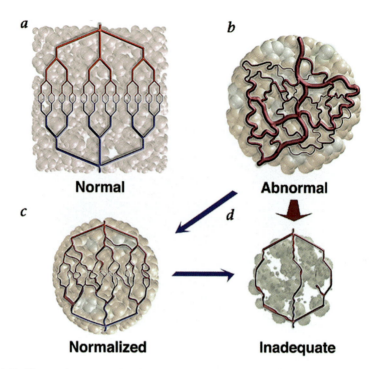

Fig. 6.10 Changes in tumor vasculature during the course of anti-angiogenic therapy. (**a**) Normal vasculature, with well-organized vascular architecture and mature vessels that are less susceptible to damage by anti-angiogenic therapy. (**b**), Abnormal tumor vasculature, composed largely of immature vessels with increased permeability, uneven vessel diameter, tortuous vessel patterns, chaotic vascular architecture, and increased interstitial fluid pressure. The delivery of therapeutics and nutrients is inefficient. (**c**), Judiciously applied anti-angiogenic therapies may prune immature vessels, leading to more "normalized" tumor vasculature. This network should be more efficient for the delivery of therapeutics and nutrients, so a tumor of this stage may benefit the most from conventional cytotoxic treatments. (**d**), Rapid pruning or stenosing of tumor vasculature might reduce the vasculature to the point that it is inadequate to support tumor growth and might lead to tumor regression or dormancy. This is the ultimate goal of anti-angiogenic/anti-vascular therapy. From (Jain 2001)

to this hypothesis by demonstrating significant benefits from combining anti-angiogenic agents with traditional cyotoxic (chemo- or radiation) therapies (Hurwitz et al. 2004, Tong et al. 2004, Winkler et al. 2004, Jain et al. 2006, Sandler et al. 2006, Willett et al. 2006).

In many ways, tumors are good examples of how not to construct a solid tissue. The problems caused by a dysfunctional tumor vasculature as well as the improvements seen when some of the vascular defects are corrected or "normalized", further underscore the importance of having proper vascular architecture and function in any engineered/regenerated tissue.

6.8.2 Psoriasis

Psoriasis is a chronic immune-mediated disease that affects the skin and joints. Recently there has been increasing evidence for a pathogenic role for angiogenesis in the etiology of psoriasis. Pro-angiogenic mediators, especially VEGF, are elevated in psoriasis patients, both in the systemic circulation and in the psoriatic plaques. These pro-angiogenic molecules are produced mainly by activated keratinocytes in psoriatic lesions, and lead to unregulated angiogenesis. As a consequence, psoriatic lesions feature prominent dermal microvascular expansions, with blood vessels that are dilated, elongated, tortuous, and hyperpermeable (Creamer et al. 2002). These structural and functional vascular defects contribute directly to the redness and swelling that are typical of psoriatic skin. In addition, blood vessels in psoriatic lesions express adhesion molecules (such as E-Selectin) that, mediate leukocyte recruitment (Griffiths et al. 1989, Groves et al. 1991, 1993), which in turn contributes to the overactive inflammatory response in psoriasis.

6.8.3 Ocular Neovascularization

Many of the examples of pathological angiogenesis given in this chapter involve vessel growth associated with diseases that are rooted in other, non-vascular causes. Ocular neovascularization (excessive angiogenesis in the eye) is the exception where the surplus, angiogenic blood vessels themselves are the cause of diseases such as proliferative diabetic retinopathy and age-related macular degeneration (see also Chapter 11). These diseases are the leading cause of severe visual impairment world wide. Angiogenesis can occur in the retina, choroid, or cornea, where the excessive and dysfunctional new vessels may impair vision by causing retinal detachment, hemorrhage, edema, or loss of corneal transparency (Kvanta 2006). These diseases have been traditionally treated by using laser or photodynamic therapy to coagulate or ablate the excessive blood vessels. These approaches have been almost completely replaced by novel anti-angiogenic therapeutics. Agents such as the anti-VEGF antibody fragment ranibizumab have shown excellent efficacy in clinical trials for age-related wet macula degeneration, and have emerged as the treatment of choice for diseases caused by ocular neovascularization (van Wijngaarden et al. 2005).

6.8.4 Atherosclerosis

Since the vascular endothelium is directly and constantly in contact with flowing blood, it comes as no surprise that the endothelium is susceptible to damage caused by exposure to plasma contents (including infectious agents and metabolic byproducts such as cholesterol), as well as to hemodynamic stress created by the blood flow itself. Alteration of endothelial functions due to such damage, leading to inflammatory changes and accumulation of macrophages, is the ultimate cause of atherosclerosis (Gimbrone et al. 2000). A detailed discussion of the mechanistic causes of atherosclerosis is beyond the scope of this chapter, but the fact that improper hemodynamic stress can lead to atherosclerosis underscores the importance of having the appropriate architecture in any new or regenerated vascular network.

Atherosclerotic plaques are avascular in the beginning, but similar to the growth of any tissue, the growth of plaques beyond a certain size limit (in the case of plaques this limit is 350–500 µm of intimal thickness) requires the development of new blood vessels by angiogenesis (Geiringer 1951). Blood vessels in plaques have characteristics that resemble tumor blood vessels, such as disorganized architecture and immature, leaky linings. Besides enabling plaque enlargement, these new blood vessels can contribute to plaque pathology in other ways. Vessels in a plaque can attract, transport, and provide anchorage for immune cells, thereby exacerbating inflammatory damage. Hemorrhage from plaque capillaries can also lead to build-up of erythrocyte membranes, leading to enlargement of the necrotic core, and eventually to plaque rupture and thrombosis (Virmani et al. 2005).

Atherosclerosis (and related thrombosis) presents an intriguing dilemma – while neovascularization within plaques is a key contributor to disease progression and should be prevented, neovascularization is desirable in the regions rendered ischemic by the atherosclerostic plaque. Judicious application of both pro- and anti-angiogenic therapies will likely be required for optimal treatment of atherosclerosis, especially in the advanced stages (Jain et al. 2007a–c).

6.9 Tools to Study Angiogenesis

Over the last 30 years, many different tools have been developed to study angiogenesis, both in vitro and in vivo. In vitro assays are most commonly used to study the migration and proliferation of endothelial cells, as well as the formation of vascular tubes in gels. Compared to in vivo assays, in vitro assays are generally easier and cheaper to perform, and also allow for greater control on experimental variables. However, there is a large degree of inherent heterogeneity in the reagents commonly used in vitro assays (such as the origin and passage number of endothelial cells, collagen or Matrigel substrate, growth media, etc.), which makes comparing results from different labs problematic. In addition, results obtained in vitro are not always good predictors of what would happen in vivo. Therefore, in vitro assays are most useful as a preliminary screening tool, and any results obtained must be validated in vivo. In vivo assays are much more costly and labor-intensive to perform than in vitro assays, but the results are more biologically relevant.

There are three general categories of in vivo angiogenesis assays:

1) in situ preparations (e.g., corneal pockets)
2) exterior tissue preparations (e.g., hamster cheek pouch, rodent liver)
3) chronic transparent chambers

Several commonly used assays are discussed below. Interested readers are encouraged to consult comprehensive reviews of angiogenesis assays in the literature (Jain et al. 1997a–c, 2002, Auerbach et al. 2003).

6.9.1 Matrigel Tube Formation Assay

Matrigel is a gelatinous extract obtained from the basement membrane of EHS mouse sarcoma. Endothelial cells that are cultured on matrigel-coated surfaces form tubular networks (Auerbach et al. 2003). This characteristic feature is often used to affirm endothelial cell identity in harvested primary endothelial cells as well as pluripotent cells that have been induced to differentiate towards an endothelial phenotype.

6.9.2 Cornea Pocket Assay

The cornea pocket assay is often used to screen for pro- or anti- angiogenic properties of cells, chemicals, or biologic agents (Kenyon et al. 1996). As the name implies, pellets containing the agent to be tested are implanted into small pockets in the animals' corneas (Fig. 6.11). Since the cornea is normally avascular, it provides an easily accessible and relatively "noise-free" background to study angiogenesis. However, the fact that the cornea is normally avascular also reduces the physiological relevance of findings obtained in the cornea pocket model.

6.9.3 Chamber Models

Chamber models are animal models with surgically implanted, windowed chambers (Fig. 6.12). Common sites for chamber implantation include the dorsal skin (dorsal skinfold chamber), top of the skull (cranial window), and the ear of rabbits (rabbit ear chamber) (Jain et al. 1997 a–c). The chambers allow materials (cells, drugs, etc.) to be implanted into the animal, whereas the window allows for direct observation of the implanted materials and the host's responses to them. These models allow for long-term in situ monitoring of cells and blood vessels, and are uniquely suited for detailed measurements of vascular parameters such as vascular permeability, blood flow velocity, etc. (Brown et al. 2001). Chamber models are susceptible to confounding factors

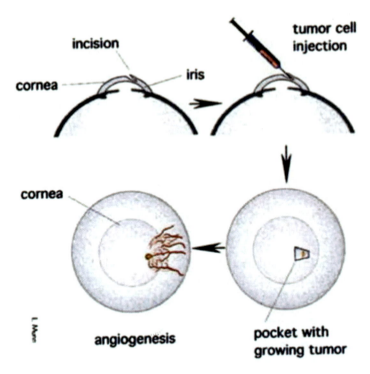

Fig. 6.11 Corneal micropocket assay. An incision is made on the cornea to create a small pocket, then cells/chemical agents are injected into the pocket, and the resultant angiogenic response is observed. From (Jain et al. 1997)

inherent to all animal models, such as inflammation and competing, or compensatory responses by the animal towards experimental treatments.

6.10 Angiogenesis in Regeneration

Although diseases are seldom classified based on vascular characteristics, deficient angiogenesis also plays a critical part in a wide range of diseases, particularly those caused by insufficient blood flow (e.g., ischemic heart disease and stroke, (Carmeliet and Jain 2000, Brown et al. 2001, Carmeliet 2003). Not surprisingly, vascular remodeling is also usually involved in the innate repair mechanisms that our bodies use to deal with these diseases. Examples include the development of collateral circulation in coronary artery disease, and the angiogenic response following ischemic stroke and myocardial infarction. Unfortunately, these remodeling responses are often inefficient to overcome the disease. Not surprisingly, the extent of angiogenesis after the initial ischemic stroke is correlated with improved survival (Krupinski et al. 1994). However,

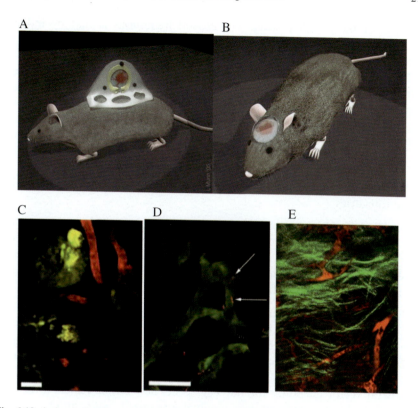

Fig. 6.12 Animal chamber models. (**A**) Illustration of the dorsal chamber model. The dorsal chamber consists of two flaps of skin on the back of the animal that are brought together and secured with a scaffold (usually made of light materials such as titanium). A circular window is cut through one flap of skin, exposing the inside surface on the opposite skin flap. Cells and chemicals can be implanted onto this exposed inside surface, and a glass slip is used to cover up this surface. (**B**) Illustration of the cranial window model. A bone flap on the top of the skull is removed, the dura mater within that opening is also removed. Cells and chemicals can be implanted, then the opening is sealed with a glass cover. Both these window models allow for direct observation of the contents of the chamber over time, without further surgical incision or sacrificing the animal. (**C–E**) Examples of observations obtained using the dorsal chamber or cranial window models. (**C**) Tumor cells that lack the HIF1-α gene (*yellow*) are seen to cluster away from blood vessels (*red*). (**D**) After delivery to a tumor, liposomes (*red*) tended to extravasate out of the tumor blood vessels (*green*) at the $\sim$25 μm looped vessel (*arrows*). (**E**) Imaging of fibrillar collagen (green) in a tumor using second harmonic generation. Blood vessels are labeled in red. Illustrations in (**A**) and (**B**) are by Dr. L.L. Munn. (**C**) and (**D**) are adapted from (Brown et al. 2001). (**E**) is adapted from (Brown et al. 2003)

this post-ischemic angiogenic response is transient and may be completely abrogated within a few weeks (Kolonin et al. 2004). In the case of myocardial infarction (MI), a significant tissue damage occurs but the residual viable endothelial cells may assist the affected tissue to regain function if perfusion is restored. It is this concept that led to the development of pro-angiogenic therapy for the post-MI heart. The central hypothesis is that by inducing angiogenesis in the damaged tissue, tissue repair is facilitated, allowing the partial restoration of normal function.

Many methods designed to increase the degree of vascularization are under active investigation as treatment for ischemic diseases such as myocardial infarction (Fukuda et al. 2004) and stroke (Shimamura et al. 2004, Taguchi et al. 2004), with the ultimate goal of relieving ischemia by producing long-lasting, functional blood vessels at the infarct site. Ideally, the new vessels would form rapidly (allowing for quick improvement in perfusion), and would resemble normal vessels in both structure and function.

Most of the approaches in pro-angiogenic therapy can be grouped into two broad categories:

1) growth-factor based approaches where angiogenic growth factors, such as members of the VEGF or FGF families, are used to induce angiogenesis from the host; or
2) cell-based approaches where cells forming blood vessels such as endothelial cells are seeded onto scaffolds and grown under various conditions.

Each of these approaches has advantages and drawbacks.

Angiogenic factors such as VEGF can induce angiogenesis when the *Vegf* gene or VEGF protein is delivered, and can potentially be used easily "off-the-shelf". However, growth factors alone may not be sufficient to create mature and stable vasculature. In a study with injection into normal tissue of an adenoviral vector encoding for the *Vegf* gene, new blood vessels formation was induced, but these vessels were highly disorganized, leaky, and hemorrhagic (Pettersson et al. 2000). Furthermore, VEGF can potentiate inflammation by increasing adhesion molecules or releasing chemokines (Melder et al. 1996, Heil et al. 2000, Lee et al. 2002, Shin et al. 2004). Sequentially controlled delivery of VEGF and PDGF-BB is able to induce mature vascular networks, but this process requires a relatively long time for host endothelial and perivascular cell infiltration, which may limit the amount of recoverable tissue (Richardson et al. 2001).

Cellular therapy is being employed in the experimental and clinical settings for treatment of myocardial infarction. Adult stem cells are isolated from either peripheral blood or bone marrow and the cells are then injected into the infracted region of the heart. The rationale behind such therapy is that the injected cells can potentially improve cardiac function by two ways:

1) by directly differentiating into the cellular component (cardiac tissue, and blood vessel) of the heart; or
2) by indirectly inducing angiogenesis through the release of growth factors.

A number of studies, in both animal models and human trials, have demonstrated improvement in cardiac functions post-MI by direct injections of adult stem cells (Kocher et al. 2001, Orlic et al. 2001, Davani et al. 2003, Mangi et al. 2003, Stamm et al. 2003, Yeh et al. 2003). While these results are encouraging, cellular therapies suffer from the scarcity of cell source, as well as potential unpredictability in the implanted cells' behavior.

Irrespective of the approach, new blood vessel formation will be essential since blood flow is required for tissue survival. The ideal pro-angiogenic therapy would induce rapid formation of vessel thus, allowing for quick resumption of blood flow. The new vessels should mature to resemble normal vessels in such properties as permeability, vessel geometry, vessel diameter distribution, etc.

6.10.1 Growth Factors Based Pro-angiogenic Therapy

Due to the critical role VEGF plays in angiogenesis, this growth factor has been actively investigated for the treatment of ischemic tissues, for which no effective therapy has been established to date. A variety of animal models has been developed to test for the efficacy of VEGF in enhancing tissue perfusion (Ferrara and Alitalo 1999). Two commonly used preclinical models include ischemic heart and ischemic limb that are induced by the ligation of the coronary or femoral artery, respectively. In some studies, recombinant VEGF was administered directly while in others the *Vegf* gene transfer was achieved by either naked DNA or viral vectors. In a rabbit model of hindlimb ischemia, a single administration of recombinant human $VEGF_{165}$ enhanced tissue perfusion and the development of collateral vessels (Takeshita et al. 1994). Similar results were obtained with gene transfer of cDNA encoding for VEGF isoforms (Takeshita et al. 1996). Based on these encouraging preclinical data, VEGF administration has been tested in patients with limb or myocardial ischemia. The results from double-blind randomized placebo-controlled trials have proven so far that protein infusion or gene transfer with VEGF is safe and generally well-tolerated but none of the studies showed significant improvement in tissue perfusion or cardiac function with treatment (Losordo et al. 2002, Henry et al. 2003, Kastrup et al. 2005, Fuchs et al. 2006).

The failure of VEGF therapy in clinical trials of therapeutic angiogenesis could be related to a number of issues including the choice and pharmacokinetics of biological agents, responsiveness of tissue to exogenous angiogenic factors, and the selection of patients (Simons 2005). All the clinical trials to date have used single agents and this fails to take into account of the complexity of the myriad molecular pathways governing vessel growth. VEGF is a key initiator of angiogenesis since it guides the sprouting of endothelial tip cells from pre-existing vessels but an orchestra of molecular signals is necessary for the process of vessel maturation (Gerhardt et al. 2003, Jain 2003). The growth factors may need to be supplied in a temporally and spatially controlled manner (Lindblom et al. 2003, Lee et al. 2004). Almost invariably, the preclinical animal models have been tested on young and healthy animals that respond robustly to the angiogenic factors. In contrast, the clinical trials were performed on elderly and diseased patients that are refractory to maximum tolerated medical treatments (Fuchs et al. 2007). Vascular response to angiogenic factors is reduced with age, hyperglycemia and atherosclerosis (Simons 2005). Patients with diabetes mellitus or elevated homocysteine exhibited impaired angiogenic response (Duan et al. 2000, Waltenberger 2001). All the factors discussed above may affect the clinical efficacy is the use of VEGF as single agent in enhancing development of collateral vessels for treatment of ischemic tissues.

6.10.2 Cell-based Pro-angiogenic Therapy

The formation of new blood vessels in adult was thought to occur exclusively through the process of angiogenesis, where the newly formed vessels are derived from pre-existing blood vessels (Folkman and Shing 1992, Sato et al. 1995). The alternate process, vasculogenesis, defined as the de novo formation of blood vessel from endothelial progenitor cells, was thought to occur only during development (Risau et al. 1988). The discovery of circulating endothelial and smooth muscle progenitor cells along with the subsequent in vivo studies of their functions caused a paradigm shift (Asahara et al. 1997, Simper et al. 2002)

The presence of endothelial cells in circulation has been speculated for almost 100 years, but only recently has the existence of such cells been established with the aid of modern techniques (Hueper and Russell 1932, Parker 1933). In 1997, Asahara et al. isolated circulating endothelial progenitor cells from human peripheral blood and demonstrated in vitro and in vivo their endothelial characteristics (Asahara et al. 1997). These $CD34^+$ mononuclear blood cells, sorted from peripheral blood and cultured on fibronectin-coated flasks, gave rise to adherent cell clusters with cobblestone morphology. The attached cells were characterized as endothelial cells by their ability to take up acetylated low-density lipoprotein (AcLDL) in their cytoplasm, and the expression of endothelial markers including CD31, VEGFR2, and Tie2. To test the in vivo function of the circulating EPCs, freshly isolated $CD34^+$ hMNBC were injected intravenously into mice with an ischemic limb and the $CD34^+$ cells were shown to participate in active angiogenesis in the ischemic tissue. Questions remain on the contribution and duration of the injected cells to the vascular network.

The tissue and lineage origin of circulating EPCs remain controversial. EPCs have been defined as cells derived from bone marrow and that they actively participate in forming new blood vessels in vivo. Bone marrow-isolated EPCs are characterized by their expression of CD133, CD34, and VEGFR, their binding to Ulex Europeus lectin, and their uptake of AcLDL. However, other populations of endothelial cells, such as the circulating endothelial cells (CECs), can be detected in the blood along with $CD34^+CD133^+$ progenitor cells (Bertolini et al. 2006, Duda et al. 2006, 2007). CECs have markers specific to mature endothelial cells, and some of them may have been shed into the blood stream under pathological settings. Moreover, recent studies have also suggested that circulating monocytes, when cultured in vitro, may upregulate the expression of endothelial markers and downregulate the expression of hematopoietic markers (Rehman et al. 2003, Rohde et al. 2006). In fact, a very recent hierarchical characterization of EPCs postulates that the CD34$^+$ EPC colonies as described by Asahara et al. (Asahara et al. 1997) and later by Hill et al. (Hill et al. 2003) are monocytic cells, and that the rare circulating EPCs derive from pre-existing endothelium (Yoder et al. 2007).

These reports have raised many unanswered questions. First, how readily do the ex vivo expanded EPCs form functional vessels when implanted in vivo? Second, if the ex vivo expanded EPCs can form functional vessels in vivo, how well do these vessels perform? Third, EPCs have been isolated and cultured from both adult peripheral blood (PB) and cord blood (CB); which source of EPCs is better at vascularizing tissue? Are monocytic cells able to form vessels or induce angiogenesis? Answering these questions is both necessary and required before EPCs are to be widely used in regenerative medicine.

An alternative, renewable and potentially unlimited source of endothelial cells is human embryonic stem cells (hESCs). Human embryonic stem cells are pluripotent, capable of differentiating into all cell types in a body. They also have the advantage of having unlimited proliferation potential. However, there are significant scientific and ethical challenges in using hESCs in the clinic. The first scientific challenge is to establish the appropriate conditions (i.e., growth factors, oxygen tension, extracellular matrix, etc.) for differentiation of the hEPCs into endothelial and perivascular cells. Second, teratoma formation by undifferentiated hESCs must be avoided. We and others have shown that some of these challenges may be overcome by deriving endothelial cells from hESCs and demonstrating that the hESC-derived endothelial cells have the ability to form functional blood vessels in vivo whitout causing teratoma formation (Lu et al. 2007, Wang et al. 2007). We observed that some of the hESC-derived blood vessels persisted for more than 150 days as functional, blood-perfused vessels.

Although endothelial precursor cells can initiate the process of neovascularization, they cannot complete by themselves this process. Endothelial cells initially form a naked tube that is highly permeable and unstable (Pettersson et al. 2000). For vessel maturation and stabilization, the naked endothelial tube undergoes a series of steps to recruit and to differentiate perivascular cells to line the outer layer of the vessels (Hirschi and D'Amore 1996, Gerhardt et al. 2003). As discussed in previous sections, perivascular cells such as vascular smooth muscle cells and pericytes are thought to provide structural integrity to the vessels, lay down basement membrane, and provide necessary survival factors to the endothelial cells (Hirschi et al. 2002). The importance of perivascular cells can be seen in a number of diseases that are caused by the lack or dysfunction of perivascular cells. Loss of pericytes can lead to leaky vessels and vessel aneurysm. In diabetic retinopathy, injury to pericytes leads to secondary changes in the endothelium that induces pathological angiogenesis (Hirschi and D'Amore 1996). Damage to smooth muscle cells due to chronic amyloid deposition is also believed to be an early step that ultimately leads to hemorrhagic stroke (Christie et al. 2001).

To date the emphasis in therapeutic angiogenesis has been almost exclusively on studies of endothelial cells, with perivascular cells either totally neglected, or treated simply as a muscular tube to maintain the blood vessel's structure. Engineered blood vessels have often been found to be immature and unstable (Schechner et al. 2000), and the lack of proper perivascular cell support is likely to be one of the main reasons for this instability. Work in our lab has have shown that perivascular cells must be appropriately incorporated into the design of engineered vessels in order to form long lasting, stable vasculature (Koike et al. 2004). By co-implanting human umbilical-vein endothelial cells (HUVECs) with 10T1/2 cells, a mesenchymal precursor cell line that is capable of differentiating into perivascular cells through heterotypic interaction with endothelial cells (Gerhardt et al. 2003), we were able to produce small diameter blood vessels that remained patent for over 1 year with good functionality, and displayed proper coverage of perivascular-like cells. While these results were encouraging, this model is not immediately applicable in the clinic because HUVECs and 10T1/2 cells are not immunocompatible with human patients. Recently we have been able to reproduce these results using human bone marrow derived mesenchymal stem cells (hMSCs) instead of the 10T1/2 cell line (Au et al. Submitted), and using endothelial progenitor cells from umbilical cord blood instead of HUVECs (Au et al. Submitted). By using cell sources that are compatible with the human body, these studies have brought our method one step closer towards possible clinical utilization.

6.11 Concluding Remarks

Angiogenesis is a complex process, which requires a tightly regulated molecular signaling in order to result in stable, mature and functional vascular networks. In adults, abnormal angiogenesis is a feature of many diseases and has become a validated target in diseases such as cancer or macula degeneration. "Normalization" of the aberrant vasculature may have beneficial effects. The potential of promoting angiogenesis for regenerative medicine and tissue engineering is yet to be fulfilled but is an area of intense investigation.

Questions/Exercises

1. What are perivascular cells (also known as mural cells)? What role do they play in blood vessels?
2. What is the difference between "vasculogenesis" and "angiogenesis"?
3. Name two structural differences between capillaries and arterioles.
4. Do tip cells and stalk cells respond the same way to VEGF? If not, how do they respond differently?
5. What are the differences between "normal" blood vessels and tumor blood vessels in the follow aspects: i) pericyte coverage; ii) vessel permeability; iii) blood flow velocity?

6. Are oxygenation and pH levels the same between tumors and normal tissues? If not, how are they different?
7. What are some putative sources of endothelial progenitor cells in adults?
8. Angiogenesis occurs predominantly during the fetal and childhood stages where the body is experiencing active growth. What are some examples of non-pathological angiogenesis that continue to occur in adults?
9. Endothelial cells in different organs have different properties. Name one difference between the endothelium of the brain compared to that of the glomerulus in the kidney and describe how this difference relates to the respective functions of blood vessels in these two locations.
10. Anti-angiogenic therapies were initially developed to treat solid tumors, but more recently their use has been expanded to treat other disorders that have vascular causes, such as diabetes retinopathy and psoriasis. Can you think of other diseases that may benefit from correcting the underlying vascular disorders?

References

Abbott NJ, Ronnback L et al. (2006) Astrocyte-endothelial interactions at the blood-brain barrier. Nat Rev Neurosci 7:41–53

Arap W, Kolonin MG et al. (2002) Steps toward mapping the human vasculature by phage display. Nat Med 8:121–127

Asahara T, Murohara T, et al. (1997) Isolation of putative progenitor endothelial cells for angiogenesis. Science 275:964–967

Au P, Daheron LM, et al. (Submitted)Differential in vivo potential of endothelial progenitor cells from human umbilical cord blood and adult peripheral blood to form functional long-lasting vessels.

Au P, Tam J et al. (Submitted)PDGF-BB overexpression in endothelial cells leads to rapid regression of engineered blood vessels in vivo.

Au P, Tam J, et al. (Submitted)Bone marrow derived mesenchymal stem cells stabilize engineered blood vessels by differentiating into perivascular cells.

Auerbach R, Lewis R, et al. (2003) Angiogenesis assays: a critical overview. Clin Chem 49:32–40

Baffert F, Le T, et al. (2006) Cellular changes in normal blood capillaries undergoing regression after inhibition of VEGF signaling. Am J Physiol Heart Circ Physiol 290:H547–559

Bertolini F, Shaked Y et al. (2006) The multifaceted circulating endothelial cell in cancer: towards marker and target identification. Nat Rev Cancer 6:835–845

Betsholtz C (2004) Insight into the physiological functions of PDGF through genetic studies in mice. Cytokine Growth Factor Rev 15:215–28

Bouloumie A, Lolmede K et al. (2002) Angiogenesis in adipose tissue. Ann Endocrinol (Paris) 63:91–95

Brizel DM, Klitzman B, et al. (1993) A comparison of tumor and normal tissue microvascular hematocrits and red cell fluxes in a rat window chamber model. Int J Radiat Oncol Biol Phys 25:269–76

Brown E, McKee T, et al. (2003) Dynamic imaging of collagen and its modulation in tumors in vivo using second-harmonic generation. Nat Med 9:796–800

Brown EB, Campbell RB, et al. (2001) In vivo measurement of gene expression, angiogenesis and physiological function in tumors using multiphoton laser scanning microscopy. Nat Med 7:864–868

Brown JM, Giaccia AJ (1998) The unique physiology of solid tumors: opportunities (and problems) for cancer therapy. Cancer Res 58:1408–1416

Bruick RK, McKnight SL (2001) A conserved family of prolyl-4-hydroxylases that modify HIF. Science 294:1337–1340

Cao L, Jiao J, et al. (2004) VEGF links hippocampal activity with neurogenesis, learning and memory. Nat Genet 36:827–835

Carmeliet P (2003) Blood vessels and nerves: common signals, pathways and diseases. Nat Rev Genet 4:710–720

Carmeliet P, Jain RK (2000) Angiogenesis in cancer and other diseases. Nature 407:249–57

Carmeliet P, Ng YS, et al. (1999) Impaired myocardial angiogenesis and ischemic cardiomyopathy in mice lacking the vascular endothelial growth factor isoforms VEGF164 and VEGF188. Nat Med 5:495–502

Carmeliet P, Tessier-Lavigne M (2005) Common mechanisms of nerve and blood vessel wiring. Nature 436:193–200

Chi JT, Chang HY, et al. (2003) Endothelial cell diversity revealed by global expression profiling. Proc Natl Acad Sci U S A 100:10623–10628

Choi K, Kennedy M, et al. (1998) A common precursor for hematopoietic and endothelial cells. Development 125:725–732

Christie R, Yamada M, et al. (2001) Structural and functional disruption of vascular smooth muscle cells in a transgenic mouse model of amyloid angiopathy. Am J Pathol 158:1065–1071

Chung YS, Zhang WJ, et al. (2002) Lineage analysis of the hemangioblast as defined by FLK1 and SCL expression. Development 129:5511–5520

Crandall DL, Hausman GJ, et al. (1997) A review of the microcirculation of adipose tissue: anatomic, metabolic, and angiogenic perspectives. Microcirculation 4:211–232

Creamer D, Sullivan D, et al. (2002) Angiogenesis in psoriasis. Angiogenesis 5:231–236

Davani S, Marandin A, et al. (2003) Mesenchymal progenitor cells differentiate into an endothelial phenotype, enhance vascular density, and improve heart function in a rat cellular cardiomyoplasty model. Circulation 108:II253–8

Dewhirst MW (1998) Concepts of oxygen transport at the microcirculatory level. Semin Radiat Oncol 8:143–150

Dor Y, Djonov V, et al. (2002) Conditional switching of VEGF provides new insights into adult neovascularization and pro-angiogenic therapy. Embo J 21:1939–1947

Drixler TA, Vogten MJ, et al. (2002) Liver regeneration is an angiogenesis- associated phenomenon Ann Surg 236:703–711; discussion 711–712

Duan J, Murohara T, et al. (2000) Hyperhomocysteinemia impairs angiogenesis in response to hindlimb ischemia. Arterioscler Thromb Vasc Biol 20:2579–2585

Duda DG, Cohen KS, et al. (2006) Differential CD146 expression on circulating versus tissue endothelial cells in rectal cancer patients: implications for circulating endothelial and progenitor cells as biomarkers for antiangiogenic therapy. J Clin Oncol 24:1449–1453

Duda DG, Cohen KS, et al. (2007) A protocol for phenotypic detection and enumeration of circulating endothelial cells and circulating progenitor cells in human blood. Nat Protoc 2:805–810

Duda DG, Fukumura D, et al. (2004) Differential transplantability of tumor-associated stromal cells. Cancer Res 64:5920–5924

Eichmann A, Yuan L, et al. (2005) Vascular development: from precursor cells to branched arterial and venous networks. Int J Dev Biol 49:259–267

Endrich B, Reinhold HS, et al. (1979) Tissue perfusion inhomogeneity during early tumor growth in rats. J Natl Cancer Inst 62:387–395

Eremina V, Sood M, et al. (2003) Glomerular-specific alterations of VEGF-A expression lead to distinct congenital and acquired renal diseases. J Clin Invest 111:707–716

Esser S, Wolburg K, et al. (1998) Vascular endothelial growth factor induces endothelial fenestrations in vitro. J Cell Biol 140: 947–959

Ferrara N Alitalo K (1999) Clinical applications of angiogenic growth factors and their inhibitors. Nat Med 5:1359–1364

Folkman J (2007) Angiogenesis: an organizing principle for drug discovery? Nat Rev Drug Discov 6:273–286

Folkman J, Shing Y (1992) Angiogenesis. J Biol Chem 267:10931–10934

Fuchs S, Battler A, et al. (2007) Catheter-based stem cell and gene therapy for refractory myocardial ischemia. Nat Clin Pract Cardiovasc Med 4:S89–S95

Fuchs S, Dib N, et al. (2006) A randomized, double-blind, placebo-controlled, multicenter, pilot study of the safety and feasibility of catheter-based intramyocardial injection of AdVEGF121 in patients with refractory advanced coronary artery disease. Catheter Cardiovasc Interv 68:372–378

Fukuda S, Yoshii S, et al. (2004) Angiogenic strategy for human ischemic heart disease: brief overview. Mol Cell Biochem 264:143–149

Fukumura D, Ushiyama A, et al. (2003) Paracrine regulation of angiogenesis and adipocyte differentiation during in vivo adipogenesis. Circ Res 93:e88–97

Fukumura D, Yuan F, et al. (1997) Effect of host microenvironment on the microcirculation of human colon adenocarcinoma. Am J Pathol 151:679–688

Geiringer E (1951) Intimal vascularization and atherosclerosis. J Pathol Bacteriol 63:201–211

Gerety SS, Wang HU, et al. (1999) Symmetrical mutant phenotypes of the receptor EphB4 and its specific transmembrane ligand ephrin-B2 in cardiovascular development. Mol Cell 4:403–414

Gerhardt H, Golding M, et al. (2003) VEGF guides angiogenic sprouting utilizing endothelial tip cell filopodia. J Cell Biol 161:1163–1177

Gersh I, Still MA (1945) Blood vessels in fat tissue. Relation to problems of gas exchange. J. Exp. Med 81:219–232

Gimbrone MA Jr, Topper JN, et al. (2000) Endothelial dysfunction, hemodynamic forces, and atherogenesis. Ann NY Acad Sci 902:230–239; discussion 239–240

Griffiths CE, Voorhees JJ, et al. (1989) Characterization of intercellular adhesion molecule-1 and HLA-DR expression in normal and inflamed skin: modulation by recombinant gamma interferon and tumor necrosis factor. J Am Acad Dermatol 20:617–629

Groves RW, Allen MH, et al. (1991) Endothelial leucocyte adhesion molecule-1 (ELAM-1) expression in cutaneous inflammation. Br J Dermatol 124:117–123

Groves RW, Ross EL, et al. (1993) Vascular cell adhesion molecule-1: expression in normal and diseased skin and regulation in vivo by interferon gamma. J Am Acad Dermatol 29:67–72

Heil M, Clauss M, et al. (2000) Vascular endothelial growth factor (VEGF) stimulates monocyte migration through endothelial monolayers via increased integrin expression. Eur J Cell Biol 79:850–857

Heldin CH, Westermark B (1999) Mechanism of action and in vivo role of platelet-derived growth factor. Physiol Rev 79:1283–1316

Helmlinger G, Sckell A, et al. (2002) Acid production in glycolysis-impaired tumors provides new insights into tumor metabolism. Clin Cancer Res 8:1284–1291

Helmlinger G, Yuan F, et al. (1997) Interstitial pH and pO2 gradients in solid tumors in vivo: high-resolution measurements reveal a lack of correlation. Nat Med 3:177–182

Henry TD, Annex BH, et al. (2003) The VIVA trial: Vascular endothelial growth factor in Ischemia for Vascular Angiogenesis. Circulation 107:1359–1365

Hill JM, Zalos G, et al. (2003) Circulating endothelial progenitor cells, vascular function, and cardiovascular risk. N Engl J Med 348:593–600

Hirschi KK, D'Amore PA (1996) Pericytes in the microvasculature. Cardiovasc Res 32:687–698

Hirschi KK, Rohovsky SA, et al. (1998) PDGF, TGF-beta, and heterotypic cell-cell interactions mediate endothelial cell-induced recruitment of 10T1/2 cells and their differentiation to a smooth muscle fate. J Cell Biol 141:805–814

Hirschi KK, Skalak TC, et al. (2002) Vascular assembly in natural and engineered tissues. Ann NY Acad Sci 961:223–242

Hoch RV, Soriano P (2003) Roles of PDGF in animal development. Development 130:4769–4784

Hueper WC, Russell MA (1932) Capillary-like formations in tissue culture of leukocytes. Arch Exp Zellforsch 12:407–424

Hurwitz H, Fehrenbacher L, et al. (2004) Bevacizumab plus irinotecan, fluorouracil, and leucovorin for metastatic colorectal cancer. N Engl J Med 350:2335–2342

Jain RK (2001) Delivery of molecular and cellular medicine to solid tumors. Adv Drug Deliv Rev 46:149–168

Jain RK (2001) Normalizing tumor vasculature with anti-angiogenic therapy: a new paradigm for combination therapy. Nat Med 7:987–989

Jain RK (2003) Molecular regulation of vessel maturation. Nat Med 9:685–693

Jain RK (2004) Vascular and Interstitial Biology of Tumors. In: Clinical Oncology Abeleff M, Armitage J, Niederhuber J, Kastan M, McKenna G (eds).. Elsevier, Philadelphia, PA, 153–172

Jain RK (2005) Normalization of tumor vasculature: an emerging concept in antiangiogenic therapy. Science 307:58–62

Jain RK, di Tomaso E, et al. (2007a) Angiogenesis in brain tumours. Nat Rev Neurosci 8:610–622

Jain RK, Duda DG, et al. (2006) Lessons from phase III clinical trials on anti-VEGF therapy for cancer. Nat Clin Pract Oncol 3:24–40

Jain RK, Finn AV, et al. (2007b) Antiangiogenic therapy for normalization of atherosclerotic plaque vasculature: a potential strategy for plaque stabilization. Nat Clin Pract Cardiovasc Med 4:491–502

Jain RK, Munn LL, et al. (2002) Dissecting tumour pathophysiology using intravital microscopy. Nat Rev Cancer 2:266–276

Jain RK, Schlenger K, et al. (1997) Quantitative angiogenesis assays: progress and problems. Nat Med 3:1203–1208

Jain RK, Tong RT, et al. (2007c) Effect of vascular normalization by antiangiogenic therapy on interstitial hypertension, peritumor edema, and lymphatic metastasis: insights from a mathematical model. Cancer Res 67:2729–2735

Kalaria RN (2002) Small vessel disease and Alzheimer's dementia: pathological considerations. Cerebrovasc Dis 13:48–52

Kamba T, Tam BY, et al. (2006) VEGF-dependent plasticity of fenestrated capillaries in the normal adult microvasculature. Am J Physiol Heart Circ Physiol 290:H560–576

Kastrup J, Jorgensen E, et al. (2005) Direct intramyocardial plasmid vascular endothelial growth factor-A165 gene therapy in patients with stable severe angina pectoris A randomized double-blind placebo-controlled study: the Euroinject One trial. J Am Coll Cardiol 45:982–988

Kenyon BM, Voest EE, et al. (1996) A model of angiogenesis in the mouse cornea." Invest Ophthalmol Vis Sci 37:1625–1632

Kim I, Kim JH, et al. (2000) Angiopoietin-2 at high concentration can enhance endothelial cell survival through the phosphatidylinositol 3'-kinase/Akt signal transduction pathway. Oncogene 19:4549–4552

Kocher AA, Schuster MD, et al. (2001) Neovascularization of ischemic myocardium by human bone-marrow-derived angioblasts prevents cardiomyocyte apoptosis, reduces remodeling and improves cardiac function Nat Med 7:430–436

Koike N, Fukumura D, et al. (2004) Tissue engineering: creation of long-lasting blood vessels. Nature 428:138–139

Kolonin MG, Saha PK, et al. (2004) Reversal of obesity by targeted ablation of adipose tissue. Nat Med 10:625–632

Krupinski J, Kaluza J, et al. (1994) Role of angiogenesis in patients with cerebral ischemic stroke. Stroke 25:1794–1798

Kvanta A (2006) Ocular angiogenesis: the role of growth factors. Acta Ophthalmol Scand 84:282–288

Lambrechts D, Storkebaum E, et al. (2003) VEGF is a modifier of amyotrophic lateral sclerosis in mice and humans and protects motoneurons against ischemic death. Nat Genet 34:383–394

Lammert E, Cleaver O, et al. (2001) Induction of pancreatic differentiation by signals from blood vessels. Science 294:564–567

Lee CW, Stabile E, et al. (2004) Temporal patterns of gene expression after acute hindlimb ischemia in mice: insights into the genomic program for collateral vessel development. J Am Coll Cardiol 43:474–482

Lee S, Jilani SM, et al. (2005) Processing of VEGF-A by matrix metalloproteinases regulates bioavailability and vascular patterning in tumors. J Cell Biol 169:681–691

Lee TH, Avraham H, et al. (2002) Vascular endothelial growth factor modulates neutrophil transendothelial migration via up-regulation of interleukin-8 in human brain microvascular endothelial cells. J Biol Chem 277:10445–10451

Leunig M, Yuan F, et al. (1992) Angiogenesis, microvascular architecture, microhemodynamics, and interstitial fluid pressure during early growth of human adenocarcinoma LS174T in SCID mice. Cancer Res 52:6553–6560

Li X, Tjwa M, et al. (2005) Revascularization of ischemic tissues by PDGF-CC via effects on endothelial cells and their progenitors. J Clin Invest 115:118–127

Lijnen HR, Christiaens V, et al. (2006) Impaired adipose tissue development in mice with inactivation of placental growth factor function. Diabetes 55:2698–2704

Lindblom P, Gerhardt H, et al. (2003) Endothelial PDGF-B retention is required for proper investment of pericytes in the microvessel wall. Genes Dev 17:1835–1840

Losordo DW, Vale PR, et al. (2002) Phase 1/2 placebo-controlled, double-blind, dose-escalating trial of myocardial vascular endothelial growth factor 2 gene transfer by catheter delivery in patients with chronic myocardial ischemia. Circulation 105:2012–2018

Louissaint A Jr, Rao S, et al. (2002) Coordinated interaction of neurogenesis and angiogenesis in the adult songbird brain. Neuron 34:945–960

Maisonpierre PC, Suri C, et al. (1997) Angiopoietin-2, a natural antagonist for Tie2 that disrupts in vivo angiogenesis. Science 277:55–60

Maloney CT Jr, Wages D, et al. (2003) Free omental tissue transfer for extremity coverage and revascularization. Plast Reconstr Surg 111:1899–1904

Mangi AA, Noiseux N, et al. (2003) Mesenchymal stem cells modified with Akt prevent remodeling and restore performance of infarcted hearts. Nat Med 9:1195–1201

Matsumoto K, Yoshitomi H, et al. (2001) Liver organogenesis promoted by endothelial cells prior to vascular function. Science 294:559–563

Medvinsky A, Dzierzak E (1996) Definitive hematopoiesis is autonomously initiated by the AGM region. Cell 86:897–906

Melder RJ, Koenig GC, et al. (1996) During angiogenesis, vascular endothelial growth factor and basic fibroblast growth factor regulate natural killer cell adhesion to tumor endothelium. Nat Med 2:992–997

Miyazawa-Hoshimoto S, Takahashi K, et al. (2003) Elevated serum vascular endothelial growth factor is associated with visceral fat accumulation in human obese subjects. Diabetologia 46: 1483–1488

Motley J.L (1861) History of the United Netherlands: from the death of William the Silent to the twelve years' truce - 1609, Harper & Brothers, New York.

Moyon D, Pardanaud L, et al. (2001) Plasticity of endothelial cells during arterial-venous differentiation in the avian embryo. Development 128:3359–3370

Olsson AK, Dimberg A, et al. (2006) VEGF receptor signalling - in control of vascular function." Nat Rev Mol Cell Biol 7:359–371

Orlic D, Kajstura J, et al. (2001) Bone marrow cells regenerate infarcted myocardium. Nature 410:701–705

Palmer TD, Willhoite AR, et al. (2000) Vascular niche for adult hippocampal neurogenesis. J Comp Neurol 425 479–494

Park JA, Choi KS, et al. (2003) Coordinated interaction of the vascular and nervous systems: from molecule- to cell-based approaches. Biochem Biophys Res Commun 311:247–253

Parker RC (1933) The development of organized vessels in cultures of blood cells. Science 77: 544–546

Pettersson A, Nagy JA, et al. (2000) Heterogeneity of the angiogenic response induced in different normal adult tissues by vascular permeability factor/vascular endothelial growth factor. Lab Invest 80:99–115

Richardson TP, Peters MC, et al. (2001) Polymeric system for dual growth factor delivery." Nat Biotechnol 19:1029–1034

Risau W, Sariola H, et al. (1988) Vasculogenesis and angiogenesis in embryonic-stem-cell-derived embryoid bodies. Development 102:471–478

Ross MA, Sander CM, et al. (2001) Spatiotemporal expression of angiogenesis growth factor receptors during the revascularization of regenerating rat liver. Hepatology 34:1135–1148

Rupnick MA, Panigrahy D, et al. (2002) Adipose tissue mass can be regulated through the vasculature. Proc Natl Acad Sci U S A 99:10730–10735

Sandler A, Gray R, et al. (2006) Paclitaxel-carboplatin alone or with bevacizumab for non-small-cell lung cancer. N Engl J Med 355:2542–2550

Sato TN, Tozawa Y, et al. (1995) Distinct roles of the receptor tyrosine kinases Tie-1 and Tie-2 in blood vessel formation. Nature 376:70–74

Schechner JS, Nath AK, et al. (2000) In vivo formation of complex microvessels lined by human endothelial cells in an immunodeficient mouse. Proc Natl Acad Sci U S A 97:9191–9196

Semenza GL (1999) Regulation of mammalian O2 homeostasis by hypoxia-inducible factor 1. Annu Rev Cell Dev Biol 15:551–578

Shen Q, Goderie SK, et al. (2004) Endothelial cells stimulate self-renewal and expand neurogenesis of neural stem cells. Science 304:1338–1340

Shim WS, Ho IA, et al. (2007) Angiopoietin: a TIE(d) balance in tumor angiogenesis. Mol Cancer Res 5:655–665

Shimamura M, Sato N, et al. (2004) Novel therapeutic strategy to treat brain ischemia: overexpression of hepatocyte growth factor gene reduced ischemic injury without cerebral edema in rat model. Circulation 109:424–431

Shin M, Matsuda K, et al. (2004) Endothelialized networks with a vascular geometry in microfabricated poly(dimethyl siloxane). Biomed Microdevices 6:269–278

Simons M (2005) Angiogenesis: where do we stand now? Circulation 111:1556–1566

Simper D, Stalboerger PG, et al. (2002) Smooth muscle progenitor cells in human blood. Circulation 106:1199–1204

Smalley KS, Brafford PA, et al. (2005) Selective evolutionary pressure from the tissue microenvironment drives tumor progression. Semin Cancer Biol 15:451–459

Stamm C, Westphal B, et al. (2003) Autologous bone-marrow stem-cell transplantation for myocardial regeneration. Lancet 361:45–46

Storkebaum E, Lambrechts D, et al. (2004) VEGF: once regarded as a specific angiogenic factor, now implicated in neuroprotection. Bioessays 26:943–954

Suri C, Jones PF, et al. (1996) Requisite role of angiopoietin-1, a ligand for the TIE2 receptor, during embryonic angiogenesis. Cell 87:1171–1180

Taguchi A, Soma T, et al. (2004) Administration of CD34+ cells after stroke enhances neurogenesis via angiogenesis in a mouse model. J Clin Invest 114:330–338

Takahashi H, Shibuya M (2005) The vascular endothelial growth factor (VEGF)/VEGF receptor system and its role under physiological and pathological conditions. Clin Sci (Lond) 109:227–241

Takakura N, Watanabe T, et al. (2000) A role for hematopoietic stem cells in promoting angiogenesis. Cell 102:199–209

Takeshita S, Tsurumi Y, et al. (1996) Gene transfer of naked DNA encoding for three isoforms of vascular endothelial growth factor stimulates collateral development in vivo. Lab Invest 75:487–501

Takeshita S, Zheng LP, et al. (1994) Therapeutic angiogenesis. A single intraarterial bolus of vascular endothelial growth factor augments revascularization in a rabbit ischemic hind limb model. J Clin Invest 93:662–670

Thomlinson RH, Gray LH (1955) The histological structure of some human lung cancers and the possible implications for radiotherapy. Br J Cancer 9:539–549

Tong RT, Boucher Y, et al. (2004) Vascular normalization by vascular endothelial growth factor receptor 2 blockade induces a pressure gradient across the vasculature and improves drug penetration in tumors. Cancer Res 64:3731–3736

Torres Filho, IP, Leunig M, et al. (1994). Noninvasive measurement of microvascular and interstitial oxygen profiles in a human tumor in SCID mice. Proc Natl Acad Sci U S A 91:2081–2085

Ueno H, Weissman IL (2006). Clonal analysis of mouse development reveals a polyclonal origin for yolk sac blood islands. Dev Cell 11:519–533

van Wijngaarden P, Coster DJ, et al. (2005). Inhibitors of ocular neovascularization: promises and potential problems. JAMA 293:1509–1513

Virmani R, Kolodgie FD, et al. (2005) Atherosclerotic plaque progression and vulnerability to rupture: angiogenesis as a source of intraplaque hemorrhage. Arterioscler Thromb Vasc Biol 25:2054–2061

Waltenberger J (2001) Impaired collateral vessel development in diabetes: potential cellular mechanisms and therapeutic implications. Cardiovasc Res 49:554–560

Wang HU, Chen ZF, et al. (1998) Molecular distinction and angiogenic interaction between embryonic arteries and veins revealed by ephrin-B2 and its receptor Eph-B4. Cell 93:741–753

Weinstein BM.(2005) Vessels and nerves: marching to the same tune." Cell 120:299–302

Willett CG, Kozin SV, et al. (2006) Combined vascular endothelial growth factor-targeted therapy and radiotherapy for rectal cancer: theory and clinical practice." Semin Oncol 33:S35–40

Winkler F, Kozin SV, et al. (2004) Kinetics of vascular normalization by VEGFR2 blockade governs brain tumor response to radiation: role of oxygenation, angiopoietin-1, and matrix metalloproteinases. Cancer Cell 6:553–563

Wong AL, Haroon ZA, et al. (1997) Tie2 expression and phosphorylation in angiogenic and quiescent adult tissues." Circ Res 81:567–574

Yamashita J, Itoh H, et al. (2000) Flk1-positive cells derived from embryonic stem cells serve as vascular progenitors. Nature 408:92–96

Yeh ET, Zhang S, et al. (2003) Transdifferentiation of human peripheral blood CD34 + − enriched cell population into cardiomyocytes, endothelial cells, and smooth muscle cells in vivo. Circulation 108:2070–2073

Yoder MC, Mead LE, et al. (2007) Redefining endothelial progenitor cells via clonal analysis and hematopoietic stem/progenitor cell principals. Blood 109:1801–1809

Yuan F, Salehi HA, et al. (1994) Vascular permeability and microcirculation of gliomas and mammary carcinomas transplanted in rat and mouse cranial windows. Cancer Res 54:4564–4568

Chapter 7
Tissue Engineering of Small– and Large– Diameter Blood Vessels

Dörthe Schmidt and Simon P. Hoerstrup

Contents

7.1   Introduction   231
7.2   Historical Overview – Development of Artificial Vascular Grafts   232
7.3   State of the Art – Currently Used Synthetic Vascular Grafts   233
    7.3.1   Dacron (PET)   233
    7.3.2   Expanded Polytetrafluorethylene (ePTFE)   233
    7.3.3   Polyurethanes (PU)   234
    7.3.4   Limitations of the Currently Used Vascular Grafts   234
7.4   Requirements for the "Ideal" Vascular Replacement   236
7.5   Tissue Engineering – A Promising Concept for "Ideal" Vascular Replacements   238
    7.5.1   The Golden Standard – Architecture and Characteristics of Native Blood
        Vessels   238
    7.5.2   Tissue Engineering of Vascular Grafts – Strategies and Approaches   240
    7.5.3   How to Match the Mechanical Requirements of Native Tissue   246
    7.5.4   Tissue Engineering of Small-Diameter Vessels – In Vitro and In Vivo
        Studies   247
    7.5.5   Tissue Engineering of Large-diameter Vessels – In Vitro and In Vivo
        Studies   250
    7.5.6   First Clinical Experiences   251
7.6   Limitations and Future Perspectives   252
7.7   Summary   253
Questions/Exercises   254
References   254

7.1 Introduction

Today, atherosclerotic vascular diseases such as coronary and peripheral vascular disease are one of the leading causes of death in the western world (www.americanheart.org/presenter.jhtml?identifier = 4440). The prevalence of arterial

D. Schmidt (✉)
Division of Regenerative Medicine (Tissue Engineering and Cell Transplantation),
Department of Surgical Research and Clinic for Cardiovascular Surgery, University
and University Hospital Zurich, Rämistrasse 100, 8091 Zurich, Switzerland
e-mail: doerthe.schmidt@usz.ch

M. Santin (ed.), *Strategies in Regenerative Medicine*,
DOI 10.1007/978-0-387-74660-9_7, © Springer Science+Business Media, LLC 2009

disease is increasing in the aging society (Baguneid et al. 1999) and is currently estimated up to 20% (AHA 2006).

Many of the patients with atherosclerotic vascular diseases suffer from angina or myocardial infarction or critical limb ischemia. If left untreated, the patient's quality of life will be restricted due to the symptoms of the advanced state of the disease and major amputation might be required. Advanced treatment and therapies such as endovascular or minimally invasive procedures and surgical reconstruction significantly increase the quality of life (www.vascularweb.org/_CONTRIBUTION_PAGES/Patient_Information/NorthPoint/Endovascular_Stent_Graft.html). For lower limb artery bypass the autologous saphenous vein represents the accepted "gold standard" vascular graft (Taylor et al. 1990), while the internal mammary artery is the preferred graft for coronary artery bypass (Cameron et al. 1996). A major drawback and cause of failure of venous grafts is occlusion (stenosis) that is a consequence of the systemic pressure-induced tissue degeneration. One-third of vein grafts are occluded within 10 years and half of those show marked atherosclerotic changes (Raja et al. 2004).

Furthermore, 30% of patients undergoing lower limb bypass do not have suitable veins (Veith et al. 1979) and suitable arterial grafts for coronary bypass (Seifalian et al. 2002) because of pre-existing diseases or because they have already been used in previous treatments (Clayson et al. 1976, Darling and Linton 1972). This shortage restricts this therapeutic approach. Thus, an alternative vascular replacement material with native analogous characteristics is highly needed, especially in the light of the increasing multi-morbidity of aging patient population. Currently, most coronary artery bypass grafting operations are performed using combinations of internal mammary artery and saphenous vein.

Beside a need for small diameter grafts such as for the coronary arteries or peripheral blood vessels, there is also a lack of appropriate materials for the replacement of large diameter vessels such as for a diseased aorta or for the repair of congenital cardiovascular malformations. In order to overcome these limitations, the search for alternative materials leading to biocompatible and anti-thrombogenic vascular substitutes has been undertaken by many research groups worldwide. Many types of vascular substitutes have been proposed. These include fresh or cryopreserved allografts, decellularized tissues, and artificial synthetic grafts.

7.2 Historical Overview – Development of Artificial Vascular Grafts

The first vascular prostheses were fabricated from metals, glass, or ivory tubes that rapidly and invariably became clotted by blood (Hess 1985) followed by first attempts using polymeric materials such as polyethylene and methacrylate

(Hufnagel 1947, Moore 1950). The observation of Vorhess in 1952 that a silk suture exposed for several months to flowing blood in the right ventricle became covered by a "glistening film" of tissue free of microscopic thrombi (Vorhess et al. 1952) brought further interest to the use of synthetic materials for vascular replacements. The idea that a woven rather than a smooth material would provide a non-thrombogenic surface that might be suitable for vascular substitutes was born and the first successful clinical use of a hand-woven fabric based on Vinyon N was reported (Blakemore and Vorhess 1952). Stimulated by these studies, a variety of materials including Nylon, Dacron (polyethylene terephthalate (PET)), Teflon (expanded polyetetra-fluoroethy-lene (ePTFE)), Orlon, and polyurethane were investigated as porous woven fabrics identifying Dacron as the most promising material (Deterling and Bohnslay 1955).

7.3 State of the Art – Currently Used Synthetic Vascular Grafts

Since the first clinical experiences with synthetic vascular grafts have been reported, Dacron (PET) and Teflon (ePTFE) have been established as standard materials for routine clinical use.

7.3.1 Dacron (PET)

PET (www.polymersdatabase.com/) was first introduced in 1939 and further developed as Dacron fiber by DuPont in 1950 (Friedman et al. 1994). The first implantations of Dacron based vascular grafts were performed by Julian in 1957 and DeBakey in 1958 (Hess 1985). For the fabrication of vascular grafts, Dacron can be processed in different modes. Both knitted and woven techniques are applied whereas knitting results in a higher porosity and radial distensibility of the graft. The high porosity of the knitted grafts necessitates a coating for example such as the use of gelatin (Vascutek, Renfrewshire, Scotland), collagen (Boston Scientific, Oakland, NJ), or albumin (Bard Cardiovascular, Billerica, Maas), sometimes cross-linked by low concentrations of formaldehyde (Jonas et al. 1988, Scott et al. 1987, Cziperle et al. 1992). In order to avoid kinking and possible mechanical compression, prosthetic rings or coils can be positioned on the external graft surface.

7.3.2 Expanded Polytetrafluorethylene (ePTFE)

PTFE was patented in 1937 by DuPont and later, in the 1960s, it was introduced into medical application as artificial heart valves. In 1996, Gore patented expanded PTFE, a microporous material that is known as Gore-tex and that is used for the fabrication of vascular grafts (www.gore.com/en_xx/products/ medical/surgical/index.html). The PTFE molecule itself is biostable and the surfaces of PTFE based materials are electronegative minimizing the reaction of blood components when exposed to blood flow. Furthermore, the materials do not undergo biological deterioration when implanted in vivo.

ePTFE based substitutes are produced by stretching a melt-extruded solid polymer tube cracked into a non-woven porous tube. The main structure is formed by solid nodes connected through fine fibrils with an internodal distance of 30 μm for a standard graft.

7.3.3 Polyurethanes (PU)

PU (www.polyurethane.org/s_api/sec.asp?CID = 824&DID = 3437) were initially developed in the 1930s as surface coatings, foams, and adhesives (Batich and DePalma 1992). PU are based on three different monomers the chemical characteristics of which provide the material with its mechanical properties: (i) a hard domain derived from diisocyanate is mainly responsible for mechanical strength, (ii) a soft domain, most commonly based on polyol, making the material flexible, and (iii) a chain extender. The properties of the material can be modified depending on the composition and selection of the three monomers. This makes available a high variety of PU-based materials with different mechanical characteristics attractive for biomedical applications.

Particularly, its elastic properties are interesting for the fabrication of vascular grafts enabling a radially compliant substitute. The first vascular grafts were manufactured from polyester PU (Vascugraft by B.Braun, Melsungen, Germany) followed by polyether-based PU grafts such as Pulse-Tec (Newtec Vascular Products of North Wales, UK) or Vectra (Thorathec Laboratories Corporation, Pleasanton, CA, USA). The latter one is fabricated with an average pore seize of 15 μm and a non-porous layer under the luminal surface, which makes it impervious to liquids (Eberhart et al. 1999). A new generation of PU grafts is based on polycarbonate PU that eliminate most ether linkages and thus are hydrolytically and oxidatively stable and more resistant to biodegradation (Tanzi et al. 2000).

7.3.4 Limitations of the Currently Used Vascular Grafts

Despite of initial good results and experiences with the above mentioned materials long-term studies revealed limitations and disadvantages of the materials. One major problem that occurred when synthetic materials were used as vascular substitutes is the patency of the grafts. Thrombogenicity and graft occlusion represent a major limitation of currently used vascular grafts. When implanted into the cardiovascular system and exposed to the circulating blood flow, a dynamic protein adsorption/desorption to the surfaces of the synthetic material can be detected. This so-called Vroman effect (Vroman and Adams 1969) follows a platelet adhesion, inflammatory cell infiltration, and endothelial and smooth muscle cell migration (Xue and Greisler 2000, Greisler 1995). Then, during the first hours to days, a coagulum is built up by fibrin, platelets, and blood cells forming a compact layer in a period of 6–18 months (Davids et al. 1999).

7.3.4.1 Patency Rates

An overview of the patency rates of currently used vascular substitutes is given in Table 7.1. Dacron based grafts demonstrated a five-year patency rate about 93% when implanted as aortic bifurcation replacements (Friedman et al. 1994), but only 43% for above-knee femoropopltiea bypass grafts (Green et al. 2000) and even lower for below-knee grafts.

In order to enhance the patency rate, modifications on the material with particular respect to the surfaces has been performed. The surface properties were changed by protein coating but without any impact on the patency rates (Jonas et al. 1988, De Mol Van Otterloo et al. 1991, Prager et al. 2001). When heparin was bound primarily through Van der Waals bonds to the synthetic fibers of a Dacron graft (available at InterVascular, La Ciotat, France) pre-treated with the cationic agent tridodecil-methyl-ammonium chloride (TMAC) and coating of the external graft wall with collagen to prevent blood extravasation (Lambert et al. 1999) the patency rates were slightly increased.

Table 7.1 Patency rates of the currently clinical used vascular prostheses

Material	Implantation site	Patency rate%	Years	Reference
Dacron	Aortic bifurcation	93	5	Friedmann et al. (1994)
Dacron	femoropoplitea	43	5	Green et al. (2000)
Dacron coated with heparin and TMAC	femoropoplitea	70	1	Devine and McCollum (2001)
Dacron coated with heparin and TMAC	femoropoplitea	63	2	Devine and McCollum (2001)
Dacron coated with heparin and TMAC	femoropoplitea	55	3	Devine and McCollum (2001)
Eptfe	aorta	91–95%	5	Friedmann et al. (1994) Alimi et al.(1994)
ePTFE	femoropoplitea	61 45	3–5	Post et al. (2001) Green et al. (2000)
Autologous vein grafts	femoropoplitea	77	5	Taylor et al. (1990)
Autologous vein grafts	femoropoplitea	50	10	Donaladson et al. (1991)
PU based Vascugraft	femoropoplitea below-knee	occlusion	1	Zhang et al. (1997)

After 1, 2, and 3 years the patency rates of coated grafts were 70, 63, and 55% respectively whereas the rates of untreated grafts were 56, 46, and 42%, respectively (Devine and McCollum 2001).

Similar to Dacron grafts, ePTFE grafts used as aortic substitutes demonstrated (www.biomaterialsvideos.org/displayVideo.php?id = 36) good five-year patency rates ranging from 91 to 95% (Friedman et al. 1995, Alimi et al. 1994). To the contrary, in femoropopliteal position, the 3–5 years patency rates were only 61% (Post et al. 2001) and 45% (Green et al. 2000) respectively, whereas the autologous vein grafts showed 5- and 10 year cumulative patency rates of 77% (Taylor et al. 1990) and 50% (Donaldson et al. 1991), respectively. Modifications on the basic ePTFE grafts such as changes of the internodal distances (Kohler et al. 1992), carbon-coating (Akers et al. 1993, Tsuchida et al. 1992, Bacourt 1997), heparin-bounding or growth factors (Greisler 1995) have improved the long-term patency-rates marginally.

Currently, no long-term results with PU graft exist. A clinical trial using Vascugrafts as below-knee bypass grafting was aborted in the first year due to occlusion of the implanted grafts (Zhang et al. 1997). In a clinical multicenter study, the polyetherurethaneurea-based Vectra graft with a pore size of 15 μm and a non-porous layer under the luminal surface was compared with ePTFE-based substitutes showing no differences in patency rates (Glickman et al. 2001). Currently, there is no evidence that PU-based grafts have significant advantages over Dacron or ePTFE based grafts.

7.3.4.2 Potential for Regeneration, Remodeling, and Growth

Beside limited patency rates of the vascular substitutes currently available in clinic, an additional major drawback is caused by the lack of tissue regeneration, repair, and growth capabilities. Indeed, synthetic materials are non-living replacements lacking the potential for promoting tissue self-renewing and growth.

Furthermore, it has been shown that the current substitutes are associated with low-level foreign body reaction and chronic inflammation (Greisler 1990) and an increased risk for microbial infections (Mertens et al. 1995).

7.4 Requirements for the "Ideal" Vascular Replacement

Many criteria have to be considered to engineer the optimal vascular replacement. The main requirements are summarized in Table 7.2.

A functional vessel substitute should be biocompatible, that is, non-thrombogenic, non-immunogenic, and resistant to infection. Furthermore, after implantation in vivo, the integration of the graft to the host tissue is crucial. Thus, the graft must induce an acceptable physiologically similar healing response that does not result in chronic inflammation, hyperplasia, or fibrous capsule formation. Ideally, the healing process would lead to a complete integration of the vascular substitute into the cardiovascular system becoming

Table 7.2 Requirements for the ideal vascular substitute

Properties of the ideal vascular replacement

- biocompatiple
- non-thrombogenic
- non-immunogenic
- matching mechanical properties of native counterpart
- capacity to remodel and regenerate
- nature-analogous healing process
- resistant to infection after implantation
- causing no inflammation
- no hyperplasia
- integration into surrounding tissue
- physiological responses such as vasoconstriction and relaxation

Table 7.3 Landmarks as to vascular tissue engineering

Engineered Vascular Graft	Year	Reference
Small diameter vascular grafts based on bovine SMC/EC and Dacron	1986	Weinberg and Bell *Science*
Small diameter blood vessels based on cell sheets fabricated from human fibroblasts/SMC/EC	1998	L'Heureux N et al. *FASEB*
Small diameter blood vessels based on bovine SMC/EC and PGA exposed to cyclic strain	1999	Niklason LE et al. *Science*
Aortic replacement based on vascular-derived fibroblasts/ SMC/EC and PGA/polyhyhroxyalkanoate	1999	Shum-Tim D et al. *Ann Thorac Surg*
Small diameter vascular grafts based on ovine blood-derived EPC and PGA	2001	Kaushal S et al. *Nature Medicine*
Autologous pulmonary artery based on vascular-derived cells and polycaprolactone–polylactic acid copolymer reinforced with woven polyglycolic acid implanted into a four-year-old girl	2001	Shin'oka T et al. *NEJM*
Small diameter blood vessels based on immortalized human cells and PGA	2005	Poh M et al. *Lancet*
pediatric blood vessels based on human umbilical cord myofibroblasts and blood-derived endothelial cells and PGA/P4HB	2006	Schmidt D et al. *Ann Thorac Surg*
Growing pulmonary arteries based on ovine vascular-derived myofibroblasts/EC and PGA/P4HB implanted into an autologous animal model	2006	Hoerstrup SP et al. *Circulation*
Strong small diameter blood vessels based on cell sheets fabricated from skin-derived human myofibrobalsts implanted into a canine model	2006	L'Heureux N et al. *Nature Med*

indistinguishable from the surrounding and connecting nature vessels. Additionally, the ideal substitute would exhibit mechanical properties similar to its native counterparts including appropriate physiological compliance and tissue strength in order to withstand hemodynamic pressure changes without failure. Moreover, responses to physiological changes and integration into the complex cardiovascular system by means of adequate vasoconstriction and relaxation pose an additional challenge for the ideal vascular replacement. For pediatric applications, it would be crucial to provide a vascular substitute able to grow with the infant body thus avoiding recurrent operations; an issue nowadays associated with increased patients' mortality and morbidity. Therefore, a living autologous vascular prosthesis with growth, remodeling, and repair capability would be the ideal substitute.

7.5 Tissue Engineering – A Promising Concept for "Ideal" Vascular Replacements

As the currently used materials do not fulfil the above mentioned requirements for an ideal vascular substitute, tissue engineering has emerged as a promising approach to address these shortcomings. Different approaches have been attempted to meet the criteria to develop the ideal vascular prosthesis. Although different strategies have been used in order to create an ideal replacement, it is widely established that three components are required: (i) a non-activated, confluent endothelium that is able to prevent thrombosis; (ii) a biocompatible component with high tensile strength to provide mechanical support (collagen fibers or their analogous); and (iii) a biocompatible elastic component to provide recoil and prevent aneurysm formation (elastin fibers or their analogous) (Mitchell and Niklason 2003).

The emerging interdisciplinary field of tissue engineering aims to develop autologous, living vascular substitutes by mimicking nature blood vessels in architecture, characteristics, and properties. As viable structures, tissue engineered blood vessels are responsive and self-renewing tissues with the inherent potential of growth, healing, and remodeling. Most of the tissue engineering approaches rely on bioabsorbable synthetic or natural scaffold materials which are able to provide temporary mechanical strength until sufficient extracellular matrix is produced; the final tissue formation would resemble histological and biomechanical characteristics similar to those of the native tissue. This is particularly important as hemodynamic competence and mechanical requirements for tissue-engineered vascular grafts are critical for their full functionality.

7.5.1 The Golden Standard – Architecture and Characteristics of Native Blood Vessels

7.5.1.1 General structure

The architecture of blood vessels differs depending on the vessel size (arterioles vs. arteries) and vessel type (arteries vs. veins). However, there is a common

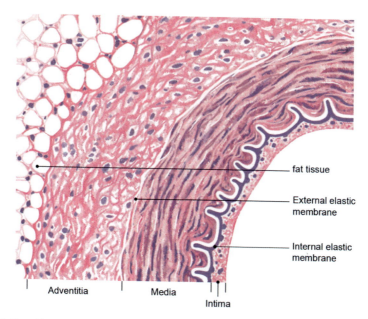

Fig. 7.1 Graphic of the nature vessel wall

structural pattern that can be found in all blood vessels independent of size and type, except capillaries (See also Chapters 2 and 6). Figure 7.1 shows a scheme of the general characteristics of the blood vessel architecture.

Native blood vessels have a three-layered tissue architecture comprising (i) an endothelium lined intima, (ii) a media containing smooth muscle cells, and (iii) an adventitia with connective tissue.

(i) As the inner layer, the intima delimits the vessel wall towards the lumen. It is formed by a functional endothelium having anti-thrombogenic properties. Because of tight intercellular connections it works as a barrier. Beneath the endothelium an associated connective tissue is present followed by an internal elastic membrane that separates the intima from the media.

(ii) As the middle layer, the media consists of one or multiple layers of smooth muscle cells (SMC) and elastic lamellas with variable amounts of connective tissue. Each layer of SMC and elastic lamella form a lamellar unit. The external elastic membrane is located beneath the outmost lamellar unit and separates the media from the adventitia.

(iii) As the outer layer, the adventitia consists of blends of connective and fat tissue anchoring the vessel with the surrounding neighbor-tissue.

The extracellular matrix of blood vessels comprises mainly collagen, glycoaminoglycans, and elastin enabling the mechanical properties of the vessels, that is, elasticity, tissue strength, and flexibility.

7.5.1.2 Arteries

Arteries can be classified into three different types: (i) elastic arteries such as the large diameter vessels originating from the pulmonary trunk, the pulmonary trunk, and the aorta, (ii) muscular arteries that are medium-sized arteries such as the carotid arteries, and (iii) the arterioles representing the small arteries with a diameter less than 2 mm. The main histological characteristic of different sized arteries represents the thickness of the different layers. The intima of elastic arteries is much thicker compared to that of arteries from the muscular type. In contrast, the media of muscular arteries is dominated by numerous concentric layers of smooth muscle cells with fine elastic fibers, whereas the media of arteries from the elastic type comprises about 50 elastic lamellae.

7.5.1.3 Veins

In contrast to artery walls, those of veins are thinner while their diameter is larger. Furthermore, the stratified architecture of vein walls is not very distinct. The intima is thin, internal, and external elastic membranes are absent or tenuous. The media appears thinner than the adventitia and the two layers tend to blend into each other. Moreover, valves, essentially endothelial folds, are found in many veins, particularly in veins of the extremities.

7.5.2 Tissue Engineering of Vascular Grafts – Strategies and Approaches

Many strategies have been developed to fabricate living vascular substitutes with anti-thrombogenic properties. Early approaches focused on surface coating of synthetic grafts with endothelial cells prior to implantation. First results were reported by Herring et al. (1978) using a single-stage technique for seeding prosthetic materials. Therefore, endothelial cells were obtained from vein segments using steel-wood pledgets and seeded directly onto the vascular prosthesis. As this method was prone to contamination, other techniques were introduced for endothelial cell harvest including a period of cell culturing and cell proliferation before seeding (Schmidt et al. 1984, Jarrell et al. 1986, Williams et al. 1994). First clinical studies performed by Deutsch et al. in 1999 (Deutsch et al. 1999) demonstrated a 68% patency of vascular-derived endothelial cell coated ePTFE grafts as synthetic substrates.

Although these reports have demonstrated promising initial results, synthetic grafts still induce low-level foreign body reaction and chronic inflammation (Greisler 1990) and as artificial materials they are associated to an increased risk of microbial infections (Mertens et al. 1995). In the attempt to overcome these limitations, more recent strategies focused on the creation of complete autologous, living vascular substitutes using a three-dimensional temporary vehicle seeded with autologous cells. Figure 7.2 shows the concept of vascular

Fig. 7.2 Concept of vascular TE: Autologous cells are obtained from the patient and isolated. After in vitro expansion (1) cells are seeded onto vascular scaffolds (2) and cultured into a biomimetic system. (3) There, tissue formation and maturation take place and after a several time tissue engineered constructs are ready for implantation

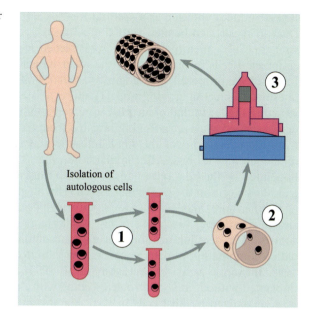

Isolation of autologous cells

tissue engineering. Ideally, autologous cells are harvested and cultivated. After proliferation in sufficient numbers, cells are seeded onto the three-dimensional scaffolds which could be synthetic or natural and exposed to a physiological in vitro environment mimicked by the conditions of a bioreactor system. There, tissue formation and maturation take place and after several weeks the vascular substitutes are ready for implantation.

Other strategies focus on a non-scaffold based vascular tissue engineering concept by using cell sheets (Ye et al. 2000). There, smooth muscle cell sheets of low passage are grown on culture plates. After sufficient time, the sheet is removed and wrapped around a porous, tubular mandrel in order to mould a media. In the same fashion, a sheet of myofibroblasts is grown and wrapped around the media to fabricate an adventitia. This step is followed by the endothelialization of the lumen that is performed after the removal of the mandrel (L'Heureux et al. 2001).

7.5.2.1 Scaffolds for the Engineering of Vascular Grafts

Various biomaterials have been investigated as scaffolds for vascular replacements including natural, permanent synthetic, biodegradable synthetic scaffolds, and biopolymer scaffolds.

Natural scaffolds

The use of natural scaffolds was first reported by Rosenberg et al. by using tanned bovine carotid arteries (Rosenberg et al. 1966). Later, Weinberg and

Bell introduced collagen as a scaffold for vascular tissue engineering by using Dacron meshes embedded into the collagen providing a necessary tensile strength. Subsequently, the grafts were seeded with smooth muscle cells and endothelial cells in order to line the inner lumen (Weinberg and Bell 1986). Lantz et al. proposed a biological vascular graft material fabricated from intestine submucosa in a dog model (Lantz et al. 1993) resulting in a histological structure similar to that of an adjacent vessel after 90 days. Similar results were demonstrated when a scaffold, fabricated from a collagen biomaterial derived from the submucosa of the small intestine and type 1 bovine collagen was implanted into rabbits. After implantation for 13 weeks, the grafts had been populated by the host cells, namely smooth muscle cells in the graft wall and endothelial cells in the graft inner lumen (Huynh et al. 1999).

Furthermore, many studies were performed using decellularized tissues fabricated from either vascular or non-vascular (e.g., small intestinal submucosa) sources. The decellularization is usually accomplished by treating the tissue with a combination of detergents, enzyme inhibitors, and buffers (Bader et al. 1998, Bishopric et al. 1999, Lively et al. 1994). Typically, these grafts are implanted without any in vitro cell seeding with the assumption that they will be recelluarized by host cells in vivo. Significant shrinkage was observed in decelluarized vessels as a result of proteoglycans being removed from the tissues during the decelluarization process (Courtman et al. 1994). However, on the example of decellularized heart valves, it was demonstrated that the implantation of decelluarized materials resulted in an adverse host response either as immunological or foreign body reaction (Simon et al. 2003). Other groups reported aneurysm formation, infection, and thrombosis after implanting decelluarized xenografts (Teebeken and Haverich 2002).

Permanent Synthetic Scaffolds

Encouraged by the favorable results reported by Weinberg and Bell enhancing the mechanical stability of the tissue engineered grafts with Dacron meshes (Weinberg and Bell 1986) other permanent synthetic scaffolds were investigated for the tissue engineering approach such as PU-based scaffolds (Ratcliffe 2000, Miwa and Matsuda 1994) and loosely woven, relatively elastic, Dacron based scaffolds (Baguneid et al. 2004). As the major limitation of these materials is the lack of compliance, poly(carbonate-urea)urethane based scaffolds were later developed (Salacinski et al. 2001, Kannan et al. 2005).

Biodegradable Synthetic Scaffolds

Many studies focused on biodegradable polymers as temporary mechanical support for in vitro-generated tissues. An example of a scaffold for the fabrication of vascular replacements is shown in Fig. 7.3. The unique feature of biodegradation makes these scaffold materials an attractive tool for the engineering of autologous vascular grafts. An ideal biodegradable synthetic

Fig. 7.3 Example of a biodegradable scaffold for vascular tissue engineering (Schmidt D et al. *Ann Thorac Surg* 2004)

material serves as a temporary scaffold guiding tissue growth and formation until the neo-tissue demonstrates sufficient mechanical properties. Then, ideally, the scaffold will be degraded resulting in a total autologous vascular graft. Particularly for pediatric applications, the use of biodegradable polymers represents an attractive strategy as it enables tissue growth and regeneration. The degradation has to be controllable and proportional to the tissue development. Furthermore, an ideal biodegradable scaffold should be at least 90% porous (Agrawal and Ray 2001) and must possess an interconnected pore network that is essential for cell growth, nutrient supply, and removal of metabolic waste products.

The two most investigated biodegradable polymers are polyglycolic acid (PGA) and polylactid acid (PLA). Both PGA and PLA are FDA accepted materials and represent therewith an attractive biodegradable material. PGA, a highly crystalline and hydrophilic polymer, has already been used to produce absorbable suture material (Frazza and Schmitt 1991). After 2–4 weeks in vivo, sutures lose their mechanical strength due to polymer hydrolysis. In contrast, PLA is a more hydrophobic material than PGA because of the presence of an additional methyl group in the lactide molecule limiting water uptake and resulting in a lower hydrolysis rate. The rapid degradation times are associated with a decrease of mechanical properties. When used for vascular tissue engineering, shrinkage and contraction of the material can be observed resulting in shape and size changes of the tissue engineered constructs. In order to overcome these problems, PGA has been successfully formulated as co-polymers by a combination with materials such as poly-L-lactid acid (Kim et al. 1999, Mooney et al. 1996), polyhydroxyalkanoate (Shum-Tim et al. 1999), poly-4-hydroxybutyrate (Hoerstrup et al. 2001, 2002, 2006, Schmidt et al. 2006), polycaprolactone-co-polylactic acid (Shin'oka et al. 2001, Watanabe et al. 2001), and polyethylene glycol (Wake et al. 1996). When seeded with cells, these materials demonstrated excellent biocompatible features and, when they are used for the fabrication of tissue engineered vascular grafts, a native analogous architecture of the resulting blood vessels was observed (Hoerstrup et al. 2006, Schmidt et al. 2006).

7.5.2.2 Cells

Choosing the right cell source for the fabrication of vascular tissue engineering products is crucial as the quality of the constructs depends on the quality and properties of the cells. Therefore, several aspects have to be considered: (i) easy cell harvest by a minimally invasive technique, (ii) phenotypes and properties of cells matching the characteristics of their native counterparts, and (iii) cells having the potential of regeneration and self-renewal. The cell harvest should not or at least minimally harm intact donor structures and has to be performed with acceptable risks for the patients. Particularly for pediatric patients, cell harvest without harming infant structures resulting in malfunction is important. With respect to these points several cell sources have been investigated including vascular structures, bone marrow, peripheral blood, umbilical cord tissue, umbilical cord blood, and prenatal progenitor cells.

Vascular-Derived Cells

Different types of animal or human vascular-derived cells have been used in order to engineer a vascular substitute. Particularly attractive are smooth muscle cells as they demonstrate contractile properties and contribute substantially to changes in vessel tonus and diameter responding to physiological and pathological impulses (Niklason et al. 1999). Alternatively, interesting results with vascular-derived myofibroblasts have been reported (Hoerstrup et al. 2006, L'Heureux et al. 2006) as myofibroblasts demonstrate contractile elements such as alpha smooth muscle actin in their cytoskeleton. Furthermore, vascular endothelium-derived endothelial cells have been successfully used for the endothelialization of vascular grafts (Deutsch et al. 1999). However, vascular-derived cells require an invasive surgical procedure associated with the sacrifice of intact vascular donor tissue. This may restrict such cell sources to adult applications.

Bone Marrow-Derived Cells

An attractive cell source enabling cell harvest without a major invasive procedure represents the bone marrow. It can be harvested easily and without high risks for the patient by a puncture, for example, of the hip (iliac crest). The bone marrow itself consists of different cell types such as endothelial cells, hematopoietic progenitor cells, fat cells, and marrow stromal cells, also termed mesenchymal stem cells. First results using bone marrow-derived cells for the engineering of vascular grafts have been reported by Noishiki et al. (1996).

Lately, evidence could be demonstrated that autologous canine bone marrow cells home into, adhere to, proliferate in, and differentiate in a scaffold in order to construct components of a new vessel wall (Matsumura et al. 2003).

Blood-Derived Cells

Recently, circulating endothelial progenitor cells (EPCs), a subpopulation of stem cells present in human peripheral blood have been introduced as a new cell source for tissue engineering. EPCs are a unique circulating subtype of bone marrow cells differentiated from hemangioblasts, a common progenitor for both hematopoetic and endothelial cells. These cells have the potential to differentiate into mature endothelial cells. Currently, EPCs have been investigated for the repair of injured vessels, neovascularization or regeneration of ischemic tissue (Kawamoto et al. 2002, Kocher et al. 2001, Assmus et al. 2002, Pesce et al. 2003), coating of vascular grafts (Shirota et al. 2003, Schmidt et al. 2004), endothelialization of decellularized grafts in an animal model (Kaushal et al. 2001), and seeding of hybrid grafts (Shirota et al. 2003).

Umbilical Cord-Derived Cells

As for pediatric applications, cell harvest without harming intact structures is crucial in order to avoid malfunction of the growing infant structures. The umbilical cord represents an attractive cell source. It provides different cell types including fibroblast-myofibroblast like cells, endothelial, and progenitor cells. Currently, only little experience using umbilical cord-derived cells for pediatric vascular tissue engineering exists. In 1996, Sipehia et al. reported the use of human umbilical cord vein derived endothelial cells. In 2002, human umbilical cord derived myofibroblasts were established as a new cell source for pediatric cardiovascular tissue engineering and used for pulmonary conduits (Hoerstrup et al. 2002, Kadner et al. 2002) In 2004, Koike et al. created long lasting blood vessels with endothelial cells from umbilical cord vein in a three-dimensional fibronectin- type I collagen gel connected to the mouse circulatory system. Recently, the fabrication of blood vessels based on Wharton's jelly – derived myofibroblast and endothelial cells derived from umbilical cord blood endothelial progenitors has been demonstrated (Schmidt et al. 2006). An example of the resulting blood vessel is shown in Fig. 7.4. Although reporting favorable results and achievements, the engineering of autologous pediatric cardiovascular tissues from umbilical cord is still in an early stage of development and a number of issues remain to be addressed such as the prenatally harvest of cells by, for example, cordocentesis having the vascular graft ready for use at birth and preventing secondary damage to the immature heart.

Fig. 7.4 Example of a tissue engineered blood vessel based on PGA/P4HB and umbilical cord-derived myofibroblasts and endothelial progenitor cells (Schmidt D et al. *Ann Thorac Surg* 2006)

7.5.3 How to Match the Mechanical Requirements of Native Tissue

Mechanical matching between vascular grafts and native tissue represents an important aspect and numerous studies have been performed addressing this issue (See also Chapter 2). The mechanical properties of native vessels are mainly determined by its extracellular matrix components, particularly collagen and elastin (Armentano et al. 1991, Barra et al. 1993, Bank et al. 1996) produced by the smooth muscle cells in the medial layer of the vessel wall. Tensile strength and maintenance of the structural integrity of a vessel is mainly determined by collagen. The transcription of collagen can be increased by growth factors (Varga et al. 1987) such as ascorbic acid (Geesin et al. 1988) and amino acids or cyclic strains (Leung et al. 1979, Mol et al. 2003). There, cross-link formation between and within collagen fibers stabilizes the proteins (Eyre 1984) resulting in a less susceptible structure for matrix metalloproteinase and other enzymes. Furthermore, cross-link formation may be controlled by modulating the activity of the enzyme lysyl oxidase. This enzyme catalyzes cross-link formation and is activated by copper (Rayton and Harris 1979) and TGF-b (Shanley et al. 1997) in vitro.

In order to match the mechanical properties of a native vessel and therewith to stimulate collagen synthesis and cross-link formation in vitro, biomimetic systems (bioreactors) exposing tissue engineered vascular constructs to mechanical stimuli have been developed (Niklason et al. 1999, Hoerstrup et al. 2001). When ovine cell based tissue engineered blood vessels were allowed to mature in a biomimetic system, the mechanical strength of the constructs improved consistently with the culture time as demonstrated in a detailed analysis of the burst pressures. After three weeks, no burst pressure could be measured; after five weeks, the burst pressure was 570 mmHg and after eight weeks, it reached values up to 2150 mmHg (Niklason et al. 2001). This phenomenon has also been reported by Hoerstrup et al. (Hoerstrup et al. 2001). An increase of burst strength up to 326 mmHg could be detected in ovine cell based tissue-engineered vascular grafts exposed to pulsatile flow for four weeks. In contrast, statically cultured controls demonstrated burst strength of 50 mmHg only. These results underline the importance of mechanical in vitro stimulation in tissue engineering of vascular grafts in order to match the mechanical strength of native vessels.

However, high burst strength is often associated with compliance mismatch (L'Heureux et al. 1998) which might lead to an intimal hyperplasia at the suture line (Bassiouny et al. 1992).

Interestingly, constructs that demonstrated compliance within a physiological range, have lacked high burst strength (Girton et al. 2000). Elastin, that is a key factor in elasticity detection and mainly responsible for elastic properties, has only been reported using the self-assembly approach, and in association with fibroblasts (L'Heureux et al. 1998). This might be due to a downregulation of elastogenesis in smooth muscle cells (Johnson et al. 1995, McMahon et al. 1985) and represents still an unsolved problem for the tissue engineering of native-analogous vascular grafts.

7.5.4 Tissue Engineering of Small-Diameter Vessels – In Vitro and In Vivo Studies

Many favorable studies have been performed focusing on tissue engineering of small-diameter blood vessels. A particular challenge for the tissue engineering of small diameter vessels represents a lower flow velocity compared to large-diameter vessels. As to the law of Hagen-Poiseuille, describing the flow characteristics of voluminal laminar stationary flows of incompressible uniform viscous liquids through cylindrical tubes with constant circular cross-sections, the volume of the flow is highly dependent on the radius of the tube (voluminal laminar stationary flow $= \pi R^4 / 8\eta \times \Delta P/L$; $R =$ internal radius of the tube, $\eta =$ dynamic fluid viscosity; $P =$ pressure difference between the two tube ends, $L =$ total length of the tubes in the x direction). This necessitates an excellent anti-thrombogenic layer and excellent scaffold design in order to engineer a functional vessel.

Fascinating results as to functional vessels growth in vitro have been reported by Niklason et al. (1999). In this study, non-cross-linked PGA mesh scaffolds, of which the surfaces had been chemically modified by sodium hydroxide, were sewn into tubular forms and positioned into a bioreactor. After ethylene oxide sterilization, the scaffolds were seeded with bovine smooth muscle cells. The bioreactor was filled with nutrient medium. Pulsatile radial stress was applied to the vessels at 165 beats per minute and 5% radial distension (strain). By these culture conditions, fetal development in large mammals should be mimicked and approximated. After eight weeks of culture time in the bioreactor, the silicon tubes were removed and the pulsatile flow directed through the vessels. Then, a suspension containing endothelial cells was injected into the lumen in order to create an anti-thrombogenic surface. Histologies of the resulting vessels and functional assessment demonstrated native analogous characteristics. When tissue engineered bovine vessels were implanted into a xenogenic model, favorable results could be achieved.

In order to transfer the methodology to human application and to overcome lifespan limitations of differentiated cells of elderly patients, recently, Poh et al. (Poh et al. 2005) demonstrated the use of immortalized human cells from donors (age 47–74 years). Therefore, cells were infected with a retrovirus carrying the plasmid pBABE (vector cells) or pBABE Hygro-FLAG-hTERT (hTERT cells). Ectopic expression of the reverse transcriptase (hTERT) in somatic cells has been shown to restore telomerase activity, to stop telomere shortening, and to overcome senescence in some cell types (Jiang et al. 1999). When using these cells for the engineering of vessels, mechanical characteristics did not reach those needed for implantation in vivo. However, this approach demonstrated the feasibility of engineering vessels also for elderly patients by lifespan extension via telomerase expression as a new strategy.

Currently, L'Heureux et al. (2006) reported the feasibility of tissue engineered blood vessels with good mechanical properties based exclusively on cultured human cells without any synthetic material. Human myofibroblasts were isolated from skin biopsies of patients who had undergone vascular bypass surgery and cell sheets were produced. After eight weeks, the fibroblast sheets were wrapped around a Teflon-coated stainless-steel temporary support tube for three revolutions. After 10 weeks of maturation, the individual plies were fused together and formed a homogenous tube. This structure were dehydrated in order to provide a decellularize substrate for endothelialzation. An adventitia was formed by wrapping a living cell sheet around the internal membrane. After a second maturation period, the stainless-steel tube was removed and the inner lumen seeded with autologous saphenous vein-derived endothelial cells. Afterwards, the constructs were exposed to pulsatile flow that increased from 3 ml/ min to 150 ml/min over three-days preconditioning period. These vascular grafts were implanted in an immunosuppressed canine model and monitored for up to 225 days. The grafts were incorporated into the surrounding tissue and demonstrated a smooth lumen as well as formation of *vasa vasorum*. Notably, the internal membrane was largely intact and still acellular. When implanted into primates up to eight weeks, all vessels were patent and demonstrated constant diameters. However, the total engineering time was approx. 28 weeks and might limit the use of these vessels in clinical reality.

With respect to pediatric cardiovascular tissue engineering, recently, tissue engineered small-diameter vessels were fabricated from human umbilical cord-derived progenitor cells (Schmidt et al. 2006). Biodegradable tubular scaffolds with an inner diameter of 0.5 cm were produced by heat application welding technique from a non-woven PGA mesh dip-coated with poly-4-hydroxybutyric acid. After sterilization, human umbilical cord-derived myofibroblasts were seeded onto the inner surfaces of the vascular tubes using fibrin as a cell carrier and cultured in nutrient medium. After 14 days, vessels were endothelialized with differentiated cord blood-derived EPCs, kept at humidified incubator conditions (37°C, 5% CO_2) for additional 24 h and implanted into a pulse duplicator system as described before (Hoerstrup et al. 2001). The pulse duplicator system (Fig. 7.5) consisted of a bioreactor connected via silicon tubing to a medium reservoir. The

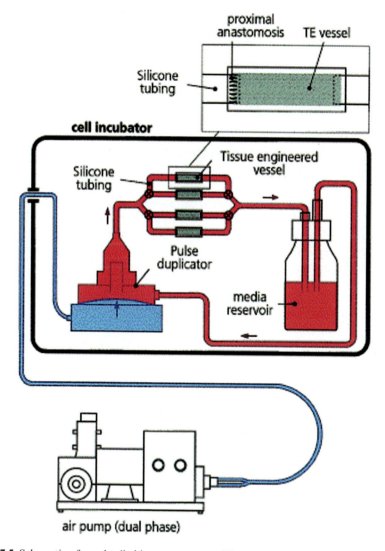

Fig. 7.5 Schemetic of a pulstaile bioreactor system (Hoerstrup SP et al. *Eur J Cardiothorac Surg* 2001)

pulsatile flow of the nutrient medium was generated by periodic expansion of a highly elastic membrane, de- and inflated using an air pump. Tissue engineered vessels were connected in parallel to the medium reservoir-bioreactor circulation. The pulsed flow of nutrient medium was directed through their inner lumen exposing them to flow and shear stress (125 ml/min at 30 mmHg) for seven subsequent days. Analysis of the tissue engineered vessels revealed a native analogous layered architecture with an intact intima (Fig. 7.6), a middle layer containing alpha smooth muscle actin positive cells and an adventitial-like outer layer. Furthermore, a physiological response to TNF-α in differentiated endothelial progenitor cells demonstrating their functional plasticity could be shown. These results indicate that umbilical cord-derived progenitor cells obtainable before or at birth combined with the possibility to manufacture scaffolds of any diameter will allow the fabrication of living autologous tissue engineered blood vessels tailored to the anatomy of the patient. This approach may enable the realization of the tissue engineering concept for pediatric applications.

7.5.5 Tissue Engineering of Large-diameter Vessels – In Vitro and In Vivo Studies

In 1999 Shum-Tim et al. demonstrated the use of large diameter vascular grafts by fabricating autologous aorta in a sheep model. Biodegradable PGA-polyhydroxyalkanoate copolymers were moulded into tubular structures and subsequently seeded with a mixed cell population of smooth muscle cells, fibroblasts, and endothelial cells obtained from carotid arteries of 20 days old

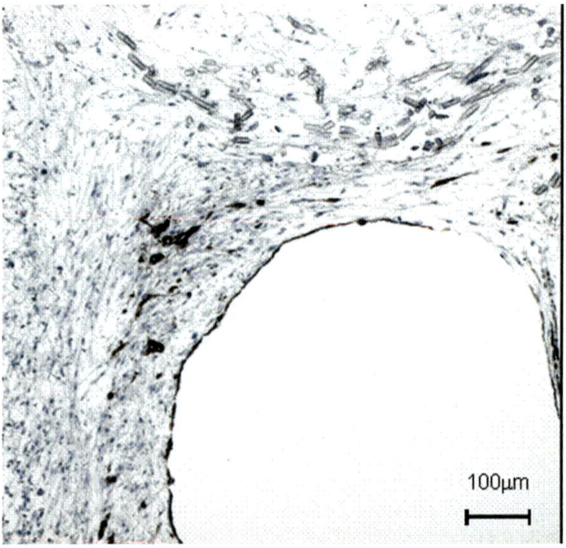

Fig. 7.6
Immunohistochemistry of tissue engineered vascular blood vessel. The CD31 staining demonstrates an endothelial cell lining in the inner surface corresponding to an intima of native vessels and potential vasa vasorum in the middle and outer layers (Schmidt D et al. *Ann Thorac Surg* 2006)

100µm

lambs. After seven days in vitro culture, seeded autologous constructs of 3–4 cm were implanted into the animals ($n = 7$) as an abdominal aortic segment, whereas 4 lambs received an acellular tubular conduit as controls. Assessment of the patency rate by periodic Doppler ultrasound and angiography displayed patent grafts and no aneurysms up to 150 days. In contrast, the four control conduits became occluded at 1, 2, 55, and 101 days, respectively. Furthermore, the mechanical properties and histological features demonstrated characteristics resembling the native aorta.

Recently, Hoerstrup et al. (2006) demonstrated the successful use of autologous tissue engineered large diameter arterial replacements with functional growths potential in a long-term animal study. Pulmonary arteries were engineered based on autologous vascular-derived cells seeded on a PGA/P4HB based scaffold. After implantation in a large animal model (lambs), the animals were monitored up to two years reflecting the full biological growth. During the growth period, an increase of 30% in diameter and of 45% in length of the tissue engineered pulmonary arteries without any indications for aneurysm formation or thrombo-embolic events could be detected. Furthermore, the tissue engineered grafts showed good functional performance in vivo as displayed by echocardiography and computed angiography. Thus, for the first time the systematic evidence of growth in living functional pulmonary arteries engineered from vascular cells in a full growth animal model could been demonstrated.

7.5.6 First Clinical Experiences

To date, only little clinical experience with tissue engineered vascular grafts exists. Particularly, limited long-term experiences have been reported. Some preliminary clinical studies have been initiated and recently, first long-term results have been reported by Shin'oka et al. (2005). In 2000, following the successful reconstruction of a peripheral pulmonary artery in a 4-year-old girl using an autologous tissue engineered construct (Shin'oka et al. 2001), 24 vascular grafts have been implanted as an extra-cavopulmonary connection in children or young adults with an age ranging from 1 to 24 years (Shin'oka et al. 2005). Those vascular grafts were fabricated from copolymers composed of l-lactide and ε-caprolactone (50:50). Furthermore, the tubes were reinforced using PLLA and PGA. After seeding of autologous bone marrow-derived cells that were obtained from the anterior superior spine, the outer layer was sprayed with fibrin glue. The seeded grafts were kept for 2–4 h in the patients own serum in order to enable cell attachment and were then implanted into the patients. A follow-up was performed for 1.3 months and up to 31.6 months. During this time no graft-specific lethal complications were observed. Furthermore, no indications for graft rupture, thromboembolism, aneurysm formation, or calcification was noted. The diameters of the implanted grafts increased about 110 $\pm$ 7% over time. These data demonstrate the successful transfer and application of one vascular tissue engineering process in a clinical scenario and confirm the potential of autologous tissue engineered vascular grafts as an alternative for currently used vascular prosthesis. However, this approach is based on a very slowly degradable scaffold material and a long time-follow-up is necessary to confirm the durability of this approach as well as carefully evaluation of the engineered tissues after implantation.

7.6 Limitations and Future Perspectives

Although many favorable studies have been performed and first clinical data have been reported, the fabrication of an autologous tissue engineered vascular replacements meeting all criteria of an ideal vascular graft remains unmatched. Achievement of the mechanical requirements of their human native counterpart still is a major drawback. Particularly, matching the mechanical properties of large-diameter vessels for the replacement of the aorta that is exposed to high pressure changes represents a high challenge. Excellent mechanical properties of engineered vessels could be obtained only in long in vitro culture times making clinical application almost impossible. However, it should be reconsidered whether an increase of strength should be associated with an elongated culture time as this approach restricts the application to non-urgent patients as particularly patients with an aortic disease often require a therapy within a few weeks. Furthermore, long in vitro culture time might also increase the potential risks for infections and cell de-differentiation. Using bioreactors with a well tuned conditioning protocol providing a physiological in vitro environment that allows tissue maturation and in vitro remodeling might enable the fabrication of strong vascular replacements in a shorter time and therewith even the applications for more urgent patients. Furthermore, an unsolved problem in vascular tissue engineering represents the production of elastin in vitro. The lack of elastin might be a risk for pathological creeping and for aneurysm formation if the in vivo remodeling process and the production of elastin will be delayed. However, future research on new biomaterials may solve this problem by developing an intelligent degradable biomaterial matching the mechanical properties and guiding and stimulating tissue formation as long as the neo-tissue has completed its remodeling process towards its native counterparts.

On the other hand, new cell sources such as progenitor or stem cells representing not-fully differentiated cells with high proliferation capacities might offer new strategies as to cellular engineering resulting in fetal-like tissues with a high potential to remodel, regenerate, and grow. The development of those tissues might be more physiological and nature-like since the undifferentiated cells are running through a development process which might closely match the natural embryonic tissue development process. Currently, several promising cell sources are under investigation. As a new cell source for cardiovascular tissue engineering prenatal progenitor cells have been recently introduced. In this study, progenitor cells were isolated prenatally from chorionic villi and used for the fabrication of cardiovascular replacements. In addition, it could be shown that cell phenotypes do not undergo any changes after cryopreservation. Combined with cell banking technology prenatal progenitor cells might be an attractive cell source for future tissue engineering approaches.

However, differentiation processes of progenitor cells guided by mechanical and/or biochemical stimuli or cell-to-cell and cell-to-scaffold interaction is unexplored. In general, a major obscurity in tissue engineering is the tissue development in vitro and its remodeling processes when once implanted in vivo. Furthermore, the fate that cells undergo in vitro after seeding and during tissue development is still unclear. The major obstacle to elucidate these aspects represents the current use of invasive or destructive methods. Once the tissue is fixed or destroyed only one state in tissue development can be displayed. Monitoring tissue growth variability such as cellularity, phenotypical characteristics, cell differentiation, localized extra cellular matrix composition, microstructure, and cell differentiation, or material and mechanical properties over time is highly desirable. The development of imaging methods, for example the field of molecular imaging, may enable a non-invasive monitoring of tissue growth and development. Particularly, the application of traceable nano-scaled particles incorporated into the cells or attached to their surfaces might open up new strategies and might bring important and valuable information as well as a better understanding of the tissue engineering process and remodeling phenomena.

7.7 Summary

Currently used synthetic vascular prostheses are associated with limitations such as increased risks for graft occlusion and thrombosis. In order to overcome these limitations, the search for alternative materials leading to biocompatible and anti-thrombogenic vascular substitutes has been initiated. Many significant advances as to the engineering of large- and small diameter vascular grafts have been demonstrated in the last years. Several strategies were applied in order to fabricate vascular substitutes matching the properties in functionality and architecture of the natural tissues. Nevertheless, autologous tissue engineered grafts have not been adopted in routine clinical practice and do not yet represent prostheses available off the shelf. Many questions with regard to quality improvements such as mechanical and structural properties as well as functional aspects remain to be answered. However, the requirements of the increasing aging population necessitate an alternative vascular substitute as the currently used vascular replacements are burdened with several disadvantages. Moreover, for pediatric applications living autologous vascular replacements would be crucial in order to avoid recurrent operations increasing the risks for mortality and morbidity. Thus, the challenge of vascular tissue engineering continues and the "ideal" vascular graft might be enabled by future novel technologies stemming from interdisciplinary efforts.

Questions/Exercises

1. Which are the two natural grafts used in bypass applications? What are their tissues of origin, how well do they match the vessel to be replaced, and how do they perform in clinics?
2. What are the main polymeric biomaterials used for synthetic graft manufacture and the principal method of engineering? Describe their biocompatibility and clinical performance in the light of their physico-chemical properties.
3. Describe the ideal properties of a vascular graft and compare them with the features of a typical, commercially available device.
4. Critically analyze two main tissue engineering approaches for small-diameter blood vessel regeneration.
5. Which are the main cell types to be used in the assembling of a tissue engineering construct for vessel replacement?
6. Name the main sources of stem cells for tissue engineering of blood vessels and analyze advantages and disadvantages in their use.
7. What are the biomolecular bases of the blood vessel mechanical properties? Can they be reproduced by synthetic grafts and tissue engineering constructs?
8. Give an example of a bioreactor system for tissue engineering of blood vessel. Describe the biomaterial, cell, and environmental conditions used and the rationale underpinning their choice.
9. What are the principal animal models employed for the in vivo assessment of blood vessel grafts and tissue engineering constructs?
10. Provide an overview of the clinical experience for blood vessel tissue engineering and compare it with the performance of traditional grafts and bypasses.

References

Agrawal CM, Ray RB (2001) Biodegradable polymeric scaffolds for musclsckeletal tissue engineering. J Biomedic Mat Res 55:41–150
Akers DL, Du YH, Kempczinski RF (1993) The effect of carbon coating and porosity on early patency of expanded polytetrafluoroethylene grafts: an experimental study. J Vasc Surg 18:10–15
Alimi Y, Juhan C, Morati N, girard N, Cohen S (1994) Dilatino of woven and knitted aortic prostic grafts: CT scan evaluation. Ann Vasc Surg 8:238–242

Armentano RL, Levenson J, Barra JG, Fischer Cabrera EI, Breitbart GJ, Pichel RH, Simon A (1991) Assessment of smooth muscle contribution to aortic elasticity in conscious dogs. Am J Physiol 260:H1870–1877

Assmus, B, Schachinger, V, Teupe, et al. (2002) Transplantation of progenitor cells and regeneration enhancement in acute myocardial infarction. Circulation 106:3009–3017

Bacourt F (1997) Prospective randomized study of carbon-impregnated polytetrafluoroethylene grafts for below-knee popliteal and distal bypass, results at 3 years. Ann Vasc Surg 11:569–603

Bader A, Schilling T, Teebken OE, Brandes G, Herden T, Steinhoff G, Haverich A (1998) Tissue engineering of heart valves-human endothelial cell seeding of detergent acellularized porcine valves. Eur J Cardiothorac Surg 14:279–284

Baguneid M, Murray D, Salacinski HJ, Fuller B, Hamilton G, Walker M et al. (2004) Shear stress preconditioning and tissue-engineering-based paradigms for generating arterial substitutes. Biotechnol Appl Biochem 39:151–157

Baguneid MS, Fulford PE, Walker MG (1999) Cardiovascular surgery in elderly. J R Coll Surg Edinb 44:216–221

Bank AJ, Wang H, Holte JE, Mullen K, Shammas R, Kubo SH (1996) Contribution of collagen, elastin, and smooth muscle to in vivo human branchial artery wall stress and elastic modulus. Circulation 94:3263–3270

Barra JG, Armentano RL, Levenson J, Fischer Cabrera EI, Pichel RH, Simon A (1993) Assessment of smooth muscle contribution to descending thoracic aortic elastic mechanics in conscious dogs. Circ Res 73:1040–1050

Bassiouny HS, White S, Glagov S, Choi E, Giddens DP, Zarins CK (1992) Anastomotic intimal hyperplasia: mechanical injury or flow induced. J Vasc Surg 15:708–716

Batich C, DePalma D (1992) Materials used in breast implants: silicones and polyurethanes. J Long-term Effects Med Implants 1:255

Bishopric NH, Dousman L, Yao Y-MM, inventors: St Jude Medical Inc, assignee (1999). Matrix substrate for a viable body tissue-derived prosthesis and method for making the same. St Paul, Minn: St Jude Inc.

Blakemore AH, Vorhess AB (1952) The use of tubes constructed from Vinyon "N" cloth in bridging arterial defects. Experimental and clinical. Ann Surg 140:324–334

Cameron A, Davis KB, Green G, Scaff HV (1996) Coronary bypass surgery with internal-thoracic-artery grafts- effects on survival over a 15-year period. N Eng J Med 334:216–219

Clayson KR, Edwards WH, Allen TR, Dale WA (1976) Arm veins for peripheralarterial reconstruction. Arch Surg 111:1276–1280

Courtman DW, Pereira CA, Kashef V, McComb D, Lee JM, Wilson GJ (1994) Development of a pericardial acellualr matrix biomaterial: biochemical and mechanical effects of cell extraction. J Biomed Mater Res 28:655–666

Cziperle DJ, Joyce KA, Tattersall CW, et al. (1992) Albumin impregnated vascular grafts: albumin resorption and tissue reactions. J Cardiovasc Surg 33:407–414

Darling RC, Linton RR (1972) Durability of femoropopliteal reconstructions. Endarterytomy versus vein bypass grafts. Am J Surg 123:472–479

Davids L, Dower T, Zilla P (1999) The lack of healing in conventional vascular grafts. In Zilla P, Greisler HP, editors. Tissue engineering of vascular prosthetic grafts. Austin (Texas): R.G. Landes Company; p. 3–44.

De Mol Van Otterloo JC, Van Bockel JH, Ponfoort ED, et al. (1991) Systemic effects of collagen-impregnated aortoiliac Dacron vasular prostheses on platelet activationand fibrin formation. J Vasc Surg 14:59–66

Deterling RA, Bohnslay SB (1955) An evaluation of synthetic materials and fabrics suitable for blood vessel replacement. Surgery 38:71–89

Deutsch M, Meinhart J, Fischlein T, Preiss P, Zilla P (1999) Clinical autologous in vitro endothelialization of infrainguinal ePTFE grafts in 100 patients: a 9-year experience. Surgery 126:847–855

Devine C, McCollum C (2001) Heparin-bounded Dacron or polytetrafluoroethylene for femoropopliteal bypass: a prospective randomized multicentre trial. J Vasc Surg 33:533–539

Donaladson MC, Mannick JA; Whittermore AD (1991) Femoral-distal bypass with in situ greater saphenous vein. Ann Surg 213:457–465

Eberhart A, Zhang Z, Guidoin R, Laroche G, Guay L, de La Faye D, Batt M, King MW (1999) A new generation of polyurethane vascular prosthese: rara avis or ignis fatuus? J Biomed Mater Res 48:546–458

Eyre D (1984) Cross-linking in collagen and elastin. Annu Rev Biochem 53:717–748

Frazza EJ, Schmitt EE (1991) A new absorbable suture. J Biomed Mater Res Symp 1:43–58

Friedman SG, Lazzaro RS, Spier LN, Moccio c, Tortolani AJ (1995) A prospective randomized comparison of Dacron and polytetrafluoroethylene aortic bifurcation grafts. Surgery 117:7–10

Friedman SG, Lazzoro RS, Spier LN, Moccio C, Tortolani AJ (1994) A prospective randomized comparison of Dacron and polytetrafluoroethylene aortic bifurcation grafts. Surgery 117:7–10

Friedmann DW, Orland PJ, Greco RS (1994) Biomaterials: a historical perspective. In Ralph S. Greco, editor. Implantation biology: the host response and biochemedical devices. Baco raton (Fla): CRC press. p 1–14.

Geesin JC, Darr D, Kaufman R, Murad S, Pinnel SR (1988) Ascorbic acid specifically increases type I and III procollagen messenger RNA levels in human skin fibroblasts. J Invest Dermatol 90:429–424

Girton TS, Oegma TR, Grassl ED, Isenberg BC, Tranquillo RT (2000) Mechanisms of stiffening and strengthening in media-equivalents fabricated using glycation. J Biomed Eng 122:216–223

Glickman MH, Stokes GK, Ross JR, Schuman ED, Sternbergh WC 3rd, Lindberg JS, Money SM, Lorber MI (2001) Multicenter evaluation of a polytetrafluoroethylene vascular access graft in hemodialysis applications. J Vasc Surg 34:465–472

Green RM, Abbott WM, Matsumoto T, Wheeler JR, Miller N, Veith FJ, Money S, Garrett HE (2000) Prosthetic above-knee femoraopopliteal bypass grafting:five-year results of a randomized trial. J Vasc Surg 31:417–425

Greisler HP (1995) Characteristics and healing of vascular grafts. In: Callow AD, Ernst CB, (eds). Vascular surgery: theory and practice. Appleton & Lange, Stamford (Conn), p. 1181–1212

Greisler HP (1990) Interactions at blood/material interface. Ann Vasc Surg 4:98–103

Herring M, Gardener A, Glover J (1978) A single-staged technique for seeding vascular grafts with autologenous endothelium. Surgery 84:498–504

Hess F (1985) History of (micro) vascular surgery and the development of small-caliber blood vessel prothesies (with some notes on patency rates and re-endothelialization). Microsurgery 6:59–69

Hoerstrup SP, Cumming I, Lachat M, Schoen FJ, Jenni R, Leschka S, Neuenschwander S, Schmidt D, Mol A, Gunter C, Gossi M, Genoni M, Zund G (2006) Functional growth in tissue engineered vascular grafts: follow-up in a large animal model. Circulation 114:159–66

Hoerstrup SP, Kadner A, Breymann C, Maurus CF, Guenter CI, Sodian R, Visjager JF, Zund G, Turina MI (2002) Living, autologous pulmonary artery conduits tissue engineered from human umbilical cord cells. Ann Thorac Surg 74:46–52

Hoerstrup SP, Zund G, Sodian R, Schnell AM, Grunenfelder J, Turina MI (2001) Tissue engineering of small caliber vascular grafts. Eur J Cardiothorac Surg 20:164–169

Hufnagel CA (1947) Permanent intubation of the thoracic aorta. Arch Surg 54:382–389

Huynh T, Abraham G, Murray J, Brockbank K, Hagen PO, Sullivan S (1999) Remodeling of an acellular collagen graft into a physiologically responsive neovessel. Nat Biotechnol 17:1083–1086

Jarrell BE, Williams SK, Stokes G, Hubbard FA, Carabasi RA, Koolpe E, Greener D, Pratt K, Moritz MJ, Radomski J (1986) Use of freshly isolated capillary endothelial cells for the immediate establishment of monolayer on a vascular graft at aurgery. Surgery 100:392–399

Jiang XR, Jimenez G, Chang E, Frolkis M, Kusler B, Sage M, Beeche M, Bodnar AG, Wahl GM, Tlsty TD, Chiu CP (1999) Telomerase expression in human somatic cells does not induce changes associated with a transformed phenotype. Nat Genet 21:111–114

Johnson DJ, Robson P, Hew Y, Keeley FW (1995) Decreased elastin synthesis in normal development and in long-term aortic organ and cell cultures is related to rapid and selective destabilization of mRNA for elastin. Circ Res 77:1107–1113

Jonas RA, Ziemer G, Schoen FJ, Britton L, Castaneda AR (1988) A new sealant for knitted Dacron prostheses: minimal cross-linked gelatine. J Cardiovasc Surg 7:414–419

Kadner A, Hoerstrup SP, Tracy J, Breymann C, Maurus CF, Melnitchouk S, Kadner G, Zund G, Turina M (2002) Human umbilical cord cells: a new cell source for cardiovascular tissue engineering. Ann Thorac Surg 74:1422–1428

Kannan RY, Salacinski HJ, Butler PE, Hamilton G, Seifalian AM (2005) Current status of prostethic bypass grafts:a review. J Biomed Mater Res Appl Biomater 74:570–581

Kaushal S, Amiel GE, Guleserian KJ, et al. (2001) Functional small-diameter neovessels created using endothelial progenitor cells expanded in vivo. Nat Med 7:1035–1040

Kawamoto A, Gwon HC, Iwaguro, H (2002) Therapeutic potential of ex vivo expanded endothelial progenitor cells for myocardial ischemia. Circulation 103:634–637

Kim BS, Nikolovski J, Bonadio J, Smiley E, Mooney DJ (1999) Engineered smooth muscle tissues: regulating cell phenotype with the scaffold. Exp Cell Res 251:318–328

Kocher AA, Schuster MD, Szabolcs MJ, Takuma S, Burkhoff D, Wang J, Homma S, Edwards NM, Itescu S (2001) Neovascularisation of ischemic myocardium by human bone-marrow-derived angioblasts prevents cardiomyocyte apoptosis reduces remodeling and improves cardiac function. Nat Med 7:430–436

Kohler TR, Stratton JR, Kirkman TR, Johansen KH, Zierler BK, Clowes AW (1992) Conventional versus high-porosity polytetrafluoroethylene grafts: clinical evaluation. Surgery 112:901–907

Koike N, Fukumura D, Gralla O, Au P, Schechner JS, Jain RK (2004) Tissue engineering: creation of long-lasting blood vessels. Nature 1:138–139

L'Heureux N, Stoclet JC, Auger FA, Lagaud GJL, Germain L, Andriantsitohaina R (2001) A human tissue-engineered vascular model for pharmacological studies of contractile responses, FASEB 15:515–524

L'Heureux N, Paquet S, Labbe R, Germain L, Auger FA (1998) A completely biological tissue-engineered human blood vessel. FASEB J 12:47–56

L'Heureux N, Dussere N, Koenig G, Victor B, Keire P, Wight TN, Chronos NA, Kyles AE, Gregory CR, Hoyt G, Robbins R, McAllister TN (2006) Human tissue-engineered blood vessles for adult arterial revascularization. Nat Med 12:361–365

Lambert AW, Fox AD, Williams DJ, Hoorocks M, Budd JS (1999) Experience with heparin-bounded collagen-coated grafts for infrainguinal bypass. Cardiovasc Surg 7:491–7494

Lantz GC, Badylak SF, Hiles MC, Coffey AC, Geddes LA, Kokini K et al. (1993) Small intestinal submucosa as a vascular graft: a review. J Invest Surg 6:297–310

Leung DY, Glagov S, Mathews MB (1979) Cyclic stretching stimulates synthesis of matrix components by arterial smooth muscle cells in vitro. Science 191:475–477

Lively SA, del campo AA, Nag A, Nichols KB, Coleman C, inventors; LifeCell Corp, assignee. Method processing and preserving collagen-based tissues for transplantation. US patent 5 336 616. Branchburg, NJ; LifeCell Corp; 1994.

Matsumura G, Miyagawa-Tomita S, Shin'oka T, Ikada Y, Kurosawa H (2003) First evidence that bone marrow cells contribute to the construction of tissue-engineered vascular auto-grafts in vivo. Circulation 108:1729–1734

McMahon MP, Faris B, Wolfe BL, Brown KE, Pratt CA, Toselli P, Franzblau C (1985) Aging effects on the elastin composition in the extracellualr matrix of cultured rat aortic smooth muscle cells. In Vitro Cell Dev Biol 21:674–680

Mertens RA, Ohara PJ, Hertzer NR, Krajewski LP, Beven EG (1995) Surgical management of infrainguinal arterial prosthetic graft infections: review of a thirty-five-year experience. J Vasc Surg 21:782–791

Mitchell SL, Niklason LE (2003) Requirements for growing tissue-engineered vascular grafts. Cardiovasc Pathol 23:59–64

Miwa H, Matsuda T (1994) An integrate approach to the design and engineering of hybrid arterial prostheses. J Vasc Surg 19:658–667

Mol A, Bouten CV, Zund G, Gunter CI, Visjager JF, turina MI, Baaijens FP, Hoerstrup SP (2003) The relevance of large strains in functional tissue engineering of heart valves. Thorac Cardiovasc Surg 51:78–83

Mooney DJ, Mazzoni CL, Breuer C, McNamara K, Hern D, Vacanti JP, Langer R (1996) Stabilized polyglycolic acid fibre-based tubes for tissue engineering. Biomaterials 17:115–124

Moore HD (1950) The replacement of blood vessels by polyethylene tubes. Surg Gynecol Obstet 91:593–600

Niklason LE, Abbott W, Gao J, Klagges B, Hirschi KK, Ulubayram K, Conroy N, Jones R, Vasanawala A, Sanzgiri S, Langer R (2001) Morphologic and mechanical characteristics of engineered bovine arteries. J Vasc Surg 33:628–638

Niklason LE, Gao J, Abbott WM, Hirschi KK, Houser S, Marini R, Langer R (1999) Functional arteries grown in vitro. Science 284:489–493

Noishiki Y, Tomizawa Y, Yamane Y, Matsumoto A (1996) Autocrine angiogenic vascular prosthesis with bone marrow transplantation. Nat Med 2:32–34

Pesce M, Orlandi A, Iacchininoto MG, Straino S, Torella AR, Rizzati V, Pompilio G, Scambia G, Caporossi MC (2003) Myoendothelial differentiation of human umbilical cord blood-derived stem cells in ischemic limb tissues. Circ Res 93:e51–62

Poh M, Boyer A, Dahl SLM, Pedrotty D, Banik SSR, McKee JA, Klinger RY, Counter CM, Niklason LE (2005) Blood vessels engineered from human cells. Lancet 365:2122–2224

Post S, Kraus T, Mueller-Reinartz U, Weiss C, Kortmann H, Quentenmeier A, Winkler M, Husfeldt KJ, Allenberg JR (2001) Dacron vs polytetrafluoroethylene grafts for femoro-poplieal bypass: a prospective randomized multicentre trial. Eur J Vasc Endovasc Surg 22:226–231

Prager M, Polterauer P, Böhmig HJ, Wagner O, Fugl A, Kretschmer G, Plohner M, Nanobasvilli J, Huk I (2001) Collagen versus gelatine-coated Dacron versus stretch polytetrafluoroethylene in abdominal aortic bifurcation graft surgery: results of a seven-year prospective, randomized multicenter trial. Surgery 130:408–414

Raja SG, Haider Z, Ahmad M, Zaman H (2004) Saphenous vein grafts: to use or not to use? Heart Lung Circ 13:403–409

Ratcliffe A (2000) Tissue engineering of vascular grafts. Matrix Biol 19:658–667

Rayton JK, Harris ED (1979) Induction of lysyl oxidase with copper. J Biol Chem 254:621–626

Rosenberg N, Martinez A, Sawyer PN, Wesolowski SA, Postlethwait RW, Dillon ML Jr. (1966) Tanned collagen arterial prosthesis of bovine carotid origin in man. Preliminary studies of enzyme-treated heterografts. Ann Surg 164:247–256

Salacinski HJ, Punshon G, Krijgsman B, Hamilton G, Seifalian AM (2001) A hybrid compliant vascular graft seeded with microvascular endothelial cells extracted from human omentum. Artif Organs 25:974–982

Schmidt D, Asmis LM, Odermatt B, Kelm J, Breymann C, Gössi M, Genoni M, Zund G, Hoerstrup SP (2006) Engineered living blood vessels: functional endothelia generated from human umbilical cord-derived progenitors. Ann Thorac Surg 82:1465–1471

Schmidt D, Breymann C, Weber A, Guenter CI, Neuenschwander S, Zund G, Turina M, Hoerstrup SP (2004) Umbilical cord blood derived endothelial progenitor cells for tissue engineering of Vascular Grafts. Ann Thorac Surg 78:2094–98

Schmidt D, Mol A, Breymann C, Achermann J, Odermatt B, Gössi M, Neuenschwander S, Prêtre R, Genoni M, Zund G, Hoerstrup SP (2006) Living autologous heart valve engineered from human prenatally harvested progenitors. Circulation 114:I125–131

Schmidt SP, Hunter TJ, Sharp WV, Malindzak GS, Evancho MM (1984) Endothelial cells seeded four millimetre Dacron vascular grafts. J Vasc Surg 1:434–441

Scott SM, Gaddy LR, Sahmel R, Hoffmann H (1987) A collagen coated vascular prosthesis. J Cardiovasc Surg (Torino) 28:498–504

Seifalian AM, Tiwari A, Hamilton G, Salacinski HJ (2002) Improving the clinical patency of prosthetic vascular and coronary bypass grafts: the role of seeding and tissue engineering. Artif Organs 26:307–320

Shanley CJ, Gharee-Kermani M, Sarkar R, Welling TH, Kriegel A, Ford JW, Stanley JC, Phan SH (1997) Transforming growth factor-$\beta 1$ increases lysyl oxidase enzyme activity and mRNA in rat aortic smooth muscle cells. J Vasc Surg 25:446–452

Shin'oka T, Imai Y, Ikada Y (2001) Transplantation of a tissue-engineered pulmonary artery. N Eng J Med 344:532–533

Shin'oka T, Matsumura G, Hibino N, Naito Y, Watanabe M, Konuma T, Sakamoto T, Nagatsu M, Kurosawa H (2005) Midterm clinical result of tissue-engineered vascular autografts seeded with autologous bone marrow cells. J Thorac Cardiovasc Surg 129:1330–1338

Shirota T, Hongbing HE, Yasui H, Matsuda T (2003) Human endothelial progenitor cell-seeded hybrid graft: proliferative and antithrombogenic potentials in vitro and fabrication processing. Tissue Eng 9:127–136

Shirota T, Yasui H, Shimokawa H, Matsuda T (2003) Fabrication of endothelial progenitor cell (EPC)-seeded intravascular stent devices and in vitro endothelialization on hybrid vascular tissue. Biomaterial 24:2295–302

Shum-Tim D, Stock U Hrkach J, Shinoka T, Lien J, Moses MA, Stamp A, Taylor G, Moran AM, Landis W, Langer R, Vacanti JP, Mayer JE Jr. (1999) Tissue engineering of autologous aorta using a new biodegradable polymer. Ann Thorac Surg 68:2298–2305

Simon P, Kasimir MT, Seebacher G et al. (2003) Early failure of the tissue engineered porcine heart valve SYNERGRAFT in pediatric patients. Eur J Cardiothorac Surg 23:1002–1006

Sipehia R, Martucci G, Lipscombe J (1996) Enhanced attachment and growth of human endothelial cells derived from umbilical veins on ammonia plasma modified surfaces of PTFE and ePTFE synthetic vascular graft biomaterials. Artif Cells Blood Substit Immobil Biotechnol 24:51–63

Tanzi MC, Fare S, Petrini P (2000) In vitro stability of polyether and polycarbonate urethanes. J Biomater Appl 14:325–348

Taylor LM Jr, Edwards JM; Porter JM (1990) Present status of reserved vein bypass grafting: five-years results of a modern series. J Vasc Surg 11:193–205

Teebeken OE, Haverich A (2002) Tissue engineering of small diameter vascular grafts. Eur J Vasc Endovasc Surg 12:59–64

Tsuchida H, Cameron BL, Marcus CS, Wilson SE (1992) Modifies polytetrafluoroethylene: indium 111-labeled platelet deposition on carbon.lined and high porosity polytetrafluoroethylene grafts. J Vasc Surg 16:643–649

Varga J, Rosenbloom J, Jimenz SA (1987) Transforming growth factor β (TGFβ) causes a persistent increase in steady-state amounts of type I and type III collagen and fibrinectin mRNAs in normal human dermal fibroblasts. Biochem J 247:597–604

Veith FJ, Moss CM, Sprayregen S, Montefusco C (1979) Preoperative saphenous venography in arterial reconstructive surgery of the lower extremity. Surgery 85:253–256

Vorhess ABJ, Jaretzki AI, Blakemore AH (1952) The use of tubes constructed from Vinyon "N" cloth in bridging arterial defects. A preliminary report. Ann Surg 135:332–336

Vroman L, Adams AL (1969) Identification of rapid changes at plasma-solid interfaces. J Biomed Mater Res 3:43–67

Wake MC, Gupta PK, Mikos AG (1996) Fabrication of pliable biodegradable polymer foams to engineer soft tissues. Cell Transplant 5:465–473

Watanabe M, Shin'oka T, Tohyama S, Hibino N, Konuma T, Matsumura G, Kosaka Y, Ishida T, Imai Y, Yamakawa M, Ikada Y, Morita S (2001) Tissue-engineered vascular autograft: inferior vena cava replacement in a dog model. Tissue Eng 7:429–439

Weinberg CB and Bell E (1986) A blood vessel model constructed from collagen and cultured vascular cells. Science 231:397–400

Williams SK, Rose DG, Jarrell BE (1994) Microvascular endothelial cell seeding of ePTFE vascular grafts: improved patency and stability of the cellular lining. J Biomed Mat Res 28:203–212

Xue L, Greisler HP (2000) Blood vessels. In: Lanza RP, Langer R, Vacanti J, editors. Principle of tissue engineering 2nd edition.Academic Press, San Diego (Calif), p. 427–446

Ye Q, Zund G, Jockenhoevel S, Hoerstrup SP, Schoeberlein A, Grunenfelder J, Turina M (2000) Tissue engineering in cardiovascular surgery: new approach to develop completely human autologous tissue. Eur J Cardiothorac Surg 17:449–454

Zhang Z, Marois Y, Guidoin RG, Bull P, Marois M, How T, Laroche G, King MW (1997) Vasculgraft polyurethane arterial prosthesis as femorao-peroneal bypass in humans: pathological, structural and chemical analyses of four excised grafts. Biomaterials 18:113–124

Chapter 8
Pancreas Biology, Pathology, and Tissue Engineering

Wendy M. MacFarlane, Adrian J. Bone, and Moira Harrison

Contents

8.1 Introduction 261
8.2 Pancreas Biology 263
    8.2.1 Islets of Langerhans 263
8.3 Pathology 264
    8.3.1 Pathogenesis and β-cell Dysfunction in Diabetes 264
    8.3.2 Pathogenesis of Diabetes and Mechanisms of β-Cell Death 265
8.4 Tissue Engineering: Generating Replacement β–cells 269
    8.4.1 Islet Transplantation.. 269
    8.4.2 Engineering Replacement β-cells 270
    8.4.3 Adult Stem Cells: Pancreas.................................... 271
    8.4.4 Liver Cells 273
    8.4.5 Bone Marrow 274
    8.4.6 Embryonic Stem Cells....................................... 275
8.5 Summary 277
Questions/Exercises... 277
References 278

8.1 Introduction

Diabetes is a chronic condition that is increasing at an alarming rate. In 1985, there were approximately 30 million people with diabetes worldwide. Today the International Diabetes Federation estimates that more than 230 million people around the world have diabetes and this total is expected to rise to 350 million by 2025. Not only is diabetes expensive to treat but also the complications of the condition account for 5–10% of total healthcare spending globally. The complications are numerous and devastating, causing blindness, cardiovascular disease, limb amputations, kidney failure, and many more. Until the discovery of insulin in 1921 by Dr. Frederick Grant Banting, Charles Best, James Bertram

W.M. MacFarlane (✉)
School of Pharmacy & Biomolecular Sciences, University of Brighton, Cockcroft
Building Lewes Road, Brighton BN2 4GJ, UK
e-mail: w.m.macfarlane@brighton.ac.uk

M. Santin (ed.), *Strategies in Regenerative Medicine*,
DOI 10.1007/978-0-387-74660-9_8, © Springer Science+Business Media, LLC 2009

Collip, and Dr. James Macleod, diabetes was a death sentence. Although the discovery of insulin lead to people surviving where previously they would have died, it was certainly not a cure. It was only after many years of treatment that it became evident that while people could be kept alive on insulin, it was difficult if not impossible to achieve the tight glucose control required to avoid the risk of complications.

There are two main types of diabetes, Type 1 and Type 2. Type 1 diabetes Mellitus is generally associated with a sudden onset of symptoms and is characterized by the cellular-mediated autoimmune destruction of the insulin containing β-cells within the pancreatic islets of Langerhans and commonly develops during childhood (The Expert Committee on the Diagnosis and Classification of Diabetes Mellitus, 2003; Atkinson and McClaren, 1994). Type 1 diabetes (T1D) develops when the body is unable to maintain circulating glucose concentrations within the normal physiological range due to a reduction or complete absence of circulating insulin. A person with T1D will suffer acute symptoms including weight loss, extreme thirst, increased urination, and if left untreated, will result in coma and death. The only current treatment for T1D is daily administration of insulin either via injection or subcutaneous insulin infusion (pump therapy). Treatment of diabetes with insulin carries an associated risk of hypoglycaemia. People who experience this sudden drop in blood glucose may find it not only causes physical symptoms but also the very fear of it occurring may impact negatively on their lives and their diabetes control.

Type 2 diabetes (T2D) is often, mistakenly, considered to be a milder form of diabetes as it does not present with the same acute symptoms. Rather than an absolute deficiency of insulin, there may be high circulating levels of insulin but cellular insensitivity combined with signaling defects prevent the cells from taking up glucose. Despite the less obvious onset, poor glycemic control in T2D has exactly the same consequences in terms of complications as Type 1. Often because the onset usually occurs later in life and presents with symptoms such as fatigue, thirst, or recurrent infections, it may go undiagnosed for a many years. People with T2D may be diagnosed due to the complications they present with rather than an initial diagnosis of diabetes. There are many therapies available for the treatment of T2D but compliance is poor and these, like insulin, are also not a cure.

Although the discovery of insulin and improved drug therapies have saved many lives, there is currently no cure for diabetes. The ultimate goal of diabetes research is to address the loss of β-cells in an enduring and physiologically accurate way so that a cure can be affected and complications prevented. Tissue engineering offers the potential to do this. In order to understand what is required of an "engineered tissue", it is first important to have a good working knowledge of what is required from such a tissue. The following sections will describe the pancreas biology and pathology before addressing the strategies for tissue engineering.

8.2 Pancreas Biology

The pancreas is an organ with an uninspiring appearance, but a complex and vital role to play in the digestion of most foods (www.pancreas.org). Not only does it produce the enzymes (exocrine function) necessary to break up and digest food, its endocrine function tightly regulates blood glucose, through the production of insulin and glucagon predominantly. Regarded by the ancient Greeks as a delicacy, the word pancreas literally means "all meat". It is a small elongated organ with a head, body, and tail and is located in the abdomen in a position posterior to the stomach and in close association with the duodenum. The pancreatic head abuts the second part of the duodenum while the tail extends towards the spleen.

As previously mentioned, the pancreas has two major functions, exocrine and endocrine. The exocrine tissue, or acinar, of the pancreas produces digestive enzymes and the endocrine (islets of Langerhans) produces the hormones that regulate normal glucose homeostasis. The exocrine function is extremely important to digestion but in terms of diabetes, it is the endocrine function that is of interest, as it is the loss or dysregulation of β-cell function that results in this condition. Digestive enzymes such as amylase, lipase, and trypsin are secreted via the pancreatic duct which in turn joins the common bile duct and consequently into the duodenum. Initially inactive pancreatic enzymes become activated only upon arrival in the digestive tract.

The endocrine, or insulin-producing, blood-glucose regulatory function of the pancreas is the main target for research and tissue engineering strategies. There are other diseases that can affect the pancreas such as cystic fibrosis and pancreatitis, but it is diabetes that presents a significant and rapidly increasing global health threat.

8.2.1 Islets of Langerhans

There are four distinct cell types contained within the islets of Langerhans, alpha (α) cells, beta (β) cells, pancreatic polypeptide (PP), and somatostatin-producing cells. Unlike the digestive enzymes, hormones secreted by the pancreatic islets are released into the bloodstream via a highly vascularized network. The three hormones produced are insulin (β-cells), glucagon (α-cells), and somatostatin. These three hormones work together to regulate blood glucose levels which are normally between 4 and 7 mM/l in people without diabetes. The β-cells release insulin to lower blood glucose levels by facilitating the removal of glucose from the blood to cells. Conversely, the α-cells produce glucagon which can elevate blood glucose by mobilising glycogen stores in the liver when blood glucose levels drop below acceptable levels. Somatostatin can prevent both insulin and glucagon from being released.

8.3 Pathology

8.3.1 Pathogenesis and β-cell Dysfunction in Diabetes

Failure to maintain a sufficient mass of properly functioning β-cells is a common determining factor in the development of both T1D and T2D. T1D is an autoimmune disease, which occurs as a result of the targeted destruction of the insulin producing-cells (Fig. 8.1a–d). T2D, on the other hand, is classically described as a condition characterized by β-cell dysfunction and tissue insulin resistance. There is, however, a growing body of evidence to suggest that the progressive decline in β-cell mass, possibly involving common distinct death effector pathways (Mathis and Benoist 2006, Butler et al. 2003, Kolb and Mandrup-Poulsen 2005), may be a major causative factor that connects T1D and T2D. This suggestion, not only serves to challenge our current thinking that categorizes human diabetes into two separate disease states, demonstrating distinct pathologies, but it could also have major therapeutic implications.

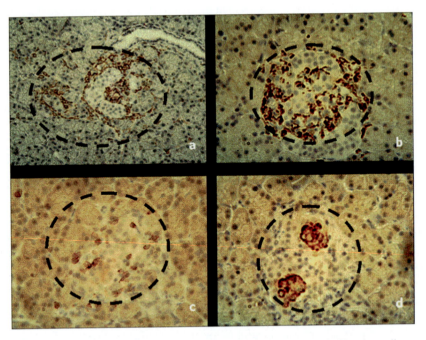

Fig. 8.1 Human pancreatic islet cross sections. (**a–c**) staining for the infiltrating cells commonly found during the development of Type 1 diabetes [(**a**) macrophages, (**b**) CD2, and (**c**) CD8 positiveT-cells]. (**d**) illustrates the residual beta cells remaining following autoimmune destruction mediated by infiltrating T-cells and macrophages. Type 1 diabetes usually becomes symptomatic following destruction of 90% of the beta cells

In overt T1D, the treatment options are to either give exogenous insulin or replace the lost β-cells. Whilst in T2D, and perhaps early T1D, preventive therapies may be able to sustain β-cell function and arrest the progressive decline in β-cell mass leading to secondary failure and insulin dependence. Moreover, recent research trials involving the incretin hormones have suggested the additional exciting possibility of restoring β-cell mass in patients with established diabetes, by enhancing new islet formation (Ratner 2005, Perfetti 2004).

It is important, therefore, to understand the pathophysiological mechanisms of disease progression in both T1D and T2D diabetes if we are to achieve the common therapeutic goal of effective β-cell mass preservation or regeneration. Thus, in this part of the chapter will briefly discuss pathogenesis of diabetes, mechanisms of β-cell death, and replacement and subsequent disease progression.

8.3.2 Pathogenesis of Diabetes and Mechanisms of β-Cell Death

8.3.2.1 Type 1 diabetes

Type 1 diabetes [T1D] is an autoimmune disease that culminates in a marked loss of β-cells following widespread pancreatic infiltration (Tisch and McDevitt 1996, Boitard et al. 1997). The precise nature of the inflammatory infiltrate [insulitis] in humans has not been well documented but studies in animal models of T1D have proved extremely useful in characterizing the infiltrating immune cells (Fig. 8.1a–c).

Thus, studies in the non-obese diabetic [NOD] mouse, BB rat, and the Lewis.*iddm* rat identified various cells of the immune system [macrophages, natural killer [NK] cells, and T lymphocytes] within the infiltrated islets (Ridgway 2003, Lally et al. 2001, Lenzen et al. 2001) at different stages of the disease process. However, these models are based on the selective breeding of animals having an elevated incidence of T1D and, as such; they are unlikely to accurately mirror the disease process in man. A limited number of studies have been undertaken to examine disease progression in humans but this is extremely difficult to achieve for two main reasons. Firstly, methods to study the disease process in situ are not available, and secondly, access to the pancreas at the time of diagnosis of diabetes is not routinely possible in humans. An early investigation did, however, successfully analyze islet antigen expression in an archive collection of pancreas samples, recovered *postmortem* from human patients who died within approximately one year of diagnosis of T1D (Foulis et al. 1997).

Two principal models of disease pathogenesis have been proposed on the basis of animal studies and in vitro work with isolated human islets. The first proposes that diabetes is caused by infiltration of islets by various cells of the immune system (including macrophages and CD4$^+$ T cells) culminating in the recruitment and activation of CD8$^+$ T cells. The CD8$^+$ T cells then destroy the β-cells by the targeted release of cytolytic molecules (such as perforin and granzymes) with resultant β-cell death either by apoptosis or necrosis (Yoon et al. 1998, Kreuwel et al. 1999, Kagi et al. 1996, Kay et al. 2000). The alternative model suggests that islets are infiltrated by inflammatory cells which release a cocktail of cytokines [e.g., interleukin-1β (IL-1β), interferon-γ (IFN-γ), tumor necrosis factor-α (TNF-α)] which in turn induce the expression of a "death receptor", Fas, on the β-cell surface (Stassi et al. 1997, Moriwaki et al. 1999, Chervonsky et al. 1997, Loweth et al. 1998, Amrani et al. 2000, Suarez-Pinzon et al. 1999). Since some infiltrating cells express the ligand for this receptor (FasL), engagement of FasL with the newly expressed receptor consequently leads to β-cell apoptosis. In addition, it has also been proposed that the β-cells themselves may express FasL, either spontaneously or in response to specific stimuli (Loweth et al. 1998, Loweth et al. 2000, Hanke 2000) and that up-regulation of Fas expression generates conditions under which they can undergo apoptosis by binding of FasL expressed on adjacent endocrine cells.

These models differ both in terms of the likely mechanism of β-cell death [necrosis or apoptosis] and in the roles played by the immune cells. It is also unclear why specific subsets of immune cells are recruited to the islets during the initiation of the pathogenic process and whether different mechanisms are involved in recruitment and in subsequent islet infiltration. Additionally, it is not known whether the variable degree of insulitis seen in different regions of the diabetic pancreas (Gepts 1965) reflects an inefficient process of insulitis or whether it results from active resistance to immune attack occurring at the level of the β-cells. Although it might seem straightforward to distinguish between these various possibilities, in practice this has not been possible because of the lack of availability of human pancreas samples in which the critical processes can be studied. Indeed, no studies in human pancreas have been able to define the factors responsible for immune cell recruitment to inflamed islets and only two groups have attempted to examine the immunopathology of human T1D in situ (Stassi et al. 1997, Moriwaki et al. 1999, Imagawa et al. 2001). Moreover, the findings of these studies differ in a number of key respects. In particular, Stassi et al. found Fas expression on islet β-cells but failed to detect it on α-cells, whereas Moriwaki et al. (1999) reported Fas expression at this location. This is an important issue since evidence of the presence of Fas on α-cells would not be consistent with the selectivity of β-cell destruction. A further critical issue that remains unresolved is the mode of β-cell death in T1D. It has been shown very convincingly that human islets can undergo apoptosis when incubated under appropriate conditions in vitro (Loweth et al. 1998, Maedler et al. 2001, Paraskevas et al. 2000, Eizirik and Darville 2002) and it has been reported that both early and sustained apoptotic β-cell death may play an important role in T1D (Meier et al. 2005, Meier et al. 2006). However, while strong evidence to support this view has been obtained in animal models (Lally et al. 2001), the limited studies undertaken with pancreata from human patients have yielded conflicting data. Thus, Moriwaki et al. (1999) failed to detect any apoptotic β-cells in pancreatic biopsy specimens taken from T1D patients whereas Stassi et al. (1997) observed a remarkable number of such cells in autopsy material.

8.3.2.2 Type 2 diabetes

T2D is a heterogeneous disorder exhibiting a natural history of deteriorating glucose control, increasing insulin resistance and subsequent β-cell failure. Although T2D is characterized both by peripheral insulin resistance and loss of β-cell function, for the purposes of this chapter we will focus exclusively on aspects of β-cell mass regulation.

The natural history of T2D demonstrates a progressive decline in β-cell function with resultant deterioration in glycemic control and development of impaired glucose tolerance. The loss of glucose-stimulated insulin secretion in T2D patients has been shown to be associated with a reduction in β-cell mass (Butler et al. 2003, Kahn 2003). The questions still remain, however, whether or not the loss of functional β-cell mass in T2D patients is due to increased rates of β-cell destruction, reduced β-cell regeneration or a combination of both. Evidence from studies in animal models and isolated human islets have highlighted a number of key potential mediators capable of producing the progressive decline in β-cell mass seen in T2D. Thus, the sustained periods of hyperglycemia and associated dyslipidemia observed in T2D increase β-cell apoptosis (Kaiser et al. 2003, Donath and Halban 2004). The precise mechanisms leading to β-cell glucotoxicity and lipotoxicity have been the subject of considerable investigation and are described in some detail in a recent review (Cnop et al. 2005). In addition, the characteristic deposits of islet amyloid seen in T2D (Kahn et al. 1999) have been shown to similarly cause apoptosis of β-cells (Janson et al. 1999). A compensatory increase in β-cell mass has been observed in response to deteriorating metabolic control in human obesity (Kloppel et al. 1985). Whether such an adaptive response exists in T2D is not known. However, it has only recently been confirmed that β-cell mass is decreased in T2D as a result of increased rates of apoptosis (Butler et al. 2003). The same study showed that the reduction in β-cell mass was due to the rates of apoptosis exceeding the normal capacity for β-cell renewal and replication. This important observation, combined with very recent evidence for islet regeneration in long-standing disease (Meier et al. 2005), opens up exciting new therapeutic approaches to the treatment of diabetes.

The new class of antidiabetic drugs, the thiazolidinediones [TZDs] and the incretin hormone Glucagon Like Peptide-1 [GLP-1] have the potential to stimulate β-cell function and mass and prevent disease progression (Finegood et al. 2001, Uwaifo and Ratner 2005).

8.3.2.3 Thiazolidinediones (TZDs)

This new class of antidiabetic drug produces a gradual improvement in glycemic control by amplifying insulin action in adipose tissue, skeletal muscle and liver.

However, it is becoming apparent, that TZDs may also act directly on the insulin producing pancreatic β-cell. Studies of rosiglitazone treatment in animal models of T2D have shown a lowering of blood glucose and an amelioration of disease associated with an increase in functional β-cell mass (Finegood et al. 2001). The combined ability of TZDs to enhance insulin action and stimulate β-cell function/mass indicates a potential for delaying disease progression in T2D. This suggestion gained support from the TRIPOD randomized, controlled trial which showed a reduction in diabetes incidence in women with gestational diabetes treated with Troglitazone (Buchanan et al. 2002). A subsequent small cohort study using rosiglitazone and pioglitazone was able to confirm a similar disease reduction in patients with impaired glucose tolerance (Durbin 2004).

8.3.2.4 GLP-1 and Dipeptidyl Peptidase IV (DPP-IV) Inhibitors

The incretin hormone GLP-1 is secreted from the intestine in response to food digestion and potentiates glucose-stimulated insulin secretion and insulin biosynthesis. GLP-1 is a product of the preproglucagon gene, which is cleaved by prohormone convertase 1 to its truncated, biologically active form, which in turn, is rapidly inactivated (half-life <2 min) by the enzyme DDP-IV. GLP-1 has been shown to exert a number of β-cell specific antidiabetic effects, which suggest it could have considerable therapeutic potential for treating T2D (Drucker 1998). Within the β-cell, GLP-1 acts on specific G-protein receptors to potentiate nutrient-induced insulin release, and in vivo is able to normalize patterns of insulin secretion in patients with T2D (Juhl et al. 2000). Furthermore, the properties of GLP-1 have been shown to extend to a modulatory role in β–cell mass regulation via direct effects on β-cell growth and apoptosis (Farilla et al. 2002). Taken together, the physiological properties of GLP-1 make it a very attractive starting point for the production of a new class of antidiabetic agents. In order to achieve this aim, much attention within the pharmaceutical industry has focused on developing therapeutic approaches involving non-degradable GLP-1 analogues and DPP-IV inhibitors.

Early studies were performed with exendin-4, a naturally occurring long-acting analogue of GLP-1 isolated from the venom of the Gila monster, which was able to improve glucose tolerance and increase the β–cell mass in animal models of T2D (Uwaifo and Ratner 2005, Buchanan et al. 2002). There are currently a number of GLP-1 analogues or "incretin mimetics" under clinical investigation and going through the clinical trials process [reviewed in (Uwaifo and Ratner 2005)].

The alternative approach to promote the antidiabetic effects of GLP-1 has been to develop inhibitors of DPP-IV in order to protect and stabilize endogenous incretin hormone activity. Although not as advanced in their development as the GLP-1 analogues, several DPP-IV inhibitors have shown promising antidiabetic effects in preliminary clinical trials [reviewed in Mest and Mentlein (2005)].

Thus, clinical evidence is accumulating that these new treatment agents are able to improve glycemic control in T2D and it is interesting to speculate on the therapeutic potential of their islet growth promoting properties. The hope would be that by maintaining the mass of functioning β-cells above the critical threshold level, it may be possible to prevent the progression to secondary β-cell failure and insulin/β-cell replacement therapy.

8.4 Tissue Engineering: Generating Replacement β–cells

Recent improvements in the success rate of islet transplantation have provided critical proof of principle that restoring the functional β-cell mass in patients with T1D can successfully restore normal control of whole body glucose metabolism. The development of the Edmonton Protocol in 2000 (Shapiro et al. 2000) revolutionized the therapeutic options for patients with T1D, allowing isolated islets of Langerhans from donated pancreases to safely replace lost β-cell function in these patients and establishing tissue replacement therapy as an attractive and viable alternative to life long glucose monitoring and daily insulin injections. However, the availability of islet transplantation is severely limited by crippling shortages in donated pancreatic tissue. As a result, a worldwide research effort is underway to engineer new sources of replacement insulin-producing cells. Here we review the latest development in the search for a viable and limitless supply of replacement insulin-producing cells for the treatment of diabetes.

8.4.1 Islet Transplantation

Unlike whole organ pancreas transplants, islet transplantation therapy involves the separation of the insulin-producing islets of Langerhans from the pancreases of cadaveric organ donors by a complex digestion and purification process (Shapiro et al. 2000). Purified islets are then transplanted to the liver of T1D recipients through delivery into the hepatic portal vein where the islets become engrafted in the vessel wall. The advent of the Edmonton protocol, a glucocorticoid-free immunosuppressive regime combined with an optimal islet engraftment mass, has revolutionized transplantation therapy for T1D. The Edmonton protocol has now been implemented in over 50 transplantation centers world wide, with over 500 islet transplants now successfully completed (Lakey et al. 2006). However, limitations on the availability of human pancreatic tissue mean that this figure represents less than 0.5% of T1D patients that could benefit from this treatment. To compound this problem, several donors are currently required to generate sufficient isolated islets for a single recipient to remain free from insulin injections. This restricted availability, combined with a lifelong requirement for immunosuppression, has led to a worldwide research effort aimed at engineering new and abundant sources of insulin-producing cells.

Islet transplantation therapy has provided critical proof of principle that restoration of functional β-cell mass in patients with T1D can restore normal glucose metabolism, freeing patients from the burden of insulin injections and delaying, or even preventing the onset of diabetic complications (Paty et al. 2006). It has been shown that transplanted islets can sustain functionality up to five years. Longer functionality seems to be inhibited by a substantial cell loss during the early phases of implantation. It has been shown that during the first two weeks after transplantation over 70% of the pancreatic islets lose viability (Shapiro et al. 2000). This is due to the slow vascularization process occurring in the portal vein wall a process leading to cell hypoxia. In addition, the immune reaction to the transplanted islet graft induces the synthesis and secretion of pro-inflammatory cytokines by the immune cells which trigger a programmed cell death (apoptosis) of the β-cells. The use of immunosuppressants has also been shown to induce islet viability loss. This significant cell loss is partially compensated with the transplantation of over $>10,000$ islets/kg of body with obvious problems related to donor shortage. Furthermore, the β-cells of the pancreas are extraordinary cells, unique in human physiology, and to engineer fully functional replacement β-cells is extremely challenging. The β-cells of the pancreas are exquisitely sensitive to changes in blood glucose concentration. Only β-cells transcribe, translate, process, and release insulin in response to these changes in glucose concentration, a set of abilities dependent on the expression of a unique complement of glucose-sensing genes within these specialized cells of the pancreas. The challenges facing scientists aiming to engineer replacement β-cells are, therefore, extensive.

8.4.2 Engineering Replacement β-cells

The successful engineering of replacement β-cells requires the generation of human cells that, as closely as possible, mimic normal pancreatic β-cell physiology and metabolism. Firstly, replacement β-cells must accurately sense fluctuations in glucose concentrations within the physiological range, transcribing, translating, processing, and secreting insulin in a manner comparable to a normal β-cell. This means that the generated cells must express the key range of glucose transporters, glycolytic enzymes, cell signaling molecules, and transcription factor proteins which form the glucose-sensing mechanism of normal β-cells (MacDonald et al. 2005). Secondly, replacement β-cells must be capable of expansion to large enough populations to be clinically relevant. It is estimated that between 2×10^8 and 5×10^8 β-cells are required for a single patient to receive enough insulin-producing cells to remain free from insulin injections (Kim et al. 2006). Thirdly, having expanded the cell population to the required extent, replacement β-cell populations must have a controlled growth and proliferation status, to avoid the potential danger of tumor formation upon transplantation. And lastly, these replacement β-cells must be safe. This means that the cells must be generated and maintained under GMP (good microbiological practice) conditions, and, ideally, must avoid provoking an increased immune response from recipients. Clearly, these desired characteristics represent an extraordinary challenge to researchers in this field. However, many advanced and innovative approaches are currently being utilized in the quest for limitless sources of these ideal replacement β-cells.

The generation of insulin-producing cells has been attempted now from a broad range of starting cell types, through both differentiation and transdifferentiation techniques in vitro. Cells of the liver and bone marrow have been utilized from adult donors. The extraordinary potential of embryonic stem cells has also been the source of huge international research interest and exciting progress has been made using these pluripotent immature cells as a starting point. However, we begin with the utilization of the donated pancreatic organs themselves, as increasing evidence suggests that the pancreas itself may hold the potential to generate new and more abundant sources of insulin-producing cells.

8.4.3 Adult Stem Cells: Pancreas

Like many other tissues in the body, the pancreas undergoes a constant process of tissue turnover and renewal. It is estimated that the lifespan of an adult β-cell is approximately 50 days, after which time β-cells are replaced by neogenesis from existing pancreatic precursor cells within the adult pancreas (Bouwens et al. 1997). Much debate has surrounded the potential existence of adult stem cells within the pancreas. The existence of such cells would represent an exciting potential source of many more β-cells. So, where do new β-cells within the pancreas arise from? Several lines of evidence suggest that the β-cells themselves replicate (Dor 2006). However, the replication rate of β-cells in vitro is too low to be useful in terms of the generation of our ideal replacement β-cells. Of greater interest in our quest for an abundant supply of insulin-producing cells are recent studies highlighting the potential of two other cell types within the islet: the acinar cells and the cells of the pancreatic duct.

Observation of the co-expression of β-cell transcription factors such as PDX1 in cells with pancreatic ductal markers such as cytokeratin 19 (CK19) initially led to the hypothesis that increases in β-cell mass occur through the expansion of the ductal cells within the pancreas (Bouwens and Rooman 2005, Dor 2006). Elegant lineage tracing experiments have recently demonstrated that within the developing and the adult pancreas, new β-cells can arise from the pancreatic ducts. Studies of the process of islet neogenesis have also shown that new islets "bud" from the pancreatic duct following pancreatectomy (partial removal of the pancreatic tissue), or in response to neogenic stimuli (including partial duct occlusion, cellophane wrapping) (Bouwens and De Blay 1996). In vitro studies on transformed ductal cell lines have shown that treatment with growth factors activin A and betacellulin can stimulate the transdifferentiation of ductal cells to an insulin-producing β-cell phenotype (Deramaudt et al. 2006). The same effect can be produced by the over-expression of key developmental transcription factor PDX1, a homeodomain protein required for normal β-cell development during embryogenesis (Deramaudt et al. 2006). Hence, the ductal cells have become an area of intense study. Several recent reports describe the process of β-cell neogenesis as a combination of firstly the de-differentiation of the adult pancreatic cells towards a ductal phenotype, with subsequent re-differentiation to a β-cell phenotype (Ogata et al. 2004). Clearly, the plasticity of adult pancreatic cells plays a fundamental role in the neogenesis of β-cells in the adult pancreas, and the accurate mimicking of this process in vitro represents an important goal.

The many varied animal models of islet neogenesis and β-cell regeneration currently being studied point to the ductal cells of the pancreas as the main source of precursor cells able to transdifferentiate to β-cells and hence provide compensatory β-cell expansion under conditions of β-cell stress in vivo. However, whilst it is clear that ductal cells may play this role in neogenic islets, several other cell types are currently under investigation for their ability to generate functional β-cells in vitro. These include the dominant cell type in the adult pancreas, the pancreatic acinar cells.

Pancreatic acinar cells form the dominant mass of the adult pancreas, surrounding individual islets. Like the ductal cells, these cells are currently discarded during the islet isolation procedure. However, like the ductal cells, recent studies indicate that the acinar cells retain the capacity to transdifferentiate towards an insulin-producing β-cell phenotype. Again, over-expression of developmental transcription factors such as PDX1 and Pax4 has been shown to provoke the transdifferentiation of acinar to β-cells (Gao et al. 2005). Recent optimized protocols have utilized a combination of EGF (epidermal growth factor) and LIF (leukaemia inhibitory factor) to stimulate the formation of β-cells from acinar starting material (Rooman et al. 2000). In vitro and in vivo testing of the derived β-cell populations suggests that these cells are functional glucose-sensing insulin-secreting cells, an exciting step forward.

The potential of acinar and ductal cells of the pancreas lies not only in the seemingly accurate formation of functional β-cells from these cell types, but in their availability from human donors. At present, islet isolation using the optimized Edmonton protocol utilizes only 5% of the adult pancreas (the volume of the pancreas made up of the islets of Langerhans), with the other 95% of the pancreas being discarded. This process is inefficient on two levels. Firstly, loss of islets during the digestion and purification procedure means that combined islets from several different donated organs are required to generate enough functional islets for transplantation. Secondly, however, the 95% of each organ currently discarded includes the acinar and the ductal cells of the pancreas, potential sources of millions of new β-cells. If optimized transdifferentiation protocols can be applied to the acinar and ductal cells of the pancreas, it may be possible to generate sufficient β-cells from a single organ to not only successfully transplant a single recipient, but to transplant several recipients from a single donor.

Optimization of the transdifferentiation protocols described above also creates the potential for cells from a patient with T1D to be utilized in the generation of replacement β-cells for that individual. As described earlier, during the autoimmune destruction of the pancreas in T1D the β-cells are specifically destroyed. The acinar and the ductal cells of the pancreas, however, remain largely intact. Optimized transdifferentation protocols may allow these cells to be harvested, transdifferentiated towards a β-cell phenotype, and then returned to the patient. This is an attractive methodology, as it may avoid the current requirement for immunosuppression, since the cells would not be recognized as "non-self" by the recipient's immune system. The avoidance of life long immunosuppression would be a hugely significant step forward in the improvement of cell replacement therapy for the treatment of T1D.

Cells of the pancreas represent an exciting resource in the quest for new replacement β-cells, with these new cells being generated through the process of transdifferentiation. Transdifferentiation is a coordinated alteration in gene expression and cellular phenotype which allows the conversion of one adult cell type into another adult cell type. The amazing plasticity of adult cells is only now becoming clear. Significant literature now suggests that the generation of functional β-cells from adult precursor cells is not limited to the ductal cells, or merely to cells of the pancreas.

8.4.4 Liver Cells

In general, cells which are capable of interconversion/transdifferentiation arise from adjacent regions of the developing embryo (Baeyens et al. 2006). Hence, some of the early regulatory events and stimuli driving the formation of these cells are common to both adult cell types. Such is the case for the liver and pancreas, which are both derived from the endoderm of the developing embryo, and which share some adult characteristics, such as glucose sensitivity and pathways of protein processing and secretion. Transdifferentiation of liver to pancreas (and of pancreas to liver) is a phenomenon naturally observed in rare human disease states, as well as in well characterized animal models (Shen et al. 2003). Hepatic *foci* have been observed in the pancreas of humans, rodents, and simian models. Indeed, the appearance of hepatic cells in the adult pancreas has been studied for over 20 years. Whether the hepatic cells observed in the pancreas in vivo arise directly from transdifferentiation of adult pancreatic cells remains controversial; however, in vitro studies with transformed cell lines have shown that the transdifferentiation from liver to β-cell and from β-cell to liver is possible, given the right experimental conditions, or the transgenic expression of key transcription factors such as PDX1 (Koizunmi et al. 2004). Several of the hepatic nuclear factor family of transcription factors have also been shown to be critically important to the transdifferentiation of liver to pancreas and *vice versa*. These include HNF1α, FoxA2, HNF4α, and HNF6 (Rao et al. 1996, Dabeva et al. 1995). Although transdifferentiation of liver to pancreas occurs in disease states such as hepatic cirrhosis (Wolf et al. 1990), and in several animal models and in vitro systems (Shen et al. 2003), the role of these transdifferentiation events in normal β-cell turnover and adult pancreatic β-cell function has not been established. It seems unlikely that these events have a major role to play in the normal processes of β-cell regeneration and neogenesis; however, these processes may allow the very effective generation of replacement β-cells in vitro. Optimized transdifferentiation protocols have now been utilized to generate glucose-responsive insulin producing cells from liver cells both in vitro and in vivo (Fodor et al. 2006). The lifespan and viability of these newly generated cells remains under investigation. Hence, although these initial data are very exciting, the potential of liver cells in the generation of replacement β-cells still requires extensive study.

8.4.5 Bone Marrow

Studies investigating the potential use of adult tissues such as pancreatic ductal/ acinar cells and liver cells as the source of clinically relevant transplant material are clearly still at an early stage. However, there are some adult stem cells that have been in safe and effective clinical use for several years: bone marrow derived stem cells. Stem cells isolated from donated human bone marrow samples have been successfully utilized in the effective treatment of leukaemia, multiple myeloma, lymphoma, and severe combined immunodeficiency (SCID) (Tuch 2006). The potential of bone marrow derived stem cells to generate replacement β-cells is only now becoming clear.

Early studies in animal models reported the formation of new insulin-producing cells in the pancreas from transplanted bone marrow stem cells (Ianus et al. 2003). However, doubt was cast on the authenticity of these reported trandifferentiation events when later studies revealed that the new "β-cells" had emerged through the fusion of bone marrow derived stem cells with exisiting rodent β-cells (Lechner et al. 2004). Despite these early set backs, however, bone marrow derived stem cells have recently been reported to form insulin producing cells in vitro in isolation (Sun et al. 2006). Improved trans-differentiation studies driven by developmental transcription factors have reported the formation of insulin-secreting cells. Chromatin remodelling factors have also been shown in vitro to drive the formation of glucose-sensing insulin secreting cells from bone marrow in vitro (Tayaramma et al. 2006). The resultant cells formed islet-like clusters, reported authentic stimulus-secretion coupling, and ultrastructure studies revealed the presence of insulin secretory granules comparable to those observed in normal isolated β-cells. Hence, work on bone marrow derived cells is at an exciting stage. Given the relative simplicity of bone marrow donation, these cells are an extremely attractive starting point for the generation of new and abundant sources of replacement β-cells in vitro.

8.4.6 Embryonic Stem Cells

The tissue engineering approaches described above detail the potential of adult cells to be utilized as a source of replacement β-cells. Each of these cell types requires the donation of primary material and hence these approaches are, like islet transplants, naturally limited through the availability of donor material. The ultimate goal of diabetes research laboratories around the world is to develop a limitless source of β-cells, allowing the effective "one size fits all" treatment of patients with T1D. This may be possible if we can harness the potential of embryonic stem cells (ESC).

ESC are derived from the isolated inner cell mass of blastocysts grown in the laboratory to day 6-8 of gestation (Stojkovic et al. 2004a, b). These pluripotent cells represent an extraordinary opportunity for many research groups focussed on tissue engineering, since, under carefully optimized growth conditions, these cells retain the capacity to develop into all three germ layers and every individual adult cell type (Stojkovic et al. 2004a, b). Capable of proliferation and self-renewal in culture, the extraordinary plasticity of these cells represents a unique opportunity to produce a potentially limitless source of replacement β-cells. Many novel and innovative protocols have been utilized to attempt to drive the differentiation of human ESC towards a β-cell phenotype.

Early differentiation studies utilized the over-expression of key developmental transcription factors (PDX1, Pax4, Isl1) to drive the differentiation process forward (Blyszczuk et al. 2003). Later studies began to develop more advanced protocols, initially based on the formation firstly of embryoid bodies (through changes in serum growth conditions) and then through a selection step narrowing the cell population based on the expression of the neural marker protein nestin (Lumelsky et al. 2001). However, the selection of nestin positive cells remains a controversial methodology, as later studies suggested that the resultant cell populations were more closely linked to the neural lineage than to the pancreatic lineage (Blyszczuk et al. 2004). These early studies had several limitations. In addition to concern regarding the potential neural origin of the cells, the derived cell populations had very low numbers of insulin-positive cells (typically less than 10% insulin positive). In addition, insulin-positive cells were often reported to lack glucose-regulated insulin secretion, a key requirement of an effective replacement β-cell.

The inefficient nature of the early ESC differentiation protocols was most likely the result of in vitro protocols which bypassed key stages of normal β-cell development. Our understanding of the developmental cues and transcription factor cascades driving the formation of the β-cells during normal pancreas development has increased remarkably over the last five years (Servitja and Ferrer 2004). Pancreatic development begins in the endodermal region of the primitive foregut and is regulated by the hierarchical expression of a specific complement of transcription factor proteins, reviewed Edlund (1998). Beginning in cells expressing the homeodomain transcription factor PDX1 (pancreatic/duodenal homeobox 1) and repressing sonic hedgehog, both endocrine and exocrine cells of the adult pancreas arise from endodermal cells expressing PDX1 (Jonsson et al. 1994), and Ptf1a (Kawaguchi et al. 2002). Proliferation of progenitor cells is stimulated by the fibroblast growth factor family of proteins (Elghazi et al. 2002) and greatly influenced by the production of specific inductive stimuli from the surrounding developing mesenchyme (Scharfmann 2000). Transient expression of the paired domain transcription factor Pax4 in ngn3 expressing cells drives the formation of the β-cells, with the final β-cell transcription factor complement being driven by PDX1, Pax6, Nkx2.2, and Nkx6.1 (Habener et al. 2005). This is the point at which β-cell expansion begins.

It has become clear from recent reports that to generate an "insulin-positive cell" from an ES cell is relatively straightforward. However, to generate a fully functional, glucose-sensing, insulin secreting β-cell is extraordinarily difficult. This is because a true β-cell (and an ideal replacement β-cell) must have all of the characteristics and all of the key proteins, and attributes described at the beginning of this section. To achieve this during normal embryogenesis, β-cell development is a highly regulated and elegantly coordinated process. To truly capitalize on the potential of ESC, the current challenge is to accurately reproduce all of the developmental cues and hierarchical transcription factor expression patterns seen during normal development.

Several studies have reported attempts to mimic the normal developmental process (Rolletschek et al. 2006). These culminated in a recent study which reported the formation of insulin-producing cells from ESC in culture (D'Amour et al. 2006). However, unlike previous studies, this ground-breaking report utilized novel optimized protocols which orchestrated the differentiation of human ESC to endocrine hormone expressing cells through five recognizable endodermal intermediates, an extraordinary feat. Initial Oct 4 and Nanog positive ESC were driven to form accurate definitive endoderm, endocrine precursors showing expression of PDX1, and then the temporal and spatial expression of the key developmental transcription factors Ngn3, Pax4, Nkx6.1, and Nkx2.2 was reproduced. This ingenius protocol resulted in the formation of all of the major hormone-producing cell types of the pancreas from human ESC in culture and stands as an elegant demonstration of how closely the normal development process may be mimicked given the thorough optimization and testing of differentiation protocols. The insulin-producing cells generated showed an insulin secretion pattern similar to that observed in isolated foetal islets. These cells represented between 3-12% of the cell populations generated.

Hence, even with this optimized protocol, the cells generated were not capable of glucose-sensitive insulin-secretion that mimicked normal adult β-cells. However, the accurate generation of foetal β-cells using a protocol which so closely mimics normal embryonic development still represents an extraordinary step forward, and exciting proof of principle that if researchers can accurately mimic the normal developmental process in its entirety, the potential of ESC can be realized. Realization of this potential may allow the generation of a limitless source of replacement β-cells which could be utilized in transplantation therapy for the treatment of T1D.

8.5 Summary

The development of the Edmonton Protocol has provided vital proof of principle that tissue replacement therapy can be used effectively for the treatment of T1D. However, to truly capitalize on these advances, new and abundant sources of insulin-producing cells are required. As discussed above, cells of the pancreas, liver, and bone marrow are being extensively studied for their potential to generate accurate glucose-sensing, insulin-producing cells. There has been exciting progress made with all of these cell types and accumulating evidence suggests that the plasticity of donated adult cells may allow the generation of new insulin-producing cells. However, the potential of embryonic stem cells perhaps holds the greatest promise, since these cells could provide a limitless supply of replacement β-cells. Successful generation of accurate adult β-cells from ESC remains the ultimate goal of many research laboratories world wide. Extraordinary progress has been made in accurately mimicking the normal developmental pathways leading to the emergence of a β-cell phenotype. Significant challenges remain; however, since the generation of β-cells from ESC does not require tissue or organ donation, successful optimization of this differentiation process is a tantalizing prospect, since it may lead to a limitless source of β-cells available to all who might benefit from β-cell replacement therapy.

Questions/Exercises

1. Describe the cell mechanisms leading to T1D.
2. Highlight the molecular mechanisms underpinning T2D and its possible links with T1D.
3. Describe the main histological features of the pancreas and their pathological alteration following T1D insurgence.
4. Highlight the main biochemical and cellular prerequisites in pancreas tissue engineering.

5. Present the main clinical procedure for pancreatic islet transplantation and alternative suggested protocols; a critical discussion of their advantages and disadvantages.
6. List the factors leading to β cell apoptosis following pancreatic islet transplantation and explore the main strategies adopted to prevent it.
7. Speculate on the physico-chemical and biocompatibility properties of an ideal biomaterial scaffold for pancreas tissue engineering.
8. Analyze the molecular and cell biology links between diabetes and other pathologies such as cardiovascular diseases.
9. Discuss in a critical manner the main protocols used for in vitro pancreatic islet regeneration.
10. Analyze the current limitations in using embryonic stem cells for pancreatic islet regeneration and indicate the future perspectives.

References

Amrani A et al. (2000) IL-1, IL-1 and IFN- mark -cells for Fas-dependent destruction by diabetogenic CD4+ T lymphocytes. J Clin Invest 105:459–468

Atkinson MA, Maclaren NK (1994) Mechanisms of Disease – The Pathogenesis of Insulin-Dependent Diabetes-Mellitus. New England Journal of Medicine 331: 1428–1436

Baeyens L et al. (2006) Ngn3 expression during postnatal in vitro beta cell neogenesis induced by the JAK/STAT pathway. Cell Death Differ 13:1892–1899

Blyszczuk P et al. (2003) Expression of Pax4 in embryonic stem cells promotes differentiation of nestin-positive progenitor and insulin-producing cells. Proc Natl Acad Sci USA 100:998–1003

Blyszczuk P et al. (2004) Embryonic stem cells differentiate into insulin-producing cells without selection of nestin-expressing cells. Int J Dev Biol 2004 48:1095–1104

Boitard C et al. (1997) Immune mechanisms leading to type 1 insulin-dependent diabetes mellitus. Horm Res 48:58–63

Bouwens L, De Blay E, (1996) Islet morphogenesis and stem cell markers in rat pancreas. J Histochem Cytochem 44:947–951

Bouwens L, Rooman I (2005) Regulation of pancreatic beta-cell mass. Physiol Rev 85:1255–1270

Bouwens L, Lu GW, De Krijger R (1997) Proliferation and differentiation in the human fetal endocrine pancreas. Diabetologia 40:398–404

Buchanan TA, Xiang AH, Peters RK (2002) Preservation of pancreatic B-cell function and prevention of type 2 diabetes by pharmacological treatment of insulin resistance in high-risk Hispanic women. Diabetes 51: 2796–2803

Butler AE et al. (2003) Beta-cell deficit and increased beta-cell apoptosis in humans with Type 2 diabetes. Diabetes 52: 102–110

Chervonsky AV et al. (1997) The role of Fas in autoimmune diabetes. Cell 89:17–24

Cnop M et al. (2005) Mechanisms of pancreatic beta-cell death in Type 1 and Type 2 diabetes. Diabetes, Suppl. 2:S97–S107

Dabeva MD, Hurston E, Sharitz DA (1995) Transcription factor and liver-specific mRNA expression in facultative epithelial progenitor cells of liver and pancreas. Am J Pathol 147:1633–1648

D'Amour KA et al. (2006) Production of pancreatic hormone-expressing endocrine cells from human embryonic stem cells. Nat Biotechnol 24:1392–1401

Deramaudt TB et al. (2006) The PDX1 Homeodomain transcription factor negatively regulates the pancreatic ductal cell-specific keratin 19 promoter. J Biol Chem 281:38385–38395

Donath MY, Halban PA (2004) Decreased B-cell mass in diabetes: significance, mechanisms and therapeutic implications. Diabetologia 47:581–589

Dor Y (2006) β-Cell proliferation is the major source of new pancreatic beta cells. Nat Clin Pract Endocrinol Metab 2:242–243

Drucker DJ (1998) Glucagon-like peptides. Diabetes 47:159–169

Durbin RJ (2004) Thiazolidinedione therapy in the prevention/delay of type 2 diabetes in patients with impaired glucose tolerance and insulin resistance. Diabetes Obes Metab 6:280–285

Edlund H (1998) Transcribing pancreas. Diabetes 47:1817–1823

Eizirik DL, Darville MI (2002) Beta cell apoptosis and defense mechanisms: lessons from type 1 diabetes. Diabetes 50:S64–S69

Elghazi L et al. (2002) Role for FGFR2IIIb-mediated signals in controlling pancreatic endocrine progenitor cell proliferation. Proc Natl Acad Sci USA 99:3884–3889

Farilla L, Lui H, Bertolotto C (2002) Glucagon-like peptide-1 promotes islet cell growth and inhibits apoptosis in Zucker diabetic rats. Endocrinology 143:4397–4408

Finegood DT, McArthur MD, Kojwang D (2001) Beta-cell mass dynamics in Zucker diabetic fatty rats. Rosiglitazone prevents the rise in net cell death. Diabetes 50:1021–1029

Fodor A et al. (2006) Adult rat liver cells transdifferentiated with lentiviral IPF1 vectors reverse diabetes in mice: an ex vivo gene therapy approach. Diabetologia [Epub ahead of print].

Fodor et al (2007) Jan;50(1):121–30. Epub 2006 Nov 28

Foulis AK et al. (1997) A search for evidence of viral infection in pancreases of newly diagnosed patients with IDDM. Diabetologia 401:53–56

Gao R et al. (2005) In vitro neogenesis of human islets reflects the plasticity of differentiated human pancreatic cells. Diabetologia 48:2296–2304

Gepts W (1965) Pathologic anatomy of the pancreas in juvenile diabetes mellitus. Diabetes 14:619–633

Habener JF, Kemp DM, Thomas MK (2005) Minireview: transcriptional regulation in pancreatic development. Endocrinology 146.

Hanke J (2000) Apoptosis and occurrence of Bcl-2, Bak, Bax, Fas and FasL in the developing and adult endocrine pancreas. Anat Embryol 202:303–312

Ianus A et al. (2003) In vivo derivation of glucose-competent pancreatic endocrine cells from bone marrow without evidence of cell fusion. J Clin Invest 111:843–850

Imagawa A et al. (2001) Pancreatic biopsy as a procedure for detecting in situ autoimmune phenomena in type 1 diabetes: close correlation between serological markers and histological evidence of cellular autoimmunity. Diabetes 50:1269–1273

Janson J et al. (1999) The mechanism of islet amyloid polypeptide toxicity is membrane disruption by intermediate-sized toxic amyloid particles. Diabetes 48:491–498

Jonsson J et al. (1994) Insulin-promoter-factor 1 is required for pancreas development in mice. Nature 371: 606–609

Juhl CB, Schmitz O, Pincus S (2000) Short-term treatment with GLP-1 increases pulsatile insulin secretion in type II diabetes with no effect on orderliness. Diabetologia 43:583–588

Kagi D et al. (1996) Development of insulitis without diabetes in transgenic mice lacking perforin-dependent toxicity. J Exp Med 183:2143–2152

Kahn SE (2003) The relative contributions of insulin resistance and B-cell dysfunction to the pathophysiology of type 2 diabetes. Diabetologia 46:3–19

Kahn SE, Andrikopoulos S, Verchere CB (1999) Islet amyloid: A long recognized but under-appreciated pathological feature of Type 2 diabetes. Diabetes 48:241–253

Kaiser N, Leibowitz G, Nesher R (2003) Glucotoxicity and B-cell failure in Type 2 diabetes mellitus. J Pediatr Endocrinol Metab 16:5–22

Kawaguchi Y, et al. (2002) The role of the transcriptional regulator Ptf1a in converting intestinal to pancreatic progenitors. Nat Genet 3. 2:128–134

Kay TWH, et al. (2000) The beta cell in autoimmune diabetes: many mechanisms and pathways of loss. Trends Endocrinol Metab 11:11–15

Kim S et al. (2006) Analysis of donor and isolation-related factors of successful isolation of human islet of langerhans from human cadaveric donors. Transplantation Proceedings 37:3402–3403

Kloppel G et al. (1985) Islet pathology and the pathogenesis of Type 2 diabetes revisted. Surv Synth Pathol Res 4:110–125

Koizumi M, et al. (2004) Hepatic regeneration and enforced PDX-1 expression accelerate transdifferentiation in liver. Surgery 136:449–457

Kolb H, Mandrup-Poulsen T (2005) An immune origin of type 2 diabetes. Diabetologia 48:1038–1050

Kreuwel HT et al. (1999) Comparing the relative role of perforin/granzyme vs Fas/Fas ligand cytotoxic pathways in CD8+ T cell-mediated insulin dependent diabetes mellitus. J Immunol 163:4355–4351

Lakey JR, Mirbolooki M, Shapiro AM (2006) Current status of clinical islet cell transplantation. Methods Mol Biol. 2006:47–104

Lally FJ, Ratcliffe H, Bone AJ (2001) Apoptosis and disease progression in the spontaneously diabetic BB/S rat. Diabetologia 44:320–324

Lechner A et al. (2004) No evidence for significant transdifferentiation of bone marrow into pancreatic beta-cells in vivo. Diabetes 53:616–623

Lenzen S et al. (2001) The LEW.1AR1/Ztm-iddm rat: a new model of spontaneous insulin-dependent diabetes mellitus. Diabetologia 44:1189–1196

Loweth AC, et al. (1998) Human islets of Langerhans express Fas-Ligand and undergo apoptosis in response to interleukin-1ß and Fas ligation. Diabetes 47:727–732

Loweth AC, et al. (2000) Dissociation between Fas expression and induction of apoptosis in human islets of Langerhans. Diabetes, Obesity and Metabol 2:57–60

Lumelsky N, et al. (2001) Differentiation of embryonic stem cells to insulin-secreting structures similar to pancreatic islets. Science 292:1389–1394

MacDonald PE, Joseph JW, Rorsman P (2005) Glucose-sensing mechanisms in pancreatic beta-cells. Philos Trans R Soc Lond B Biol Sci 360:2211–2225

Maedler K et al. (2001) Glucose induces ß-cell apoptosis via upregulation of the Fas receptor in human islets. Diabetes 50:1683–1690

Mathis D, Benoist C (2006) Beta cell death during progression to diabetes. Nature 414:792–798

Meier JJ, et al. (2005) Sustained beta cell apoptosis in patients with long-standing type 1 diabetes: indirect evidence for islet regeneration? Diabetologia 48:2221–2228

Meier JJ, et al. (2006) Increased vulnerability of newly forming beta cells to cytokine-induced cell death. Diabetologia 49:83–89

Mest H, Mentlein R (2005) Dipeptidyl peptidase inhibitors as new drugs for the treatment of type 2 diabetes. Diabetologia 48:616–620

Moriwaki M, et al. (1999) Fas and Fas ligand expression in inflamed islets in pancreas sections of patients with recent-onset Type 1 diabetes mellitus. Diabetologia 42:1332–1340

Ogata T, et al. (2004) Reversal of streptozotocin-induced hyperglycemia by transplantation of pseudoislets consisting of beta cells derived from ductal cells. Endocr J. 51:381–386

Paraskevas S, et al. (2000) Cell loss in isolated human islets occurs by apoptosis. Pancreas 20:270–276

Paty BW, et al. (2006) Assessment of glycemic control after islet transplantation using the continuous glucose monitor in insulin-independent versus insulin-requiring type 1 diabetes subjects. Diabetes Technol Ther. 8:165–173

Perfetti R (2004) The role of GLP-1 in the regulation of islet cell mass. Medscape Diabetes & Endocrinology 6(2)

Rao MS, et al. (1996) Expression of transcription factors and stem cell factor precedes hepatocyte differentiation in rat pancreas. Gene Expr 6:15–22

Ratner RE (2005) Therapeutic role of incretin mimetics. Medscape Diabetes & Endocrinology 7(1)

Report of the Expert Committee on the Diagnosis and Classification of Diabetes Mellitus (2003) Diabetes Care 26:S5–20S

Ridgway WM (2003) The non obese diabetic (NOD) mouse: a unique model for understanding the interaction between genetics and T cell response. Rev Endocr Metab Disord 4:263–269

Rolletschek A, Kania G, Wobus AM (2006) Generation of pancreatic insulin-producing cells from embryonic stem cells – 'Proof of principle', but questions still unanswered. Diabetologia 49:2541–2545

Rooman I, et al. (2000) Modulation of rat pancreatic acinoductal transdifferentiation and expression of PDX-1 in vitro. Diabetologia 43:907–914

Scharfmann R (2000) Control of early development of the pancreas in rodents and humans: implications of signals from the mesenchyme. Diabetologia 43:1083–1092

Servitja JM, Ferrer J (2004) Transcriptional networks controlling pancreatic development and beta cell function. Diabetologia 47:597–613

Shapiro AM, et al. (2000) Islet transplantation in seven patients with type 1 diabetes mellitus using a glucocorticoid-free immunosuppressive regimen. N Engl J Med 343:230–238

Shen CN et al. (2003) Transdifferentiation of pancreas to liver. Mech Dev 120:107–116

Stassi G et al. (1997) Nitric oxide primes pancreatic -cells for Fas-mediated destruction in insulin-dependent diabetes mellitus. J Exp Med 186:1193–1200

Stojkovic M et al. (2004a) Derivation of human embryonic stem cells from day-8 blastocysts recovered after three-step in vitro culture. Stem Cells 22:790–797

Stojkovic M et al. (2004b) Derivation, growth and applications of human embryonic stem cells Reproduction 128:259–267

Suarez-Pinzon W et al. (1999) ß-cell destruction in NOD mice correlates with Fas (CD95) expression on ß-cells and proinflammatory cytokine expression in islets. Diabetes 48:21–28

Sun J et al. (2006) Expression of Pdx-1 in bone marrow mesenchymal stem cells promotes differentiation of islet-like cells in vitro. Sci China C Life Sci 49:480–489

Tayaramma T et al. (2006) Chromatin-remodeling factors allow differentiation of bone marrow cells into insulin-producing cells. Stem Cells 24:2858–2867

Tisch R, McDevitt H (1996) Insulin-dependent diabetes mellitus. Cell 85:291–297

Tuch BE (2006) Stem cells–a clinical update. Aust Fam Physician 35:719–721

Uwaifo GI, Ratner RE (2005) Novel pharmacologic agents for type 2 diabetes. Endocrinol Metab Clin North Am. 34:155–197

Wolf HK et al. (1990) Exocrine pancreatic tissue in human liver: a metaplastic process? Am J Surg Pathol 14:590–595

Yoon JW, Jun H-S, Santamaria P (1998) Cellular and molecular mechanisms for the initiation and progression of -cell dstruction resulting from the collaboration between macrophages and T cells. Autoimmunity 27:109–122

Chapter 9
The Holy Grail of Hepatocyte Culturing and Therapeutic Use

Andreas K. Nussler, Natascha C. Nussler, Vera Merk, Marc Brulport, Wiebke Schormann, Ping Yao, and Jan G. Hengstler

Contents

9.1 Introduction 284
9.2 Legal Aspects of Cell Therapy in Future 284
9.3 Hepatocyte Isolation and Culture 285
9.4 Clinical Hepatocyte Transplantation 289
9.5 Hepatocytes Used in Drug Development – Metabolism and Toxicology 291
9.6 Hepatocytes Generated by Stem Cell Technology 292
     9.6.1 Can Human Hepatocytes Be Produced by Stem Cell Technology? An Overview 292
     9.6.2 Transdifferentiation of Extrahepatic Cells to Hepatocytes? First Evidence Pointing towards Bone Marrow Cells 294
     9.6.3 Cell Fusion or Transdifferentiation? A Question of Potential Clinical Relevance 295
     9.6.4 The Discovery of "Fusogenic Cells" in Mouse Bone Marrow 296
     9.6.5 Elegant Reporter Studies Differentiate Between Cell Fusion and Real Differentiation 297
     9.6.6 Fate of Human Stem and Precursor Cells in Livers of Mice 300
     9.6.7 Transdifferentiation of Extrahepatic Human Stem Cells to Hepatocytes In Vitro? 302
     9.6.8 Quantitative Analyses Comparing Stem Cell Derived Cells with Primary Hepatocytes 306
     9.6.9 5-Years Perspective 308
Questions/Exercises 312
Ten Key Publications 313
References 314

A.K. Nussler (✉)
Technical University Munich, Department of Traumatology, Ismaningerstr. 11, 81829 Munich, Germany
e-mail: andreas.nuessler@gmail.com

M. Santin (ed.), *Strategies in Regenerative Medicine*,
DOI 10.1007/978-0-387-74660-9_9, © Springer Science+Business Media, LLC 2009

9.1 Introduction

In the past 30 years, orthotopic liver transplantation became a routine procedure, to cure irreversible liver damage (Starzl and Lakkis 2006). However, the number of organs needed is much higher than the number of organs available, and thus since the late 1980s and early 1990s alternative therapeutic strategies have been envisaged (www.nhsdirect.nhs.uk/articles/article.aspx?articleId = 233). These strategies include bioreactors with and without hepatocytes (Rifai et al. 2005, Gerlach 2006, Mitzner et al. 2006) hepatocytes of various species and cell lines (Enosawa et al. 2001, Donini et al. 2001, Gerlach et al. 2003) , split liver and living related liver transplantation (Otte et al. 1990, Strong 2006), as well as xenogenic organ transplantation (Starzl et al. 1993).

Bioreactor systems with primary human hepatocytes have in common that they are exclusively applied in acute liver damage, underlining their use only as a temporary bridge until a graft is available. The main problem with these systems is cell availability and the stand-by costs of the bioreactor, resulting in the phenomenon that either the bioreactor or the patient is available. Various systems are actually under investigation at different clinical phases (Pless et al. 2006, Chiu and Fan 2006, Gerlach 2006).

Since Strom et al. (1982) reported that human hepatocytes isolated from split livers or livers declined for transplantation could be transplanted into rodents, many clinicians and researchers have tried to establish successful programs aiming to treat patients with isolated liver cells for various liver diseases of monogenetic, autoimmune, systemic, or infective origin (for review see Nussler et al. 2006). Although experimental progress has been achieved in the past years, a routine application of hepatocyte transplantation must still be established. Apart from the isolation process, many new legal and ethical aspects have to be considered in the European Community and will come into place in the upcoming years (Directive 2001/83/EC and Regulation (EC) No 726/2004. COM567final). Since the isolation of human liver cells is also a problem of availability, a number of investigators have recently focused on the establishment of stem or precursor cells to overcome organ and cell deficits. We here present the current literature on hepatocyte isolation and the possibility of using isolated human hepatocytes and alternative hepatocyte-like cells for the treatment of liver diseases and to identify areas of future pre-clinical and clinical research.

9.2 Legal Aspects of Cell Therapy in Future

The Directive 2001/83/EC and Regulation (EC) No 726/2004 of European Community is a discussion paper which aims to improve the patients' safety and access to advanced therapies by increasing the research, development and authorization of gene therapy, somatic cell therapy, and tissue engineered products. This regulation plans to lay down a legal and regulatory framework for the evaluation and authorization of these advanced therapy products. As a consequence, these products will be subject to a centralized EU approval procedure in the near future (See Section 16.4).

Apart from a number of important regulations, it states that human tissue and cell based products should be founded on the philosophy of voluntary and unpaid donation and that clinical trials on advanced therapy medicinal products should be conducted in accordance with the overarching principles and the ethical requirements of each member state *and* the European law. Actually this directive is still under discussion; however, the commission's proposal foresees exempting hospitals from complying with the new centralized regulation. This fact would create a two-tier system: companies would be obliged to take their products to the European Medical Agency, while hospitals and research institutions could routinely produce products without being obliged to follow European law.

However, so far all clinical trial applications using cell therapy products will still fall under each member state's national legislation.

9.3 Hepatocyte Isolation and Culture

The liver histological and physiological features are exhaustively described at www.vivo.colostate.edu/hbooks/pathphys/digestion/liver/histology.html and will not be discussed in this chapter. Since the introduction of the two-step in situ collagenase perfusion to isolate hepatocytes in different species, protocols have been optimized and improved in many laboratories (see for reference Seglen 1976, Dorko et al. 1994, Berry et al. 1991, Hengstler et al. 2000a, b, Gebhardt et al. 2003). In 1976, Bojar and co-workers showed that a large number of viable human hepatocytes can be isolated from the whole liver and Guillouzo et al. showed that perfusion of a single lobe can also give satisfactory results. However, the limited availability of liver tissue for studying pathological phenomena demands the development of techniques that enable isolation of large numbers of viable human hepatocytes. In particular, the possible use of hepatocytes for temporary life support and transplantation requires large amounts of viable hepatocytes. The perfusion technique introduced in the late 1960s involves perfusion through the liver vessels, resulting in dissociation of the liver tissue and leading to isolated hepatocytes. This technique has also been successfully applied for hepatocyte isolation from human liver resections (Strom et al. 1982, Nussler et al. 1992). The perfusion of whole livers or liver segments is fairly expensive due to the high amount of collagenase needed. Additionally, single catheters in large vessels such as the portal vein and the vena cava do not guarantee an even distribution of collagenase in the entire specimen. Similarly, implantation of a single perfusion tube into one vessel does result in a homogenous perfusion of the whole resected liver lobe. Here we present a routine protocol, which has been successfully applied in our laboratory (Dorko et al. 1994, Strom et al. 1998).

Human liver samples can be obtained from tissue organ banks, organ recovery centers or directly from the surgical ward. It is mandatory to seek approval according to institutional guidelines of all participating institutions before any isolation of human hepatocytes from tissue takes place. In addition, all tissue specimens must be tested negative for hepatitis virus B and C as well as for HIV in agreement with the national guidelines. Similarly, livers infected with *Entamoeba histolytica* should not be processed for cell isolation. The reasons why an organ is not transplantable and is, therefore, given to organ tissue banks for research activities include macro and micro steatosis of more than 50%, cirrhosis, necrosis, trauma, abnormal anatomy, or the presence of a hepatic or extrahepatic malignant tumor. It should also be emphasized that hepatocytes isolated from tumor patients or from patients previously treated because of a tumor must not be used for cell therapy, since a possible transplantation of tumor cells can not be excluded.

Surgical preparation of the liver specimen for perfusion is an important step. Using whole or partial organs, the general technique involves perfusion of the liver through the existing vasculature. The initial step, however, requires correct specimen orientation and determination of landmark structures. In the case of whole livers, the organ should first be placed with the inferior surface facing the investigator. This is accomplished by placing the liver onto the falciforme ligament, a structure running from anterior to posterior, which divides the larger right lobe containing the gall bladder or gall bladder fossa in case of prior cholecystectomy. The right lobe lays now on the left side and the left lobe on the right side (Fig. 9.1). At least two catheters are placed into the vena cava and collagenase is infused into the hepatic veins which drain directly into the vena cava. The large hepatic veins allow perfusion of the entire liver parenchyma and multiple canulas can easily be placed. After placement of all canulas, the vena cava is sutured in order to prevent leakage of collagenase perfusion. Prior to perfusion, the portal vein is ligated with a ligature. The hepatic artery remains open to allow for egress of the perfusate. Be careful not to canulate the bile duct.

Working with resected liver tissue (10–50 g) (Fig. 9.2A), a maximum of three catheters are placed into the largest vessels (Fig. 9.2B) and are fixed by tissue glue. Smaller arteries are sealed using tissue glue in order to avoid leakage (Fig. 9.2C). Then, the tubing is placed into a peristaltic pump, and the tissue is perfused to remove blood and/or preservation solution from the specimen (Fig. 9.2D). The ischemia time should not exceed 1½ hours if the resected specimens are obtained directly from the operating room. Once the remaining blood or preservation solution has been flushed from the specimen with buffer I, the samples are perfused with perfusion Buffer II containing 4.8 mM calcium

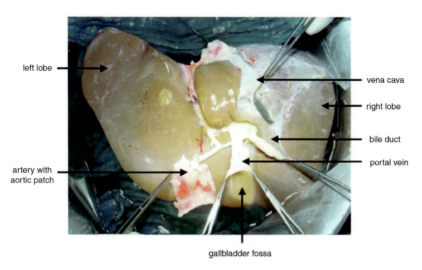

left lobe

vena cava

right lobe

bile duct

portal vein

artery with
aortic patch

gallbladder fossa

Fig. 9.1 Whole liver specimen prior to isolation. The liver is placed onto the falciforme ligament and the inferior surface is facing upwards. Vena cava, hepatic artery, and portal vein are identified for placement of isolation catheters

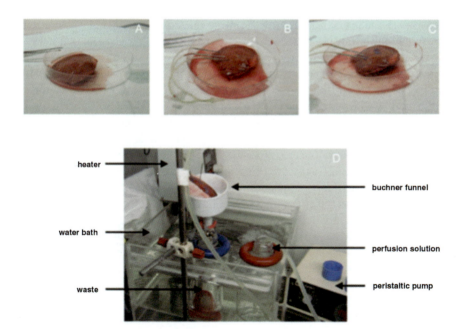

heater

buchner funnel

water bath

perfusion solution

waste

peristaltic pump

Fig. 9.2 Isolation steps for hepatic specimens obtained after liver resection. Using this technique, specimens should not exceed 50 g (**A**). Up to three catheters are placed into the largest vessels (**B**). Tissue glue is applied to fix catheters and to seal the open vessels (**C**). Tubing is connected with peristaltic pump and perfusion is started with perfusion solution I (Table 9.1). Perfusion should be performed at 37°C (**D**)

Table 9.1 Perfusion Buffers used in the hepatocyte isolation process

Buffer I		Buffer II	Buffer III	
NaCl	143.0 mM	Buffer I + EGTA 1 mM	NaCl	67.0 mM
KCl	6.7 mM		KCl	6.7 mM
HEPES	10.0 mM		CaCl$_2$	4.8 mM
Albumin	0.5%		HEPES	25.0 mM
			Collagenase P	*
			Albumin	0.5%

Perfusion buffers for hepatocyte isolation are adjusted to pH 7.6. * The enzymatic activity of collagenase is variable. Useful concentrations range from 200 to 400 mg/liter.

(Table 9.1) at a flow rate of app. 25 ml/min. Then tissue samples are perfused depending on the activity of the collagenase and the size of the specimen for 15–45 min at 37°C (Buffer III). It is mandatory that the perfusion solutions has a temperature of app. 37°C at the outflow of the tubing (just before it enters the tissue) to guarantee optimal collagenase enzyme activity. After perfusion, the tissue is transferred into a flat Petrie dish with buffer I and the liver tissue is disassociated by scraping the samples with a glass pipette or by finely chopping the specimen with a pair of scissors (Dorko et al. 1994, Strom et al. 1998). Thereafter, the cell suspension is poured through a gauze-lined funnel into 50 ml centrifuge tubes. Remember to keep the cell suspension at 4°C on ice at all times after perfusion. Hepatocytes are isolated by differential centrifugation at 50 × g for 4 min at 4°C (to remove non-parenchymal cells that remain in the supernatant). Subsequently, the supernatant is poured off and fresh cold perfusion solution I w/o EGFTA is added. Then the cell suspension is passed over a 30% Percoll gradient to eliminate damaged hepatocytes and cell fragments (Kreamer et al. 1986). At the end, the supernatant is discarded and the pellet is resuspended in hepatocyte culture medium, washed twice (50 g, 5 min, 4°C) and cell number, viability and purity using Trypan blue are determined.

Common problems with human hepatocyte isolation include the concentration of collagenase, the maintenance of the temperature during perfusion (37°C), as well as to keep cells at 4°C once they are isolated. The viability is also influenced by the Percoll gradient and the proper washes after the gradient. However, even when all technical aspects have been followed, one can still fail to have viable cells. In our hands, livers with bile duct tumors or cirrhosis show a very poor outcome in cell isolation. Another important point of failure is a too long ischemia time and subsequent hypoxia of the tissue in operation room before the liver piece is handed over to the pathologist or the person in charge of the isolation.

9.4 Clinical Hepatocyte Transplantation

The progress made in orthotopic liver transplantation has revolutionized the treatment of liver diseases. Hepatocyte transplantation has been used to bridge patients waiting for an organ transplantation (Strom et al. 1997, Bilir et al. 2000), to decrease mortality in acute liver failure (Fisher et al 2000, Ott et al. 2000), and for treatment of metabolic liver disease (Strom et al. 1999, Fox et al. 1998, Sokal et al. 2003, Muraca et al. 2002, Horslen et al. 2003, Dhawan et al. 2004, Ambrosino et al. 2005, Hughes et al. 2005). Up to know most of the published articles have reported a positive impact of hepatocyte transplantation in human studies. Despite these reports, application of hepatocyte transplantation in humans is limited to less than 100 cases. The main reason is the limited availability of human hepatocytes. Since all reasonable donor livers are used for organ transplantation, only limited numbers and/or quality of hepatocytes isolated from non-transplantable livers can be used in hepatocyte transplantation. This situation will remain unaltered unless an alternative to primary hepatocytes is available. As soon as a cell source equivalent to primary hepatocytes becomes available, treatment of acute liver failure, acquired or monogenetic metabolic liver diseases and perhaps of end-stage liver disease will might be in the focus of future cell therapy. This optimistic view is based on the fact that we recently were able to show that monocyte-derived liver-like cells could improve the survival of rats suffering from an extended liver-resection (Glanemann et al. 2008).

Acute liver failure is characterized by rapid deterioration of liver function and a high mortality. Other acute forms are caused by viral hepatitis (www.he patitis-central.com/), idiosyncratic drug reactions, acetaminophen, and mushroom ingestion. Hepatic encephalopathy, brain edema, coagulopathy, septicemia, and multi-organ failure are critical key events (Meier et al. 2005, Jalan 2005) during the course of the disease. Cell therapy of acute liver failure should provide rapid support for the failing liver by providing metabolism of liver toxins, the secretion of proteins such as clotting factors or albumin and stabilization of hemodynamic parameters. In several studies, allogeneic primary hepatocytes isolated from cadaver livers have been applied (Mito et al. 1992, Strom et al. 1997, Bilir et al. 2000). Improvements in ammonia levels encephalopathy scores, and prothrombin time levels were reported. Despite the positive results, it should be considered that the evaluation of therapies in acute liver failure might be difficult because of the inclusion of adequate controls. In this respect, studies in patients with inherited metabolic diseases are easier to interpret.

The results of hepatocyte transplantation for many metabolic liver diseases have been encouraging (for review see: Burlina 2004). For instance, a therapeutic benefit has been reported in a girl with Crigler-Najjar Syndrome Type I, which is a recessively inherited metabolic disorder characterized by severe unconjugated hyperbilirubinaemia (Fox et al. 1998). Isolated hepatocytes were infused through the portal vein, which partially corrected plasma bilirubin levels for more than 11 months (Fox et al. 1998). Similarly, a 9-year-old boy received 7.5×10^9 hepatocytes infused via the portal vein, which resulted in a decrease in bilirubin level from 530 ± 38 μmol/L (Mean $\pm$ SD) to 359 ± 46 μmol/L (Ambrosino et al. 2005). Hughes et al. (2005) also reported a 40% reduction in bilirubin levels in a Crigler-Najjar Syndrome Type I patient following transplantation with hepatocytes. Although these data demonstrate efficacy and safety, a single course of cell application does not seem sufficient to correct Crigler-Najjar Syndrome Type I completely.

Promising results have also been obtained in a 47 year old woman suffering from glycogen storage disease type 1a, an inherited disorder of glucose metabolism resulting from mutations in the gene encoding of the hepatic enzyme glucose-6-phosphatase (Muraca et al. 2002). 2×10^9 ABO-compatible hepatocytes were infused into the portal vein. Nine months after cell transplantation, her metabolic situation had clearly improved. Successful hepatocyte transplantation has also been achieved in a 4 year old girl with infantile Refsum disease, an inborn error of peroxysome metabolism, leading to increased levels of serum bile acids and the formation of abnormal bile acids (Sokal et al. 2003). A total of 2×10^9 hepatocytes from a male donor were given during eight separate intraportal infusions. Abnormal bile acid production (for instance, pipecholic acid) decreased by 40% after 18 months.

Recently, hepatocyte transplantation was used successfully to treat inherited factor VII deficiency (Dhawan et al. 2004). Two brothers (aged three months and three years) received infusions of 1.1 and 2.2×10^9 AB0-matched hepatocytes into the inferior mesenteric vein. Transplantation clearly improved the coagulation defect and decreased the necessity for exogenous factor VII to approximately 20% prior to cell therapy. As with the other metabolic liver diseases, hepatocyte transplantation has been shown to provide a partial correction of urea cycle defects. Patients showed clinical improvement, reduced ammonia levels, and increased production of urea (Strom et al. 1997, Horslen et al. 2003, Mitry et al. 2004, Stephenne et al. 2005).

Relatively, little is known about long-term engraftment of transplanted hepatocytes. An important observation of Dhawan et al. (2004) is an increased requirement for recombinant factor VII six months after cell therapy, suggesting loss or decreased function of the transplanted hepatocytes. Therefore, it will be important to obtain further information of whether loss of transplanted allogeneic hepatocytes is inevitable or a consequence of suboptimal immuno-suppression.

Cell therapy of end-stage liver disease is more problematic. Besides, loss of functional hepatocyte abnormalities of the hepatic architecture contributes to the decrease in liver function. Intrahepatic portal-to-portal venous shunts may prevent an efficient exchange between hepatocytes and blood plasma. In this situation, the benefit of additionally transplanted hepatocytes into the liver without restoring the normal liver architecture may be questionable. An alternative strategy may be the transplantation of hepatocytes into other sites, for example, the spleen, peritoneum, or omentum, to support metabolic function and regeneration.

The response to hepatocyte transplantation in humans with end-stage liver disease has not resulted in the same degree of improvement compared to experimental animal studies (Mito et al. 1992, Strom et al. 1999). One explanation may be that the hepatocytes in clinical studies were delivered into the splenic artery and not into the splenic pulp. From an immunological point of view, the spleen represents the "lion's den", where transplanted cells could possibly cause greater immune response than in most other ectopic transplantation sites. Application of cells in patients with decompensated chronic liver disease has resulted in some improvement of laboratory parameters, but was not able to change the natural cause of the disease (Ott M., unpublished). Encapsulation of hepatocytes (Aoki et al. 2005, Ringel et al. 2005) or their attachment to microcarriers may be an alternative method to improve efficacy of intraperitoneal cell transplantation. Implantable hepatocyte-based devices may represent another alternative for the treatment of end-stage liver disease (Chan et al. 2004). However, these approaches are only conceptual and presently will not pass any legal hurdle to be used in a routine clinical application.

9.5 Hepatocytes Used in Drug Development – Metabolism and Toxicology

The liver is the target organ for metabolism of endogenous and xenogenic compounds. It performs basic functions, which include biosynthesis, biotransformation, biodegradation, and detoxification. Xenobiotics and environmental pollutions are metabolized in the liver via either phase I and/or phase II enzymes (Berthou et al. 1989, Seddon et al. 1989). Cytochrome P450 are a group of primary enzymes responsible for the oxidative mechanism of xenobiotics, many of which are converted to more hydrophilic molecules in order to be easily eliminated from the body. The cytochrome P450 (CYP) isoenzymes are a group of heme containing enzymes embedded primarily in the lipid layer of the endoplasmatic reticulum of hepatocytes. The enzymatic complexes consist of two functional units: A cytochrome P450 mono-oxygenase (mixed function oxygenase) and a NADPH cytochrome P450 reductase (Nelson et al. 1996, Wieczorek and Tsongalis 2001). The major human drug metabolizing CYPs belongs to the families 1, 2, and 3. Among the numerous human enzymes identified, the major human CYPs isoforms involved in drug metabolism are CYP1A2, CYP2A6, CYP2B6, CYP2C9, CYP2C19, CYP2P6, CYP2E1, and CYP3A4.

Phase II enzymes are located in the cytoplasm and in the microsomes. These enzymes primarily include glutathione S-transferases, UDP-glucoronyl transferases (UGTs), N-acetyl transferases, methyl transferases, and amino acid transferases. Phase II metabolism leads to conjugation with a cofactor and usually increases water solubility. Reactions include conjugation with UDP-glucuronic acid, glutathione, sulphate, acetate, and amino acids (Parkinson 1996). The majority of cells in the liver are hepatocytes.

In recent years, some drugs were taken off the market with disturbing regularity, because of late discovery of hepatotoxicity. Drug-drug as well as drug-xenobiotics interactions are a major source of this clinical problem. Therefore, it appears to be particularly important to evaluate the drug-drug interactions of new drug candidates, as well as obtaining a more complete picture of the metabolic fate of what may happen in humans and to avoid inappropriate costs during clinical introduction of a new drug candidate (Li 2001). Primary cultured hepatocytes are considered an invaluable tool for these kinds of studies. Unlike immortal cell lines, freshly isolated hepatocytes are able to retain liver specific functions, including cytochrome P450 inducibility in culture. However, it is well known that primary cultured hepatocytes tend to loose expression of specific liver functions, as they adapt to culture conditions over time. Thus cultured liver cells have a limited useful life span. This fact is especially critical in the case of human hepatocytes, where availability of human tissue is irregular and unpredictable. It is well recognized that human hepatocytes are increasingly the model of choice for predicting drug hepatic events in humans, since species differences make extrapolation from animal models to humans unreliable (Bowen et al. 2000). The demand for human liver cells, either for therapeutic intervention, such as artificial livers, fundamental research, or for the study of drug metabolism during pre-clinical development (e.g., drug interactions, drug metabolism, or drug metabolic profiles) is constantly increasing.

9.6 Hepatocytes Generated by Stem Cell Technology

9.6.1 Can Human Hepatocytes Be Produced by Stem Cell Technology? An Overview

As discussed in previous paragraphs, primary human hepatocytes are well established in drug metabolism research (Hengstler et al. 2000a, b, 2003, Li et al. 1999, Najibulla et al. 2000, Ringel et al. 2002, 2005, Reder-Hilz et al. 2004, Carmo et al. 2005, 2004). In addition, transplantation of isolated hepatocytes represents a promising strategy for treatment of acute and chronic liver disease (von Mach 2002, von Mach et al. 2005, Nussler et al. 2006). However, availability of human hepatocytes is limited. Therefore, it is intriguing that numerous articles between 1999 and 2006 reported the generation of hepatocytes from different types of extrahepatic stems or precursor cells. This seems to open exciting new possibilities for pharmacology and toxicology, as well as for cell therapy. Considering these examples, one might also ask why we still use the delicate and extremely expensive primary human hepatocytes for drug metabolism research and cell therapy. In the present article, we focus on the generation of liver cells or "hepatocyte-like cells" from extrahepatic human stem cells. These studies reported hepatocyte marker expressions, including albumin, CK18, c-met, alpha-fetoprotein, and cytochrome P450s after transplantation of different types of human stem cells into the liver of laboratory animals or in vitro after incubation with cytokines. These intriguing observations have prompted scientists to classify stem cell derived cell populations as hepatocytes, "liver cells", or more carefully "hepatocyte-like cells". However, this conclusion might be premature. This is well illustrated by the example of albumin expression, which has been analyzed in most studies. It is known that the hepatocyte is the only cell type secreting albumin. But is it adequate to conclude that a cell type secreting albumin has to be a hepatocyte? Indeed, this assumption might be a far too optimistic oversimplification of the real facts and obtained experimental data. For example, it can not be excluded that incubation of precursor cells with some specific cytokines may on the one hand activate transcription factors which finally lead to albumin mRNA expression, whereas on the other hand further factors usually expressed in hepatocytes (such as cytochrome P450s) remain silent. In fact, it has been observed that some types of human stem cell derived "hepatocyte-like cells" express albumin but not CYP3A4. For this reason, it is to be assumed that cytokines can induce expression of a limited number of hepatocyte marker genes in non-hepatic cell types. To conclude on the grounds of a limited number of markers that these cells are true hepatocytes is definitely not indicated. In this case, one should, in addition, demonstrate that the majority of hepatocyte-specific genes are indeed expressed in these cells, which can be achieved by gene array analysis. Besides having an overview over the gene expression pattern, a second aspect is crucial, namely the quantity of the expressed hepatocyte factors compared to primary hepatocytes (obtained from liver tissue). It is essential to understand that the definition of a bona fide hepatocyte should not be limited to qualitative assays, but has to include a quantitative analysis of mRNA and protein expression as well as enzymatic activities, thus allowing direct quantitative comparison to primary hepatocytes. For instance, it is not sufficient demonstrating expression of albumin, alpha-fetoprotein, and a number of further "hepatocyte markers" by RT-PCR or immunostaining. Even if these analyses are all positive, they do not exclude that expression levels are orders of lower magnitude lower compared to hepatocytes ex vivo. In the next sections, we critically discuss published phenotypes of stem cells after application of various differentiation protocols aimed at generating human hepatocytes. In addition, the criteria needed for defining a true hepatocyte are suggested (Hengstler et al. 2005). We conclude that the "stem cell derived human hepatocyte" does not yet exist. Nevertheless, there is a good chance that this aim might be achieved in future.

9.6.2 *Transdifferentiation of Extrahepatic Cells to Hepatocytes? First Evidence Pointing towards Bone Marrow Cells*

The first report on bone marrow derived hepatocytes was published in 1999 (Petersen et al. 1999). Petersen et al. (1999) used three approaches to demonstrate that bone marrow cells contribute to liver cells: (i) Female rats were lethally irradiated and transplanted with bone marrow from a male rat. Engrafted females were treated with CCl_4 and 2-acetylaminofluorene (2-AAF) to simultaneously induce hepatotoxicity and block endogenous hepatocyte proliferation. Under these conditions, Y-chromosome positive hepatocytes were observed in the female recipients. (ii) Using the same protocol, bone marrow cells from $DPPIV^+$ F-344 male rats were injected into $DPPIV^-$ F-344 female rats, resulting in DPPIV expression in bile canalicular sites between hepatocytes of the $DPPIV^-$ F-344 recipients. (iii) Livers from Lewis rats expressing the major histocompatibility complex class II L21-6 isozyme were transplanted into Brown-Norway rats that do not express L21-6. After the CCl_4/2-AAF protocol, the recipients (Lewis rats) showed positive L21-6 staining in livers.

In contrast to Petersen et al. (1999) inducing massive liver damage by the CCl_4/2-AAF protocol, Theise et al. (2000) performed an experiment inducing only mild liver damage. B6D2F1 female mice were lethally irradiated and received bone marrow cells from male donors. A fraction of about 2.2% Y-chromosome positive hepatocytes were observed in all female recipients.

To study the role of purified hematopoietic stem cells, Lagasse et al. (2000) applied the fumarylacetoacetate hydrolase-deficient mouse model ($FAH^{-/-}$), an animal model of fatal tyrosinemia type 1. $FAH^{-/-}$-mice suffer from severe liver damage as a consequence of accumulation of hepatotoxic metabolite fumarylacetoacetate and its precursor maleylacetoacetate (Lagasse et al. 2000). Due to deterioration of hepatocytes, FAH-deficient mice cannot survive unless they are treated with the drug 2-(2-nitro-4-trifluoro-methylbenzyol)-1, 3-cyclohexanedione (NTBC) which prevents production of the toxic metabolites. Due to permanent deterioration of hepatocytes, the $FAH^{-/-}$-mice represent an animal model with an extremely high selection pressure for wild-type (i.e., $FAH^{+/-}$ or $FAH^{+/+}$) hepatocytes. Lagasse et al. intravenously transplanted c-$kit^{high}Thy^{low}Lin^-SCA$-$1^+$ hematopoietic stem cells from bone marrow of male ROSA26/BA mice into lethally irradiated, female $FAH^{-/-}$-mice. NTBC was administered in the drinking water for two months. Subsequently, NTBC was (periodically) withdrawn to induce liver damage and livers were analyzed after four months. Beta-galactosidase positive nodules with 0.5–4 mm diameter were observed with Y-chromosome positive cells co-expressing FAH and albumin.

Further studies have been performed in which extrahepatic stem cells from mice were transplanted to study their fate in the liver (Fujii et al. 2002, Mallet et al. 2002, Wang et al. 2002, Vassilopoulos et al. 2003, Alvarez-Dolado et al. 2003, Kanazawa and Verma 2003, Terai et al. 2003, Alison et al. 2004). What emerged from these studies was that specific markers, which were used for tagging the transplanted stem cells, such as Y-chromosomes, EGFP, or β-galactosidase were subsequently detected in the recipient's livers. A convincing proof of hepatocyte functionality after transplantation with bone marrow cells is survival of $FAH^{-/-}$ mice. However, as discussed in the next section, this does not automatically mean that the transplanted cells have transdifferentiated to become true hepatocytes.

9.6.3 Cell Fusion or Transdifferentiation? A Question of Potential Clinical Relevance

The mechanism underlying the conversion of stem cells has been heavily debated. Meanwhile, it is accepted that some stem cell types can fuse with the recipient's cells leading to cytoplasmic mixing and reprogramming of cell fate. Alternatively, it seems that stem cells can be "instructed" by factors of the host's microenvironment to adopt a hepatocyte fate. Differentiation between cell fusion and transdifferentiation is of fundamental importance and could have possible clinical implications: For instance, "fusogenic" cells might be used as vehicles for gene therapy (although this has not yet been introduced in clinical studies). Good candidates for "fusogenic cell therapy" are monogenetic liver diseases (or liver diseases caused by defects in a limited number of genes), such as Wilson's disease (caused by a defect in a copper transporting ATPase (ATP7B) leading to an accumulation of copper in the liver) or hereditary hemochromatosis characterized by hepatic iron overload due to a mutated hemochromatosis gene (HE) (Howell and Mercer 1994, Tomatsu et al. 2003). If stem cells deliver the needed wild type gene (such as ATP7B or HE) and the latter becomes expressed in hepatocytes after cell fusion, the "repaired" hepatocytes might gain a healthy phenotype and a selection advantage over hepatocytes which did not undergo cell fusion. Other monogenetic diseases, such as hemophilia, might also be treated by "fusogenic cells", but "repaired" hepatocytes will not gain a selection advantage. Secretion of clotting factors is important for the organism, but factor VIII deficiency does not influence survival of hepatocytes. (In contrast, defects in ATP7B or HE will lead to deterioration of hepatocytes.) Therefore, the outcome of "fusogenic therapy" will probably depend on the relevance of the defect genes. If delivery of a defect gene does not improve hepatocyte survival (as in the case of hemophilia), it can be expected that higher numbers of cells have to be transplanted compared to a situation where gene transfer leads to an improved survival of hepatocytes after fusion with transplanted cells. Thus, diseases caused by gene defects that do not induce deterioration of hepatocytes are only "second best candidates" for "fusogenic cell therapy". However, these are speculations on future developments. In contrast to treatment of genetic diseases, it is obvious that drug metabolism studies do not require fusogenic cells or hybridomas of fused stem cells and hepatocytes, but instead require the establishment of in vitro cultures of stem cell derived hepatocytes or at least appropriate hepatocyte-like cells.

9.6.4 The Discovery of "Fusogenic Cells" in Mouse Bone Marrow

In 2002, two studies raised the question if cell fusion between stem cells and the host's cells could explain cell fate transition (Terada et al. 2002, Ying et al. 2002). Terada et al. demonstrated that mouse bone marrow cells can fuse spontaneously with cultured embryonic stem cells. Fused bone marrow cells can subsequently adopt the phenotype of the recipient cells, which could be misinterpreted as transdifferentiation (Terada et al. 2002). Ying et al. co-cultured cells from mouse forebrains with embryonic stem cells. Following selection for a transgenic marker carried only by the brain cells (puromycin resistance), cells which carry a transgenic marker were obtained and chromosomes were derived from the embryonic stem cells. The frequency of such cell fusion events was between 10^{-4} and 10^{-5} per brain cell plated (Ying et al. 2002).

In the subsequent two years, three articles (from two independent groups) convincingly demonstrated that in the $FAH^{-/-}$ mouse model the transplanted stem cells fuse with the host's hepatocytes leading to liver regeneration (Vassilopoulos et al. 2003, Wang et al. 2003, Willenbring et al. 2004). Wang et al. transplanted bone marrow cells from female FAH wild-type lacZ transgenic mice into male $FAH^{-/-}$ recipients. Cytogenetic analysis demonstrated 80, XXXY karyotypes, indicating cell fusion between two diploid cells. Similarly, 120, XXXXYY karyotypes demonstrated cell fusion events between a tetraploid recipient's hepatocyte to a diploid donor bone marrow cell (Wang et al. 2003). Vassilopoulos et al. also used the $FAH^{-/-}$ mouse model (Vassilopoulos et al. 2003). The authors analyzed genomic DNA of FAH-expressing liver nodules after transplantation of $FAH^{+/+}$ bone marrow cells. Interestingly, the nodules contained more mutant (host) than wild-type (donor) FAH alleles. If donor bone marrow had transdifferentiated into hepatocytes, the FAH-expressing liver nodules should have contained mostly donor DNA. Two further intriguing observations suggest that reprogramming of cell nuclei after fusion of bone marrow cells with hepatocytes is not a symmetric matter, but the hepatocyte seems to be the "dominant" cell type: (i) the pan-hematopoietic surface marker CD45 initially expressed in the transplanted bone marrow cell fraction became silenced after fusion with hepatocytes and (ii) FAH mRNA was not formed in bone marrow cells, but was activated after fusion with hepatocytes (Vassilopoulos et al. 2003). Although examples of only two genes are not sufficient for generalizing, the data suggest in essence that the hepatocyte might reprogram the bone marrow cell and not vice versa. This seems to be plausible since a hepatocyte contains much more cytoplasm than a bone marrow cell. Willenbring et al. concentrated on the cell type responsible for therapeutic cell fusion (Willenbring et al. 2004). They demonstrated that differentiated macrophages obtained from the mononuclear fraction of ROSA26$^{+/-}$ (R26R), FAH$^{+/+}$-mice could act as fusion partners for FAH$^{-/-}$ hepatocytes. In contrast, hepatocyte to hepatocyte fusion did not occur or was extremely rare, which has been demonstrated by transplantation of FAH-wild-type hepatocytes into FAH$^{-/-}$ mice.

One might assume that the extreme selective pressure in FAH$^{-/-}$ livers might create a situation facilitating cell fusion. Even if cell fusion would occur with an extremely low frequency, a single fusion event might lead to a large FAH positive nodule or eventually even to repopulation of large parts of the liver. In addition, it has been suggested that the disrupted microanatomy and cellular integrity prevents transdifferentiation and facilitates cell fusion, for instance, by trapping of transplanted stem cells in the disrupted sinusoids and mechanical compression against hepatocytes with membrane instabilities (Theise 2003). Although it cannot be excluded that such a mechanism indeed facilitates cell fusion in FAH$^{-/-}$ mice, it has recently been demonstrated that cell fusion also occurs in normal livers (Alvarez-Dolado et al. 2003). These results have been obtained using elegant reporter constructs as described in the next paragraph.

9.6.5 Elegant Reporter Studies Differentiate Between Cell Fusion and Real Differentiation

The principle of cre recombinase (cre)-based reporter systems for cell fusion is that cre activates a reporter gene, for instance, EGFP or lacZ. This can be achieved by cre-mediated excision of a loxP flanked STOP cassette (Fig. 9.3). Cre is expressed in cell type A, the reporter in cell type B (or vice versa). Only if active cre is present in the cell carrying the reporter (for instance, by cell fusion) the latter will be activated. In 2003 and 2004, two groups (Alvarez-Dolado et al. and Harris et al.) used this elegant reporter system to study possible cell fusion between bone marrow cells and hepatocytes. Both studies are highly elegant, seemed well prepared, and were published in high-ranking journals (Nature and Science). Amazingly, they produced completely controversial results. In order to understand possible reasons for this discrepancy, some experimental details must be reviewed. Alvarez-Dolado et al. transplanted bone marrow from β-actin-Cre-EGFP donor mice into lethally irradiated R26R recipient mice (Alvarez-Dolado et al. 2003). The donor mice constitutively express Cre recombinase and EGFP. In the R26R recipient mice, the LacZ reporter gene is only expressed after the excision of a loxP-flanked stop cassette by Cre-mediated recombination. Only when β-actin-Cre-EGFP donor cells fuse with R26R cells the reporter gene LacZ will be expressed (Fig. 9.3). Using this elegant technique, Alvarez-Dolado et al. analyzed livers of R26R mice two and four months after transplantation with β-actin-Cre-EGFP bone marrow cells and consistently observed LacZ expressing cells (which had hepatocellular morphology

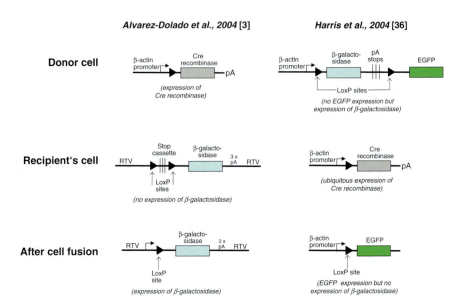

Fig. 9.3 Two different strategies to study cell fusion (Alvarez-Dolado et al. 2003, Harris et al. 2004). Alvarez-Dolado et al. transplanted Cre-expressing bone marrow cells into recipients expressing the reporter gene ß-galactosidase that is transcriptionally activated by Cre-recombinase (Alvarez-Dolado et al. 2003). In contrast, Harris et al. transplanted bone marrow cells expressing the reporter EGFP into Cre-recombinase positive recipients (**B**) (Harris et al. 2004)

and expressed albumin). Interestingly, no LacZ negative EGFP expressing hepatocytes were observed, suggesting that cell fusion and not transdifferentiation is the major mechanism. In conclusion, the experiments of Alvarez-Dolado demonstrated that fusion between bone marrow cells and hepatocytes also occurs without strong selective pressure, because the recipient mice had normal livers. As expected, no nodules of fused cells were observed (as in the $FAH^{-/-}$ mouse model). However, a relatively small number of single hybrid cells ranging between 7 and 37 fused cells per liver section were detected.

Only six months after the Alvarez-Dolado article, a second research group (Harris et al. 2004) published a controversial conclusion, although at first glance similar techniques were used. In this publication, Harris et al. (2004) also made use of a Cre/loxP recombinase-based reporter system (Harris et al. 2004). They show that transplantation of bone marrow cells into lethally irradiated mice led to bone marrow derived epithelial cells in the liver of recipients without cell fusion. Without additional experiments it seems impossible to explain this discrepancy. Nevertheless, differences in the experimental setup give a hint how the difference could be explained: Alvarez-Dolado et al. transplanted Cre-expressing bone marrow cells into recipients, harboring the reporter gene, which was transcriptionally activated by Cre-recombinase (Fig. 9.3A). In contrast, Harris et al. transplanted bone marrow cells expressing the reporter into constitutively Cre-recombinase expressing recipients (Fig. 9.3B). At a first glance, this difference may seem insignificant, since after cell fusion both, Cre-recombinase and the reporter, are unified in one cell. However, this is incorrect, since at least two differences may be relevant: (i) If Cre-recombinase is expressed under control of a constitutively active promoter in the donor cell, active cre is already present and will immediately activate the reporter after cell fusion (Fig. 9.3, left side). The recipient cell has been engineered for expression of the reporter construct. The only step required for a positive result is floxing-out the STOP cassette. Therefore, this system actually detects protein (in this case: cre) transfer from the donor into the host cell. (ii) If the reporter is in the donor cell (Fig. 9.3, right side), additional steps are required in order to obtain a positive result. After cell fusion, Cre will also flox-out the DNA sequence between the loxP sites (Fig. 9.3, right side). However, it should be considered that cell fusion between a bone marrow cell and the much larger hepatocyte does not occur on an equal footing. Bearing in mind that hepatocytes are the "dominant" cell type that can silence gene expression in the fused bone marrow cell as described above: (Vassilopoulos et al. 2003), this difference in experimental design may be critical. Silencing of the bone marrow delivered Cre-recombinase after cell fusion in Alvarez-Dolado's experiment will not be critical, because a short period of Cre activity after cell fusion is sufficient to irreversibly activate the R26R reporter construct (Fig. 9.3A). In contrast, the design of Harris et al. (2004) may lead to silencing of the EGFP reporter derived from the bone marrow cells (Fig. 9.3B). This might be misinterpreted as lack of cell fusion. This possibility has not been excluded by positive controls, which could be done by generating hybridomas of bone marrow cells and hepatocytes in vitro. Instead bone marrow from Z x CF1 mice expressing both, the reporter and Cre-recombinase, was used as positive control. In addition, the relatively low cell numbers studied by Harris et al. (2004) may be critical, since only 18 donor-derived hepatocytes were evaluated (in total 36 000 hepatocytes were examined, but only 0.05% of them contain Y chromosomes) (Harris et al. 2004). Nevertheless, it would be extremely helpful to have the data of both the Harris and the Alvarez-Dolado experiments with interchanged donors and recipients. As long as these or similar data are not available, the cell fusion story remains enigmatic.

Besides studies demonstrating cell fusion, there have been several papers published in previous years, which did not report cell fusion as a main mechanism of conversion (for instance, Jang et al. 2004, Kogler et al. 2004, Newsome 2003). It should also be considered that Terada et al. (2002) observed cell fusion between mouse embryonic stem cells and whole mononuclear mouse bone marrow cells. In contrast, the purified $SCA1^+Lin^-$ hematopoietic stem cell fraction from mouse bone marrow did not increase the frequency of the hybrid cells (Terada et al. 2002). Perhaps the extent of cell fusion depends on both, the type of host and donor cells. However, the situation remains obscure. More research is needed to identify the key factors controlling fusion between the transplanted and the host cells.

9.6.6 Fate of Human Stem and Precursor Cells in Livers of Mice

Human hepatocytes have been shown to survive and maintain some hepatocellular functions after transplantation into mouse livers (Dandri et al. 2006, Petersen et al. 2004, Dandri et al. 2001). Transplantation of normal human hepatocytes into the livers of immunodeficient urokinase-type plasminogen activator (uPA)/recombinant activation gene-2 (RAG-2) mice led the production of human albumin for at least two months after transplantation. Therefore, it seems to be a reasonable approach to transplant human precursor cells into livers of mice or other laboratory animals in order to study their capacity for hepatocellular differentiation. Meanwhile, a relatively large number of independent groups have studied the fate of different types of human stem and precursor cells in livers of experimental animals. Usually recipients were immunodeficient mice, whereas two studies performed intra-fetal injections in sheep.

Although different human cell types, different routes of injection and different recipients (mice and sheep) have been tested, very similar experimental results were obvious. Eight of the 11 studies observed human albumin positive cells in the recipient's livers. In three of the studies listed in Table 9.2, human

Table 9.2 Substrates recommended for comparison of the metabolic capacity of stem cell derived cells with primary hepatocytes

Phase I – metabolism

CYP 1A1/2, 2A6, 2B6, 2C8/9/19, 2D6, 2E1, 3A4/5 (ng/h/mg protein or ng/h/ 10^6 cells)

Examples for adequate substrates:

Coumarin (CYP 2A6)
Caffeine (CYP 1A2)
Diclofenac (CYP 2C9)
Bupropion (CYP 2B6)
S-Mephenytoin (CYP 2C19)
Dextromethorpan (CYP 2D6)
Verapamil (CYP 3A4, CYP 2C8)

Phase II – metabolism

UDP-Glucuronosyltransferase (UDPGT), glutathione-S-transferase (GST), sulfotransferase (ST), (ng/h/mg protein or ng/h/ 10^6 cells)

Examples for adequate substrates:

4-Methylumbelliferon (UDPGT)
1-chloro-2,4-dinitrobenzene (GST)
2-Naphthol and p-nitrophenol (ST)
Dextromethorphan (UDPGT, ST)

Enzyme induction

Inducer

Rifampicin (CYP 3A4)
3-Methylcholanthrene (CYP 1A1 and 1A2)
Phenobarbital (CYP2B)

albumin was not analyzed, but positive immunostaining with antibodies specific for "human hepatocyte specific antigen" (such as HSA or HepPar1, clone OCH15E) was observed (Newsome et al. 2003, Danet et al. 2002, Ishikawa et al. 2003). Most of these studies also confirmed their positive immunostaining data by RT-PCR analysis (Wang et al. 2003, Beerheide et al. 2002, Danet et al. 2002, Kollet et al. 2003, Kakinuma et al. 2003, Turrini et al. 2005). Newsome et al. (2003) examined whether unsorted mononuclear cell preparations of human cord blood would fuse with hepatocytes after transplantation into SCID/NOD mice (Newsome et al. 2003). No evidence of cell fusion was found in the study using FISH analysis and probes of human and mouse DNA. Similarly, no evidence of cell fusion was observed in two studies using fetal sheep as recipients. Thus, human stem cells might be different to their murine counterparts where hepatocyte cell fusion has been demonstrated (Vassilopoulos et al. 2003, Alvarez-Dolado et al. 2003, Wang et al. 2003, Willenbring et al. 2004). Nevertheless, it may be premature to postulate a general difference between stem cells of human and mouse origin with respect to their "fusogenic" potency. Further studies are needed. To our knowledge, the sensitive Cre-reporter system (originally described by Alvarez-Dolado et al. 2003) has not yet been tested with human stem cells expressing Cre-recombinase.

Without doubt, the observation of human albumin positive cells in livers of animals after transplantation of human stem cells is intriguing. But are these cells indeed human hepatocytes? Some authors concluded that the "human albumin positive cells" should not yet be considered as human hepatocytes (Beerheide et al. 2002, von Mach et al. 2004). However, this may be discussed controversially, since several experts have classified similar cell types as human hepatocytes. Examples are the interpretation of human stem cell derived cells as "functional hepatocytes" (Kakinuma et al. 2003) for "human hepatocytes" (Almeidia-Porada et al. 2004) after detection of human albumin and "human hepatocyte specific antigen". Kogler et al. transplanted adherently proliferating cells isolated from human cord blood into livers of fetal sheep (Kogler et al. 2004). The authors observed expression of albumin and "human hepatocyte specific antigen" in livers of sheep and concluded that the human cord blood cells differentiated to "human parenchymal hepatic cells". Newsome demonstrated expression of the HepPar1 human hepatocyte specific antigen and concluded that cells from human cord blood "become mature hepatocytes" in livers of SCID/NOD mice (Newsome et al. 2003). Finally, Ishikawa detected human albumin and the HepPar1 antigen in livers of immunodeficient mice, postulating that the engrafted cells from human cord blood "functioned as hepatocytes" (Ishikawa et al. 2003). Obviously the controversy is a consequence of different conceptions of the term "hepatocyte". It should be considered that parameters used to classify the stem cell derived hepatocytes as "human hepatocyte" in these studies are: (i) albumin, (ii) c-met, and (iii) antibodies against the human hepatocyte antigens HepPar1 or OCH15E. Other parameters, including activities of drug metabolizing enzymes, clotting factors, complement, etc. were not tested in these experiments. It is well known that under normal physiological conditions hepatocytes are the only cell type secreting albumin. On the other hand, is it possible to conclude that a cell type secreting albumin must be a hepatocyte? This may be a premature conclusion if a tight definition for hepatocytes is applied. In this regard, it is tempting to assume that the liver microenvironment is capable of inducing a limited number of genes including albumin, but evidence is lacking that these cells also express the majority of the several hundreds of proteins needed to make up a hepatocyte.

A strategy to elucidate the (albumin, c-met, or HepPar1 positive) cells real identity is isolation from the recipient's liver and quantitative analysis of activities including primary human hepatocytes as controls, which has not yet been done. This will be difficult with mice, especially since the fraction of human albumin positive cells was small, usually ranging between 0.01 and 1%. Recently, Kogler et al. (2004) reported that their sheep model resulted in 'more than 20% albumin-producing human parenchymal hepatic cells' (Kogler et al. 2004). Such a high percentage is surprising, considering the much lower numbers in mice and the fact that no liver damage has been induced in the sheep. Therefore, it will be interesting to see whether this result can be reproduced by independent groups. From a technical point of view, it should be easy to isolate the hepatocytes from sheep livers by standard EGTA/collagenase perfusion, enrich the human cells, for instance, by FACS and analyze how these cells compare to primary human hepatocytes.

9.6.7 Transdifferentiation of Extrahepatic Human Stem Cells to Hepatocytes In Vitro?

Recently, three articles reporting transdifferentiation of extrahepatic stem cells to hepatocytes or "functional hepatocyte-like cells" have attracted considerable attention. Schwartz et al. reported that "multipotent adult progenitor cells differentiate into functional hepatocyte-like cells" (MAPCs) and "had phenobarbital-inducible cytochrome P450" (Schwartz et al. 2002, Jiang et al. 2002). The authors claim that progenitor cells (MAPCs) derived from human, rat, or mouse bone marrow can "differentiate into most mesodermal and neuroectodermal cells in vitro and into all embryonic lineages in vivo." Since 2002, these two articles have been cited more than 700 times. Referring to Schwartz et al. and Jiang et al., several reviews came to the conclusion that multipotent adult progenitor cells "differentiate into mature hepatocytes with apparently fully functional properties" (Fausto 2004) and "have multilineage differentiation potential including differentiation into hepatocytes" (Gilgenkrantz 2004). On the other hand, one might be surprised that MAPCs are not yet used in drug metabolism research, although they have been described more than four years ago. An important step during drug development is elucidation of the metabolism of candidate pharmaceutical agents. Since the liver is a key site for drug metabolism, a stem cell derived "fully functional hepatocyte" would be immediately welcome, especially since primary human hepatocytes are always rare. To understand the discrepancy between the enthusiastic celebration of "MAPC derived hepatocytes" and their absence in drug metabolism research, we will next discuss the article of Schwartz et al. (2002) applying the tight hepatocyte conception relevant for the drug metabolism research community. Schwartz et al. (2002) isolated $CD44^-$, $CD45^-$ HLA class I^- and II^-, as well as c-kit$^-$ adherently growing cells from bone marrow of human, rat (Sprague Dawley) and mouse (C57Bl/6), termed hMAPS, rMAPCs, or mMAPCs. These cells were cultured with 1% Matrigel in a DMEM-based medium containing 10 ng/ml FGF-4 and 20 ng/ml HGF. After 14 days in culture, approximately 60% of the mMAPCs or rMAPCs and 90% of hMAPCs expressed albumin. These cells have been used to study induction of pentoxyresorufin-O-dealkylase activity (PROD) by phenobarbital. Dealkylation of pentoxyresorufin is predominantly catalyzed by cytochrome P450 2B-isoforms. This is a relevant experiment, since in hepatocytes PROD activity is strongly induced by phenobarbital. Stem cell derived hepatocytes that respond similar to enzyme inducers as primary hepatocytes are urgently needed in drug development, because new drugs should be tested for their capacity to cause enzyme induction and to avoid dangerous drug-drug interactions. Schwartz et al. incubated mMAPC and hMAPC derived cells as well as primary rat hepatocytes for four days with 1 mM phenobarbital and reported a 1.3-, 1.4- and 1.4-fold increase in PROD activity, respectively. This extremely weak increase in PROD activity has not been confirmed by a statistical test (Schwartz et al. 2002). Based on these data, Schwartz et al. concluded that MAPCs 'differentiate into functional hepatocyte-like cells' and 'had phenobarbital inducible cytochrome P450'. It is not surprising that these data could not convince the drug metabolism research community. It is important to consider that in conventional cultures with primary hepatocytes, PROD activity is induced more than 10-fold after incubation with phenobarbital (Fig. 9.2) (Hengstler et al. 2000a, b, Ringel et al. 2002, Wortelboer et al. 1991). Thus, a possible 1.3-fold increase in PROD activity as reported by Schwartz et al. should not be interpreted as a 'functional characteristic consistent with hepatocyte activities' (p. 1300). In contrast, the extremely low (if not absent) inducibility of PROD activity shows a clear difference in functionality between the progenitor cell derived "hepatocyte-like cells" of Schwartz et al. and primary hepatocytes. Interpreting the reported 1.3-fold increase in MAPCs by phenobarbital, Schwartz et al. wrote that 'a number of different cells have CYP activity, but CYP activity upregulation by phenobarbital is seen only in hepatocytes' (p. 1298). This is not correct, since, for instance, phenobarbital induces CYP2B isoforms also in small intestine epithelial cells (Zhang et al. 2003), white adipose tissue (Yoshinari et al. 2004) and increases several phase I and II xenobiotic metabolizing enzymes in the colonic mucosa (Baijal et al. 1997). In principle, PROD induction by phenobarbital is a good experiment to find out whether a candidate cell is similar to hepatocytes. However, this requires determination of absolute PROD activities expressed as pmol resorufin formed per minute per mg of protein (or per million of cells) (Ringel et al. 2002, Hengstler et al. 2000a, b) and not by semiquantitative techniques. It is necessary to compare induction factors and also basal activities to those of primary hepatocytes. Our experience has been that stem cell derived cells often show lower basal activities than primary hepatocytes. Finally, it raised problems when Schwartz et al. (2002) argued that 'CYP2B1 activity in rats, CYP2B9 and CYP2B13 in mice and CYP2B6 in humans is considered relatively hepatocyte specific' (p. 1300). In contrast, rat CYP2B1 is a dominating enzyme in the rat lung (Skarin et al. 1999) and also found in the rat brain (Miksys et al. 2000). CYP2B9 is also expressed in mouse small intestinal epithelial cells (Zhang et al. 2003), and CYP2B6 is furthermore found in human nasal mucosa, trachea, and lung (Ding and Kaminsky 2003, Hukkanen et al. 2000). Thus, using a tight hepatocyte definition, a detailed evaluation of Schwartz et al.'s (2002) data suggested that some conclusions of this study should be treated with caution. Despite of the expression of some markers usually found in hepatocytes, such as albumin and CK18, the proximity of the MAPC derived "functional hepatocyte-like cells" is likely to be overestimated.

Jang et al. reported that hematopoietic stem cells 'become liver cells when cocultured with damaged liver separated by a barrier' (Jang et al. 2004). The authors used a relatively complex method of isolating hematopoietic stem cells, which include three steps: (i) isolation of a small-sized cell population from male C57Bl6 mice by counter-flow elutriation of bone marrow cells, (ii) depletion of lineage-positive cells, (iii) labeling of the resulting cell fraction with the red fluorescence dye PKH26 and isolation after injection into lethally irradiated female C57Bl6 mice. The authors exposed these rigorously purified stem cells to damaged liver tissue of mice in transwell dishes. After just 48 h, albumin and CK18 were detectable in 2–3% of the stem cells. Several liver transcription factors and cytoplasmic proteins expressed during the differentiation of liver (α-FP, GATA4, HNF4, HNF3β, HNF1α, and C/EBPα) and in mature hepatocytes (CK18, albumin, fibrinogen, transferrin) were analyzed in the hematopoietic stem cell derived cell population. The expression of all markers increased over time, with the exception of α-FP, which initially increased and later decreased again, indicating a possible maturation (Jang et al. 2004). A limitation of this promising study is the relatively small percentage of cells in the stem cell derived cell population expressing the above-mentioned markers (ranging between 2% for albumin and transferrin and 17% for HNF3β). Therefore, an interesting next step would be isolation of the positive cells and analysis of drug metabolizing enzyme activities compared to primary hepatocytes.

Experiments aimed at differentiating human extra-hepatic stem cells to hepatocytes have been performed by a relatively large number of groups (Zulewski et al. 2001, Avital et al. 2001, Fiegel et al. 2003, Lee et al. 2004, Lavon et al. 2004). Here, we give a brief overview of the phenotype of cells obtained after in vitro differentiation. Zulewski et al. cultured nestin-positive adherently proliferating cells from human pancreatic islets with 10 µM dexamethasone in the absence of serum and demonstrated expression of hepatocyte growth factor, c-met, α-fetoprotein, E-cadherin, and transthyretin by RT-PCR. Avital et al. (2001) isolated β_2 microglobulin$^-$ Thy1$^+$ cells from bone marrow (Zulewski et al. 2001, Avital et al. 2001). In a trans-well culture system, the β dtgs$_2$ microglobulin$^-$ Thy1$^+$ cells were cocultured with hepatocytes isolated from cholestatic rat livers (induced by ligation of the common bile duct) in the presence of 5% "cholestatic" serum on matrigel. The β_2 microglobulin$^-$ Thy1$^+$ cells differentiated to a cell type that metabolized ammonia into urea, expressed albumin (IHC, RT-PCR), CAM5.2 (IHC), C/EBPβ (RT-PCR), HNF1 (RT-PCR), and HNF-4 (RT-PCR). Fiegel et al. separated CD34$^+$ bone marrow cells and cultured these cells on collagen coated plates (Fiegel et al. 2003). After exposure to HGF, EGF and insulin, expression of albumin, and CK19 was observed by RT-PCR. Lee et al. (2004) used adherently proliferating cells from human bone marrow and umbilical cord blood (Lee et al. 2004). These cells were serum deprived for two days in the presence of EGF and bFGF prior to induction with HGF, bFGF, and nicotinamide for seven days followed by subsequent exposure to oncostatin M, dexamethasone, and ITS. This procedure resulted in a cell population expressing albumin, α-fetoprotein, glucose-6-phosphatase, tyrosine-aminotransferase, CK18, tryptophan 2,3-dioxygenase and CYP2B6 as evidenced by RT-PCR. In addition, cells showed albumin production, urea secretion, and uptake of low-density lipoprotein. However, these data were not quantitatively compared to primary hepatocytes. Using a semiquantitative assay, the authors observed induction of PROD. As discussed above, classification of a cell type as a hepatocyte will require also a quantitative analysis. Human embryonic stem cells derived from the inner cell mass of blastocyst stage embryos should in principle be able to serve as a source for hepatocytes. On the other hand, it might turn out to be extremely difficult to establish adequate differentiation protocols. Lavon et al. revealed a subset of fetal hepatic enriched genes in human embryonic stem cells upon their differentiation to embryonic bodies (Lavon et al. 2004). The authors introduced an albumin promoter sequence with EGFP as a reporter and isolated EGFP positive cells. These cells are promising candidates for further in vitro differentiation experiments. However, differentiation of this embryonic stem cell derived precursors to cells expressing hepatocyte-like enzyme activities has to our knowledge not yet been achieved.

Recently, three studies have reprogrammed human peripheral blood monocytes and differentiated them to hepatocytes (Ruhnke et al. 2005a, Zhao et al. 2003, Ruhnke et al. 2005b). Zhao et al. (2003) isolated monocytes from buffy coats by adherence to plastic dishes (Zhao et al. 2003). Incubation of monocytes with macrophage colony-stimulating growth factor (M-CSF) for 14 days resulted in formation of fibroblast-like cells. Subsequent incubation with HGF for 5–7 days led to a round cell morphology, whereby 75 and 81% of the cells stained positive for albumin and α-fetoprotein. Ruhnke et al. (2005a, b) used a protocol culturing human blood monocytes with β-mercaptoethanol, M-CSF, and IL-3 for six days followed by a 14 day incubation period with FGF-4 (Ruhnke et al. 2005b, a). The resulting cell type (termed NeoHep cell) showed a hexagonal shape with diameters between 54 and 112 μm, which is larger compared to cultured primary human hepatocytes. RT-PCR was positive for albumin, CYP3A4, CYP2B6, CYP2C9, coagulation factor VII, and asia-loglycoprotein receptor 1 and 2. Positive immunostaining was obtained with antibodies to albumin, α_1-antitrypsin, CYP2D6, and CYP2C9. However, detailled experiments aimed at determination of the proximity of NeoHep cells to primary hepatocytes still have to be performed.

In conclusion, numerous studies have demonstrated that extrahepatic stem cells can be induced to express several factors that are found in hepatocytes. In the vast majority of experiments, these markers have been demonstrated using IHC and/or RT-PCR. Quantitative analysis of enzyme activities and direct comparison of the stem cell derived cell types with primary hepatocytes is not yet available. Thus, the published studies do not establish if extrahepatic stem cells can indeed generate hepatocytes.

9.6.8 Quantitative Analyses Comparing Stem Cell Derived Cells with Primary Hepatocytes

In principle, it should not be necessary to define the term "hepatocyte", since this cell type is well known and most of its activities (including the large ranges of interindividual differences) have been qualitatively and quantitatively described (Hengstler et al. 2000b, Gebhardt et al. 2003, Steinberg et al. 1999). However, in previous years some confusion about the definition of the term "hepatocyte" has occurred. In a number of different publications, stem or precursor cells were exposed to specific substances or defined microenviron-ments. These studies reported on "hepatocyte markers" such as albumin or cytokeratin 18, as well as induction of an epithelial phenotype. Without a doubt, de novo expression of previously silent hepatocyte markers in stem cells is an intriguing observation. Therefore, it is understandable that many scientists termed their stem cell derived cell types, for instance, "hepatocyte" or "parenchymal hepatic cell", or "functional hepatocyte" or "liver cell" (whereby the latter definition is imprecise, because it is relatively easy to differentiate between hepatocytes and the various non-parenchymal liver cell types). How-ever, to use the term hepatocyte for a cell displaying some molecular hallmarks of a true hepatocyte is misleading. For the vast majority of scientists working on toxicological or pharmacological problems, hepatocytes are used as a valuable experimental tool for identifying metabolites, toxic effects, and to study enzyme induction. Obviously, if stem cell derived hepatocytes are not fulfilling, a minimal set of defined properties consequences to patient's health can not be excluded if these cells are used for toxicological experiments. An example illustrating this important point is a scenario where toxic metabolites will be missed by virtue of an inadequate cell type. Therefore, a clear and reason-able definition of the term "hepatocyte" is needed particularly for scientists interested in drug metabolism. From a toxicologist's point of view, properties such as epithelial morphology and expression of some hepatocyte markers are not sufficient to consider a cell as a hepatocyte. Independent from the confusion caused by different notions of the term "hepatocyte", some conclusions drawn from the expression of hepatocyte markers are of interest. An example is albumin expression. As a matter of fact, hepatocytes are the only cell type secreting albumin. However, the conclusion that any albumin secreting cell (or even a cell positive for albumin in immunostaining or RT-PCR) necessarily represents a hepatocyte is highly problematic. It is possible that stem cell derived cell types express for example albumin together with a limited number of hepatocyte markers, but this does not mean that they also express the necessary set of hundreds of genes that make up a true hepatocyte. For instance, several cell lines obtained from hepatomas or even from pancreatic carcinomas are known to secrete albumin, but these lines also lack the majority of drug metabolizing enzymes. Therefore, it seems reasonable to introduce additional criteria to define if a cell is a true hepatocyte or only shares several characteristics. In this context it is also important to understand what properties a hepatocyte or the surrogate hepatocyte-like cell should have to be appropriate for the planned assay or experimental question.

In order to obtain an overview whether a stem cell derived cell type is really hepatocyte-like gene array analysis is an adequate approach. Cells from at least five donors should be compared. Using, for instance, principle component analysis it will be possible to determine to which degree gene expression patterns of stem cell derived "hepatocyte-like cells" resemble primary hepatocytes. It will also be possible to analyze to which degree mRNA expression patterns of the stem cells approach the primary hepatocytes during the differentiation process. To our knowledge, this type of analysis has not yet been published for "hepatocyte-like cells" obtained from stem cells in vitro. In addition to an overview over gene expression patterns functional analysis is mandatory. From a toxicological point of view any candidate hepatocyte-like cell type should mirror a minimal set of functions of a true hepatocyte, namely (i) metabolism of xenobiotics and endogenous substances (hormones and ammonia), (ii) synthesis and secretion of albumin, clotting factors, complement, transporter proteins, bile, lipids and lipoproteins, and (iii) storage of glucose (glycogen), fat soluble vitamins A, D, E and K, folate, vitamin B_{12}, copper, and iron. A stem cell derived hepatocyte candidate cell should be thoroughly compared to cultured primary hepatocytes by using a battery of quantitative functional assays. A reasonable choice of parameters will allow an experimental evaluation with a very good probability to decide if the selected cell is appropriate for the planned assay. In case of drug metabolism, a candidate cell is practically indistinguishable from primary hepatocytes if all criteria in Table 9.2 are fulfilled, that is, activities of the surrogate cell are equal or very similar to the enzymatic activities of primary hepatocytes. In this regard, it is crucial to determine properly normalized activities, for instance, expressed as "nmol product formed per minute per mg protein". This normalization allows to directly compare the measured activities to standard reference data (Hengstler et al. 2000a, b, Gebhardt et al. 2003). Some of the enzyme activity suggested in Table 9.2 shows a relatively large interindividual variability as expected from described enzyme polymorphisms (Hengstler et al. 1999). Nevertheless, the range of activities in the human population is known and activities of stem cell derived hepatocytes should be within these ranges.

RNA or protein analysis alone is not sufficient, since these parameters do not prove any functionality. Experiments from our laboratory aimed at differentiating stem or precursor cells to hepatocytes suggested that there might be a hierarchy in the above-mentioned parameters and useful criteria for classifying candidate hepatocytes. For example, it seems to be relatively easy to induce an epithelial morphology, expression of albumin, or synthesis of urea in stem cells. However, this does not automatically mean that other hepatocyte-specific parameters will likewise be functional. Studies performed in our laboratory point to the cytochrome P450 3A4 as an important marker for hepatocytes. In cultured primary human hepatocytes, basal CYP3A4 activities range between 300 and 700 pmol 6β-hydroxytestosterone/min/mg protein using an HPLC standard assay determining 6β-hydroxylation of testosterone (Oesch et al. 1992). After induction with rifampicin (50 µM) for 72 h, CYP3A4 activity of primary human hepatocytes increased to 1.500–14.000 pmol 6α-hydroxytestosterone/min/mg protein. To our knowledge, this hepatocyte criterion has not yet been met by any extrahepatic human stem cell derived candidate and might, therefore, be adequate for testing the hepatocyte potential of any stem cell derived hepatocyte-like cell type.

When stem cell derived cell types are characterized by their capacity to show enzyme induction, it is critical to not only determine induction factors, but also to always measure the (absolute) basal activities. This is crucial, since cytochrome P450 is not exclusively limited to hepatocytes. Indeed, cytochrome P450 induction was also reported for lung, colon and small intestine epithelial cells, white adipose tissue, and several other cell types (Vassilopoulos et al. 2003, Yoshinari et al. 2004, Baijal et al. 1997). In terms of drug metabolism it is, therefore, possible to unequivocally define whether a candidate cell is a hepatocyte or not. Central to this definition is of course both a qualitative (presence/absence of hepatocytic markers) together with a qualitative (enzyme activity) evaluation (Table 9.3).

9.6.9 5-Years Perspective

Presently primary human hepatocytes isolated from resected liver tissue are the gold standard in pharmacology, toxicology, and cell transplantation. A relatively large number of groups have demonstrated that extrahepatic stem cells can be induced to express several hepatocyte-specific markers. Expression of these markers was achieved after transplantation of human stem cells into the

Table 9.3 Markers of hepatocyte differentiation and cell sources

Human cell type	Route and number of transplanted cells	Liver injury	Observations	Recipient mice (m), sheep (s)	Reference
Adherently proliferating cells from cord blood	Injection of 2×10^5 cells into the left liver lobe	None	Expression of human albumin in the left liver lobe of SCID mice 7 and 21 days after transplantation (IHC, RT-PCR)	m	Beerheide et al. (2002)
$Lin^-CD38^-CD34^-ClqR_p^+$ and $Lin^-CD38^-D34^+ClqR_p^+$ cells isolated from cord blood and bone marrow; no culture	Tail vein injection of 500 to 7×10^4 cells	3.75 cGy	Expression of "human hepatocyte specific antigen" and human c-met 8–10 weeks after transplantation (IHC), human albumin (RT-PCR). Mouse liver suspensions showed a rare population of human HLA- ABC$^+$ and CD45$^-$ cells (flow cytometry)	m	Danet et al. (2002)
Unsorted mononuclear cell preparations of human cord blood	Tail vein infusion of 50×10^6 cells	2.5 Gy	Expression of the HepPar1 antigen (IHC) in livers 4, 6, and 16 weeks after transplantation into SCID/ NOD mice. No evidence for cell fusion (FISH)	m	Newsome et al. (2003)
$CD34^+$ or $CD34^+$, $CD38^-$, $CD7^-$ cells isolated by from cord blood; no cell culture	Tail vene injection of 0.02 $CD34^+$ or 1×10^5 $CD34^+$, $CD38^-$, $CD7^-$ cells	3 Gy CCl$_4$. 0.4 ml/kg	Expression of human albumin (WB, RT-PCR, IHC) in livers of SCID/NOD mice 5 and 30 days after liver injury. Positive RT-PCR for CK19. Negative results without CCl$_4$ induced liver injury	m	Wang et al. (2003)

Table 9.3 (continued)

Human cell type	Route and number of transplanted cells	Liver injury	Observations	Recipient mice (m), sheep (s)	Reference
CD34$^+$ cells isolated from cord blood and from peripheral blood	Tail vein injection of 2×10^5 cells	3.75 Gy CCl$_4$, 10,15, or 30 μl/ mouse	Human albumin positive cells preferentially around bile ducts (IHC, RT-PCR, WB) in livers of SCID/NOD mice. Neutralization of the SDF-1 receptor CXCR4 abolished homing of human stem cells to the mouse liver, whereas local injection of SDF-1 into the mouse liver increased homing	m	Kollet et al. (2003)
Adherently proliferating cells from cord blood	Injection of 10×10^6 cells into the portal vein	2-AAF and one.third hepat-ectomy	Expression of human albumin (RT-PCR, IHC) and "human hepatocyte antigen" (IHC) in liver, human albumin detection in serum (WB) and detection of human X chromosome centromeres (FISH) in liver of SCID mice	m	Kakinuma et al. (2003)
CD34$^+$ and CD45$^+$ cells from cord blood	Tail vein injection of 1×10^5 cells	5-fluoro-uracil and anti mouse c-kit	Expression of human albumin (RT-PCR), HepPar1 antigen (IHC) and human centromeres (FISH) in livers of NOD/SCID/BMG^{null} mice	m	Ishikawa et al. (2003)
Nestin positive islet-derived adherently proliferating precursor cells	Injection of 0.15, 1.5, and 7.5×10^5 cells into the left liver lobe	None	Expression of human albumin but not of mouse albumin in individual cells (IHC, RT-PCR) of the left liver lobe of SCID/NOD mice after 3 and 12 weeks. Negative results	m	Von Mach et al. (2004)

Table 9.3 (continued)

Human cell type	Route and number of transplanted cells	Liver injury	Observations	Recipient mice (m), sheep (s)	Reference
			after injection of 0.15×10^5 cells, but positive for 1.5 and 7.5×10^5 injected cells		
CD34$^+$ cells isolated from cord blood	$3-5 \times 10^5$ for intra-fetal and 15–20 cells for intra-blastocyst injection	None	Human albumin, HepPar1 antigen, and human α1-antitrypsin (IHC, RT-PCR) were expressed in livers of immunocompetent C57Bl6 and CD1 mice 1 and 4 weeks after birth	m	Turrini et al. (2004)
CD34+ Lin-CD38- cells from human bone marrow, cord blood and mobilized peripheral blood	Intra-fetal injection of 2×10^4 cells	None	Expression of human albumin (IHC, ELISA), "human hepatocyte antigen" (IHC), and detection of Alu-sequences (FISH)	s	Almeida-Porada et al. (2004)
Adherently proliferating cells from cord blood	Intra-fetal injection of 1 500 cells in sheep	None	Expression of human albumin and "human hepatocyte specific antigen" in livers of sheep (IHC, WB)	s	Kogler et al. (2005)

liver of laboratory animals or by incubation with cytokines. These intriguing observations prompted quite a number of scientists to classify these stem cell derived cell types as hepatocytes. This may create the impression that a generation of human hepatocytes from stem cells is already state of the art, which is factually wrong. For further progress, it is important to clearly define activities of stem cell derived "hepatocyte-like cells" that closely resemble those of primary hepatocytes, and it is similarly important to identify those which are not hepatocyte-like. Mechanisms responsible for lack of expression of hepatocyte factors may be loss of transcription factor expression, methylation of genes and promoters, alterations in histone acetylation status. Elucidation of these mechanisms is a precondition for further progress. It may be extremely difficult to differentiate stem cells to a cell type that resembles primary hepatocytes in all aspects. However, complete conformity with hepatocytes is not necessary in order to be a valuable tool for specific applications, for example, in drug metabolism or cell therapy. For instance, a stem cell derived cell type with inducible CYP3A4, CYP2B6, CYP1A1/2, and CYP2E1 that responds to all substances that can cause induction of these enzymes in primary hepatocytes would be a valuable tool. Although the "stem cell derived human hepatocyte" does not yet exist, there is a good chance to generate stem cell derived cell types that may be useful for specific applications in drug development or cell therapy.

Questions/Exercises

1. Explain the actual legal situation to perform hepatic cell isolation and transplantation.
2. Why is it not allowed to use isolated liver cells for cell transplantation from patients that underwent tumor therapy?
3. Describe the isolation process of liver cells from large and small liver specimens and describe possible differences and similarities of both procedures.
4. Which types of bioreactors/artificial liver devices exist? Explain their advantages and disadvantages.
5. Describe the advantages to use human liver cells instead of non-human cells in drug developments.
6. Give at least five examples for human hepatocyte transplantation and explain different clinical diseases in which hepatocyte transplantation has been applied.
7. In the chapter about "reporter studies for cell fusion", we describe two studies resulting in contradictory conclusions (Alvarez-Dolado versus Harris). Design a specific experiment that could dissolve the discrepancy.
8. A scientist presents a new stem cell derived "hepatocyte-like cell type". He demonstrates expression of 17 "hepatocyte-factors" by conventional RT-PCR and 12 by immunostaining and concludes: 'The high number of hepatocyte factors expressed in our stem cell derived cells proves that they are

indeed hepatocytes.' Do you agree? Which additional experiments would you suggest to determine the proximity of these cells to genuine hepatocytes? How can interindividual differences between hepatocytes from different donors be considered?

9. Can unmodified embryonic stem cells be used for cell therapy of metabolic liver diseases? (answer: no; risk of teratoma formation).
10. Give typical characteristics of a human hepatocyte-like cell and discuss differences and similarities with "native" human hepatocytes.

Ten Key Publications

1. Dorko K, Freeswick PD, Bartoli F, Cicalese L, Bardsley BA, Tzakis A, Nussler AK (1994) A new technique for isolating and culturing human hepatocytes from whole or split livers not used for transplantation. Cell Transplant 3:387–395
2. Hengstler JG, Brulport M, Schormann W, Bauer A, Hermes M, Nussler AK, Fandrich F, Ruhnke M, Ungefroren H, Griffin L, Bockamp E, Oesch F, von Mach MA (2005) Generation of human hepatocytes by stem cell technology: definition of the hepatocyte. Expert Opin Drug Metab Toxicol 1:61–74
3. Li AP (2001) Screening for human ADME/Tox drug properties in drug discovery. Drug Discov Today 6:357–366
4. Michalopoulos GK, DeFrances M (2005) Liver regeneration. Adv Biochem Eng Biotechnol 93:101–134
5. Michalopoulos GK, DeFrances MC (1997) Liver regeneration. Science 276:60–66
6. Nussler A, Konig S, Ott M, Sokal E, Christ B, Thasler W, Brulport M, Gabelein G, Schormann W, Schulze M, Ellis E, Kraemer M, Nocken F, Fleig W, Manns M, Strom SC, Hengstler JG (2006) Present status and perspectives of cell-based therapies for liver diseases. J Hepatol 45:144–159
7. Petersen BE, Bowen WC, Patrene KD, Mars WM, Sullivan AK, Murase N, Boggs SS, Greenberger JS, Goff JP (1999) Bone marrow as a potential source of hepatic oval cells. Science 284:1168–1170
8. Ruhnke M, Ungefroren H, Nussler A, Martin F, Brulport M, Schormann W, Hengstler JG, Klapper W, Ulrichs K, Hutchinson JA, Soria B, Parwaresch RM, Heeckt P, Kremer B, Fandrich F (2005) Differentiation of in vitro-modified human peripheral blood monocytes into hepatocyte-like and pancreatic islet-like cells. Gastroenterology 128:1774–1786
9. Seglen PO (1976) Preparation of isolated rat liver cells. Methods Cell Biol. 13:29–83
10. Strom SC et al. (1998) Large scale isolation and culture of human hepatocytes. In: Franco D, Boudjema K, Varet K (eds) Ilots de Langerhans et hepatocytes: Vers une utilisation therapeutique, pp 195–205

References

Alison MR, Vig P, Russo F, Bigger BW, Amofah E, Themis M, Forbes S (2004) Hepatic stem cells: from inside and outside the liver? Cell Prolif. 37:1–21

Almeida-Porada G, Porada CD, Chamberlain J, Torabi A, Zanjani ED (2004) Formation of human hepatocytes by human hematopoietic stem cells in sheep. Blood 104:2582–2590

Alvarez-Dolado M, Pardal R, Garcia-Verdugo JM, Fike JR, Lee HO, Pfeffer K, Lois C, Morrison SJ, Alvarez-Buylla A (2003) Fusion of bone-marrow-derived cells with Purkinje neurons, cardiomyocytes and hepatocytes. Nature 425:968–973

Ambrosino G, Varotto S, Strom SC, Guariso G, Franchin E, Miotto D, Caenazzo L, Basso S, Carraro P, Valente ML, D'Amico D, Zancan L, D'Antiga L (2005) Isolated hepatocyte transplantation for Crigler-Najjar syndrome type 1. Cell Transplant 14:151–157

Aoki T, Jin Z, Nishino N, Kato H, Shimizu Y, Niiya T, Murai N, Enami Y, Mitamura K, Koizumi T, Yasuda D, Izumida Y, Avital I, Umehara Y, Demetriou AA, Rozga J, Kusano M (2005) Intrasplenic transplantation of encapsulated hepatocytes decreases mortality and improves liver functions in fulminant hepatic failure from 90% partial hepatectomy in rats. Transplantation 79:783–790

Avital I, Inderbitzin D, Aoki T, Tyan DB, Cohen AH, Ferraresso C, Rozga J, Arnaout WS, Demetriou AA (2001) Isolation, characterization, and transplantation of bone marrow-derived hepatocyte stem cells. Biochem Biophys Res Commun 288:156–164

Baijal PK, Fitzpatrick DW, Bird RP (1997) Phenobarbital and 3-methylcholanthrene treatment alters phase I and II enzymes and the sensitivity of the rat colon to the carcinogenic activity of azoxymethane. Food and Chemical Toxicology 35:789–798

Beerheide W, von Mach MA, Ringel M, Fleckenstein C, Schumann S, Renzing N, Hildebrandt A, Brenner W, Jensen O, Gebhard S, Reifenberg K, Bender J, Oesch F, Hengstler JG (2002) Downregulation of beta2-microglobulin in human cord blood somatic stem cells after transplantation into livers of SCID-mice: an escape mechanism of stem cells? Biochem Biophys Res Commun 294:1052–1063

Berry M et al. (1991) Biochemical properties. In: Berry M, Edwards AM, Barritt GJ (eds) Isolated hepatocytes: Preparation, properties and applications, pp 121–178

Berthou F, Ratanasavanh D, Riche C, Picart D, Voirin T, Guillouzo A (1989) Comparison of caffeine metabolism by slices, microsomes and hepatocyte cultures from adult human liver. Xenobiotica 19:401–417

Bilir BM, Guinette D, Karrer F, Kumpe DA, Krysl J, Stephens J, McGavran L, Ostrowska A, Durham J (2000) Hepatocyte transplantation in acute liver failure. Liver Transpl 6:32–40

Bowen WP, Carey JE, Miah A, McMurray HF, Munday PW, James RS, Coleman RA, Brown AM (2000) Measurement of cytochrome P450 gene induction in human hepatocytes using quantitative real-time reverse transcriptase-polymerase chain reaction. Drug Metab Dispos 28:781–788

Burlina AB (2004) Hepatocyte transplantation for inborn errors of metabolism. J Inherit Metab Dis 27:373–383

Carmo H, Hengstler JG, de Boer D, Ringel M, Carvalho F, Fernandes E, Remiao F, dos Reys LA, Oesch F, de Lourdes BM (2004) Comparative metabolism of the designer drug 4-methylthioamphetamine by hepatocytes from man, monkey, dog, rabbit, rat and mouse. Naunyn Schmiedebergs Arch Pharmacol 369:198–205

Carmo H, Hengstler JG, de Boer D, Ringel M, Remiao F, Carvalho F, Fernandes E, dos Reys LA, Oesch F, de Lourdes BM (2005) Metabolic pathways of 4-bromo-2,5-dimethoxyphenethylamine (2C-B): analysis of phase I metabolism with hepatocytes of six species including human. Toxicology 206:75–89

Chan C, Berthiaume F, Nath BD, Tilles AW, Toner M, Yarmush ML (2004) Hepatic tissue engineering for adjunct and temporary liver support: critical technologies. Liver Transpl 10:1331–1342

Chiu A and Fan ST (2006) MARS in the treatment of liver failure: controversies and evidence. Int.J.Artif.Organs 29:660–667

Dandri M, Burda MR, Torok E, Pollok JM, Iwanska A, Sommer G, Rogiers X, Rogler CE, Gupta S, Will H, Greten H, Petersen J (2001) Repopulation of mouse liver with human hepatocytes and in vivo infection with hepatitis B virus. Hepatology 33:981–988

Dandri M, Lutgehetmann M, Volz T, Petersen J (2006) Small animal model systems for studying hepatitis B virus replication and pathogenesis. Semin.Liver Dis 26:181–191

Danet GH, Luongo JL, Butler G, Lu MM, Tenner AJ, Simon MC, Bonnet DA (2002) C1qRp defines a new human stem cell population with hematopoietic and hepatic potential. Proc. Natl.Acad.Sci.U.S.A 99:10441–10445

Dhawan A, Mitry RR, Hughes RD, Lehec S, Terry C, Bansal S, Arya R, Wade JJ, Verma A, Heaton ND, Rela M, Mieli-Vergani G (2004) Hepatocyte transplantation for inherited factor VII deficiency. Transplantation 78:1812–1814

Ding X and Kaminsky LS (2003) Human extrahepatic cytochromes P450: function in xenobiotic metabolism and tissue-selective chemical toxicity in the respiratory and gastrointestinal tracts. Annu.Rev.Pharmacol.Toxicol 43:149–173

Donini A, Baccarani U, Risaliti A, Sanna A, Degrassi A, Bresadola F (2001) In vitro functional assessment of a porcine hepatocytes based bioartificial liver. Transplant.Proc 33:3477–3479

Dorko K, Freeswick PD, Bartoli F, Cicalese L, Bardsley BA, Tzakis A, Nussler AK (1994) A new technique for isolating and culturing human hepatocytes from whole or split livers not used for transplantation. Cell Transplant 3:387–395

Enosawa S, Miyashita T, Fujita Y, Suzuki S, Amemiya H, Omasa T, Hiramatsu S, Suga K, Matsumura T (2001) In vivo estimation of bioartificial liver with recombinant HepG2 cells using pigs with ischemic liver failure. Cell Transplant 10:429–433

EU Directive 2001/83/EC and Regulation (EC) No 726/2004. COM567final.

Fausto N (2004) Liver regeneration and repair: hepatocytes, progenitor cells, and stem cells. Hepatology 39:1477–1487

Fiegel HC, Lioznov MV, Cortes-Dericks L, Lange C, Kluth D, Fehse B, Zander AR (2003) Liver-specific gene expression in cultured human hematopoietic stem cells. Stem Cells 21:98–104

Fisher RA, Bu D, Thompson M, Tisnado J, Prasad U, Sterling R, Posner M, Strom S (2000) Defining hepatocellular chimerism in a liver failure patient bridged with hepatocyte infusion. Transplantation 69:303–307

Fox IJ, Chowdhury JR, Kaufman SS, Goertzen TC, Chowdhury NR, Warkentin PI, Dorko K, Sauter BV, Strom SC (1998) Treatment of the Crigler-Najjar syndrome type I with hepatocyte transplantation. N.Engl.J.Med 338:1422–1426

Fujii H, Hirose T, Oe S, Yasuchika K, Azuma H, Fujikawa T, Nagao M, Yamaoka Y (2002) Contribution of bone marrow cells to liver regeneration after partial hepatectomy in mice. J.Hepatol 36:653–659

Gebhardt R, Hengstler JG, Muller D, Glockner R, Buenning P, Laube B, Schmelzer E, Ullrich M, Utesch D, Hewitt N, Ringel M, Hilz BR, Bader A, Langsch A, Koose T, Burger HJ, Maas J, Oesch F (2003) New hepatocyte in vitro systems for drug metabolism: metabolic capacity and recommendations for application in basic research and drug development, standard operation procedures. Drug Metab Rev. 35:145–213

Gerlach JC (2006) Bioreactors for extracorporeal liver support. Cell Transplant. 15 Suppl 1:S91–103

Gerlach JC, Zeilinger K, Grebe A, Puhl G, Pless G, Sauer I, Grunwald A, Schnoy N, Muller C, Neuhaus P (2003) Recovery of preservation-injured primary human hepatocytes and non-parenchymal cells to tissuelike structures in large-scale bioreactors for liver support: an initial transmission electron microscopy study. J.Invest Surg. 16:83–92

Gilgenkrantz H (2004) Mesenchymal stem cells: an alternative source of hepatocytes? Hepatology 40:1256–1259

Glanemann M, Gaebelein G, Nussler N, Hao L, Kronbach Z, Shi B, Neuhaus P, Nussler AK (2008) Transplantation of monocyte-derived hepatocyte-like cells (NeoHeps) improves survival in a model of acute liver failure. Ann Surg. In press

Harris RG, Herzog EL, Bruscia EM, Grove JE, Van Arnam JS, Krause DS (2004) Lack of a fusion requirement for development of bone marrow-derived epithelia. Science 305: 90–93

Hengstler JG, Bogdanffy MS, Bolt HM, Oesch F (2003) Challenging dogma: thresholds for genotoxic carcinogens? The case of vinyl acetate. Annu.Rev.Pharmacol.Toxicol. 43: 485–520

Hengstler JG, Brulport M, Schormann W, Bauer A, Hermes M, Nussler AK, Fandrich F, Ruhnke M, Ungefroren H, Griffin L, Bockamp E, Oesch F, von Mach MA (2005) Generation of human hepatocytes by stem cell technology: definition of the hepatocyte. Expert.Opin.Drug Metab Toxicol. 1:61–74

Hengstler JG, Ringel M, Biefang K, Hammel S, Milbert U, Gerl M, Klebach M, Diener B, Platt KL, Bottger T, Steinberg P, Oesch F (2000a) Cultures with cryopreserved hepatocytes: applicability for studies of enzyme induction. Chem.Biol.Interact. 125:51–73

Hengstler JG, Utesch D, Steinberg P, Platt KL, Diener B, Ringel M, Swales N, Fischer T, Biefang K, Gerl M, Bottger T, Oesch F (2000b) Cryopreserved primary hepatocytes as a constantly available in vitro model for the evaluation of human and animal drug metabolism and enzyme induction. Drug Metab Rev. 32:81–118

Hengstler JG, Van der BB, Steinberg P, Oesch F (1999) Interspecies differences in cancer susceptibility and toxicity. Drug Metab Rev. 31:917–970

Horslen SP, McCowan TC, Goertzen TC, Warkentin PI, Cai HB, Strom SC, Fox IJ (2003) Isolated hepatocyte transplantation in an infant with a severe urea cycle disorder. Pediatrics 111:1262–1267

Howell JM and Mercer JF (1994) The pathology and trace element status of the toxic milk mutant mouse. J.Comp Pathol. 110:37–47

Hughes RD, Mitry RR, Dhawan A (2005) Hepatocyte transplantation for metabolic liver disease: UK experience. J.R.Soc.Med. 98:341–345

Hukkanen J, Lassila A, Paivarinta K, Valanne S, Sarpo S, Hakkola J, Pelkonen O, Raunio H (2000) Induction and regulation of xenobiotic-metabolizing cytochrome P450s in the human A549 lung adenocarcinoma cell line. Am.J.Respir.Cell Mol.Biol. 22:360–366

Ishikawa F, Drake CJ, Yang S, Fleming P, Minamiguchi H, Visconti RP, Crosby CV, Argraves WS, Harada M, Key LL, Jr., Livingston AG, Wingard JR, Ogawa M (2003) Transplanted human cord blood cells give rise to hepatocytes in engrafted mice. Ann.N.Y. Acad.Sci. 996:174–185

Jalan R (2005) Acute liver failure: current management and future prospects. J.Hepatol. 42 Suppl:S115–S123

Jang YY, Collector MI, Baylin SB, Diehl AM, Sharkis SJ (2004) Hematopoietic stem cells convert into liver cells within days without fusion. Nat.Cell Biol. 6:532–539

Jiang Y, Jahagirdar BN, Reinhardt RL, Schwartz RE, Keene CD, Ortiz-Gonzalez XR, Reyes M, Lenvik T, Lund T, Blackstad M, Du J, Aldrich S, Lisberg A, Low WC, Largaespada DA, Verfaillie CM (2002) Pluripotency of mesenchymal stem cells derived from adult marrow. Nature 418:41–49

Kakinuma S, Tanaka Y, Chinzei R, Watanabe M, Shimizu-Saito K, Hara Y, Teramoto K, Arii S, Sato C, Takase K, Yasumizu T, Teraoka H (2003) Human umbilical cord blood as a source of transplantable hepatic progenitor cells. Stem Cells 21:217–227

Kanazawa Y and Verma IM (2003) Little evidence of bone marrow-derived hepatocytes in the replacement of injured liver. Proc.Natl.Acad.Sci.U.S.A 100 Suppl 1:11850–11853

Kogler G, Sensken S, Airey JA, Trapp T, Muschen M, Feldhahn N, Liedtke S, Sorg RV, Fischer J, Rosenbaum C, Greschat S, Knipper A, Bender J, Degistirici O, Gao J, Caplan AI, Colletti EJ, Almeida-Porada G, Muller HW, Zanjani E, Wernet P (2004) A new human somatic stem cell from placental cord blood with intrinsic pluripotent differentiation potential. J.Exp.Med. 200:123–135

Kollet O, Shivtiel S, Chen YQ, Suriawinata J, Thung SN, Dabeva MD, Kahn J, Spiegel A, Dar A, Samira S, Goichberg P, Kalinkovich A, Arenzana-Seisdedos F, Nagler A, Hardan I, Revel M, Shafritz DA, Lapidot T (2003) HGF, SDF-1, and MMP-9 are involved in stress-induced human CD34+ stem cell recruitment to the liver. J.Clin.Invest 112:160–169

Kreamer BL, Staecker JL, Sawada N, Sattler GL, Hsia MT, Pitot HC (1986) Use of a low-speed, iso-density percoll centrifugation method to increase the viability of isolated rat hepatocyte preparations. In Vitro Cell Dev.Biol. 22:201–211

Lagasse E, Connors H, Al Dhalimy M, Reitsma M, Dohse M, Osborne L, Wang X, Finegold M, Weissman IL, Grompe M (2000) Purified hematopoietic stem cells can differentiate into hepatocytes in vivo. Nat.Med. 6:1229–1234

Lavon N, Yanuka O, Benvenisty N (2004) Differentiation and isolation of hepatic-like cells from human embryonic stem cells. Differentiation 72:230–238

Lee KD, Kuo TK, Whang-Peng J, Chung YF, Lin CT, Chou SH, Chen JR, Chen YP, Lee OK (2004) In vitro hepatic differentiation of human mesenchymal stem cells. Hepatology 40:1275–1284

Li AP (2001) Screening for human ADME/Tox drug properties in drug discovery. Drug Discov.Today 6:357–366

Li AP, Gorycki PD, Hengstler JG, Kedderis GL, Koebe HG, Rahmani R, de Sousas G, Silva JM, Skett P (1999) Present status of the application of cryopreserved hepatocytes in the evaluation of xenobiotics: consensus of an international expert panel. Chem.Biol. Interact. 121:117–123

Mallet VO, Mitchell C, Mezey E, Fabre M, Guidotti JE, Renia L, Coulombel L, Kahn A, Gilgenkrantz H (2002) Bone marrow transplantation in mice leads to a minor population of hepatocytes that can be selectively amplified in vivo. Hepatology 35:799–804

Meier M, Woywodt A, Hoeper MM, Schneider A, Manns MP, Strassburg CP (2005) Acute liver failure: a message found under the skin. Postgrad.Med.J. 81:269–270

Miksys S, Hoffmann E, Tyndale RF (2000) Regional and cellular induction of nicotine-metabolizing CYP2B1 in rat brain by chronic nicotine treatment. Biochem.Pharmacol. 59:1501–1511

Mito M, Kusano M, Kawaura Y (1992) Hepatocyte transplantation in man. Transplant.Proc. 24:3052–3053

Mitry RR, Dhawan A, Hughes RD, Bansal S, Lehec S, Terry C, Heaton ND, Karani JB, Mieli-Vergani G, Rela M (2004) One liver, three recipients: segment IV from split-liver procedures as a source of hepatocytes for cell transplantation. Transplantation 77:1614–1616

Mitzner S, Klammt S, Stange J, Schmidt R (2006) Albumin regeneration in liver support-comparison of different methods. Ther.Apher.Dial. 10:108–117

Muraca M, Gerunda G, Neri D, Vilei MT, Granato A, Feltracco P, Meroni M, Giron G, Burlina AB (2002) Hepatocyte transplantation as a treatment for glycogen storage disease type 1a. Lancet 359:317–318

Najibulla K, Christina A, Jan-Georg H, Axel H, Augustinus B (2000) Diazepam metabolism and albumin secretion of porcine hepatocytes in collagen-sandwich after cryopreservation. Biotechnology Letters V22:1647–1652

Nelson DR, Koymans L, Kamataki T, Stegeman JJ, Feyereisen R, Waxman DJ, Waterman MR, Gotoh O, Coon MJ, Estabrook RW, Gunsalus IC, Nebert DW (1996) P450 superfamily: update on new sequences, gene mapping, accession numbers and nomenclature. Pharmacogenetics 6:1–42

Newsome PN, Johannessen I, Boyle S, Dalakas E, McAulay KA, Samuel K, Rae F, Forrester L, Turner ML, Hayes PC, Harrison DJ, Bickmore WA, Plevris JN (2003) Human cord blood-derived cells can differentiate into hepatocytes in the mouse liver with no evidence of cellular fusion. Gastroenterology 124:1891–1900

Nussler A, Konig S, Ott M, Sokal E, Christ B, Thasler W, Brulport M, Gabelein G, Schormann W, Schulze M, Ellis E, Kraemer M, Nocken F, Fleig W, Manns M, Strom SC, Hengstler JG (2006) Present status and perspectives of cell-based therapies for liver diseases. J.Hepatol. 45:144–159

Nussler AK, Di Silvio M, Billiar TR, Hoffman RA, Geller DA, Selby R, Madariaga J, Simmons RL (1992) Stimulation of the nitric oxide synthase pathway in human hepatocytes by cytokines and endotoxin. J.Exp.Med. 176:261–264

Oesch F, Wagner H, Platt KL, Arand M (1992) Improved sample preparation for the testosterone hydroxylation assay using disposable extraction columns. J.Chromatogr. 582:232–235

Ott M, Schmidt HH, Cichon G, Manns MP (2000) Emerging therapies in hepatology: liver-directed gene transfer and hepatocyte transplantation. Cells Tissues.Organs 167:81–87

Otte JB, de Ville dG, Alberti D, Balladur P, de Hemptinne B (1990) The concept and technique of the split liver in clinical transplantation. Surgery 107:605–612

Parkinson A (1996) Biotransformation of xenobiotics. In: Klaassen CD (eds) Casarett & Doull's Toxicology, pp 113–186

Petersen BE, Bowen WC, Patrene KD, Mars WM, Sullivan AK, Murase N, Boggs SS, Greenberger JS, Goff JP (1999) Bone marrow as a potential source of hepatic oval cells. Science 284:1168–1170

Petersen J, Burda MR, Dandri M, Rogler CE (2004) Transplantation of human hepatocytes in immunodeficient UPA mice: a model for the study of hepatitis B virus. Methods Mol. Med. 96:253–260

Pless G, Steffen I, Zeilinger K, Sauer IM, Katenz E, Kehr DC, Roth S, Mieder T, Schwartlander R, Muller C, Wegner B, Hout MS, Gerlach JC (2006) Evaluation of primary human liver cells in bioreactor cultures for extracorporeal liver support on the basis of urea production. Artif.Organs 30:686–694

Reder-Hilz B, Ullrich M, Ringel M, Hewitt N, Utesch D, Oesch F, Hengstler JG (2004) Metabolism of propafenone and verapamil by cryopreserved human, rat, mouse and dog hepatocytes: comparison with metabolism in vivo. Naunyn Schmiedebergs Arch.Pharmacol. 369:408–417

Rifai K, Ernst T, Kretschmer U, Hafer C, Haller H, Manns MP, Fliser D (2005) The Prometheus device for extracorporeal support of combined liver and renal failure. Blood Purif. 23:298–302

Ringel M, Oesch F, Gerl M, Klebach M, Quint M, Bader A, Bottger T, Hengstler JG (2002) Permissive and suppressive effects of dexamethasone on enzyme induction in hepatocyte co-cultures. Xenobiotica 32:653–666

Ringel M, von Mach MA, Santos R, Feilen PJ, Brulport M, Hermes M, Bauer AW, Schormann W, Tanner B, Schon MR, Oesch F, Hengstler JG (2005) Hepatocytes cultured in alginate microspheres: an optimized technique to study enzyme induction. Toxicology 206:153–167

Ruhnke M, Nussler AK, Ungefroren H, Hengstler JG, Kremer B, Hoeckh W, Gottwald T, Heeckt P, Fandrich F (2005a) Human monocyte-derived neohepatocytes: a promising alternative to primary human hepatocytes for autologous cell therapy. Transplantation 79:1097–1103

Ruhnke M, Ungefroren H, Nussler A, Martin F, Brulport M, Schormann W, Hengstler JG, Klapper W, Ulrichs K, Hutchinson JA, Soria B, Parwaresch RM, Heeckt P, Kremer B, Fandrich F (2005b) Differentiation of in vitro-modified human peripheral blood monocytes into hepatocyte-like and pancreatic islet-like cells. Gastroenterology 128: 1774–1786

Schwartz RE, Reyes M, Koodie L, Jiang Y, Blackstad M, Lund T, Lenvik T, Johnson S, Hu WS, Verfaillie CM (2002) Multipotent adult progenitor cells from bone marrow differentiate into functional hepatocyte-like cells. J Clin Invest 109:1291–1302

Seddon T, Michelle I, Chenery RJ (1989) Comparative drug metabolism of diazepam in hepatocytes isolated from man, rat, monkey and dog. Biochem Pharmacol 38:1657–1665

Seglen PO (1976) Preparation of isolated rat liver cells. Methods Cell Biol. 13:29–83

Skarin T, Becher R, Bucht A, Duvefelt K, Bohm S, Ranneberg-Nilsen T, Lilleaas EM, Schwarze PE, Toftgard R (1999) Cis-acting sequences from the rat cytochrome P450 2B1 gene confer pulmonary and phenobarbital-inducible expression in transgenic mice. Am J Respir Cell Mol Biol 21:177–184

Sokal EM, Smets F, Bourgois A, Van Maldergem L, Buts JP, Reding R, Bernard OJ, Evrard V, Latinne D, Vincent MF, Moser A, Soriano HE (2003) Hepatocyte transplantation in a 4-year-old girl with peroxisomal biogenesis disease: technique, safety, and metabolic follow-up. Transplantation 76:735–738

Starzl TE, Fung J, Tzakis A, Todo S, Demetris AJ, Marino IR, Doyle H, Zeevi A, Warty V, Michaels M, . (1993) Baboon-to-human liver transplantation. Lancet 341:65–71

Starzl TE and Lakkis FG (2006) The unfinished legacy of liver transplantation: emphasis on immunology. Hepatology 43:S151–S163

Steinberg P, Fischer T, Kiulies S, Biefang K, Platt KL, Oesch F, Bottger T, Bulitta C, Kempf P, Hengstler J (1999) Drug metabolizing capacity of cryopreserved human, rat, and mouse liver parenchymal cells in suspension. Drug Metab Dispos 27:1415–1422

Stephenne X, Najimi M, Smets F, Reding R, de Ville dG, Sokal EM (2005) Cryopreserved liver cell transplantation controls ornithine transcarbamylase deficient patient while awaiting liver transplantation. Am J Transplant 5:2058–2061

Strom SC, Chowdhury JR, Fox IJ (1999) Hepatocyte transplantation for the treatment of human disease. Semin Liver Dis 19:39–48

Strom SC et al. (1998) Large scale isolation and culture of human hepatocytes. In: Franco D, Boudjema K, Varet K (eds) Ilots de Langerhans et hepatocytes: Vers une utilisation therapeutique, pp 195–205

Strom SC, Fisher RA, Rubinstein WS, Barranger JA, Towbin RB, Charron M, Mieles L, Pisarov LA, Dorko K, Thompson MT, Reyes J (1997a) Transplantation of human hepatocytes. Transplant Proc 29:2103–2106

Strom SC, Fisher RA, Thompson MT, Sanyal AJ, Cole PE, Ham JM, Posner MP (1997b) Hepatocyte transplantation as a bridge to orthotopic liver transplantation in terminal liver failure. Transplantation 63:559–569

Strom SC, Jirtle RL, Jones RS, Novicki DL, Rosenberg MR, Novotny A, Irons G, McLain JR, Michalopoulos G (1982) Isolation, culture, and transplantation of human hepatocytes. J Natl Cancer Inst 68:771–778

Strong RW (2006) Living-donor liver transplantation: an overview. J Hepatobiliary Pancreat Surg 13:370–377

Terada N, Hamazaki T, Oka M, Hoki M, Mastalerz DM, Nakano Y, Meyer EM, Morel L, Petersen BE, Scott EW (2002) Bone marrow cells adopt the phenotype of other cells by spontaneous cell fusion. Nature 416:542–545

Terai S, Sakaida I, Yamamoto N, Omori K, Watanabe T, Ohata S, Katada T, Miyamoto K, Shinoda K, Nishina H, Okita K (2003) An in vivo model for monitoring trans-differentiation of bone marrow cells into functional hepatocytes. J Biochem (Tokyo) 134:551–558

Theise ND (2003) Liver stem cells: the fall and rise of tissue biology. Hepatology 38:804–806

Theise ND, Badve S, Saxena R, Henegariu O, Sell S, Crawford JM, Krause DS (2000) Derivation of hepatocytes from bone marrow cells in mice after radiation-induced myeloablation. Hepatology 31:235–240

Tomatsu S, Orii KO, Fleming RE, Holden CC, Waheed A, Britton RS, Gutierrez MA, Velez-Castrillon S, Bacon BR, Sly WS (2003) Contribution of the H63D mutation in HFE to murine hereditary hemochromatosis. Proc Natl Acad Sci USA 100:15788–15793

Turrini P, Monego G, Gonzalez J, Cicuzza S, Bonanno G, Zelano G, Rosenthal N, Paonessa G, Laufer R, Padron J (2005) Human hepatocytes in mice receiving pre-immune injection with human cord blood cells. Biochem Biophys Res Commun 326:66–73

Vassilopoulos G, Wang PR, Russell DW (2003) Transplanted bone marrow regenerates liver by cell fusion. Nature 422:901–904 von Mach MA (2002) Primary biliary cirrhosis in classmates: coincidence or enigmatic enviromental influence? EXCLI J 1:1–7

von Mach MA, Hengstler JG, Brulport M, Eberhardt M, Schormann W, Hermes M, Prawitt D, Zabel B, Grosche J, Reichenbach A, Muller B, Weilemann LS, Zulewski H (2004) In vitro cultured islet-derived progenitor cells of human origin express human albumin in severe combined immunodeficiency mouse liver in vivo. Stem Cells 22: 1134–1141

von Mach MA, Hermanns-Clausen M, Koch I, Hengstler JG, Lauterbach M, Kaes J, Weilemann LS (2005) Experiences of a poison center network with renal insufficiency in acetaminophen overdose: an analysis of 17 cases. Clin Toxicol (Phila) 43:31–37

Wang X, Montini E, Al Dhalimy M, Lagasse E, Finegold M, Grompe M (2002) Kinetics of liver repopulation after bone marrow transplantation. Am J Pathol 161:565–574

Wang X, Willenbring H, Akkari Y, Torimaru Y, Foster M, Al Dhalimy M, Lagasse E, Finegold M, Olson S, Grompe M (2003) Cell fusion is the principal source of bone-marrow-derived hepatocytes. Nature 422:897–901

Wieczorek SJ and Tsongalis GJ (2001) Pharmacogenomics: will it change the field of medicine? Clin Chim Acta 308:1–8

Willenbring H, Bailey AS, Foster M, Akkari Y, Dorrell C, Olson S, Finegold M, Fleming WH, Grompe M (2004) Myelomonocytic cells are sufficient for therapeutic cell fusion in liver. Nat Med 10:744–748

Wortelboer HM, de Kruif CA, van Iersel AA, Falke HE, Noordhoek J, Blaauboer BJ (1991) Comparison of cytochrome P450 isoenzyme profiles in rat liver and hepatocyte cultures. The effects of model inducers on apoproteins and biotransformation activities. Biochem Pharmacol 42:381–390

Ying QL, Nichols J, Evans EP, Smith AG (2002) Changing potency by spontaneous fusion. Nature 416:545–548

Yoshinari K, Sato T, Okino N, Sugatani J, Miwa M (2004) Expression and induction of cytochromes p450 in rat white adipose tissue. J Pharmacol Exp Ther 311:147–154

Zhang QY, Dunbar D, Kaminsky LS (2003) Characterization of mouse small intestinal cytochrome P450 expression. Drug Metab Dispos 31:1346–1351

Zhao Y, Glesne D, Huberman E (2003) A human peripheral blood monocyte-derived subset acts as pluripotent stem cells. Proc Natl Acad Sci USA 100:2426–2431

Zulewski H, Abraham EJ, Gerlach MJ, Daniel PB, Moritz W, Muller B, Vallejo M, Thomas MK, Habener JF (2001) Multipotential nestin-positive stem cells isolated from adult pancreatic islets differentiate ex vivo into pancreatic endocrine, exocrine, and hepatic phenotypes. Diabetes 50:521–533

Chapter 10
Peripheral Nerve Injury, Repair, and Regeneration

Rudolf K. Potucek, Stephen W.P. Kemp, Naweed I. Syed, and Rajiv Midha

Contents

10.1   Introduction   321
10.2   Current Clinical Methods of Nerve Repair
     and Reconstruction   323
     10.2.1   Direct Co-aption (Neurorrhaphy)   324
     10.2.2   Nerve Autografts.   325
     10.2.3   Nerve Transfers.   326
     10.2.4   Nerve Guidance Tubes   326
10.3   Emerging Technologies for Nerve Repair   328
     10.3.1   Electric Fields to Aid and Guide Nerve Growth   328
     10.3.2   Potential Cellular Machinery Affected by Electric Field Effects   329
     10.3.3   Wound Potentials and Spinal Cord Repair   330
10.4   Use of Stem Cells to Aid Regeneration   331
     10.4.1   Mechanical Guidance of Regeneration.   332
     10.4.2   Microchip Technologies Under Development   332
     10.4.3   Technologies for Aiding Repair   332
     10.4.4   Non-Repair Technologies: Prosthetics   334
10.5   Summary   335
Questions/Exercises.   335
References   336

10.1 Introduction

Injuries to the peripheral nervous system (PNS) present a serious health problem for society, affecting approximately 2.8% of all trauma cases, often resulting in poor recovery of function and subsequent impaired quality of life for the patient (McAllister et al. 1996, Noble et al. 1998, Belkas et al. 2004, Lundborg 2004). For example, approximately 360,000 people in the United States alone suffer from paralytic syndromes of the upper extremity annually, resulting in 8,648,000 and 4,916,000 restricted activity and bed/disability days

N.I. Syed (✉)
Hotchkiss Brain Institute, Faculty of Medicine, University of Calgary, Calgary,
Alberta, Canada
e-mail: nisyed@ucalgary.ca

M. Santin (ed.), *Strategies in Regenerative Medicine*,
DOI 10.1007/978-0-387-74660-9_10, © Springer Science+Business Media, LLC 2009

respectively (Kelsey et al. 1997). Approximately 100,000 patients undergo neurosurgical procedures of the PNS in the United States and Europe annually (Schlosshauer et al. 2006). The majority of patients with PNS injury has both motor and sensory deficits and often suffer from neuropathic pain. In contrast to the central nervous system (CNS), the PNS may exhibit spontaneous regeneration, albeit over shorter distances, with poor recovery of function being particularly prevalent in injuries that completely sever nerves from their targets.

The healing of nerve injuries is unique in that it is a process of cellular repair as opposed to tissue repair (Lundborg and Danielsen 1991). Transected axons (Fig. 10.1) must redeem their original axoplasmic volume by extending processes distally, concomitant with the surrounding microenvironment.

Transection of peripheral nerves results in Wallerian degeneration in all axons distal to the injury site, evidenced by the disintegration of axoplasmic microtubules and neurofilaments (Seckel 1990). The majority of axons along the distal stumps of transected nerves are reduced to granular and amorphous debris within 24 h; by 48 h, the myelin sheath transforms into short segments

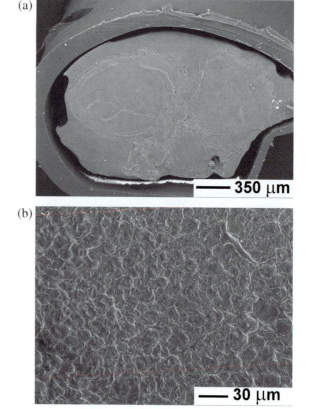

Fig. 10.1 Scanning electron microscopy of a mouse nerve. Arrows show epinerium, asterisks indicate fascicles arranged into fibres. (Photos by Dr. Antonio Merolli, Catholic University, Rome, Italy and Dr. Matteo Santin, University of Brighton, UK). (a) Nerve architecture, (b) axon bundles

to form ovoids (Chaudhry et al. 1992). Activated macrophages invade the degenerating distal nerve stump and phagocytose the disintegrating nerve fibers and myelin. There is an accompanying proliferation of Schwann cells, resulting in the formation of longitudinal Schwann cell bands (bands of Bungner), as they divide and remain within the basal lamina lined endoneurial tubes. Proximal to the injury site where the axons are still intact, both myelinated and unmyelinated fibers spontaneously sprout new axons, forming a "regenerating unit" that is surrounded by a common basal lamina (Lundborg 2004). Axonal sprouts, facilitated by laminin, proceed in a distal fashion, as they establish an intimate relationship with axons and proliferating/migrating Schwann cells (Chen et al. 2005).

Clinical treatment of peripheral nerve injuries focuses on proximally regenerating axons into distal endoneurial tubes, which ultimately leads to specific end-organ re-innervation. Unfortunately, the clinical treatment of peripheral nerve injuries has changed relatively little over the past 30 years, despite considerable advances in our understanding of the neurobiology of nerve regeneration (Belkas et al. 2004). The following sections detail current clinical methods of nerve repair and nerve reconstruction management.

10.2 Current Clinical Methods of Nerve Repair and Reconstruction

Prior to discussing the strategies for repair and reconstruction of injured nerves, it is imperative to elucidate the normal fascicular anatomy of peripheral nerves and their surrounding connective tissue layers. When examining normal peripheral nerve architecture (Figs. 10.1a and b), it is apparent that each peripheral nerve is encircled by an external epineurial sheath (Fig. 10.1, arrow), which is composed of connective tissue and longitudinal blood vessels. Each nerve contains both an external and an internal epineurium. Nerve fibers (Fig. 10.1b asterisks) occupy approximately 25–75% of the cross-sectional area of a nerve, depending on the examined nerve and its location (Matsuyama et al. 2000). In general, there is more connective tissue in the nerve at the point where it crosses the joint. In addition, nerve fibers may be of a myelinated or unmyelinated nature. The diameter of myelinated nerve fibers ranges from 2 to 25 μm (Sunderland 1968), whereas unmyelinated fibers are in the range of 0.2–3 μm (Gibbels 1989).

Each myelinated nerve fiber and groups of unmyelinated nerve fibers are closely associated and reside within endoneurial connective tissue (Fig. 10.1b), elaborated by their ensheating Schwann cell. In turn, several thousand nerve fibers are packed within connective tissue bundles called fascicles (Fig. 10.1 asterisks). The fascicles (0.04–3 mm) are surrounded by a perineurium, which is a lamellated sheath of considerable tensile strength (Lundborg 2004). From a surgical perspective, the perineurium is the smallest structure capable of accepting sutures in peripheral nerves (Sunderland 1951). Based on their fascicular structure, nerves may generally be divided into four basic patterns of intraneural architecture. Nerves containing one large fascicle are termed monofascicular (e.g., terminal branches of the digital nerves), while those containing a few or discrete number of fascicles are termed oligofascicular. However, the majority of nerves contains fascicles of varying sizes which are termed polyfascicular (e.g., the radial nerve in the upper arm).

10.2.1 Direct Co-aption (Neurorrhaphy)

The majority of peripheral nerve repair surgeries fall under two basic categories: (1) direct repair (neurorrhaphy), and; (2) bridge procedures, with autologous nerve grafts being the most common procedure to repair nerve injury gaps. Direct end-to-end repair is usually possible in the majority of clean lacerating injuries and in some cases of delayed repair when the two nerve ends can be brought together without undue tension. Neurorrhaphy can be further broken down into three distinct repair procedures: (1) epineurial repair; (2) grouped fascicular repair, and: (3) fascicular repair.

10.2.1.1 Epineurial Repair

Co-aptation of the nerve stumps by suturing the external epineurium is the traditional method of nerve repair (Diao and Vannuyen 2000, Trumble and McCallister 2000, Lundborg 2004). These repairs are most appropriate for monfascicular (e.g., digital) nerves and diffusely grouped polyfascicular (most proximal limb and plexus element) nerves. Simplistically, this method achieves continuity of the connective tissue from the proximal to the distal stumps, without tension and with appropriate alignment of both stumps. The primary goal is to achieve good co-aption of proximal and distal fascicular anatomy. Freshening of the two nerve ends to debride the nerve and remove scar tissue is, therefore, critical. Achieving appropriate nerve alignment can be aided by inspecting for longitudinal blood vessels in the epineurium as well as attending to fascicular alignment.

10.2.1.2 Grouped Fascicular Repair

Grouped fascicular repair is a potentially more accurate method than epineurial repair. One theoretical disadvantage of epineurial repair is the inability to precisely match the appropriate proximal and distal fascicles. It must be noted, however, that both animal and clinical studies have not shown a clear advantage of one repair method over the other (Posen et al. 1989, Matsuyama et al. 2000). For practical purposes, the grouped fascicular technique is especially indicated in situations where an easily identifiable part of the cross section of the nerve supplies sensory function, while another portion of the nerve supplies motor function. More distal extremity nerves, such as the elbow to wrist segments of the ulnar and median nerve, are examples of nerves that merit this type of repair. Another indication is the nerve injury requiring a split repair. In this situation, a portion of the nerve that is clearly regenerating (diagnosed by both clinical and electrical criteria) is preserved in continuity using external and internal neurolysis techniques to split apart fascicular groups, so that the groups of fascicles that are clearly neurotemetic (nerve in discontinuity or regeneration blocked by scar within nerve) undergo repair.

10.2.1.3 Fascicular Repair

In rare situations, a specific fascicular repair may be indicated. An example of where this repair would be appropriate is a clean lacerating injury of the median or ulnar nerve at the wrist, where individual motor and sensory fascicles can be defined. An additional example is where a partial lacerating injury occurs across a portion of the cross section of the nerve. In this case, individual fascicles from the proximal stump may be repaired to their reciprocal fascicles in the distal stump.

10.2.2 Nerve Autografts

Direct surgical co-aption of the ends of a severed or lacerated nerve is unfortunately not always feasible. When direct repair cannot be performed without undue tension, nerve grafting must be undertaken and is the current surgical gold standard of treatment of peripheral nerve injuries (Kline and Hudson 1995, Gordon et al. 2003). A requirement for this is that when nerve grafting is performed, there must be no tension at the proximal and distal repair sites to prevent postoperative distraction. Therefore, nerve grafting is usually performed with the extremity in full extension. In general, a nerve graft should be about 10% longer than the existing nerve gap. The majority of autologous nerve grafts are cutaneous nerves of the lower (e.g., sural nerve), and the upper extremities (Matsuyama et al. 2000). When the caliber of the host stump undergoing grafting is the same dimension as the nerve graft, a slightly larger segment than the gap is needed. Frequently, the cross sectional area of the host nerve stump will be several times larger than the diameter of the nerve graft. Multiple segments of the nerve graft will, therefore, be needed to allow coverage of the entire cross sectional surface of the host nerve stumps. Thus, it is imperative to harvest the maximal amount of nerve graft material available when harvesting the donor nerve. The primary advantage of this method of repair is that nerve grafts contain Schwann cells and basal lamina endoneurial tubes (Hudson et al. 1972, Belkas et al. 2004), which provide neurotrophic factors as well as favorable cell and endoneurial tube surface adhesion molecules to regenerating axons (Richardson 1991, Martini 1994, Fu and Gordon 1997).

10.2.3 Nerve Transfers

Nerve transfers (also known as "neurotization") involve the repair of a distal denervated nerve element using a proximal foreign nerve as the donor of neurons and their axons, which will re-innervate the distal targets. The concept is to sacrifice the function of a donor (lesser valued) muscle to revive function in the recipient nerve and muscle that will undergo re-innervation (Narakas 1984). Since their first report by Tuttle in 1913 (Tuttle 1913), and popularization by Narakas 3 decades ago (Narakas 1984), nerve transfers have become increasingly popular for the repair of brachial plexus injuries, especially where the proximal motor source of the denervated element is absent because of avulsion from the spinal cord. Increasingly advocated is the use of transfers in situations where the proximal motor source is available, but the regeneration distances are long to warrant successful re-innervation. A nerve transfer into the denervated distal nerve stump close to the motor end-organ would then restore function, which would otherwise not be possible (Nath et al. 1997). The use of nerve transfers has, therefore, been a major advance in the field of brachial plexus nerve reconstructive surgery, with many different ingenious transfers associated with improving results (Hou and Xu, 2002 Bertelli and Ghizoni 2004).

10.2.4 Nerve Guidance Tubes

Although nerve autografts remain the gold standard for peripheral nerve repair, nerve autografting is inherently flawed due to donor site morbidity, a lack of donor tissue availability, incomplete and non-specific regeneration, and subsequent modest recovery of function in both animal models and clinical cases (Dahlin and Lundborg 1998, Bellamkonda 2006, Nichols et al. 2006). However, an artificial (non-nerve) conduit or nerve guidance tube interposed between the proximal and distal nerve stumps may provide a more suitable environment for regenerating nerve fibers to sample and respond to appropriate directional (tropic) and trophic cues derived from migrating Schwann cells and soluble growth factors emanating from both nerve stumps. (Fig. 10a)

10.2.4.1 Silicone Tubes and the Development of Clinically Approved Nerve Guidance Tubes

Clinical translation of nerve guidance tubes occurred in the past two decades. In pioneering studies in the early 1990s, Lundborg demonstrated the feasibility and success of ulnar and median nerve reconstruction using short silicone conduits in a few patients (Lundborg and Danielsen 1991, Lundborg et al. 1994). However, these impermeable, non-biodegradable tubes elicited an inflammatory and fibrotic reaction and produced chronic nerve compression, requiring their removal after regeneration had occurred through them (Merle et al. 1989). Surgeons have, therefore, increasingly used biodegradable materials for clinical use, as recently reviewed (Schlosshauer et al. 2006). Based on promising primate and clinical experience, polyglycolic acid (PGA) tubes were found to be comparable to nerve autografts in the repair of digital nerves with defects up to 3 cm in a prospective randomized clinical study (Weber et al. 2000). PGA tubes (Neurotube, Neuroregen LLC, Bel Air, MD) and subsequently collagen nerve tubes (NeuraGen, Integra Life Sciences, Plainsboro, NJ) have been approved in the U.S. for the repair of peripheral nerve injuries (Archibald et al. 1991, Archibald et al. 1995). In 2001, SaluMedica (Atlanta, GA) and Collagen Matrix (Franklin Lakes, NJ) each received approval for their tubular constructs used in repairing peripheral nerves. Using a repeated freeze-thawing technique, SaluMedica produces a hydrogel tube made from polyvinyl alcohol (PVA) while Collagen Matrix has developed a collagen nerve cuff made from collagen fibers. Most recently, Polyganics (Groningen, Netherlands) employed a dip coating procedure to manufacture a resorbable poly(DL-lactide-caprolactone) tube (Neurolac). Although approved for human use, the efficacy and thus the indications for all the tubes marketed to date are limited to the repair of short defects (<3 cm) of mainly the small-caliber nerves (Meek and Coert 2002).

10.2.4.2 Future Biotechnological Applications of Nerve Guidance Tubes: Administration of Neurotrophic Factors

The creation of an artificial, growth factor augmented nerve tube, built on our fundamental knowledge of axonal guidance, may eventually provide an improved alternative to current nerve guidance tubes. For instance, it has been well established that gradients of neurotrophic factors direct and guide axonal growth during development and that the varying concentrations of growth factors can influence the differentiated state of the cells in neural tissue (Belkas et al. 2004). Similarly to the natural process of development, regeneration is regulated in a stepwise fashion by temporal and spatial molecular cues and cellular responses (Saltzman and Olbricht 2002). Although neurotrophic factors (NGF, BDNF, NT-3, NT-4/5) have not yet been approved for clinical administration to patients who experience peripheral nerve damage, they are very common in animal models of regeneration. The most common modes of employment have been through the use of osmotic pumps (Baffour et al. 1995, Nagano et al. 2003), or at the site of transection through the delivery of gelfoam (Bregman et al. 1997), fibrin glue (Cheng and Chen 2002), and genetically engineered cells (Xu et al. 1995, Nakahara et al. 1996).

An additional possibility is to incorporate neurotrophic factors within the lumen of a nerve guidance conduit, which has been used with considerable success to obtain enhanced regeneration, as compared to autografts (Seckel 1990). For example, Midha et al. (2003) showed that poly (2-hydroxyethyl methacrylate-co-methyl methacrylate (PHEMA-MMA)) tubes augmented with FGF-1 provided for improved sciatic nerve regeneration compared to either empty tubes or those containing collagen alone, and demonstrated regeneration comparable to nerve autografts. Similarly, Lee and colleagues have shown that controlled release of nerve growth factor through a novel fibrin matrix conduit enhances sciatic nerve regeneration in the rat (Lee et al. 2003). The primary goal here is for neurotrophic factors to not only favor neuronal cell survival following injury, but also to display chemotactic properties through providing appropriate directional cues to regenerating axons.

Two potential problems with this method concern the bioavailability of the neurotrophic factor being administered (NGF, BDNF, and NT-3 have half-lives of under 10 min, Poduslo and Curran 1996), and the establishment of an inadequate concentration gradient delivered across the apparatus. Recently, a new model to study and influence nerve regeneration has been developed (McDonald and Zochodne 2003), which allows for the direct manipulation of the microenvironment of regenerating peripheral axons through the use of a microinjection port (MIP) connected to a silicon based T-tube chamber. It has long been known that neurotrophic factors accumulate in vivo within silicon nerve regeneration chambers at both the proximal and distal stumps (Longo et al. 1983a, Longo et al. 1983b, Longo et al. 1984), however, a neurotrophic gradient has yet to be established experimentally in this model. The model designed by McDonald & Zochodne allows for the strategic placement of neurotrophic factor delivery at either the proximal, middle, or distal portion of the nerve conduit, thus enabling a constant gradient of neurotrophic factors within its lumen. Another promising strategy is to provide local cells within the nerve guidance tube, such as Schwann cells or undifferentiated stem cells, that are either intrinsically capable of producing neurotrophic factors or are bioengineered to do so (Nishiura et al. 2004, Keilhoff et al. 2006a). One advantage to this method is that using living cells instead of extrinsic factors may allow for the automatic adjustment of the production of neurotrophic substances in a more meticulous way than what is possible by a synthetic conduit.

Considering these exciting methods, we can, therefore, look forward to future advances in tissue engineering and biotechnology, which will provide the next generation of enhanced nerve tubes and growth factor administration. Ultimately, this research may lead to a better understanding of how soluble neurotrophic factors influence peripheral nerve regeneration and functional recovery, and will undoubtedly lead to better outcomes from nerve repairs in patients.

10.3 Emerging Technologies for Nerve Repair

10.3.1 Electric Fields to Aid and Guide Nerve Growth

It has long been known that electric fields and electric currents can influence the growth of individual neurons. In ground-breaking work, the group of Moo-Ming Poo developed a method to apply an electric field across dissociated embryonic *Xenopus* neurons. The strength of the electric field was calculated from the total current passed through the setup, voltage across the setup, the length of the current path and the resistivity of the solution (Patel and Poo 1984). Using this method they showed that the application of an external DC field of 1–10 V/cm would increase the number of sprouts initiated in the regeneration process (Patel and Poo 1982, Patel and Poo 1984). They also showed that the total number of neurons initiating sprouts would increase due to an applied DC field. Most importantly they observed that growing axons oriented in parallel to the electric field, show a preference and enhanced growth rate towards the cathode. McCaig observed that the sprout initiation was also preferred in the direction of the cathode (McCaig et al. 1994). In subsequent work, Poo et al. showed that similar effects can be achieved by using unipolar pulsed fields (Patel and Poo 1984), indicating that a time-averaged field strength of 250 mV/cm is required to achieve directed growth, while others suggest a minimum of 500 mV/cm (Hinkle et al. 1981). Studies by McCaig and colleagues also showed that the ability of either cathode or anode to influence growth was contingent upon the type of the growth permissive substrate - indicating an additional interaction between applied electric fields and substrate adhesion molecules (McCaig et al. 1994, Rajnicek et al. 1998).

As mentioned previously, the regeneration potential for CNS neurons differs considerably from their PNS counterparts. Similarly successful manipulations of embryonic neurons should not be taken to mean that such approach could also be useful in promoting directed growth from injured peripheral nerves. From these studies, it is nevertheless apparent that electrical stimulation of regenerating nerves may lead to accelerated regeneration. It has also been shown, that without stimulation, regeneration will occur in a staggered fashion with only a few axons extending processes initially. Whereas the applied electric stimulus will not only decrease this stagger but also the total regeneration time (Politis et al. 1988, Al-Majed et al. 2000, Brushart et al. 2002, Al-Majed et al. 2004). More importantly, it has also been shown, that an initial 1 h electrical stimulation of a nerve immediately after surgery will increase the specificity of re-innervation (Brushart et al. 2005). These studies show the importance of electric fields and their application during regeneration. However, both the potential target sites and the cellular mechanism by which electric fields affect growth remain poorly understood.

10.3.2 Potential Cellular Machinery Affected by Electric Field Effects

Various ideas have been brought forward to explain the effect of an electric field on the sprouting and growth of neurons. Poo and colleagues observed that the electric fields might facilitate ion channels migration within the membrane to make them more responsive to various growth specific factors (trophic factors and cell adhesion molecules) present in the extracellular milieu. Consistent with this notion are the data which show that poisoning various ion channels eliminates axonal responsiveness to external electric fields (Patel and Poo 1982). Others have observed migration of membrane-bound proteoglycans towards the source of an electric field, thus suggesting their affects on various macromolecules embedded within the regenerating fibers. This notion is supported by observations that preferred direction of neuronal outgrowth depends upon substrate polarity, for example, due to different polymer used in cell culture dishes, suggesting that electric-field sensitive membrane components could interact with the substrate. Hinkle and colleagues proposed that the electro-osmotic flow of solution might be responsible for electric filed-induced directed growth. However, they subsequently found that the directed growth would occur even in gelled solution (Hinkle et al. 1981), thus ruling out the possibility that the electric field- induced effects involve mere electro-osmotic flow movements. In contrast, Borgens observed that solution oozing from wounds and the associated wound currents observed at the injury site would provide improved regeneration (Borgens et al. 1989). Furthermore, Borgens and colleagues observed that an electric current could be detected at the injury site, which in turn facilitated improved regeneration. These researchers argued that the guidance effect may involve Ca^{2+} channel migration within the electric field, which in turn may re-organize cytoskeleton for preferred growth in the direction of the cathode (Borgens et al. 1989). Additionally, Politis et al. (1988) showed that electrical stimulation leads to generally faster and more oriented growth of the cytoskeleton. Taken together, the above studies not only demonstrate naturally existing electric fields in the biological tissue but that they can also influence nerve regeneration after injury. Moreover, the cellular mechanisms by which extrinsically applied electric fields affect nerve regeneration are also being elucidated.

10.3.3 Wound Potentials and Spinal Cord Repair

A different school of thought suggests that electric currents, or "wound currents", can be observed in regenerating tissue and that these wound currents contribute significantly to regeneration after tissue damage. This school of thought maintains that these wound currents contribute to improved regeneration of all tissues, including nerves (Borgens et al. 1989). McCaig and colleagues have observed wound currents of about 100 μA/cm² flowing through tissue with about 1 kΩ resistance, suggesting an electric field of about 100 mV/cm (McCaig et al. 2005). They also observed currents in advance of nerve development in amphibians. Following the reasoning that the flow of electric current is essential to the growth and guidance of nerves during development and regeneration, various experiments have been performed on amphibians (Borgens et al. 1989), guinea pigs (Borgens et al. 1990, Borgens and Bohnert, 1997) and dogs (Borgens et al. 1993), showing that both DC currents and very slow AC currents (15–30 min per cycle) improve nerve growth and regeneration across spinal lesions after transection. Currently, Phase I human trials are under way testing this technology for spinal cord injury repair (Shapiro et al. 2005).

No consensus appears to exist on whether electric fields alone, or aided by the diffusion and movement of nutrients underlie regeneration of CNS and PNS neurons. Nor is there consensus vis-à-vis the mechanisms by which electric field may differentially affect central verses peripheral neuronal regeneration. Even though many of the underlying mechanisms remain unknown, it is, however, clear from the literature that a significant improvement of nerve regeneration can be achieved by applying electric fields during the regeneration process.

10.4 Use of Stem Cells to Aid Regeneration

During nerve regeneration, Schwann cells proliferate to form a sheath along which the axons of the surviving neurons can extend. However, this sheath may not form fast enough or be disrupted by the healing connective tissue (neuroma) surrounding the injury site. As such, an interest exists in providing an artificial sheath for the regenerating nerve to grow in, or in seeding scaffolds with Schwann cells (*see previous section on nerve regeneration tubes*).

It has been demonstrated that such artificial sheaths of Schwann cells can be created by extracting and culturing proliferating Schwann cells from a neuroma (Keilhoff et al. 2006b). However, this approach requires two surgeries, one to extract the cells from the neuroma and a second to resect the neuroma and insert the cultured Schwann cells. Thus, in order to obtain Schwann cells or Schwann cell analogues in a less invasive manner, the use of stem cells has been investigated. It has been shown that stem cells can be differentiated into Schwann cell like cells and used in place of naturally proliferating Schwann cells (Hou et al. 2006). Stem cells or multi-potent precursor cells capable of differentiating into Schwann cells, can be obtained from various tissues (Gage et al. 1995), including gut (Kruger et al. 2002), pancreas (Seaberg et al. 2004), placental mesenchyma (Portmann-Lanz et al. 2006), bone marrow mesenchyma (Keilhoff et al. 2006b), olfactory bulb cells and olfactory epithelium (Chalfoun et al., 2006, Deumens et al. 2006, Marshall et al. 2006), and skin (Toma et al. 2001, McKenzie et al. 2006). While many different sources of stem cells can be used to aid nerve regeneration, it remains a priority to use cells from easily accessible tissues. Thus, the use of olfactory epithelium cells and skin derived precursor cells would be preferred, for example, over gut and pancreas derived stem cells. While good results have been demonstrated using Schwann cells differentiated from stem cells, a remaining problem is de-differentiation of these cells when the factors used in the differentiation procedure are no longer available after implantation (Keilhoff et al. 2006b).

10.4.1 Mechanical Guidance of Regeneration

It has been observed in vitro that the growth of regenerating neurons will align with the orientation of micro-patterned structures (Clark et al. 1993). Similarly it has been shown that proliferating Schwann cells will show 100% alignment to the orientation of geometric structures such as grooves on a surface (Miller et al. 2001). Further research in this field is currently focusing on the development of novel polymers with the aim of improving protein fouling behavior, biodegradability, and erosion characteristics. Additional work is focused on developing polymer materials that can be used to dispense growth factors to aid regeneration (Torres et al. 2007). The ultimate goal of this work would be to generate growth tubes or scaffolds that combine guidance by geometric means with the influence of electric fields and growth factors.

10.4.2 Microchip Technologies Under Development

The development of readily available micro-manufacturing techniques for semiconductors in the 1970s has led to a continuing interest in creating micro-scale devices for drug delivery, regeneration and man-machine interfacing. Today, after more than 30 years of research, the first of these devices are on the brink of clinical utilization and many show great promise in aiding nerve regeneration. Two significantly different approaches are being taken to achieve functional recovery of nerve damage. The first approach is to aid the natural recovery process by enhancing regeneration speed, and the degree of functional recovery. The second approach is to replace natural repair with prosthetics. This research is currently most advanced in the field of limb prosthetics, but some of the current work may hold promise to promote repair of long-gap peripheral nerve injury, obviating the need for autologous nerve grafting.

10.4.3 Technologies for Aiding Repair

10.4.3.1 Controlled Delivery of Trophic Factors

As mentioned previously, regeneration of neurons depends on the availability and gradient of trophic factors (*see section on administration of trophic factors above*). The most basic approach explored for the purpose of providing trophic factors is the use of a porous substrate from which the trophic factors can diffuse into the tissue. This approach has been demonstrated with porous silicon substrates developed using simple etching techniques (Anglin et al. 2004). To improve the dispensing reproducibility, silicon chips containing wells of defined dimensions were manufactured using microchip techniques (Shanley and Parker 2006).

The diffusion technique, however, does not allow controlled release of trophic factors to influence regenerating axons along the gap and thus does not easily allow setting up time-varying gradients needed for improved regeneration. Simple microfluidic pumping has been demonstrated to address the problem (Thompson et al. 2003). An alternative pumping technique that could address this problem is a pump based on a heat-expanding material under the control of electronic circuitry. Heated by an electric current the material at the bottom of a well expands, pressing the trophic factors through a narrow opening faster than natural diffusion would occur. Originally reported in a macroscopic form (Kriesel 2002), this concept has now been proven in a microchip form (Santini Jr. et al. 2003a).

Various problems associated with changing temperature within tissue have led to the development of a microchip with capped wells, in which each well can be individually opened by applying an electrical current, allowing the trophic factors to be released into the surrounding tissue. In its most basic form (Santini Jr. et al. 1998, Cho et al. 2005), each well is capped with a thin layer of gold. The gold is although stable against corrosion in the biological environment, however, in the presence of chloride and under applied current it quickly dissolves, thus removing the cap. There are no reports on the effect of the dissolved gold but to avoid problems, an alternative method has been described that uses a conductive polymer cap instead of gold. An even better alternative would be to use a micro actuator to mechanically pierce the cap without producing any corrosion products (Uhland 2006). Finally, as all of these methods do not allow re-sealing of the well, the use of a memory material has been proposed to allow opening and closing the well under electrical control (Santini Jr. et al. 2003b). At this point the technology has not been tested for nerve regeneration, but it shows promise for both the study of the time-dependent effect of growth factors and for clinical use once the desired spatio-temporal pattern of growth factor dispensing is known.

10.4.3.2 Microscopic Guidance Techniques

A field currently under development is the guidance of individual axons using micro-electrode arrays and microchips. These approaches are based on capacitive coupling between the substrate and neurons or by a scaled-down version of the electric current guidance described above. In either case, however, the nature of the substrate will affect the guided regeneration. Research to date has shown that porous silicon substrates will provide much higher adhesion for neurons (Johansson et al. 2005) and as such may guide growth in purely mechanical ways. In the case of current-based guidance, additional problems facing this research are striking the balance between using a cheap commercial manufacturing technique, which will provide unstable and corrosion prone aluminum electrodes, and a custom manufacturing or post-processing technique that provides corrosion resistant electrodes. In the case of capacitive coupling, the major problem is to achieve a very high dielectric constant of the insulating material while reducing its thickness and maintaining corrosion resistance. The most promising materials appear to be TiO_2 and Ta_2O_5 (Rose et al. 1985, Palti 1996). None of these techniques have so far proven as promising avenues in the field of nerve regeneration.

10.4.4 Non-Repair Technologies: Prosthetics

Non-invasive prosthetic techniques will only be mentioned here to point out that significant efforts are being made to improve microprocessor, or microchip technologies to decrease the size of the prosthesis and to speed up the initial relearning process involved in the use of a prosthetic device (Kajitani et al. 1999). A number of technologies are under development, which offer promise for future prosthetic development: a) capacitive interfacing with nanowires and transistors b) current based interfacing with microneedle arrays, and c) partial regeneration through sieve electrodes.

10.4.4.1 Capacitive Interfacing with Transistors and Nanowires

The idea of recording neuronal activity through an electronic device was first presented in the 1970s (Bergveld et al. 1976). In this technology a field effect transistor (FET) is used to control the amount of current flowing in a circuit based on the electric field at the transistor gate. A single neuron is placed on the gate and the change of the electric field during an action potential is recorded as a change in the transistor current. This technology was developed to maturity when it was used to record synaptic transmission between snail (*Lymnaea stagnalis*) neurons (Kaul et al. 2004). A similar technology, using nanowire transistors was recently demonstrated (Patosky et al. 2006). While this technology does not currently lend itself to mass manufacturing using semiconductor technology, the increased sensitivity of the approach makes it likely that this will be a significant area of future investigation. Neither technology has been tested on peripheral neurons or fasciculated nerves, but both hold promise for brain controlled prosthetic devices (Hochberg et al. 2006).

10.4.4.2 Current Based Interfacing

Microneedle arrays (Campbell et al. 1989) are another technology that could be used in prosthetic devices. These electrodes have originally been developed as intracortical electrodes and are now commercially available. Similar electrodes have been specifically designed with different needle lengths for interfacing with peripheral nerves (Rutten et al. 2001). Microneedle arrays have also been successfully used for recording directly from peripheral nerves to interface with a robotic arm (http://en.wikipedia.org/wiki/Kevin_Warwick). While this technology has much more obvious application in prosthetics and man-machine interfacing, there is definite promise that future development of micro-needles could lead to bio-mechanical repair methods to replace grafting.

10.4.4.3 Partial Regeneration Through Sieve Electrodes

Micromachined sieve structures were pioneered at Stanford in the early 1990s (Kovacs et al. 1992) and have since been further investigated by groups around the world (Akun et al. 1994, Zhao et al. 1997, Wallman et al. 1999, Meyer et al. 2002). To form the sieve, holes are etched into a silicon substrate and each hole is electrically contacted for interfacing with an axon. When the two ends of a dissected nerve are grafted to both sides of the sieve, the regenerating neurons will grow through the sieve and allow for the electrodes in the sieve to make contact to all the axons passing through the sieve. While this has not been demonstrated experimentally, this approach could be useful for achieving high fidelity in prosthetic devices where major nerve damage has occurred.

10.5 Summary

As presented in this chapter, the surgical nerve repair by suture or grafting, will lead to only partial success in regaining lost nerve function. However, many advances on the current state of the art technologies that combine surgical procedures with bio-medical engineering are breaking ground. From these technologies will stem regeneration tubes that employ electric fields and con-trolled delivery of trophic factors through the intelligence built within the chip circuit. The interfacing of these electronic devices with intact or regenerating neurons will enable us not only to better manage chronic pain but also to control prosthetic limbs.

Questions/Exercises

1. What are the main cell types and histological features of peripheral nerves?
2. Describe the mechanisms of degeneration and repair of peripheral nerves.
3. When direct coaption is used and which are its main procedures?
4. Which are the main features differentiation nerve grafting and nerve transfer?
5. Provide examples of the main types of biomaterials used for nerve guide manufacturing and critically discuss their clinical performance
6. List the biochemical signaling leading to peripheral nerve regeneration and its potential exploitation in surgery

7. Describe at least 2 types of nerve repair strategies which are still at research stage and critically discuss their potential transfer to clinic.
8. What are the strategies and the limitations deriving from the use of stem cells in peripheral nerve repair?
9. Are microchips able to restore nerve functionality? Provide a critical overview of the use of this technology.
10. Which are the main types of technologies adopted for nerve prosthetics?

References

Akun T, Najafi K, Bradley RM (1994) A micromachined silicon sieve electrode for nerve regeneration applications. IEEE Trans Biomed Eng 41:305–313

Al-Majed AA, Neumann CM, Brushart TM, Gordon T (2000) Brief electrical stimulation promotes the speed and accuracy of motor axonal regeneration. J Neurosci 20: 2602–2608

Al-Majed, AA, Tam, SL, Gordon T (2004) Electrical stimulation accelerates and enhances expression of regeneration-associated genes in regenerating rat femoral motoneurons. Cell Mol Neurobiol 24:379–402

Anglin E, Schwartz M, Ng V, Perelman L, Sailor M (2004) Engineering the chemistry and nanostructure of porous silicon fabry-pérot films for loading and release of a steroid. Langmuir 20:11264–11269

Archibald SJ, Krarup C, Shefner J, Li S-T, Madison RD (1991) A collagen-based nerve guide conduit for peripheral nerve repair: an electrophysiological study of nerve regeneration in rodents and nonhuman primates. J Comp Neurol 306:685–696

Archibald SJ, Shefner J, Krarup C, Madison RD (1995) Monkey median nerve repaired by nerve graft or collagen nerve guide tube. J Neurosci 15:4109–4123

Baffour R, Achanta K, Kaufman J, Berman J, Garb JL, Rhee S, Friedmann P (1995) Synergistic effect of basic fibroblast growth factor and methylprednisolone on neurologic function after experimental spinal cord injury. J Neurosurg 83:105–110

Belkas JS, Shoichet MS, Midha R (2004) Peripheral nerve regeneration through guidance tubes. Neurol Res 26:151–160

Bellamkonda RV (2006) Peripheral nerve regeneration: an opinion on channels, scaffolds and anisotropy. Biomaterials 27:3515–3518

Bergveld P, Wiersma J. et al. (1976) Extracellular potential recordings by means of field effect transistor without gate metal, called OSFET. IEEE Trans Biomed Eng 23: 136–144

Bertelli JA, Ghizoni MF (2004) Reconstruction of C5 and C6 brachial plexus avulsion injury by multiple nerve transfers: spinal accessory to suprascapular, ulnar fascicles to biceps branch, and triceps long or lateral head branch to axillary nerve. J Hand Surg 29:131–139

Borgens R, Robinson K, Vanable jr J, McGinnis M (1989) Electric fields in vertebrate repair. Alan R. Liss, Inc., New York.

Borgens R, Blight A, McGinnis M (1990) Functional recovery after spinal cord hemisection in guinea pigs: The effects of applied electric fields. J Compar Neurol 296: 634–653

Borgens R, Toombs J, Blight A, McGinnis M, Bauer M, Widmer W, Cook jr J (1993) Effects of applied electric fields on clinical cases of complete paraplegia in dogs. Restor Neurol Neurosci 5:305–322

Borgens RB, Bohnert DM (1997) The responses of mammalian spinal axons to an applied dc voltage gradient. Experimental Neurology 145:376–389

Bregman BS, McAtee M, Dai HD, Kuhn PL (1997) Neurotrophic factors increase axonal growth after spinal cord injury and transplantation in the adult rat. Exp Neurol 148:475–494

Brushart TM, Hoffman PN, Royall RM, Murinson BB, Witzel C, Gordon T (2002) Electrical stimulation promotes motoneuron regeneration without increasing its speed or conditioning the neuron. J Neurosci 22:6631–6638

Brushart TM, Jari R, Verge V, Rohde C, Gordon T (2005) Electrical stimulation restores the specificity of sensory axon regeneration. Exp Neurol 194:221–229

Campbell P, et al. (1989) A chronic intracortical electrode array: preliminary results. J Biomed Mater Res 23:245–259

Chalfoun C, Wirth G, Evans G (2006) Tissue engineered nerve constructs: where do we stand? J Cell Molecular Med 10:309–317

Chaudhry V, Glass JD, Griffin JW (1992) Wallerian degeneration in peripheral nerve disease. Neurologic Clinics 10:613–627

Chen YY, McDonald D, Cheng C, Magnowski B, Durand J, Zochodne DW (2005) Axon and Schwann cell partnership during nerve regrowth. J Neuropathol Exp Neurol 64:613–622

Cheng B, Chen Z (2002) Fabricating autologous tissue to engineer artificial nerve. Microsurgery 22:133–137

Cho ST, Cromack K, Jara-Almonte J, VerLee DJ (2005) Medicine delivery system. In: US006,953,455), vol. October 11.

Clark P, Britland S, Connolly P (1993) Growth cone guidance and neuron morphology on micropatterned laminin surfaces. Journal of Cell Science 105:203–212

Dahlin LB, Lundborg G (1998) Experimental nerve grafting: towards future solutions of a clinical problem 165–173

Deumens R, Koopmans GC, Honig WMM, Hamers FPT, Maquet V, Jerome R, Steinbusch HWM, Joosten EAJ (2006) Olfactory ensheathing cells, olfactory nerve fibroblasts and biomatrices to promote long-distance axon regrowth and functional recovery in the dorsally hemisected adult rat spinal cord. Experimental Neurology 200:89–8103

Diao E, Vannuyen T (2000) Techniques for primary nerve repair. Hand Clinics 16:53–66

Fu SY, Gordon T (1997) The cellular and molecular basis of peripheral nerve regeneration. Molecular Neurobiology 14:67–116

Gage FH, Ray J, Fisher LJ (1995) Isolation, Characterization, and use of Stem Cells from the CNS. Annual Review of Neuroscience 18:159–192

Gibbels E (1989) Morphometry of unmyelinated nerve fibers. Clinical Neuropathology 8:179–187

Gordon T, Sulaiman O, Boyd JG (2003) Experimental strategies to promote functional recovery after peripheral nerve injuries. Journal of the peripheral nervous system 8:236–250

Hinkle L, McCaig C, Robinson K (1981) The direction of growth of differentiating neurones and myoblasts from frog embryos in an applied electric field. The Journal of Physiology (London) 313:121–135

Hochberg LR, Serruya MD, Friehs GM, Mukand JA, Saleh M, Caplan AH, Branner A, Chen D, Penn RD, Donoghue JP (2006) Neuronal ensemble control of prosthetic devices by a human with tetraplegia. Nature 442:164–171

Hou S-Y, Zhang H-Y, Quan D-P, Liu X-L, Zhu J-K (2006) Tissue-engineered peripheral nerve grafting by differentiated bone marrow stromal cells. Neuroscience 140:101–110

Hou Z, Xu Z (2002) Nerve transfer for treatment of brachial plexus injury: comparison study between the transfer of partial median and ulnar nerves and that of phrenic and spinal accessory nerves. Chinese Journal of Traumatology 5:263–266

Hudson AR, Morris J, Weddell G, Drury A (1972) Peripheral nerve autografts. The Journal of Surgery Research 12:267–274

Johansson F, Kanje M, Eriksson C, Wallman L (2005) Guidance of neurons on porous patterned silicon: is pore size important? Physica Status Solidi (C) 2:3258–3262

Kajitani I, Murakawa M, Nishikawa D, Yokoi H, Kajihara N, Iwata M, Keymeulen D, Sakanashi H, Higuchi T (1999) An evolvable hardware chip for prosthetic hand controller. MicroNeuro '99. Proceedings of the Seventh International Conference on Microelectronics for Neural, Fuzzy and Bio-Inspired Systems, 1999 pp. 179–186

Kaul R, Syed N, Fromherz P (2004) Neuron-semiconductor chip with chemical synapse between identified neurons. Physical Review Letters 92: 038102

Keilhoff G, Goihl A, Stang F, Wolf G, Fansa H (2006a) Peripheral nerve tissue engineering: Autologous Schwann cells vs. transdifferentiated mesechymal stem cells. Tissue Engineering 12:1454–1465

Keilhoff G, Goihl A, Stang F, Wolf G, Fansa H (2006b) Peripheral Nerve Tissue Engineering: Autologous Schwann Cells vs. Transdifferentiated Mesenchymal Stem Cells. Tissue Engineering 12:1451–1465

Kelsey JL, Praemer A, Nelson L, Felberg A, Rice LM (1997) Upper extremity disorders. Frequency, impact, and cost. Churchill Livingstone Inc., New York.

Kline D, Hudson A (1995) Vertebral artery compression. J Neurosurg 83:759

Kovacs, GTA, Storment CW, Rosen JM (1992) Regeneration microelectrode array for peripheral nerve recording and stimulation. IEEE Transactions on Biomedical Engineering. 39:893–902

Kriesel MS (2002) Fluid delivery device with heat activated energy source. US006 485:462

Kruger GM, Mosher JT, Bixby S, Joseph N, Iwashita T, Morrison SJ (2002) Neural Crest Stem Cells Persist in the Adult Gut but Undergo Changes in Self-Renewal, Neuronal Subtype Potential, and Factor Responsiveness. Neuron 35:657–669

Lee AC, Yu VM, Lowe JB, Brenner MJ, Hunter DA, Mackinnon SE, Sakiyama-Elbert SE (2003) Controlled release of nerve growth factor enhances sciatic nerve regeneration. Experimental Neurology 295–303

Longo FM, Manthorpe M, Skaper SD, Lundborg G, Varon S (1983a) Neuronotrophic activities accumulate in vivo within silicone nerve regeneration chambers. Brain Research 109–117

Longo FM, Skaper SD, Manthorpe M, Williams LR, Lundborg G (1983b) Temporal changes of neuronotrophic activities accumulating in vivo within nerve regeneration chambers. Experimental Neurology 81:756–769

Longo FM, Hayman EG, Davis GE, Ruoslahti E, Engvall E, Manthorpe M, Varon S (1984) Neurite-promoting factors and extracellular matrix components accumulating in vivo within nerve regneration chambers. Brain Research.

Lundborg G (2004) Nerve injury and repair: regeneration, reconstruction, and cortical remodeling. Elsevier, Philadelphia.

Lundborg G, Danielsen N (1991) Injury, degeneration, and regeneration. Gelberman RH (Ed.), Operative Nerve Repair and Reconstruction. J.P. Lippincott, New York.

Lundborg G, Rosen B, Abrahamson SO, Dahlin L, Danielsen N (1994) Tubular repair of the median nerve in the human forearm. Preliminary findings. The Journal of Hand Surgery (British and European Volume) 19:273–276

Marshall C, Lu C, Winstead W, Zhang X, Xiao M, Harding G, Klueber·K, Roisen F (2006) The therapeutic potential of human olfactory-derived stem cells. Histology and Histopathology 21:633–643

Martini R (1994) Expression and functional roles of neural cell surface molecules and extracellular matrix components during development and regeneration of peripheral nerves. Journal of Neurocytology 23:1–28

Matsuyama T, Mackay MS, Midha R (2000) Peripheral nerve repair and grafting techniques: a review. Neurol Med Chir (Tokyo) 40:187–199

McAllister RMR, Gilbert SEA, Calder JS, Smith PJ (1996) The epidemiology and management of upper limb peripheral nerve injuries in modern practice. The Journal of Hand Surgery (British and European Volume) 21B:4–13

McCaig CD, Allan DW, Erskine L, Rajnicek AM, Stewart R (1994) Growing Nerves in an Electric Field. Neuroprotocols 4:134–141

McCaig CD, Rajnicek AM, Song B, Zhao M (2005) Controlling cell behavior electrically: current views and future potential. Physiological Reviews 85:943–979

McDonald DS, Zochodne DW (2003) An injectable nerve regeneration chamber for studies of unstable soluble growth factors. J Neurosci Methods 122:171–178

McKenzie IA, Biernaskie J, Toma JG, Midha R, Miller FD (2006) Skin-Derived Precursors Generate Myelinating Schwann Cells for the Injured and Dysmyelinated Nervous System. J Neurosci 26:6651–6660

Meek MF, M.D, Coert JH, M.D (2002) Clinical Use of Nerve Conduits in Peripheral-Nerve Repair: Review of the Literature. Journal of Reconstructive Microsurgery 097–110

Merle M, Dellon AL, Campbell JN, Chang PS (1989) Complications from silicon-polymer intubulation of nerves. Microsurgery 10:130–133

Meyer J-U, Stieglitz T, Ruf HH, Robitzki A, Dabouras V, Wewetzer K, Brinker T (2002) A biohybrid microprobe for implanting into the peripheral nervous system. In: 2nd Annual International IEEE-EMB Special Topic Conference on Microtechnologies in Medicine & Biology) pp. 265–268

Midha R, Munro C, Dalton P, Tator C, Shoichet M (2003) Growth factor enhancement of peripheral nerve regeneration through a novel synthetic hydrogel tube. J Neurosurg 99:555–565

Miller C, Shanks H, Witt A, Rutkowski G, Mallapragada S (2001) Oriented Schwann cell growth on micropatterned biodegradable polymer substrates. Biomaterials 22:1263–1269

Nagano M, Sakai A, Takahashi N, Umino M, Yoshioka K, Suzuki H (2003) Decreased expression of glial cell line-derived neurotrophic factor signaling in rat models of neuropathic pain. British Journal of Pharmacology 140:1252–1260

Nakahara Y, Gage FH, Tuszynski MH (1996) Grafts of fibroblasts genetically modified to secrete NGF, BDNF, NT-3, or basic FGF elicit differential responses in the adult spinal cord. Cell Transplantation 5:191–204

Narakas AO (1984) Thoughts on neurotization or nerve transfers in irreparable nerve lesions. Clinics in Plastic Surgery 11:153–159

Nath RK, Mackinnon SE, Shenaq SM (1997) New nerve transfers following peripheral nerve injuries. Operative Techniques in Plastic and Reconstructive Surgery 4:2–11

Nichols CM, Brenner MJ, Fox IK, Tung TH, Hunter DA, Rickman SR, Mackinnon SE (2006) Effects of motor versus sensory nerve grafts on peripheral nerve regeneration. Experimental Neurology 190:347–355

Nishiura Y, Brandt J, Nilsson A, Kanje M, and Dahlin LB (2004) Addition of cultured Schwann cells to tendon autografts and freeze-thawed muscle grafts improves peripheral nerve regeneration. Tissue Engineering 10:157–164

Noble J, Munro CA, Prasad VSSV, Midha R (1998) Analysis of upper and lower extremity peripheral nerve injuries in a population of patients with multiple injuries. The Journal of Trauma 45:116–122

Palti Y (1996) Implantable sensor chip US5,513,636 vol. May 7

Patel N, Poo M (1982) Orientation of neurite growth by extracellular electric fields. J Neurosci 2:483–496

Patel N, Poo M (1984) Perturbation of the direction of neurite growth by pulsed and focal electric fields. J Neurosci 4:2939–2947

Patosky F, Timko BP, Yu G, Fang Y, Greytak AB, Zheng G, Lieber CM (2006) Detection, Stimulation, and Inhibition of Neuronal Signals with High-Density Nanowire Transistor Arrays. Science 313:1100–1104

Poduslo JF, Curran GL (1996) Permeability at the blood-brain and blood-nerve barriers of the neurotrophic factors: NGF, CNTF, NT-3, BDNF. Brain Research Molecular Brain Research 36:280–286

Politis M, Zanakis M, Albala B (1988) Facilitated regeneration in the rat peripheral nervous system using applied electric fields. The Journal of Trauma 28:1375–1381

Portmann-Lanz CB, Schoeberlein A, Huber A, Sager R, Malek A, Holzgreve W, Surbek DV (2006) Placental mesenchymal stem cells as potential autologous graft for pre- and perinatal neuroregeneration. American Journal of Obstetrics and Gynecology 194:664–673

Posen JM, Phou HN, Hentz VR (1989) Fascicular tubulization: A comparison of experimental nerve repair techniques in the cat. American Journal of Plastic Surgery 22:467–468

Rajnicek AM, Robinson KR, McCaig CD (1998) The Direction of Neurite Growth in a Weak DC Electric Field Depends on the Substratum: Contributions of Adhesivity and Net Surface Charge. Developmental Biology 203:412–423

Richardson PM (1991) Neurotrophic factors in regeneration. Current Opinion in Neurobiology 1:401–406

Rose TL, Kelliher EM, Robblee LS (1985) Assessment of capacitor electrodes for intracortical stimulation. J Neurosci Methods 12:181–193

Rutten W, Mouveroux J-M, Buitenweg J, Heida C, Ruardij T, Marani E, Lakke E (2001) Neuroelectronic interfacing with cultured multielectrode arrays toward a cultured probe. Proceedings of the IEEE 89:1013–1029

Saltzman WM, Olbricht WL (2002) Building drug delivery into tissue engineering. Nature Reviews Drug Discovery 1:177–186

Santini Jr JT, Cima MJ, Langer RS (1998) Microchip drug delivery devices. US005,797,898), vol. August 25

Santini Jr JT, Cima MJ, Uhland SA (2003a) Thermally-activated microchip chemical delivery devices. US006,527,762 vol. March 4

Santini Jr JT, Cima MJ, Uhland SA (2003b) Thermally-activated microchip chemical delivery devices. In: US006,669,683 vol. December 30

Schlosshauer B, Dreesmann L, Schaller H, Sinis N (2006) Synthetic nerve guide implants in humans: a comprehensive survey. Neurosurgery 59:747–748

Seaberg RM, Smukler SR, Kieffer TJ, Enikolopov G, Asghar Z, Wheeler MB, Korbutt G, van der Kooy D (2004) Clonal identification of multipotent precursors from adult mouse pancreas that generate neural and pancreatic lineages. Nature Biotechnology 22: 1115–1124

Seckel B (1990) Enhancement of peripheral nerve regeneration. Muscle & Nerve 13:785–800

Shanley JF, Parker TL (2006) Therapeutic agent delivery device with controlled therapeutic agent release rates. US007,056,338 vol. June 6

Shapiro S, Borgens R, Pascuzzi R, Roos K, Groff M, Purvines S, Rodgers R, Hagy S, Nelson P (2005) Oscillating field stimulation for complete spinal cord injury in humans: a phase 1 trial. J Neurosurg, Spine 2:3–10

Sunderland S (1951) A classification of peripheral nerve injuries producing loss of function. Brain 74:491–516

Sunderland S (1968) Nerve and nerve injuries. Williams & Wilkins, Baltimore.

Thompson DL, Mattes MF, Larson LR, Heruth KT (2003) Single-use therapeutic substance delivery device with infusion rate control. US006,562,000, vol. May 13

Toma JG, Akhavan M, Fernandes KJL, Barnabe-Heider F, Sadikot A, Kaplan DR, Miller FD (2001) Isolation of multipotent adult stem cells from the dermis of mammalian skin. Nature Cell Biology 3:778–784

Torres MP, Determan AS, Anderson GL, Mallapragada SK, Narasimhan B (2007) Amphiphilic polyanhydrides for protein stabilization and release. Biomaterials 28:108–116

Trumble TE, McCallister WV (2000) Repair of peripheral nerve defects in the upper extremity. Hand Clinics 16:37–52

Tuttle H (1913) Exposure of the brachial plexus with nerve transplantation. The Journal of the American Medical Association 61:15–17

Uhland SA (2006) Implantable drug delivery device. US007,052,488, vol. May 30

Wallman L, Levinsson A, Schouenborg J, Holmberg H, Danielsen N, Laurell T (1999) Perforated silicon nerve chips with doped registration electrodes: in vitro performance and in vivo operation. IEEE Transactions on Biomedical Engineering 46:1065–1073

Weber RA, Breidenbach WC, Brown RE, Jabaley ME, Mass DP (2000) A randomized prospective study of polyglycolic acid conduits for digital nerve reconstruction in humans. Plastic and Reconstructive Surgery 106:1036–1045

Xu XM, Guenard V, Kleitman N, Aebischer P, Bunge MB (1995) A combination of BDNF and NT-3 promotes supraspinal axonal regeneration into Schwann cell grafts in adult rat thoracic spinal cord. Experimental Neurology 134:261–272

Zhao Q, Drott J, Laurell T, Wallman L, Lindstrom K, Bjursten LM, Lundborg G, Montelius L, Danielsen N (1997) Rat sciatic nerve regeneration through a micromachined silicon chip. Biomaterials 18:75–80

Chapter 11
Therapeutic Strategies in Ocular Tissue Regeneration: The Role of Stem Cells

K. Ramaesh, N. Stone, and B. Dhillon

Contents

11.1	Introduction	342
11.2	Historical Perspective	343
	11.2.1 Healing, Repair, Regeneration, and Reconstitution	343
	11.2.2 Evolution-Two Levels of Regeneration	343
	11.2.3 "Three Type of Cells" – The Old Theory	344
11.3	Regenerative Medicine of the Eye	344
11.4	Stem Cells	345
	11.4.1 Stem Cell Hierarchy	345
	11.4.2 Embryonic Stem Cells	345
	11.4.3 Sources of Embryonic Stem Cells	346
	11.4.4 Adult Stem Cells	347
	11.4.5 Stem Cells from Umbilical Cord Blood	347
	11.4.6 What Determines the Fate of Stem Cells?......................	347
	11.4.7 Stem Cell Niche..	348
	11.4.8 How to Identify Stem Cells? Stem Cell Markers................	349
	11.4.9 How Safe are Stem Cells for Cell Therapy?	349
11.5	Genetic Regulations of Stem Cells	349
	11.5.1 Genes Controlling Embryonic Stem Cells......................	350
	11.5.2 Regulation of Ocular Stem Cells – Pax6	350
	11.5.3 Transdifferentiation, Metaplasia, and Re-Programming of Cell Fate	351
11.6	Ocular Stem Cells	352
	11.6.1 Corneal Epithelial Stem Cells	352
	11.6.2 Endothelial Stem Cells	353
	11.6.3 Retinal Stem Cells	353
	11.6.4 Stem Cells in New Blood Vessels	353
11.7	Disorders of Stem Cells in the Eye	354
11.8	Strategies of Ocular Regeneration	354
	11.8.1 Limitation of Tissue Damage	354
	11.8.2 Ocular Regeneration with Stem Cells	357
11.9	Gene Therapy	360
11.10	The Future	361
	Questions/Exercises...	361
	References	362

K. Ramaesh (✉)
Tennent Institute of Ophthalmology Gartnavel General Hospital, Glasgow,
Scotland, UK

M. Santin (ed.), *Strategies in Regenerative Medicine*,
DOI 10.1007/978-0-387-74660-9_11, © Springer Science+Business Media, LLC 2009

11.1 Introduction

The eye is a unique window providing direct visualization of the structure and function of epithelial, vascular, and neural tissue. It thus serves as an ideal "model" for assessing cell-therapy based regenerative strategies. Loss of ocular tissues through injury and degeneration may result in visual morbidity; therefore, a need to explore possible avenues of cellular regeneration by manipulation of host and embryonic stem cells exists. In the adult eye, stem cells have been identified in various locations including the conjunctiva, corneal limbus, ciliary body, and neural retina. Recent studies have also suggested that bone marrow derived stem cells are recruited in pathological neovascularization involving both the retina and choroid.

The aim of this chapter is to translate basic scientific understanding into useful clinical application ensuring that the potential promise of stem cell studies is not "lost in translation" in the journey from laboratory to clinic.

Stem cell science underpins regenerative medicine by providing the means to potentially restore the function of pre-existing but under-performing host tissue. It also has the capacity to introduce a completely new viable stem cell pool into the host. Maximizing the potential of both approaches requires a detailed understanding of the stem cell "niche", inter-cellular dialogue, and cell traffic. Maintenance of a newly established stem cell pool also requires insight into pharmacological manipulation of the extracellular matrix and micro-environmental factors. In order to achieve successful tissue regeneration it is necessary to exploit a number of technologies embracing multiple disciplines – molecular and cell biology, gene therapy, tissue engineering, immunotherapy, and sustained-release nanotechnology.

In order to appreciate the concept of ocular tissue regeneration, an understanding of applied ocular anatomy is assumed. The eye is formed by the anterior chamber, the cornea, the conjunctiva, and the lens (www.eyeatlas.com/Eyeatlas/hom.html, www.icaen.uiowa.edu/~aip/Lectures/eye_phys_lecture.pdf). The eye is protected by the eyelid and is hosted in the skull orbital chamber. A wide variety of ocular cell types exist - epithelial cells, stromal cells, neurones, photo-receptors, blood vessels, capillaries, immunological cells, and others. Each cell type has a specific role, thus ensuring the process of vision through which light waves from an object penetrate the cornea to progress into the pupil that is the opening in the centre of the iris. In sequence, the light waves are converged towards a so called nodal point by the cornea and by the crystalline lens that is located behind the pupil. At the nodal point, the image is reversed and inverted. After this image transformation, the light progresses through the vitreous humor that is a gelatineous substance that represents approximately 80% of the eye volume. Through the vitreous humor, the light reaches the retina that is located in the eye posterior area. There, the light waves hit the macula that is a small area of the retina that is responsible for transforming the light waves into electrical signals. These signals can be captured by the optic nerve to be conveyed to the brain by the visual pathway. Because of their regenerative potential, upon damage, stem cells should be able to proliferate into one or more of the different cell types belonging to each eye tissue.

11.2 Historical Perspective

The history of medicine have seen many eras and seen many concepts and ideas how the human body functioned and repaired itself. The direction of medical approach is changing all the time and the current era of "regenerative medicine" medicine no exception. Stem cell biology and cell therapy has enabled to achieve regeneration of lost tissues and, among them, ocular tissues. To understand the cellular bases of ocular tissue regeneration, it is important to distinguish the different mechanisms triggered by insults of traumatic or pathological origin and the clinical strategies based on cell therapy while defining terms such as tissue healing repair and regeneration.

11.2.1 Healing, Repair, Regeneration, and Reconstitution

To understand the concepts underpinning stem cell science, it is important to clearly define the following terms – *healing, repair, regeneration,* and *reconstitution*. Conventionally, the term "healing" refers to the body's replacement of destroyed tissue by living tissue. This may be achieved by two different processes- regeneration or repair. In regeneration, specialized tissue is replaced by proliferation of surrounding undamaged specialized cells. In repair, lost tissue is replaced by granulation tissue which matures to form a scar. When repair occurs, there is typically loss of functional tissue. This is in contrast to regeneration where potential to maintain original function exists.

11.2.2 Evolution-Two Levels of Regeneration

It is well recognized that mammals can replace both cells and tissues, but not individual organs or limbs. In contrast, less evolutionary developed animals possess the ability to regenerate lost tissues, organs, and limbs. This process is known as reconstitution and involves a co-ordinated regeneration of multiple tissues. For example, a lizard is able to re-grow a lost tail. Reconstitution has clearly been lost with evolutionary superiority and the majority of higher animals have only very limited potential to regenerate tissue. It is theorized that mammals did not retain the ability to reconstitute as it did not confer any selective advantage – by the time full functional regeneration would have been achieved, death due to dehydration, starvation, or predation would undoubtedly have occurred. Through evolution the power to regenerate was clearly superseded by the ability to repair – fibrosis being more quickly attained in sacrifice for restoration of function.

11.2.3 *"Three Type of Cells" – The Old Theory*

To provide a full understanding of the idea of healing, it is worth revisiting some previous popular concepts. In 1894, Giulio Bizzozero, a distinguished Italian Professor of pathology noticed a link between mitosis and regenerative capacity (Bizzozero 1894). He divided mitotic cells into three categories.

> ➢ *Labile cells*, which demonstrate mitosis throughout life and regenerate incessantly- for example, surface epithelium, epidermis, and bone marrow.
> ➢ *Stable cells,* in which spontaneous mitotic activity is uncommon following birth but which does occur in response to stimuli -for example, connective tissue and liver cells.
> ➢ *Permanent* cells which are mitotic in adulthood but do not regenerate -for example, striated muscle, cardiac muscle, and neural tissue.

To some extent, these observations still hold true, but experimental biology has discovered a new dimension to regeneration. Manipulation of tissues in vitro in various culture conditions has led to the discovery of "stem cells" even amongst *permanent* cell populations. The potential of this finding is huge and generates optimism that "curing" through regeneration may be a real possibility. For example, a spinal cord injury heals by fibrosis typically leaving the victim with paralysis. If the injured neurons could be regenerated, the lost function could potentially be restored (see Chapter 10). This introduces the question – if these stem cells are already present how can they be manipulated into action? In particular, how this manipulation can be exploited for the regeneration of ocular tissues?

11.3 Regenerative Medicine of the Eye

With respect to the eye - when corneal epithelium is scrapped, the deficit is replaced with identical epithelium. This is not unsurprising since epithelium is a *labile* cell that, by definition, will be able to regenerate. However, when retinal cells are damaged or lost, repair occurs with scarring and loss of function. This bears similarities with events that follow myocardial infarction or ischemic brain injury reflecting the fact that these cells are *permanent* and lack any natural regenerative ability. Retina which is composed of neurons and retinal pigment epithelial does not regenerate in vivo. To achieve retinal regeneration, clinicians need to manipulate the healing process to replace the lost retinal tissue with identical retinal cells having active neuronal function.

Among many clinical applications, regenerative or reparative medicine has been focussed to achieve regeneration of ocular tissues through cell-based therapies. Advancement of cell culture techniques and molecular biology has made this a reality. Indeed, scientists have discovered that tissues in vitro behave differently when cultured under certain circumstances and that all tissues have a tiny population of cells that retain the ability to regenerate. These cells are largely undifferentiated and do not possess any specific markers but have the potential to differentiate into several different types of mature cell. Potentially, this small number of stem cells could be grown under culture conditions and re-introduced into damaged tissues to regenerate and restore.

11.4 Stem Cells

Stem cells are a unique population of cells which have the capacity to self-renew, express markers of primitive undifferentiated cells, and show the capacity to differentiate into a variety of specialized lineages. In the adult eye, stem cells have been identified in various locations that include the conjunctiva, corneal limbus, neural retina, and ciliary body. However, it appears that only limbal stem cells contribute to regeneration.

11.4.1 Stem Cell Hierarchy

A hierarchy of stem cell potential can be considered. That is, these cells may be totipotent, pluripotent, multipotent, or unipotent. For example, the zygote which is the earliest stage in development has the capacity to form the entire organism developing into both fetus and placenta – this is an example of totipotency. In contrast, the inner cell mass of the zygote also known as embryonic stem cells is pluripotent and can differentiate into any of the three major tissues types – endoderm, mesoderm, and ectoderm. They cannot, however, develop into a full organism. In adults, the level of cellular plasticity declines with time but some cells such as hemopoetic stem cells can still give rise to limited differentiated cell types. This is known as multipotency. Under normal circumstances, corneal stem cells can only produce corneal epithelial cells and are, therefore, considered unipotent.

11.4.2 Embryonic Stem Cells

A fertilized ovum develops from a unicellular stage by multiplication into an embryo. The zygote which is the earliest stage in development attains a cystic form known as the blastocyst. Embryologists have identified a central core in the blastocyst known as the inner cell mass which gives rise to three germ layers from which all body tissues develop. In 1981, Evans and Kaufman managed to isolate mouse inner cell mass in culture (Evans and Kaufman 1981). When incubated in vitro with appropriate signals, these cells were directed to differentiate into a variety of cell types that would form the basis of a mouse. Cells derived from the inner cell mass are known as embryonic cells. Human embryonic stem cells line was established in 1998 by Thomson et al. (1998).

11.4.3 Sources of Embryonic Stem Cells

As the name suggests, these cells are derived from embryos at the pre-implantation stage. Thus significant ethical controversy and debate surrounds stem cell work. The underlying potential use of stem cell therapies has generated great excitement and even greater expectation within the scientific world. Their potential use in a vast range of conditions has attracted countless scientists and significant funding but many religious groups strongly oppose this type of research. There is no denying that stem cells possess great potential, but how close is the scientific world to turning myth and speculation into fact? Moreover, can one really justify destroying a "life" in the hope that it might one day save another?

11.4.3.1 Blastocyst

Pre-implantation embryos are known as blastocysts and are typically available from IVF programmes, although some have been created deliberately for research purposes. Regulations regarding the use of embryonic stem cells (ES) varies from country to country. In the UK and Australia, new embryonic cell lines can be created from *unused* embryos harvested from IVF programmes (Dayton 2002a, b). But in US, federal funding for research is only available for use of lines generated before August 2001. The rationale behind this arrangement is that although these cell lines lawfully generated to that date, have pluripotency, they do not have the ability to develop into whole human being when implanted, thus the sanctity of human life is not compromised.

11.4.3.2 Somatic Cell Nuclear Transfer Techniques (SCNT)

The other potential source of ES is somatic cell nuclear transfer techniques also known as the therapeutic cloning. In this procedure, a patient's own mature cells are transferred into an oocyte to generate ES. In this way the cells generated will be histo-compatible. However, claims of successful SCNT producing human embryos have been disputed. Currently therapeutic cloning is in an extremely early and experimental stage of development.

11.4.3.3 Pre-Implantation Diagnostic Screening

Safe techniques have been developed to remove a cell from the inner cells mass of a blastocyst without harming the future development of the embryo and fetus. This technique is already used to screen the embryo for genetic disorders before it is implanted. This technique does not sacrifice the embryo and in the future may become an ethically approvable source of ES.

11.4.4 Adult Stem Cells

Adult stem cells are widely distributed in hemopoietic, neural, liver, pancreatic, epidermal skeletal, and mesenchymal tissues. Under certain circumstances, they have the capacity to develop into cells characteristic of other tissues. This ability of tissue specific stem cells to give rise to another cell type in a new location is referred as "stem cell plasticity". Recent studies have demonstrated that the adult stem cells have the ability to cut across cell boundaries with hemopoetic and neural cells appearing most versatile in this regard.

11.4.5 Stem Cells from Umbilical Cord Blood

Umbilical cord blood contains hemopoietic stem cells with high potential for proliferation. These cells, therefore, have clear therapeutic potential in bone marrow disorders. Protocols have been developed to expand these cells in vitro with subsequent transplantation. Cases have occurred where creation of a "designer baby" has taken place to allow cord blood to be used in the treatment of a sibling terminal disease. Despite this, no precedent has been set for wide-spread use of stem cells harvested in this way.

11.4.6 What Determines the Fate of Stem Cells?

Given appropriate culture conditions, ES can be propagated in vitro and be preserved indefinitely in an undifferentiated state (Bishop et al. 2002) and with the correct micro-environment and signals these cells can be made to differentiate into any desired cell type. By providing defined culture conditions, in vitro murine embryonic stem cells (mES) can be induced to differentiate into hemopoietic cells (Wiles and Keller 1991), cardiac myocytes (Klug et al. 1996), insulin secreting cells (Soria et al. 2000), neurones, and glial cells (Brustle et al. 1999, Brustle et al. 1997, Reubinoff et al. 2001). Most of this work has been pioneered in the mouse, but in 1998 pluripotent stem cells were also isolated from human embryos and grown in culture. These human embryonic stem cells have also been shown to differentiate into neurones and glial cells (Zhang et al. 2001). The early development of a stem cell can be considered both "plastic " and "directable"

(Watt and Hogan 2000, Blau et al. 2001) with its fate being controlled by both intrinsic and extrinsic factors (Watt and Hogan 2000, Blau et al. 2001). The extrinsic factors provide a surrounding micro-environment which is typically referred to as the stem cell niche (Watt and Hogan 2000, Blau et al. 2001).

The developmental pathways that pluripotent stem cells and progenitor cells follow are not irrevocably programmed and can be altered or even reversed. Cloning of "Dolly the sheep" and other animals clearly shows just how plastic mammalian development is, and demonstrates that differentiated adult nuclei can be re-programmed to become fully totipotent after transfer into a recipient oocyte (Schnieke et al. 1997). Several studies have shown that stem cell plasticity is determined by environmental factors. These include the concentration and type of cytokines present in both the culture media and the recipient implantation tissues (Watt and Hogan 2000). The environment thus strongly influences development, as demonstrated in bone marrow studies in which mesenchymal stem cells were used to create endothelium, ectodermal, and endodermal cells (Krause et al. 2001, Hovanesian et al. 2001). Adult murine bone marrow cells which have the ability to develop into epithelial cells of the liver, lung, gut have been isolated (Krause et al. 2001). When these cells were grafted into appropriate sites, they developed into functioning liver, lung, and gut epithelium (Krause et al. 2001). Further examples include cells derived from Wharton's jelly which have the capacity to give rise to neuronal cells (Mitchell et al. 2003). These experiments confirm that stem cells, both embryonic and adult are pluripotent and plastic and these properties are determined largely by external environmental factors which include extracellular matrix composition and cytokine specific profiles. Given the right niche and combination of signals stem cells can be made to follow any desired path of differentiation (Bishop et al. 2002, Watt and Hogan 2000).

11.4.7 Stem Cell Niche

In vivo stem cells are maintained in a highly regulated micro-environment usually provided by surrounding mesenchymal cells. This specialized niche is paramount in preserving the original stem cell population. When the stem cell exits its niche, the factors that maintain its pluripotency are no longer available, and the cell is likely to enter a new differentiation pathway as a result of a foreign local environment. Similarly, injuries or diseases affecting the niche can also directly or indirectly deplete the stem cell population. With respect to the eye, chronic ocular inflammation can affect the limbal stem cell niche leading to limbal stem cell deficiency (Fig. 11.1).

The ultimate consequence of this is blindness. Understanding of the structure and dynamics of the niche has obvious therapeutic importance. Transplanted stem cells will not successfully graft unless the appropriate niche is created prior to the transplantation.

Fig. 11.1 Severe stem cell deficiency due to chronic inflammation. The corneal surface is completely covered with conjunctival derived tissue

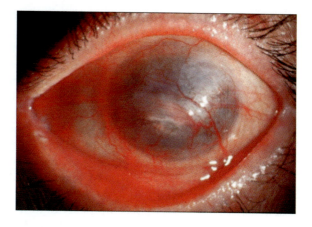

11.4.8 How to Identify Stem Cells? Stem Cell Markers

Cell therapy involves delivery of putative stem cells that have been manipulated ex vivo. One has to be certain that for maximum therapeutic efficiency, the transplanted population is purely stem cell in origin. Stem cells are undifferentiated cells and are not expected to have progenetic markers. So how can stem cell populations be identified? Certain markers to identify embryonic stem cells and hematopoietic stem cells have been discovered – these include p63, ABCG2, and other candidates including *Notch, Nanog, Wnt,* and *Sonic Hedgehog.* However, definitive markers for ocular stem cells are yet to be determined.

11.4.9 How Safe are Stem Cells for Cell Therapy?

Stem cells have high proliferative capacity and multiply significantly during in vitro culture. Any cells undergoing rapid proliferation are at risk of developing a malignant lineage. Experimental evidence shows that ES can display cellular atypia after several cell cycles thus raising concern with respect to therapeutic application.

11.5 Genetic Regulations of Stem Cells

All cell dynamics are controlled by intrinsic and extrinsic factors and stem cells are no exception to this. Various growth factors and genes have been identified in relation to cellular growth and function. Regenerative strategies depend on our ability to manipulate stem cells to perform a desired function or to direct them to develop into a particular cell type.

11.5.1 Genes Controlling Embryonic Stem Cells

The blastocyst and the cells of the inner cell mass differentiate and proliferate into different germ layers which contribute to the development of the fetus. However, when cultured in ex vivo conditions the inner cell mass remains undifferentiated, retaining pluripotency and proliferative capacity. *Oct3/4* and *Nanog* are two genes that have been identified in active ES. It is thought that a critical level in the concentration of the transcription factors for which these genes code is required before the cell commits to differentiation. Murine ES require fibroblast feeder layers and the presence of cytokines such as leukemia inhibitory factor (LIF) to maintain pluripotency in vitro. Pluripotency of stem cells is linked to coordinated activity of key genes and recent studies have shown *Nanong* has the ability to turn partially differentiated cells into pluripotent cells. This gene is fittingly named after *Tir na nOg*, the mythological Celtic land of perpetual youth.

11.5.2 Regulation of Ocular Stem Cells – Pax6

The quest to identify the genes that control ocular stem cell activity is ongoing. Although many genes such as *Pax6, Six3, Rx1, Chx10,* and *Hes1* are implicated during the process of cell differentiation the exact gene signals that maintain stem cell populations in the adult eye need to be elucidated (Ashery-Padan and Gruss 2001, Marquardt and Gruss 2002, Tropepe et al. 2000, Marquardt 2003, Marquardt et al. 2001). The authors have investigated the role of *Pax6* in the ocular surface in a murine model with a *Pax6* mutation. Human *Pax6* is located on the 11p13 chromosome segment. It codes for a paired domain, a homeodomain, a glycine-rich region, and a proline-serine–threonine rich region (Alexiades and Cepko 1997). The homeobox genes are considered "master" genes because they control the activity of many subordinate genes. The function of homeobox genes is mediated by the homeodomain, which recognizes and binds to specific DNA sequences in the subordinate genes, thereby activating or repressing their expression. The *Pax6* gene was identified as the candidate gene for aniridia and the heterozygous mutation of the *Pax6* gene was subsequently found to result in aniridia (Alexiades and Cepko 1997, Jordan et al. 1992, Hanson et al. 1994). The *Pax6* gene is, therefore, an important regulator at the top of the genetic hierarchy which controls eye morphogenesis. Regulation by the *Pax6* gene continues into adulthood as evidenced by increased expression during corneal epithelial wound healing (Koroma et al. 1997). Among the various regulatory roles in the cornea *Pax6* is essential for the expression of cytokeratin-12 (*K12*), gelatinase-B (*Gel-B*), and cell adhesion molecules (CAM) (O'Guin et al 1987, Gillett et al. 1993). Expression of cytokeratin 12 is restricted to corneal epithelium and it has been shown that expression of *Pax6* is essential for the up-regulation of the *K12* gene in human corneal epithelial cells (O'Guin et al. 1987, Gillett et al. 1993). Gelatinase-B belongs to a group of enzymes known as matrix metalloproteinases (MMP). These enzymes play an important role in wound healing. In the cornea, Gel-B, which is regulated by *Pax6* plays an important role in the metabolism of matrix proteins. Reduced Pax6 activity has been associated with down regulation of Gel-B activity (Sivak et al. 2000). Furthermore, the expression of cell-adhesion molecules which offer anchorage to adjacent cells and to the extra-cellular matrix also appear to be regulated by *Pax6* (Simpson and Price 2002, Davis et al. 2003, Stoykova et al. 1997).

Heterozygosity for *Pax6* deficiency ($Pax6^{+/-}$) results in aniridia. Aniridia affects all layers of the eye including the cornea. Corneal changes in aniridia are profound and the changes include corneal vascular pannus formation, conjunctival invasion of the corneal surface, corneal epithelial erosions, and epithelial abnormalities, which eventually result in corneal opacity and contribute to visual loss. Corneal changes in aniridia have been attributed to congenital deficiency of corneal limbal stem cells. However, current evidence based on both clinical observation and analysis of an animal model of aniridia, suggests that the proliferative potential of the corneal limbal stem cells may not be impaired. Rather the corneal changes in aniridia may in fact be related to an abnormality within the limbal stem cell niche. The mechanisms underlying progressive corneal pathology in aniridia appear to be multi-factorial. They include (1) abnormal corneal healing responses secondary to anomalous extra-cellular matrix metabolism, (2) abnormal corneal epithelial differentiation leading to fragility of epithelial cells, (3) reduction in cell adhesion molecules in the *Pax6* heterozygous state, rendering the cells susceptible to natural shearing forces, and (4) conjunctival and corneal changes leading to the presence of conjunctival derived cells on the corneal surface.

11.5.3 Transdifferentiation, Metaplasia, and Re-Programming of Cell Fate

Phenotypic changes in response to the environment are common. Cells can change their size, their "identity" surface markers and their functions. This adaptation has been recognized at two levels. The first is modulation, where by mild reversible phenotypic changes occur in response to environmental change. The second more significant level is metaplasia. This term introduced by Virchow means *the replacement of cells of one type with cells of another type* Biologists in the 1990s observed adult phenotype changes in lower animals such as zebra fish and amphibian urodele and coined the term "transdifferentiation" (McDevitt 1989). Biologically, both metaplasia and transdifferentiation encompass the same process which is mediated through gene expression. It is possible to conceive that the developmental potential of adult mammalian cells could be redirected by exposing them to specific environmental signals. Studies have shown that adult corneal epithelium can transdifferentiate into dermal cells and hair follicles under the influence of embryonic dermis (Ferraris et al. 2000). Further studies indicate that *Wnt*, *Pax6* and *Noggin* signals are involved in corneal epithelial transdifferentiation into dermal cell types (Pearton et al. 2005). Transdifferentiation in clinico-pathological context is known as metaplasia and is a common event often resulting from chronic inflammation. It occurs in cells with the ability to replicate (Lugo and Putong 1984). Various chronic ocular surface inflammatory conditions are known to cause keratinzation or epidermalization of the conjunctiva. These chronic conditions also lead to corneal limbal stem cell (LSC) deficiency that results in blindness (Colville et al. 2000, Tsai and Tseng 1995, Kenyon and Tseng 1989). Depletion of LSC and epidermalization of the conjunctiva appears to be linked and may operate through common genetic pathways. Elucidating the mechanisms which these genes control will helpfully enable us to develop treatment strategies that can prevent the development of stem cell deficiency.

11.6 Ocular Stem Cells

11.6.1 Corneal Epithelial Stem Cells

In the 1940s, Ida Mann observed that melanin pigments from the corneal limbus migrated towards a healing corneal abrasion (Espinosa-Heidmann et al. 2003). In the late 1940s, evidence for centripetal movements of corneal epithelial cells was provided by several investigators, including Maumenee and Scholz (1948) and Buschke (1949). Eventually Davanger and Evensen raised the possibility that the limbus was the source of these migrating cells (Davanger and Evensen 1971). Further experimental and clinical observations suggested that the source of proliferating cells was located in the basal layers of the corneal limbus (Cotsarelis et al. 1989, Kruse and Volcker 1997, Kruse 1994). A substantive body of evidence now indicates that a sub-population of cells at the basal layer of the corneal limbus exist in an "undifferentiated" state (Cotsarelis et al. 1989, Kruse and Tseng 1992, Ebato et al. 1987, Dua 195, Pellegrini et al. 1999). In other words, they are stem cells and have the capacity to (a) proliferate, (b) self-renew, (c) produce large numbers of terminally differentiated functional progeny, and (d) regenerate tissue after injury (Hall 1989, Lajtha 1979, Potten and Loeffler 1990).

Limbal stem cells are protected by the limbal environment or niche. The physical characteristics of the limbus together with the Palisades of Vogt provide a physically protected environment. The cells are protected from shear forces. The melanin gives a degree of protection from ultra violet rays. The anatomical closeness to the limbal blood vessels in the underlying stroma ensures that nourishment is available. Thus the underlying stroma and cells within, as well as the local blood supply, are all likely to contain factors that determine stem cell fate.

11.6.2 Endothelial Stem Cells

Corneal endothelial cells do not show any proliferative capacity in vivo or in culture. However, recent studies have shown that the human corneal endothelium contains stem cells that have the capacity to proliferate but which in normal physiology are held in a state of inhibition. The trabecular mesh work has been suggested as the probable location of these cells (Whikehart et al. 2005).

11.6.3 Retinal Stem Cells

Adult mammalian retina is part of the nervous system and totally lacks any regenerative capacity. Injured retinal tissue is replaced with fibrous tissue that results in visual impairment. The possibility of retinal stem cells was first realised in the zebra fish. In such lower life forms, the entire retina can regenerate from the ciliary body. Further studies have shown the presence of stem cells in the avian retina.

A recent study by Mayer et al. demonstrated that in culture conditions adult retina has the potential to generate primary neurospheres. These can differentiate into definitive secondary neurospheres, neurons, photoreceptors, and glial cells (Mayer et al. 2005). The failure of progenitor cells in mammals and avians to renew retinal cells in the postnatal period again suggests that the progenitor cells find themselves in an inhibitory environment.

11.6.4 Stem Cells in New Blood Vessels

Pathological neovascularization typified by diabetic retinopathy, age-related macular degeneration, and retinopathy of pre-maturity is one of the major causes of visual impairment in the developed world. This type of post-natal neovascularization has been attributed to angiogenesis, a process whereby new vessels are derived from pre-existing vasculature. The classical cascade of events are as follows – dilation and increased permeability of retinal capillaries, basement membrane lysis, migration, and proliferation of endothelial cells from pre-existing vessels, lumen formation, and finally anastomoses between adjacent new vessels to allow blood flow. However, it appears that the process of post-natal angiogenesis may not be as clear cut as originally believed. Circulating endothelial progenitor cells can be recruited to areas of neovascularization. The regulatory process involved in endothelial progenitor cell differentiation and recruitment to sites of neovascularization has yet to be fully elucidated. Vascular Endothelial Growth Factor (VEGF) appears to have an important role in this cascade of events. VEGF contributes to post-natal neovascularization by mobilizing bone-marrow derived endothelial progenitor cells (Asahara et al.

1999). These endothelial cells appear to be particularly responsive to VEGF and the expression of high levels of VEGFR-2. Endothelial progenitor cell levels in the blood are also augmented by GM-CSF and HMG-CoA reductase inhibitors (statins). Matrix proteins such as fibronectin are also likely to be important since they are required for the attachment and differentiation of endothelial progenitor cells in vitro.

Adult bone marrow derived stem cells provide the cellular components of the blood through out the life. Recent studies have show the potential of bone marrow derived stem cells homing to sites of injury and ischemia. The cells participate in the development of new blood vessels. These findings have therapeutic application in diabetic retinopathy and wet type of age related macular degeneration.

11.7 Disorders of Stem Cells in the Eye

Although a wide variety of conditions can lead to loss of vision, only a very few can be directly related to stem cell dysfunction. Clinically stem cell dysfunction is only considered in relation to corneal epithelium but stem cells may have therapeutic applications in other ocular tissues. Corneal epithelium depends on limbal stem cells for maintenance. When depleted through chemical injury or chronic inflammation the surrounding conjunctival epithelium takes over the corneal epithelium. When the limbal stem cells are depleted there will be loss of limbal anatomy, irregular epithelium, fluorescein staining of the area covered by abnormal epithelium, unstable tear film, persistent epithelial defect, scarring, keratinization, and calcification. In these situations, the LSC deficiency can be total or partial.

11.8 Strategies of Ocular Regeneration

Ocular disorder resulting from loss of tissue or cell layers requires replacement of the loss with identical and functional cells or tissues. This can be achieved through transplantation form a healthy source and cell therapy. In addition, any regenerative strategy should take account of the original mechanism of injury or inflammation and include steps to limit tissue damage.

11.8.1 Limitation of Tissue Damage

11.8.1.1 Inflammation and Tissue Damage

Injuries elicit an inflammatory response. The function of the inflammatory response is to eliminate the injurious agent and facilitate healing (Fig. 11.2).

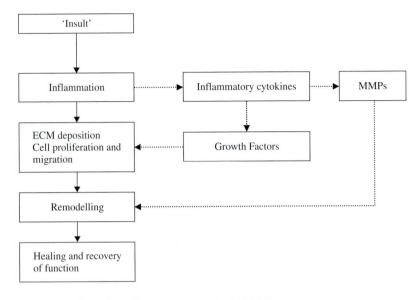

In a well coordinated healing response, the initial inflammatory
response results in proliferation, remodelling of tissue and healing

Fig. 11.2 Schematic representation of the main tissue healing and remodeling phases

The early phases of healing are characterized by deposition of extra cellular matrix and proliferation, and migration of cells within this matrix, which act as a scaffold. The proliferative phase is followed by a remodeling phase, in which the ECM is degraded by proteolytic enzymes known as matrix metalloproteinases (MMPs) (Sivak and Fini 2002, Wong et al. 2002). Irrespective of the underlying cause, inflammation typically results in tissue destruction leading to ulceration and scarring. Pathological activity of proteolytic enzymes and active pro-inflammatory cytokines are implicated in the progression of tissue destruction. Thus the therapeutic inhibition of tissue destruction is usually and is likely to result in a better visual outcome.

Initial injury often initiates release of interleukin-1 (IL-1) that in turn initiates a cascade of down stream changes. IL-1 is considered to be the master regulator of the corneal wound healing response (Wilson et al. 2001). It induces expression of keratinocyte growth factor and hepatocyte growth factor, both potent mitogens for corneal epithelial cells. IL-1 accelerates epithelial wound closure but also has destructive effects. When IL-1 is released from injured epithelium subsequent induction of the *Fas* ligand results in keratocyte apoptosis (Mohan et al. 1997, Hitchcock et al. 1996). In addition, IL-1 is known to induce corneal epithelial cells to express IL-6 and IL-8. IL-8 is a potent chemotactic agent for neutrophils and has angiogenic effects. IL-1 and IL-8 stimulate MMP expression. Cellular sources of MMPs include resident corneal cells such as epithelial cells and fibroblasts and invasive inflammatory cells (Sivak and Fini 2002a, Wong et al. 2002). Apart from inflammatory cytokines, fragments of extra cellular matrix and growth factors also up-regulate MMP gene expression. The substrates of MMP include cytokines, cell adhesion molecules, and ECM that may result in the creation of further biologically active fragments (Boudreau and Bissell 1998, Sivak and Fini 2002b, Mohan et al. 2002). During epithelial healing, up-regulation of MMP-9 also known as gelatinase-B (Gel-B) is primarily responsible for degradation of damaged matrix. The catabolism of matrix continues even after re-epithelialization is complete (Sivak and Fini 2002a, Wong et al. 2002). A well coordinated response to injury will result in healing and recovery of function of the injured tissue.

Although the release of cytokines may facilitate elimination of the injurious agent and mediate healing, a vicious cycle of tissue destruction can be initiated. The initial injury causes the expression of IL-1 that leads to further expression of other cytokines and the release of MMPs. MMPs dissolve ECM, basement membrane, and stroma, resulting in ulceration leading to the release of biologically active components initiating a "second injury". This second injury leads to another cycle of inflammatory cytokines and MMP activation thus amplifying tissue damage.

In such corneal conditions as recurrent corneal erosion, stromal ulceration, pterygium, keratoconus, corneal melting after chemical injury and persistent corneal epithelial defects, pathological activity of proteolytic enzymes, and pro-inflammatory cytokines are implicated.

11.8.1.2 Treatments to Limit Tissue Damage

Regulation of cytokine expression and inhibition of MMPs can limit tissue destruction. Therapeutically, topical and systemic steroids, cyclosporine, and FK506 may provide a means of regulating cytokine expression. Agents such doxycycline that have anti-MMP activity also have therapeutic application in managing ocular conditions where MMP activity is abnormally elevated. Oral doxycycline and topical steroids have been used successfully in the treatment of recalcitrant cases of recurrent corneal erosion (Dursun et al. 2001). This approach is based on biochemical evidence implicating increased levels of matrix metalloproteinase activity in the pathogenesis of this condition (Girard et al. 1991, Fini et al. 1998, Garrana et al. 1999). Doxycycline has been shown to specifically decrease the activity of MMP-9 derived from corneal epithelial cells (Sobrin et al. 2000).

11.8.1.3 Modulation by NF-κB

NF-κB plays a pivotal role in inducing the expression of multiple genes in immune and inflammatory responses (Baldwin 1996). Inactive NF-κB is cytoplasmic and in response to activation by oxidative stress and action of ROS, NF-κ;B undergoes phosphorylation (Baldwin 1996, Griendling et al. 2000). The active form translocates to the nucleus to initiate transcription by high affinity binding to regulatory κB motifs. κB has been identified in the promoter region of a number of genes involved in the phospholipid metabolic pathway, pro inflammatory cytokines (IL-1B, TNF a, IL-6, IL-8), MMP-9, and urokinase-type plasminogen activator. By modulating the oxidation path way, one may be able to limit tissue damage.

11.8.2 Ocular Regeneration with Stem Cells

Interest in regenerating specific ocular tissues, for example, retinal ganglion cells in glaucoma and photoreceptors in retinal dystrophies is driven by a clinical imperative to devise novel approaches to treat currently "untreatable" diseases. Restoring functionality may be achieved through transplantation of tissues from either the host source (autologous) or another individual (allogeneic) or a modified composite (bioengineered) or a combination of these approaches (Fig. 11.3).

When tissues or materials "foreign" to the host are introduced, masking the site to immunological surveillance in order to prevent rejection is of paramount importance. This might be achieved by:

1. Using autologous materials devoid of antigenic stimuli, for example, limbal transplantation in pterygium surgery
2. Genetically modifying the autologous tissue to be recognized as "self", for example, in the treatment of epidermolysis bullosa
3. Modulate the effector arm of the immune response by immune suppression, for example, using systemic steroids and cytotoxic therapies in penetrating keratoplasty

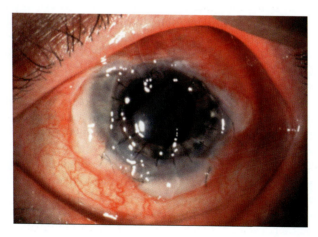

Fig. 11.3 Patient who underwent donor corneal limbal tissue transplant. The cornea is now transparent

Achieving a balance between sustained benefit of visual restoration and toxicity of long-term systemic therapies requires close monitoring. Local and systemic factors markers which may predict response to immune suppression regimens would be of therapeutic value and are likely to expand with further studies of biomarkers and pharmocogenetics. Understanding the extracellular milieu will allow strategies to be developed ensuring lasting functionality of transplanted tissues. For example, the use of growth factor-rich autologous serum and inhibitors of matrix metalloproteinase proteinases may be the key to long-term survival of limbal/corneal stem cell transplantation (see Part II for rationale of this approach). Despite the clinical success of transplantation in specific diseases, new data from DNA fingerprinting studies following kerato-limbal transplanation suggests that tissue recovery in phenotypically normal epithelial may lack evidence of donor cells. This raises the interesting possibility of regeneration through stimulation of host stem cells from a local or distant source, that is, hemopoetic. These observations underscore the need to review mechanisms of host stem cell failure which may arise from dysfunctionality rather than depletion or absence. Tracing the source and mechanism of cell fate specification in ocular development in health and disease may provide clues to new targets for intervention which may, in part, require replacement of a precursor population of cells. In this endeavor, it is worth considering the possibility of genetically modifying the target cell population, and while this may attract criticism concerning ethics and safety, it offers therapeutic potential.

11.8.2.1 Keratolimbal Transplantation

Kenyon and Tseng applied stem cell theory and adopted a direct approach to replenish the limbal stem cell population through grafting healthy limbal tissue with the resident stem cells either from the fellow healthy eye or from a donor (Kenyon 1989). Limbal transplantation is a collective term to describe various procedures that include Cadaveric Keratolimbal Allograft (KLAL), live or living-related Conjunctival Limbal Allograft (lr-CLAL), Conjunctival Limbal Autograft (CLAU). This classification considers the location as well as origin of tissue and the common goal is to restore the limbal stem cell population. Conditions such as Steven-Johnson's syndrome, cicatricial pemphigoid, and chemical burns, among others, can severely compromise the ocular surface due corneal vascularization, chronic inflammation, in-growth of fibrous tissue, and corneal opacification. The common pathological feature of this diverse group of disorders is the depletion of limbal stem cell population responsible for maintaining corneal epithelial equilibrium. In these conditions, limbal trans-plantation allows restoration of the corneal limbal stem cells and significant long term corneal epithelial healing may be achieved. Limbal transplantation requires more than a span of two to three clock hours of limbal tissue excision and may seriously compromise the living donor eye (Dua 1995). Moreover, limbal allograft, requires prolonged immunosuppression to prevent graft rejection and the systemic side effects of immunosuppressive agents remain a therapeutic concern.

11.8.2.2 Ex vivo Limbal Stem Cell expansion

Much interest has been generated by the prospect of re-implanting ex vivo expanded limbal stem cells as a technique to replenish the corneal surface (Ramaesh and Dillon 2003). In 1995, Noguchu et al. first showed tracheal epithelium could be cultured on amniotic membrane. Tsai et al. chose human amniotic membranes as a carrier to expand the limbal stem cells in vitro and transfer the expanded cells to the ocular surface (Tsai et al. 2000). Amniotic membrane (AM) is being extensively used in ocular surface disorders and this was a logical choice. Amniotic membrane serves as "transplanted basement membrane" facilitating migration of cells and reinforces epithelial cell adhesion. Amniotic membrane alone is not effective in treating total stem cell deficiency. Cultivation of limbal epithelial cells on amniotic membrane for transplantation offers the additional advantage of replenishing the stem cells and providing the potential growth factors present in amniotic membrane. This method gave an advantage of ensuring a compatible substrate for the graft. Technically, transferring the amniotic membrane-epithelial sheet composite, to the cornea was easy and was an advance over the previous methods. Tsai et al. report transplantation of autologous limbal epithelial cells cultured on amniotic membrane in successful ocular reconstruction. They achieved visual improvement in six cases of ocular surface disease secondary to chemical burns and pseudopterygium. Recently there is an encouraging steady increase in the reported success of this technique to reconstruct the ocular surface. Recent clinical trials have used a novel source of cultured cells in the reconstruction of debilitated ocular surface due to stem cell deficiency. Although the use of oral mucosal cells in the potential treatment of severe ocular surface disease with LSC deficiencies is promising and would negate the use of immunosuppression required in allogeneic transplants, these studies are only at a preliminary stage.

11.8.2.3 Regeneration of the Retina

Age related macular degeneration (ARMD) affects the elderly population and is a significant cause of visual disability and blindness worldwide. ARMD is due to degeneration of retinal pigment epithelium and overlying neural retina in the macular region. ARMD can be classified into dry and wet types depending on the absence or presence of choroidal new blood vessel formation. While there is no treatment available for the dry type wet ARMD can be treated by ablating the new blood vessels. This can be achieved by lasers to the newly forming blood vessels (photodynamic therapy). A new generation of intra-ocular anti-vascular agents which have been recently licensed have also been shown to be effective in new blood vessel regression. However, none of these treatments are effective in actually restoring the macular function.

In animal models, investigators have tried to restore lost retinal pigment epithelial and neural retinal layers by cell therapy. The basic approach is to engraft injected cultured cells into the affected retinal layers and restore anatomical integrity and function. Although there has been some success in animal models clinical application is currently a distant dream. Furthermore, the anatomical location of the retina makes this site very difficult to approach.

There is no doubt that the clinical application of stem cell therapies may hold promise for diseases affecting the retina but a number of challenges need to be addressed before translating emerging research findings into safe and effective treatments. Understanding the anatomic and functional retinal stem cell (RSC) niche in both health and disease is the key to developing replacement strategies. Our understanding of the limbal SC niche in the anterior segment of the eye may be a paradigm for parallel studies in the posterior segment. Further examination of the interdependencies between neuronal, glial and retinal pigment epithelial cells in development and maintenance of the RSC pool are required (Fischer 2005). This information would enable the potential of autologous cell replacement strategies to be realized. For example, adult iris and ciliary body can be induced to transdifferentiate into a photoreceptor phenotype by manipulating specific cell signals (Akagi et al. 2004). Maintenance of graft viability following transdifferentiation induction through ex vivo selection and culture manipulation requires a detailed understanding of RSC regulation in vivo, in which the Hedgehog pathway may be a major player (Chiou et al. 2005, Spence et al. 2004). Retinal or brain stem cell replacement strategies have met with only limited success in restoring neural connectivity. Recently, McClaren et al. have demonstrated photoreceptor precursor transplantation may be successful in improving retinal function (MacLaren et al. 2006). The stage of photoreceptor differentiation was defined by the expression of *Nrl* and illustrates the importance of defining and refining the "critical period" for donor/host integration.

Clinical experience in LSC grafts has confirmed the need for systemic and local immunosupression in order to prevent graft rejection. While the optimal regimen has yet to be established the likelihood is that a similar approach may be required to prevent autologous RSC graft rejection. In addition, clinical evaluation of RSC function used to titrate immunosuppressive therapy and monitor graft survival is likely to prove more difficult than LSC graft evaluation which is assessed by direct examination.

11.9 Gene Therapy

Stem cells are ideal targets for gene therapy since they represent "error-free" cell division, offering life-long cell renewal. The external position of the corneal and limbal epithelium offers the option of gene therapy not available to deeper tissues. It is possible to inactivate a gene producing an abnormal substance or insert a gene which has been lost due to a mutation. In theory, transfer of genes expressing anti-inflammatory proteins or growth factors could be a valuable way to prevent an inflammatory cascade or to supply components for healthy epithelial turn over (Fischer 2005). The genetic and biochemical basis of many inherited corneal epithelial dystrophies has been characterized and targets genes identified. To manipulate the tissue or cells affected by mutation, the gene delivery technique should be efficient and non-toxic and should be able to express the transfected gene for a prolonged period of time. The potential for the genetically manipulated cells to become resident cells and continue to proliferate depends on a transgene vector employed in gene manipulation. Expression of a transgene following non-viral plasmid DNA and adenoviral vectors is short lived. Retroviral vectors integrate into the host genome, providing the potential for long term passenger DNA expression. Proof of concept for this approach has been provided in a rabbit model. In this model, limbal autografts were transduced with retrovirus carrying a marker gene in vitro. The cornea in vivo continued to express the marker gene for long period of time. Similarly, ex vivo expanded limbal stem cells could be transduced with a target gene with a suitable vector before transplantation. Given the special external access afforded by the corneal limbus gene delivery through ex vivo expanded limbal cells have clear potential for therapeutic application in the cornea.

11.10 The Future

Regenerative medicine of ocular tissues is a dynamic and daring field pushing forward the boundaries of stem cell research. Its potential benefits are unparalleled. Scientists and clinicians have many hurdles to overcome before the vision of cell based therapy is realised. Unfortunately, progress is slow as the ethical issues surrounding embryonic stem cells and therapeutic cloning divides the scientific, political, and public communities. The use of adult stem cells would avoid such ethical dilemmas and accelerate much needed research in this field.

Questions/Exercises

1. Explain the role played by the main eye component in the vision process
2. List the main locations of stem cells in eye tissues.
3. Give an example of a major eye pathology and discuss its mechanism.
4. What is the cell biology mechanisms linking inflammatory response to cell transdifferentiation?
5. Describe the role of corneal epithelium in the tissue regeneration.
6. Critically discuss the treatments to prevent ocular tissue damages in the light of inflammatory response and tissue remodeling biochemistry.

7. What are the main limbal transplantation methods and which is the role played by stem cells in their clinical efficacy?
8. Which are the main substrates for ex vivo limbal cell expansion? Why these substrates are preferred for this process?
9. Describe the main clinical treatments for macula degeneration and compare their efficacy.
10. Critically discuss the main strategies for eye gene therapy.

References

Akagi T, et al. (2004) Otx2 homeobox gene induces photoreceptor-specific phenotypes in cells derived from adult iris and ciliary tissue. Invest Ophthalmol Vis Sci. 45:4570–5

Alexiades MR, Cepko CL (1997) Subsets of retinal progenitors display temporally regulated and distinct biases in the fates of their progeny. Development 124:1119–31

Asahara T, et al. (1999) VEGF contributes to postnatal neovascularization by mobilizing bone marrow-derived endothelial progenitor cells. Embo J. 18:3964–72

Ashery-Padan R, Gruss P (2001) Pax6 lights-up the way for eye development. Curr. Opin. Cell Biol. 13:706–14

Baldwin AS, Jr (1996) The NF-kappa B and I kappa B proteins: new discoveries and insights. Annu Rev Immunol. 14:649–83

Bishop AE, LD Buttery, Polak JM (2002) Embryonic stem cells. J Pathol. 197:424–9

Bizzozero G (1894) An address on growth and regeneration of organism. Br Med J. 1:728–732

Blau HM, TR Brazelton, Weimann JM (2001) The evolving concept of a stem cell: entity or function? Cell 105:829–41

Boudreau N, Bissell MJ (1998) Extracellular matrix signaling: integration of form and function in normal and malignant cells. Curr Opin Cell Biol. 10:640–6

Brustle O, et al. (1997) In vitro-generated neural precursors participate in mammalian brain development. Proc Natl Acad Sci USA 94:14809–14.

Brustle O, et al. (1999) Embryonic stem cell-derived glial precursors: a source of myelinating transplants. Science 285:754–6

Buschke W (1949) Morphologic changes in cells of corneal epithelium in wound healing. Arch Ophthalmol 41:306–316

Chiou SH, et al. (2005) A novel in vitro retinal differentiation model by co-culturing adult human bone marrow stem cells with retinal pigmented epithelium cells. Biochem Biophys Res Commun. 326:578–85

Colville D, et al. (2000) Absence of ocular manifestations in autosomal dominant Alport syndrome associated with haematological abnormalties. Ophthalmic Genet. 21:217–25

Cotsarelis G, et al. (1989) Existence of slow-cycling limbal epithelial basal cells that can be preferentially stimulated to proliferate: implications on epithelial stem cells. Cell 57:201–9

Davanger M, Evensen A (1971) Role of the pericorneal papillary structure in renewal of corneal epithelium. Nature 229:560–1

Davis J, et al. (2003) Requirement for Pax6 in corneal morphogenesis: a role in adhesion. J Cell Sci. 116:2157–67

Dayton L (2002a) Biomedical research. Australia pushes stem cell advantage. Science 296:1779–81

Dayton L (2002b) Embryonic stem cells. Australian agreement allows new lines. Science 296:238

Dua HS (1995) Stem cells of the ocular surface: scientific principles and clinical applications. Br J Ophthalmol 79:968–9

Dursun D, et al. (2001) Treatment of recalcitrant recurrent corneal erosions with inhibitors of matrix metalloproteinase-9, doxycycline and corticosteroids. Am J Ophthalmol. 132:8–13

Ebato B, Friend J, Thoft RA (1987) Comparison of central and peripheral human corneal epithelium in tissue culture. Invest Ophthalmol Vis Sci. 28:1450–6

Espinosa-Heidmann DG, et al. (2003) Bone marrow-derived progenitor cells contribute to experimental choroidal neovascularization. Invest Ophthalmol Vis Sci. 44:4914–9

Evans MJ, Kaufman MH (1981) Establishment in culture of pluripotential cells from mouse embryos. Nature 292:154–6

Ferraris C, et al. (2000) Adult corneal epithelium basal cells possess the capacity to activate epidermal, pilosebaceous and sweat gland genetic programs in response to embryonic dermal stimuli. Development 127:5487–95

Fini ME, Cook JR, Mohan R (1998) Proteolytic mechanisms in corneal ulceration and repair. Arch Dermatol Res. 290:S12–23

Fischer AJ (2005) Neural regeneration in the chick retina. Prog Retin Eye Res. 24:161–82

Garrana RM, et al. (1999) Matrix metalloproteinases in epithelia from human recurrent corneal erosion. Invest Ophthalmol Vis Sci. 40:1266–70

Gillett NA, et al. (1993) Leukemia inhibitory factor expression in human carotid plaques: possible mechanism for inhibition of large vessel endothelial regrowth. Growth Fact. 9:301–5

Girard MT, Matsubara M, Fini ME (1991) Transforming growth factor-beta and interleukin-1 modulate metalloproteinase expression by corneal stromal cells. Invest Ophthalmol Vis Sci. 32:2441–54

Griendling KK, Sorescu D, Ushio-Fukai M (2000) NAD(P)H oxidase: role in cardiovascular biology and disease. Circ Res. 86:494–501

Hall PA (1989) What are stem cells and how are they controlled? J Pathol. 158:275–7

Hanson IM, et al. (1994) Mutations at the PAX6 locus are found in heterogeneous anterior segment malformations including Peters' anomaly. Nat Genet. 6:168–73

Hitchcock PF, et al. (1996) Antibodies against Pax6 immunostain amacrine and ganglion cells and neuronal progenitors, but not rod precursors, in the normal and regenerating retina of the goldfish. J Neurobiol. 29:399–413

Hovanesian JA, Shah SS, Maloney RK (2001) Symptoms of dry eye and recurrent erosion syndrome after refractive surgery. J Cataract Refract Surg. 27:577–84

Jordan T, et al. (1992) The human PAX6 gene is mutated in two patients with aniridia. Nat Genet. 1:328–32

Kenyon KR (1989) Limbal autograft transplantation for chemical and thermal burns. Dev Ophthalmol. 18:53–8

Kenyon KR, Tseng SC (1989) Limbal autograft transplantation for ocular surface disorders. Ophthalmology. 96:709–22; discussion 722–3

Klug MG, et al. (1996) Genetically selected cardiomyocytes from differentiating embronic stem cells form stable intracardiac grafts. J Clin Invest. 98:216–24

Koroma B, Tseng S, Sundin OH (1997) Expression of the PAX6 gene in the anterior segment in human aniridia and Sey mouse model. Investigative Ophthalmol Vis Sci. 38:4405 (abstract).

Krause DS, et al. (2001) Multi-organ, multi-lineage engraftment by a single bone marrow-derived stem cell. Cell 105:369–77

Kruse FE (1994) Stem cells and corneal epithelial regeneration. Eye 8:170–83

Kruse FE, Tseng SC (1992) Proliferative and differentiative response of corneal and limbal epithelium to extracellular calcium in serum-free clonal cultures. J Cell Physiol. 151:347–60

Kruse FE, Volcker HE (1997) Stem cells, wound healing, growth factors, and angiogenesis in the cornea. Curr Opin Ophthalmol. 8:46–54

Lajtha LG (1979) Stem cell concepts. Differentiation 14:23–34

Lugo M, Putong PB (1984) Metaplasia. An overview. Arch Pathol Lab Med. 108:185–9

MacLaren RE, et al. (2006) Retinal repair by transplantation of photoreceptor precursors. Nature 444:203–7

Marquardt T, Gruss P (2002) Generating neuronal diversity in the retina: one for nearly all. Trends Neurosci. 25:32–8

Marquardt T, et al. (2001) Pax6 is required for the multipotent state of retinal progenitor cells. Cell 105:43–55

Marquardt T (2003) Transcriptional control of neuronal diversification in the retina. Prog Retin Eye Res. 22:567–77

Maumenee A, Scholz R (1948) Histopathology of the ocular lesions produced by the sulphur and nitrogen mustards. Johns Hopkins Hospital Bull. 82:121–147

Mayer EJ, et al. (2005) Neural progenitor cells from postmortem adult human retina. Br J Ophthalmol. 89:102–6

McDevitt DS (1989) Transdifferentiation in animals. A model for differentiation control. Dev Biol. (NY 1985). 6:149–73

Mitchell KE, et al. (2003) Matrix cells from Wharton's jelly form neurons and glia. Stem Cells. 21:50–60

Mohan RR, et al. (1997) Apoptosis in the cornea: further characterization of Fas/Fas ligand system. Exp Eye Res 65:575–89

Mohan R, et al. (2002) Matrix metalloproteinase gelatinase B (MMP-9) coordinates and effects epithelial regeneration. J Biol Chem. 277:2065–72

O'Guin WM, et al. (1987) Patterns of keratin expression define distinct pathways of epithelial development and differentiation. Curr Top Dev Biol. 22:97–125

Pearton DJ, Yang Y, Dhouailly D (2005) Trans differentiation of corneal epithelium into epidermis occurs by means of a multistep process triggered by dermal developmental signals. Proc Natl Acad Sci USA 102:3714–9

Pellegrini G, et al. (1999) Location and clonal analysis of stem cells and their differentiated progeny in the human ocular surface. J Cell Biol. 145:769–82

Potten CS, Loeffler M (1990) Stem cells: attributes, cycles, spirals, pitfalls and uncertainties. Lessons for and from the crypt. Development 110:1001–20

Ramaesh K, Dhillon B (2003) Ex vivo expansion of corneal limbal epithelial/stem cells for corneal surface reconstruction. Eur J Ophthalmol. 13:515–24

Reubinoff BE, et al. (2001) Neural progenitors from human embryonic stem cells. Nat Biotechnol. 19:1134–40

Schnieke AE, et al. (1997) Human factor IX transgenic sheep produced by transfer of nuclei from transfected fetal fibroblasts. Science 278:2130–3

Simpson TI, Price DJ (2002) Pax6; a pleiotropic player in development. Bioessays 24:1041–51

Sivak JM, Fini M (2002a) MMPs in the eye: emerging roles for matrix metalloproteinases in ocular physiology. Prog Retin Eye Res. 21:1–14

Sivak J Fini M (2002b) Pax-6 Deficient Mice Show Altered Phenotype and Gene Expression During Corneal Re-Epithelialization. ARVO poster. Presentation Number:4197.

Sivak JM, et al. (2000) Pax-6 expression and activity are induced in the reepithelializing cornea and control activity of the transcriptional promoter for matrix metalloproteinase gelatinase B. Dev Biol. 222:41–54

Sobrin L, et al. (2000) Regulation of MMP-9 activity in human tear fluid and corneal epithelial culture supernatant. Invest Ophthalmol Vis Sci. 41:1703–9

Soria B, et al. (2000) Insulin-secreting cells derived from embryonic stem cells normalize glycemia in streptozotocin-induced diabetic mice. Diabetes 49:157–62

Spence JR, et al. (2004) The hedgehog pathway is a modulator of retina regeneration. Development 131:4607–21

Stoykova A, et al. (1997) Pax6-dependent regulation of adhesive patterning, R-cadherin expression and boundary formation in developing forebrain. Development 124:3765–77

Thomson JA, et al. (1998) Embryonic stem cell lines derived from human blastocysts. Science 282:1145–7

Tropepe V, et al. (2000) Retinal stem cells in the adult mammalian eye. Science 287:2032–6

Tsai RJ, Tseng SC (1995) Effect of stromal inflammation on the outcome of limbal transplantation for corneal surface reconstruction. Cornea 14:439–49

Tsai R, Lin-Min, Chen J-K (2000) Reconstruction of damaged corneas by transplanation of autologous limbal epithelial cells. New Engl J Med. 343:86–94

Watt FM, Hogan BL (2000) Out of Eden: stem cells and their niches. Science 287:1427–30

Whikehart DR, et al. (2005) Evidence suggesting the existence of stem cells for the human corneal endothelium. Mol Vis. 11:816–24

Wiles MV, Keller G (1991) Multiple hematopoietic lineages develop from embryonic stem (ES) cells in culture. Development 111:259–67

Wilson SE, et al. (2001) The corneal wound healing response: cytokine-mediated interaction of the epithelium, stroma, and inflammatory cells. Prog Retin Eye Res. 20:625–37

Wong TT, et al. (2002) Matrix metalloproteinases in disease and repair processes in the anterior segment. Surv Ophthalmol. 47:239–56

Zhang SC, et al. (2001) In vitro differentiation of transplantable neural precursors from human embryonic stem cells. Nat Biotechnol. 19:1129–33

Chapter 12
Cartilage Development, Physiology, Pathologies, and Regeneration

*Xibin Wang, *Lars Rackwitz, Ulrich Nöth, and Rocky S. Tuan

Contents

12.1 Introduction 367
12.2 Articular Cartilage Development, Composition,
     and Structure 368
    12.2.1 Development 368
    12.2.2 Composition 369
    12.2.3 Structure 371
12.3 Pathology, Pathophysiology, Pharmacological Therapies of Osteoarthritis . . . 372
    12.3.1 Arthritis. 372
    12.3.2 Osteoarthritis Pathology 372
    12.3.3 Risk Factors for Osteoarthritis 373
    12.3.4 Pathophysiology of Osteoarthritis 374
    12.3.5 Pharmacological Therapies for Osteoarthritis 375
12.4 Cartilage Repair and Regeneration 375
    12.4.1 Articular Cartilage Defects and Spontaneous Repair. 375
    12.4.2 Current Operative Strategies for Articular
            Cartilage Repair 377
    12.4.3 Mesenchymal Stem Cells (MSCs) 379
    12.4.4 Chondrogenic Differentiation of MSCs 379
    12.4.5 Tissue Engineering 381
    12.4.6 Experimental Approaches for Articular Cartilage Tissue Engineering 382
12.5 Conclusions 386
Questions/Exercises. 386
References 387

12.1 Introduction

Articular cartilage covers the end of long bones. It provides a smooth surface for joint articulation and confers resilient properties for dynamic loadings. In general, cartilage defects in adult humans fail to heal spontaneously and usually lead

R.S. Tuan (✉)
Department of Health and Human Services, Cartilage Biology and Orthopaedics
Branch, National Institute of Arthritis and Musculoskeletal and Skin Diseases,
National Institutes of Health, Bethesda, MD 20892, USA
e-mail: tuanr@mail.nih.gov
*Authors contributed equally.

M. Santin (ed.), *Strategies in Regenerative Medicine*,
DOI 10.1007/978-0-387-74660-9_12, © Springer Science+Business Media, LLC 2009

to progressive cartilage destruction, which is the major feature of osteoarthritis, a debilitating joint disease prevalent in older population. Thus, repair and regeneration of articular cartilage is of great public interest. Experimental and clinical studies have shown the possibility to transplant chondrocytes into isolated cartilage defects, with or without scaffold material, to restore the articular surface. Mesenchymal stem cells (MSCs) offer a new cell source to develop sufficient quantity of cartilage analogue for the reconstruction of damaged or degenerated articular cartilage. Successful articular cartilage repair and regeneration depends critically on understanding cartilage biology. This chapter will first review cartilage development, structure, composition, and pathology, and then summarize recent advances made in the field of cartilage repair and regeneration using MSCs and tissue engineering technologies.

12.2 Articular Cartilage Development, Composition, and Structure

12.2.1 Development

Articular cartilage, together with menisci, ligaments, synovium, and the fibrous joint capsule make up a diathrodial joint. The development of such joints is an integral component of embryonic limb development. The primordial limb bud consists of mesodermal cells covered by an ectodermal ridge. An elongated cartilaginous skeleton emerges in the limb bud through a precartilage condensation process of the mesenchymal cells, giving rise to a cartilage anlage for the formation of the endochondral skeletal elements (Carter et al. 2004).

The cartilaginous rod becomes segmented at specific regions and the future joint location begins to form. A secondary remodeling step occurs at the presumptive joint location, where some cells become dedifferentiated and elongated simultaneously with the loss of collagen type II and appearance of collagen type I expression (Craig et al. 1987). This region is termed the joint interzone, which plays a critical role in the development of articular cartilage, meniscus, and ligaments (Archer et al. 2003). The interzone is composed of three layers. The central layer consists of randomly arranged cells between two denser, chondrogenic layers. In the chondrogenic layers, the cells are aligned parallel with the adjacent epiphyseal region. In the developing human knee joint, the layers follow the shape of the femoral and tibial condyles, forming a dense band of connective tissue that marks the first sign of organized articular cartilage (Merida-Velasco et al. 1997). The central layer merges laterally with the joint capsule mesenchyme, and in the developing knee joint, it is from this area that the synovium and other intracapsular structures, such as the menisci and ligaments arise. Cavitation of the joint follows, characterized by cellular apoptosis and selective, high-level synthesis of hyaluronan by interzone cells, and presumptive synovial cells (Dowthwaite et al. 2003).

In human, during the seventh embryonic week, the chondrocytes in the central region of the cartilaginous segments become hypertrophic and the local matrix begins to calcify. Capillary invasion follows and bone tissue replaces the calcified cartilage, which is described as endochondral ossification. This process forms the primary center of ossification. Chondrocytes on either side of the primary ossification center go through a sequence of proliferation, maturation, and hypertrophy to form the growth plate. The bone tissue expands longitudinally as it continuously replaces the hypertrophic cartilage. Usually postnatally, a secondary ossification center forms in the cartilage near both ends of long bones. The growth of the bone tissue from the primary and secondary ossification centers eventually replaces all the cartilage tissue in the middle of the long bone, while only the layer of cartilage covering the bone end persists throughout life and provides a smooth surface for joint articulation; hence it is termed articular cartilage. Recent molecular studies demonstrate that growth differentiation factor-5 (GDF-5), a growth factor of the bone morphogenetic protein (BMP) family, is expressed in the articular cartilage and not in the growth plate cartilage (Koyama et al. 2007). This suggests the articular cartilage is formed through a distinct developmental process, different from the growth plate.

12.2.2 Composition

Articular cartilage is composed of a single type of cells, chondrocytes, and a large amount of extracellular matrix (ECM). It is a unique tissue that is not innervated and has no blood or lymphatic vessels. The ECM of the cartilage occupies 90% or more of the total volume and consists of tissue fluid and structural macromolecules. Overall, water constitutes 60–80% of the wet weight of the cartilage. Cartilage ECM structural macromolecules are classified into three major groups: collagens, proteoglycans, and other non-collagenous proteins (Buckwalter and Mankin 1997).

Articular cartilage contains multiple types of collagens, including types II, VI, IX, X, and XI. Collagen type II is the major collagen form, accounting for 90–95% of the total collagen in the cartilage tissue (Buckwalter and Mankin 1997, Cremer et al. 1998). Collagen types II, IX, and XI form matrix fibrils, in which collagen type IX can bind covalently to fibrils on one end and to other collagen type IX molecules on the other end. This interaction enables the collagen fibrils to form a meshwork entrapping other macromolecules, such as proteoglycans, in the tissue. Collagen type VI connects chondrocytes and the surrounding extracellular matrix, while collagen type X, which plays a likely role in mineralization, is only found in hypertrophic chondrocytes in the mineralized zone of the cartilage.

Cartilage proteoglycans include aggrecan, decorin, biglycan, fibromodulin, and perlecan (Buckwalter and Mankin 1997, French et al. 2002). These molecules consist of a protein core and one or more glycosamioglycan (GAG) side chains. GAGs are polysaccharides composed of repeating disaccharide units, which are modified with amino, carboxylate, or sulfate groups. Based on the composition of the disaccharide units and modifying groups, GAGs are classified as chondroitin sulfate, keratan sulfate, heparan sulfate, and dermatan sulfate. Each proteoglycan has its own unique GAG composition. For example, aggrecan may have over 100 chondroitin sulfate chains and 30 to 60 keratan sulfate chains (Carney and Muir 1988, Kiani et al. 2002, Roughley et al. 2006), while fibromodulin has several keratan sulfate chains, and decorin and biglycan have only one and two dermatan sulfate chains, respectively (Roughley and Lee 1994). Hyaluronic acid is a high-molecular-weight GAG molecule with no core protein. Aggrecan molecules attach to hyaluronic acid and link proteins, forming large proteoglycan aggregates. Because of the hydrophilic characteristics of GAGs, the large proteoglycan aggregates help maintain the high water content in the tissue. The swelling pressure of the proteoglycans is counter-balanced by the stiffness of the collagen meshwork. Loading forces water out of the cartilage tissue, while proteoglycans help draw water back and restore the water content and tissue shape when the loading pressure is released. Therefore, proteoglycan aggregates and collagen meshwork together give the cartilage stiffness and resilience to repeated mechanical loadings. The functions of the small proteoglycans are not fully understood. It has been proposed that they may regulate collagen fibril formation or sequester growth factors (Buckwalter and Mankin 1997). For example, decorin and fibromodulin bind to collagen type II and may stabilize the collagen meshwork. These small proteoglycans also bind growth factors like transforming growth factor $-\beta$ (TGF-β), and may regulate their availability and/or activities (Hildebrand et al. 1994). Heparan sulfate-containing proteoglycans, such as perlecan, in the matrix sequester basic fibroblast growth factor (bFGF), which can be released upon cartilage cutting and loading, mediating tissue response to mechanical stimuli and damage (Vincent et al. 2002, Vincent and Saklatvala 2006).

Besides collagens and proteoglycans, cartilage matrix also contains many other non-collagenous proteins and glycoproteins, including cartilage oligomeric matrix protein (COMP), matrilins, chondroadherin, and fibronectin (Roughley 2001). These proteins play structural roles through either matrix–matrix or matrix–cell interactions. COMP and matrilin-3 interact with each other, and both can bind collagens such as collagen types II and IX (Briggs and Chapman 2002, Hecht et al. 2005). Mutations in COMP, matrilin-3, or collagen type IX collagen result in various forms of multiple epiphyseal dysplasia (Briggs and Chapman 2002, Hecht et al. 2005). COMP mutations also cause pseudoachondroplasia, in which the secretion of COMP, matrilin-3 and type IX collagen is reduced (Hecht et al. 2005). The matrix deficient in these proteins is disorganized, suggesting their important role in matrix assembly. A number of non-collagenous proteins, such as chondroadherin and fibronectin, mediate cell–matrix interaction (Roughley 2001). Fibronectin is increased in osteoarthritic cartilage, and its fragments stimulate chondrocytes to secrete anabolic growth factors as well as catabolic cytokines and enzymes (Homandberg 2001). Therefore, these structural proteins may not only provide an anchor for chondrocytes in the matrix, but also influence cell activities in different physiologic and pathologic conditions.

12.2.3 Structure

From the surface to the subchondral bone, the articular cartilage can be divided into four zones (Fig. 12.1): the superficial zone, middle zone, deep zone, and calcified zone (Buckwalter and Mankin 1997, Ge et al. 2006). These four zones differ in cell shape, composition and organization of matrix molecules, as well as chondrocyte activities. In general, cell size, diameter of collagen fibrils, and proteoglycan content increase from the superficial zone to the deep zone. In the superficial zone, the chondrocytes are flattened, assuming an ellipsoid shape. The long axis of cells and collagen fibrils lie parallel to the cartilage surface. This arrangement gives the cartilage the strength to resist the shear force produced during joint movement. In the middle zone and deep zone, chondrocyets assume a spheroidal shape and line up in columns perpendicular to the surface. The diameter of collagen fibrils in the matrix becomes larger, while the orientation of the fibrils changes from oblique to the joint surface in the middle zone to perpendicular in the deep zone. In deeper region of the articular cartilage, the calcified zone, where cells are embedded in the calcified matrix, unites the uncalcified cartilage with the subchondral bone. Chondrocytes from different zones vary in their synthetic activities. Generally, cells in the middle and deep zones are more active than those in the superficial zones (Wong et al. 1996). Chondrocyte activities respond to mechanical stimuli in a zone-specific manner (Wong et al. 1997). Thus, the zonal organization may play an important role for the cartilage to exert its normal function.

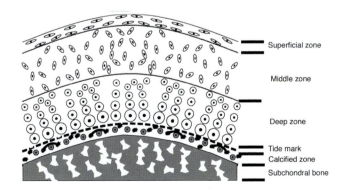

Fig. 12.1 Structure of articular cartilage

12.3 Pathology, Pathophysiology, Pharmacological Therapies of Osteoarthritis

12.3.1 Arthritis

Two major forms of arthritis, rheumatoid arthritis (RA) and osteoarthritis (OA), are the leading causes of articular cartilage destruction (www.arthritis.org/). RA is an autoimmune disease, in which the immune system attacks cartilage components and destroys cartilage and subchondral bone. In addition to joint damage, rheumatoid diseases have multisystemic manifestations, including subcutaneous nodules, vasculitis, pericarditis, pulmonary nodules, etc. The prevalence of RA in the population is between 0.5 and 1% (Sangha 2000). Osteoarthritis is a degenerative joint disease, which is characterized by progressive loss of cartilage, increased formation of subchondral bone, and new bone formation at the joint margins (osteophytes). More than 10% of the population aged 65 years and older suffers from OA, the leading cause of disability in the elderly (Sangha 2000). Both RA and OA have inflammatory components and share some common symptoms, such as joint stiffness and pain. Non-steroidal anti-inflammatory drugs (NSAIDs) and analgesics are used in both RA and OA to alleviate symptoms. However, strategies for disease modification are different in RA and OA. Besides the treatment of acute inflammation, the major approach to slow RA progression is to suppress the function of the immune system. In the case of OA, efforts have been targeted at finding new approaches to prevent the loss of cartilage and to regenerate cartilage tissue. A potentially promising approach to cartilage regeneration is the emerging technology of tissue engineering, which is discussed later.

12.3.2 Osteoarthritis Pathology

OA is mostly found in the joints of the knee, hip, hand, spine, and foot. The wrists, shoulders, and ankles are less likely affected (Arden and Nevitt 2006). The joints removed from OA patients show loss of the articular cartilage to varying degrees. Grossly, the articular surface appears uneven, fibrillated, and fissured. Total loss of the cartilage can occur, especially in the load-bearing area, where subchondral bone is exposed at the joint surface (eburnation). At the margins of the joint, new cartilage and bone (osteophytes) grow out. As subchondral bone formation and remodeling are increased, cysts and sclerosis (bone thickening) co-exist in the subchondral area. X-ray radiography of the OA joint shows the loss of cartilage as coincident with joint space narrowing. Osteophytes, subchondral cysts, and sclerosis are also useful features for OA diagnosis.

Histological observations have revealed that the normal cartilage structure is totally disrupted in the osteoarthritic tissues. The articular surface is fibrillated. Chondrocytes exist in clusters, surrounded by broad acellular areas. Loss of proteoglycan can be seen histologically.

12.3.3 Risk Factors for Osteoarthritis

A definite cause cannot be found in most cases of OA. However, a number of factors are associated with the incidence of the disease, namely, age, gender, genetics, joint injury, obesity, joint deformity, and greater bone density. OA is prevalent in the elderly population, while it is less common before 40 years of age. Females are more susceptible to OA than males at the same age. Recent studies reveal increasing number of genetic linkages to OA (Peach et al. 2005). Candidate genes include those encoding structural proteins such as collagen type II, COMP, collagen type XI, matrillin-3, and aggrecan. Mutations in some of these genes cause epiphyseal dysplasia as discussed earlier. However, the association of OA in the population with mutations or polymorphism in the structural genes has not yet been confirmed. Genetic analysis has also linked OA to interleukin-1 (IL-1), which modulates chondrocyte catabolic activities and to BMP-5, which may contribute to bone and cartilage development and repair of the joint.

Cartilage damage and factors that increase mechanical load to cartilage also contribute to the development and progression of OA. As discussed earlier, water is the main component of the ECM of human articular cartilage, and is distributed within a macromolecular network of collagen type II and large polyanionic, sulfated proteoglycans (Buckwalter and Mankin 1997). Collagen fibrils confer to the form and tensile strength of the tissue, whereas the interaction of proteoglycans with water contributes to the compression-stiffness and resilience of articular cartilage. Physiological loading of articular surfaces causes water movement within the cartilage ECM that distributes the load to the subchondral bone and dampens the impact during load (Mow and Rosenwasser 1988). If load is applied slowly, the fluid movement allows the articular cartilage to deform, and the stress force applied to the macromolecular network is decreased. On the other hand, impacts with a rapidly applied force can exceed the ability of fluid movement within the ECM, and a significant higher force is transferred to the macromolecular network of the tissue. These impacts can exceed the cohesive integrity of the network and lead to its disruption. This is followed by a loss of proteoglycans, resulting in cell damage or cell death (Donohue et al. 1983, Thompson et al. 1991, Zang et al. 1999, Loening et al. 2000). Surrounding chondrocytes are, to a certain extent, capable of detecting changes in the composition of the ECM, and through increased synthesis of ECM molecules the damage might be limited (Martin and Buckwalter 2000). However, a further loss of collagen type II and proteoglycans can be detected if the cartilage damage exceeds the limited repair capacity of the local chondrocyte population. This process results in compromised mechanical properties of the articular cartilage and alterations in the biomechanical load distribution in the joint. Furthermore, joint loading or blunt trauma can result in fibrillation and tears in the articular cartilage, which progress into larger defects and degeneration in the long-term.

Obesity increases joint load, and is positively associated with OA in the knee joint, while weight loss can reduce the risk of developing symptomatic OA

(Sangha 2000). Joint injury, joint laxity, and joint misalignment due to deformity substantially increase the risk of OA, probably by changing the distribution of the mechanical load (for example, shear forces may increase due to joint laxity). In young adults, the tear of the anterior cruciate ligament (ACL) or the meniscus is the most common factor related to OA of the knee (Roos 2005). Greater bone density is another factor changing load distribution in the joint. Bones with lower density would be more compliant to mechanical load and absorb more forces, thus reducing the burden of the cartilage. Consistent with this hypothesis, an inverse relationship is found between bone mineral density and OA (Sangha 2000).

12.3.4 Pathophysiology of Osteoarthritis

Despite the identification of a number of the risk factors for OA development, the pathophysiological mechanisms have not been fully elucidated. The articular cartilage in OA is not simply worn away by mechanical forces. Lines of evidence suggest that chondrocyte activities play an essential role in cartilage degradation. Two families of enzymes, MMPs (matrix metalloproteinases) and ADAMTS ("A disintegrin and metalloproteinase with thrombospondin motifs"), are secreted by chondrocytes and promote degradation of cartilage matrix proteins such as collagens and proteoglycans. MMP-1 (collagenase 1), MMP-3 (stromelysin 1), MMP-13 (collagenase 3), and ADAMTS-5 (aggrecanase 2) are present in normal and OA cartilage. Expression of MMP-13, which degrades collagen type II with high efficiency, is significantly increased in human chondrocytes of late stage OA (Bramono et al. 2004). Over-expressing MMP-13 in articular chondrocytes results in joint degeneration in mice similar to human OA (Neuhold et al. 2001). Studies have also demonstrated that deletion of ADAMTS-5 (aggrecanase 2) gene inhibits cartilage degradation in mouse models of OA and inflammatory arthritis (Glasson et al. 2005, Stanton et al. 2005). These data suggest that MMPs and ADAMTS are the primary enzymes responsible for cartilage degradation during the onset and progression of OA.

Multiple factors induce cartilage-degrading enzyme expression and participate in catabolic events in OA. Inflammatory cytokines (e.g., interleukin1-β, IL1-β, and tumor necrosis factor α, TNFα), cartilage degradation products (e.g., fibronectin fragments), and growth factors (e.g., bFGF) all stimulate MMP expression (Homandberg 2001, Henrotin et al. 2003, Goldring and Goldring 2004, Wang et al. 2004, Pelletier et al. 2006). Intracellular signaling molecules, for example, nuclear factor-κB (NF κB) and the mitogen-activated protein kinases (MAPKs), mediate MMP expression induced by the extracellular factors. In addition, cartilage impact injury and inflammatory cytokines also increase nitric oxide and other reactive oxygen species, which in turn cause chondrocyte death, direct matrix protein damage, inhibition of matrix synthesis, and increased matrix breakdown through induction of MMPs (Henrotin et al. 2003). These factors altering the balance of anabolic and catabolic events in the cartilage have been the targets of therapeutic development.

12.3.5 Pharmacological Therapies for Osteoarthritis

No standard pharmacological therapies have been established for OA treatment. Besides NSAIDS and analgesics, used to alleviate symptoms, most of the disease-modifying OA drugs (DMOADs) are still under investigation. The action of DMOADs is targeted towards cartilage preservation, based on current understanding of OA pathophysiology. New developments in DMOADs are reviewed recently by Abramson et al. (Abramson et al. 2006), and Pelletier et al. (Pelletier et al. 2006) and are briefly summarized here. (1) *Nutriceuticals.* Glucosamine sulfate and chondroitin sulfate are components of GAGs. Clinical studies have shown that both can prevent joint space from further narrowing in OA patients, suggesting their cartilage protection effects. Glucosamine may act via modulation of the effects of pro-inflammatory cytokines on induction of MMP and suppression of ECM genes (Derfoul et al. 2007). (2) *MMP inhibitors.* MMPs require zinc for their enzymatic activities. Tetracycline analogues, for example, doxycycline, acting as zinc chelators, have been shown to inhibit MMP activities, and prevent loss of cartilage in animal models. In human clinical trials, doxycycline slows joint space narrowing in the OA joint. (3) *Inhibitors of inflammatory pathways.* Diacerein, a chemical inhibiting IL-1 synthesis and activity, has been shown to preserve joint space in hip OA patients. Many more drugs are also under investigation in clinical trial stages or preclinical stages, which include IL-1 antagonists, inhibitors of IL-1 converting enzyme, inhibitors of inducible nitric oxide synthase, and inhibitors of intracellular signaling molecules such as MAPKs. Although the development of DMOADs has advanced dramatically in recent years, many challenges remain. First, OA is a chronic disease, which progresses over years or decades, and the possibility of drug toxicity will not be tolerated given the long time frame. Second, the cartilage protection effects of several drugs have not resulted in symptom alleviation. Clearly, a successful drug therapy should protect the structural integrity of the joint, reduce disease symptoms, and improve joint functions. Such criteria have not been met. Third, high resolution imaging technologies and high specificity biomarkers need to be developed to monitor OA progression and to validate the structure protection effects of the drugs. These difficulties in drug therapy thus point to opportunities in cartilage repair and regeneration as treatment approaches for OA.

12.4 Cartilage Repair and Regeneration

12.4.1 Articular Cartilage Defects and Spontaneous Repair

Articular cartilage defects can be divided into chondral and osteochondral lesions, – the former are confined to the chondral layer and the latter also penetrate the subchondral bone plate. Today, the classification by the "International Cartilage Repair Society" (ICRS, www.cartilage.org) (Fig. 12.2) for

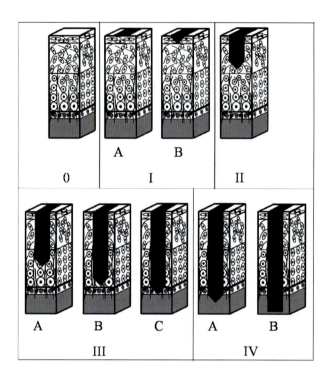

Cartilage defect grade	Defect dimension
0	Healthy articular cartilage
I	Softening of the intact cartilage surface (**A**) or superficial fibrillation/tears, <25% of the cartilage depth (**B**).
II	Cartilage lesions >25% but < 50% of the cartilage depth.
III	Cartilage lesions >50% of the cartilage depth (**A**), reaching the calcified layer (**B**) down to the subchondral bone, without penetrating (**C**).
IV	Lesions penetrating the subchon dral lamella (**A**) and/or penetrating the subchondral bone (**B**).

Fig. 12.2 Classification of cartilage defects, adapted from the International Cartilage Repair Society (www.cartilage.org)

osteo-/chondral defects is widely accepted and mainly based on the "Outer-bridge classification" (1961) (Outerbridge 1961).

Chondral lesions in adult rarely heal by themselves and usually develop into OA. Superficial defects can persist unchanged depending on their size without progression to a further stage (Mankin 1982, Hayes et al. 2001). In contrast, osteochondral lesions that penetrate the underlying bone plate can be repaired by a spontaneous mechanism. By virtue of opening bone-marrow spaces and vessels, bleeding occurs and the defect is filled with a fibrin-clot. Within the first 7 days, the fibrin-clot becomes organized and mesenchymal progenitor cells migrate from the bone marrow into the defects site. Subsequently, these cells proliferate and differentiate into cartilage- and bone-forming cells, and initiate a remodeling process within the fibrin-clot that results in the formation of a neo-tissue. After 6 to 8 weeks a cartilage-like repair tissue fills the former substantial defect. However, non-cartilage specific collagens, such as collagen types I, III, and hypertrophy-related collagen type X, are expressed within this repair tissue over time, such that this spontaneous repair process ultimately leads to the formation of a fibrous and mechanically inferior tissue (Johnson 2001, Hunt et al. 2002). It also has been shown that this neo-tissue fails to provide substantial stability under physiological load-bearing in the long term, and progressive degeneration can occur (Shapiro et al. 1993, Jackson et al. 2001).

12.4.2 Current Operative Strategies for Articular Cartilage Repair

As stated by Hunter in 1743, "from Hippocrates to the present age, ulcerated cartilage is a troublesome thing and once destroyed, is not repaired" (Hunter 1743). There is a long history of clinical trials to analyze the physiology and healing capacity of articular cartilage lesions to improve therapeutic interventions. The choice of the adequate operative strategy to treat articular cartilage defect is based on its size, grade, localization, and the age of the patient (Hunziker 2002). Arthroscopic lavage (rinsing of the joint), shaving and debridement are treatment options, which aim to remove pro-inflammatory substances and diseased cartilage. However, numerous studies have shown that these interventions cannot elicit a reactive or reparative response of the remaining cartilage. In fact, mechanical manipulation during these procedures entails cell death at the wound edge, therefore, furthering substantial loss of cartilage ECM and subsequent degeneration (Mitchell and Shepard 1987, Kim et al. 1991, Tew et al. 2000, Hunziker and Quinn 2003). Overall, the scientific data are extremely variable and inconsistent, ranging from good (> 65% of patients with pain relief) (Sprague 1981, Linschoten and Johnson 1997) to very poor (< 25% of patients with pain relief) (McGinley et al. 1999, Kruger et al. 2000) post-operative outcome.

Abrasion arthroplasty (Magnuson 1941, Rockett and Wageck 1997), Pridie-drilling (Pridie 1959), and microfracturing (Steadman et al. 1997, Steadman et al. 1999) comprise an extensive debridement of the degenerated articular cartilage to enhance the penetration of the subchondral bone plate. The difference between these techniques relates to the manner of the opening to the bone marrow spaces and vasculatures. Abrasion arthroplasty uses an automated burr, microfracture involves a small awl, and Pridie drilling stimulates bleeding by drilling the subchondral bone. All these approaches rely on the development of a fibrin clot and the formation of a fibrous neo-tissue (Steadman et al. 1999) as described above. It has been shown that the formation and composition of the repair tissue provoked by these techniques are extremely variable, inconsistent, and exhibit inferior mechanical properties compared to native articular cartilage (Furukawa et al. 1980, Shapiro et al. 1993). Nevertheless, a significantly better postoperative outcome in the short term using this technique, in comparison to a sole debridement, has been described (Schmidt and Hasse 1989). However, long-term studies rarely report a satisfactory clinical outcome (Johnson 1986, Kim et al. 1991). While the clinical outcomes after bone-marrow stimulating procedures are controversial, it should be emphasized that micro-fracturing, in particular, shows good long term result for smaller defects within a selected patient cohort (Sledge 2001, Steadman et al. 2003).

Other strategies for the reconstruction of the articular surface focus on the osteochondral transplantation of autologous or allogenic tissue grafts. The autologous osteochondral transplantation was first described by Wagner in 1972 (Wagner 1972). An autologous osteochondral cylinder of corresponding size to the defect is harvested from a non-weight bearing area of the knee and transplanted into the full-depth cartilage defect. This procedure was further developed as "Osteochondral Autologous Transplantation^{TM}" (OATS^{TM}, Arthrex) (Bobic 1999) or, when using multiple grafts, "MosaicPlasty^{TM}" (Smith & Nephew) (Hangody and Fules 2003). Postoperative results have shown decreased pain and improved joint function, mainly when small or medium sized full thickness defects were treated (Jakob et al. 2002, Hangody and Fules 2003). Regardless, transplantation of allogenic tissue is often followed by an immune response, despite the immunologically privileged environment of the knee joint (Langer et al. 1978, Stevenson 1987). Autologous tissue transplantation has to deal with the limited ability of osteochondral cylinders, incongruence of the surface, varying thickness of the cartilage layer, and extent of the donor site morbidity.

Soft tissue grafts including the transplantation of periosteum and perichondrium have been used in animal models and human clinical trials to treat cartilage defects. The periosteum exhibits a chondrogenic potential attributed to the presence of chondroprogenitor cells located within the cambial layer (Ito et al. 2001a, b). When using perichondrium compared to periosteum, no significant difference in the development of the cartilage-like repair tissue could be found, whereas varying results in the quality of the neo-tissue have been reported (Carranza-Bencano et al. 1999).

Besides tissue grafts, cells, with or without a scaffold, have been used to repair cartilage defects. Two cell sources have been experimented. One source is chondrocytes harvested from a non-weight bearing area of the cartilage in patients. This approach is referred to as Autologous Chondrocyte Implantation (ACI), which has been discussed in Chapter 13. Another cell source is mesenchymal stem cells (MSCs), which can be obtained from various adult tissues (Tuan 2006). The following section will discuss MSCs and their application in tissue engineering to treat cartilage defects.

12.4.3 Mesenchymal Stem Cells (MSCs)

MSCs are adult tissue derived progenitor cells that exhibit the potential to differentiate into cell types of multiple mesenchymal lineages, including chondrocytes, osteoblasts, adipocytes, fibroblasts, stromal cells, muscle cells, and other cells of mesenchymal origin (Tuan et al. 2003, Giordano et al. 2007, Kolf et al. 2007). MSCs can be isolated from a number of different adult tissues, and often exhibit the capacity to regenerate cell types specific for the tissue they reside in. Therefore, MSCs reflect a population of reparative cells that can be mobilized in case of injury, disease or degeneration, to induce a local repair mechanism in conjunction with secreted growth factors and resident cells. Adult MSCs reveal the unique property to replicate over a long period of time, and the primary MSCs and their daughter cells are capable of differentiating into various cell lineages of mesenchymal tissues. So far, MSCs have been isolated with different techniques from a variety of mesenchymal tissues, such as adipose, synovial membrane, muscle, blood, dermis, pericytes, trabecular bone, periosteum, and most commonly from bone marrow (Tuan et al. 2003). Interestingly, recent findings suggest the capacity of MSCs to differentiate into neuroectodermal and endodermal cell linages under certain conditions, suggesting a multipotent plasticity of this cell type (Giordano et al. 2007). At present, human bone marrow aspirate is considered the most accessible and enriched source of MSCs (Baksh et al. 2004). By virtue of their chondrogenic differentiation capability and their self-renewal capacity, MSCs have received significant attention in the emerging research field of cell- and matrix-based cartilage repair.

12.4.4 Chondrogenic Differentiation of MSCs

Chondrogenic differentiation of MSCs is reflected by the time-dependent changes in their morphology, and molecular and biosynthetic characteristics from an undifferentiated into a differentiated chondrocyte-like cell type (Tuan 2006). By analogy, during embryonic cartilage development, mesenchymal cells undergo a complex process consisting of cell condensation, proliferation and differentiation, requiring highly orchestrated cell–cell/–matrix interactions, growth-factor stimulation and mechanical activation (Shum et al. 2003, Tuan 2006). To activate and maintain chondrogenic differentiation of MSCs in vitro, some of these factors are also essential. In general, the phenotype of chondrocytes isolated from articular cartilage is unstable when cultured in a two-dimensional culture system, and time-dependent dedifferentiation into a fibroblastic cell type occurs, accompanied by the loss of the ability to produce cartilage-specific ECM, especially collagen type II and aggrecan (von der Mark 1999). This de-differentiation process is circumvented by culturing the chondrocytes in a three-dimensional environment (Holtzer et al. 1960, Manning and Bonner 1967), which is thought to promote cell–cell/–matrix interactions as observed during the embryonic development.

The principle of a three-dimensional environment is essential for the chondrogenic differentiation of MSCs as shown by Johnstone and co-workers, who established a high-density cell pellet culture system for MSCs (Johnstone et al. 1998). With this culture system, the critical cell density for chondrogenesis is achieved and cell–cell/–matrix interactions are promoted. Other factors are also needed to obtain defined, chondrogenic differentiation of MSCs in vitro. For example, chemically defined, serum free culture conditions with certain bioactive substances, which play a crucial role during induction and progression of chondrogenesis, are also necessary. Various studies have revealed the pro-chondrogenic influence of members of the TGF-ß superfamily, especially TGF-ß1, TGF-β3, BMP-2, and BMP-6 (Tuan et al. 2003). Furthermore, the pro-chondrogenic influence of dexamethasone, insulin, ascorbic acid and ITS plus alone or in combination has been widely demonstrated (Johnstone et al. 1998, Altaf et al. 2006, Derfoul et al. 2006). Mechanical stimuli also influence chondrogenic differentiation. It has been demonstrated that intermittent hydrostatic pressure alone or in combination with reduced oxygen tension leads to an increase of cartilage-specific gene expression and subsequent higher accumulation of cartilage-like ECM (Domm et al. 2000, Angele et al. 2003). Taken together, these results strongly suggest the importance of a mechano-sensitive component in the chondrogenic differentiation pathway of MSCs.

Effective chondrogenic differentiation in vitro is generally analyzed histologically and biochemically. After chondrogenesis of MSCs, a collagen- and proteoglycan-rich ECM can be identified histologically, while specific collagen types, such as collagen types II, IX, and XI, specific for adult articular cartilage (Srinivas et al. 1993, Bruckner and van der Rest 1994), may be detected by immunohistochemistry.

With the high-density pellet system as an experimental model for the chondrogenic differentiation of MSCs, the following outline of biochemical and gene expression during in vitro differentiation of MSCs is recognized, which will also be addressed later in the context of cartilage tissue engineering. Induction of chondrogenesis is followed by characteristic changes in expression of cartilage specific marker genes (Barry et al. 2001). The 21-day chondrogenic differentiation program of MSCs represents three distinct phases of gene expression activities. In the early phase (day 1–6), the expression of SOX-9, fibromodulin, collagen type IX, and COMP is upregulated. This is followed by a short intermediate phase (day 6–8) with expression of chondroadherin, versican core protein, the leucin-rich proteins decorin and biglycan, and collagen type XI. The late phase (day 8–21) is characterized by the expression of marker genes that are specific for the ECM of adult articular cartilage, namely, collagen type II and aggrecan. It should be mentioned that collagen type X is also expressed during this period. Expression of collagen type X, a marker gene for terminal chondrogenic differentiation, is associated with hypertrophic cartilage, as observed in the growth plate of long bones and osteoarthritic cartilage (Chambers et al. 2002).

12.4.5 Tissue Engineering

Tissue engineering is an ambitious, interdisciplinary research approach that combines principles and methods of engineering sciences, cell and molecular biology, and clinical medicine to develop and improve strategies to recover diseased or damaged human tissues and restore its form, structure, and function (Langer and Vacanti 1993). Tissue engineering has developed into its own discipline in the field of regenerative medicine and is potentially applicable for the reconstruction of almost every tissue of the human body (Fuchs et al. 2001).

Tissue engineering is of particular interest for the treatment of skeletal diseases, the majority of which are caused by degeneration or failure to heal. Injury, disease or degeneration of the musculoskeletal system often results in a reduction in quality of life and loss of earning capacity, and accordingly has a huge impact on public health and social security. The lack of adequate therapies that lead to a "restitutio ad integrum", especially in cartilage repair, drives the need for innovative therapies; tissue engineering thus presents significant therapeutic potential for the restoration and regeneration of diseased or degenerated tissue of the musculoskeletal system (Marler et al. 1998, Laurencin et al. 1999, Huard et al. 2003).

Chondral defects arising from trauma or degenerative diseases, such as osteoarthritis, reflect one of the most challenging fields in orthopaedic surgery. Because of the limited spontaneous repair mechanisms of chondral defects, the potential progression of these defects into extensive joint degeneration and possible necessity of a total joint arthroplasty highlights the special need for sufficient cartilage repair strategies. Tissue engineered osteo-/chondral analogs present a great potential for the improvement of current cartilage repair concepts.

Tissue engineering strategies are based on three principal components (Song et al. 2004):

1. Cell carriers (Scaffolds)
2. Cells (e.g., primary chondrocytes/MSCs)
3. Tissue-specific external stimuli (e.g., growth factors, mechano-biological stimulus)

Scaffolds used in tissue engineering strategies for cartilage repair that are in clinical or experimental stages can be grouped into native and synthetic materials. Native materials, such as collagen, fibrin, or hyaluronan, exhibit optimal biocompatibility and biodegradability, whereas synthetic polymers, such as poly (glycolic acid) (PGA), poly (lactic acid) (PLA), and their copolymer poly (lactic-co-glycolic acid) (PLGA) have higher primary stability and are more accessible for macro-/microstructure formation (Vacanti and Langer 1999). A key requirement is the porosity of the scaffold to facilitate nutrition, proliferation, and migration of cells, and to ensure cell colonization of the entire carrier (Langer and Vacanti 1993, Laurencin et al. 1999).

The application of primary, autologous chondrocytes, with or without the association of an organic scaffold for the reconstruction of isolated cartilage defects is already successfully used in clinical application (Fig. 12.3) (Brittberg et al. 1994, Schneider and Andereya 2003, Andereya et al. 2006, Marlovits et al. 2006). The utilization of autologous MSCs instead of chondrocytes marks a new promising approach for the improvement of current cartilage repair concepts. Although chondrocytes are inherently the most suitable cell source, they have limitations in practice because of the shortage of supply, loss of phenotype during expansion, and potential donor-side morbidity after harvest. On the other hand, MSCs may address these shortcomings because they are easily obtained from bone marrow aspirate, possess high expansion capacity, and are able to differentiate into chondrocytes even after a high number of population doublings (Baksh et al. 2004, Song et al. 2004). In addition, recent studies have revealed the immunosuppressive properties of MSCs in vitro and in vivo, that is, after allogenic transplantation of MSCs, a reduced immune response could be detected (Djouad et al. 2005, Caplan and Dennis 2006). The proliferation and chondrogenic differentiation of MSCs, accompanied by the synthesis of cartilage specific ECM, are crucial for the fabrication of functional cartilage constructs. Chondrogenic differentiation and maintenance of chondrocyte phenotype can be promoted and influenced by a variety of different growth factors, bioactive substances, and mechano-biological stimulations (Carver and Heath 1999, Elder et al. 2005, Stoltz et al. 2006). Cartilage tissue engineering has recently been reviewed by Kuo et al. (Kuo and Tuan 2003, Kuo et al. 2006) and Chen et al. (Chen et al. 2006).

12.4.6 Experimental Approaches for Articular Cartilage Tissue Engineering

The aim of articular cartilage repair is the fabrication of a suitable cartilage analogue that can replace the function, form, and structural integrity of the

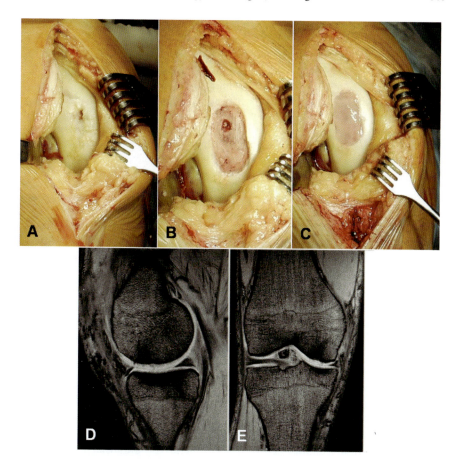

Fig. 12.3 Repair of isolated cartilage defects using primary autologous chondrocytes. The traumatic cartilage defect (III–IV ICRS classification) is located in the medial condyle of a 33 year old male, 3 x 2 cm. Cartilage defect before (**A**) and after debridement (**B**), revealing a small osteochondral lesion in the proximal region (**B**). Defect after implantation of a chondrocyte-laden collagen type I hydrogel (**C**), which is secured with fibrin glue into defect site (CaReS®, Arthro Kinetics PCL, Esslingen, Germany; www.arthro-kinetics.com). MRI images, six months after performing matrix-based ACI, reveal an optimal lateral and horizontal integration into the surrounding native cartilage (**D–E**). The dimension of chondrocyte-laden collagen hydrogel matches the height of the neighboring cartilage in the medial-lateral view (**D**) as well as in the anterior-posterior view (**E**). A small residual incongruence of the subchondral bone is seen in the anterior-posterior view (**E**)

diseased or degenerated articular cartilage. The complexity of articular cartilage tissue engineering encompasses the development of scaffolds, choices of cells, and the utilization of biological and physical stimuli.

The rudimentary function of a tissue engineering scaffold is to provide a temporary structure for the seeded cells to adhere and to synthesize new natural ECM in a shape and form guided by the scaffold. Scaffold design criteria,

Table 12.1 Requirements of scaffolding biomaterials used for cartilage tissue engineering (Adapted from Hunziker (1999))

Scaffold requirements	Biological function
Porosity	Cell penetration and migration
Carrier function	Controlled liberation of signaling molecules
Adhesion	Cell attachment
Biodegradability	Physiologic remodeling process
Volume stability	Integration, contact with surrounding tissue
Biocompatibility	Prevention of immune reaction.
Bonding	Integration and internal cohesion of the scaffolds
Internal cohesion	Structural integrity
Elasticity	Physiologic and dynamic deformation
Structural anisotropy	Physiologic formation of regenerative neo-tissue

shown in Table 12.1, include controlled biodegradability, suitable mechanical strength and surface chemistry, and the ability to regulate cellular activities, such as proliferation, interaction, and differentiation (Kuo and Tuan 2003, Chen et al. 2006, Kuo et al. 2006). A large variety of natural and synthetic scaffolds are under current investigation to determine their potential for cartilage repair. The structures of the scaffolds include hydrogels, porous sponges, woven or non-woven meshes, and fibrous meshes.

The poly (α-hydroxy esters), PGA, PLA, and their copolymer PLGA, are approved for use as biodegradable polymers for clinical use by the US Food and Drug Administration, and have been extensively studied as scaffold material for more than a decade (Cima et al. 1991, Vacanti et al. 1991). A large number of in vitro and in vivo studies have shown that chondrocytes seeded within these scaffolds maintain a stable chondrogenic phenotype and the production of cartilage-specific ECM (Cima et al. 1991, Hunter and Levenston 2004, Fuchs et al. 2005, Griffon et al. 2005, Mahmoudifar and Doran 2005, Mouw et al. 2005).

More recently, non-woven, three-dimensional meshes consisting of nanofibers with the dimentional properties of natural ECM components have been successfully electrospun from poly (α-hydroxy ester) and used as scaffolding material (Li et al. 2005). By virtue of their morphological shape and arrangement, nanofibers mimic the organization of the native ECM. The use of nanofibrous PLA or PCL scaffolds enabled chondrocytes to preserve their phenotype and to produce cartilage specific ECM (Li et al. 2003), and further promoted the chondrogenic differentiation of MSCs (Li et al. 2005). These findings, taken together with results of other investigations (Li et al. 2003, Malda et al. 2005), strongly underscore the importance of optimized scaffold architecture to obtain articular cartilage analogs.

Collagen-based matrices, such as fleeces, sponges, or hydrogels, have been widely used over the last 20 years in different tissue engineering approaches. Hydrogels are crosslinked polymer networks that are characterized by the ability to take up large amounts of water, similar to the ECM of the articular cartilage. These scaffolding materials are characterized by their unique property of gelation, and cells can, therefore, be homogeneously encapsulated within the matrix, thus promoting a spherical cell morphology. Chondrocyte-laden collagen type I hydrogels have already been successfully used in the treatment of focal articular cartilage defects in humans (Schneider and Andereya 2003, Andereya et al. 2006) (Fig. 12.3); postoperative follow-up examinations showed decreased joint pain and increased mobility and joint function. In vitro studies also demonstrated the potential use of MSCs in association with collagen I hydrogels (Noth et al. 2007). In comparison, additional stimulation with growth factors was found to be necessary to achieve chondrogenic differentiation of the encapsulated MSC.

Inductive stimuli, including growth factors, cytokines, and other bioactive agents, are commonly used to enhance synthesis of specific ECM and tissue growth in cartilage tissue engineering. A possible intrinsic biological difference between primary articular chondrocytes and MSC-derived chondrocytes has recently been reported by Mauck et al. (Mauck et al. 2006). In this study, bovine MSCs were seeded in agarose and then induced to undergo chondrogenesis with TGF-ß3 containing medium. As outcome parameters, mechanical and biochemical characteristics of the MSC-laden constructs were monitored and compared with those of chondrocyte-laden constructs, similarly cultured over a 10-week period. The results showed that while chondrogenesis occurred under the influence of TGF-ß3, the amount and mechanical properties of the accumulated matrix were reduced compared to that produced by chondrocytes cultured under the same conditions. However, it remains unclear whether MSCs, compared to primary chondrocytes, are in fact inferior in their capacity to form a functional cartilaginous ECM, or whether the in vitro conditions used did not match the optimal conditions for MSC differentiation and subsequent production of cartilage specific ECM within the construct.

A number of growth factors that initiate intracellular signal pathways via binding to cell surface receptors and instruct cells to proliferate, differentiate, and synthesize extracellular matrix proteins have been shown to be essential for the fabrication of cartilage constructs in vitro. Members of the TGF-β superfamily, insulin-like growth factors (IGFs), fibroblast growth factors (FGFs), platelet-derived growth factors (PDGFs), and epidermal growth factor (EGF) exhibit strong regulatory effects on chondrocytes and on MSC chondrogenesis. TGF-ßs (Chua et al. 2004, Indrawattana et al. 2004, Lee et al. 2004, Glowacki et al. 2005, Li et al. 2005) have been shown to have the highest chondrogenic inductive potential in MSCs, and enhance the synthesis of cartilage ECM in chondrocytes. Bone morphogenetic protein (BMP)-2 (Park et al. 2005, Noth et al. 2007), a member of the TGF-β superfamily, and FGF-18 have also been shown to promote chondrogenesis of MSCs and limb bud mesenchymal cells, respectively (Davidson et al. 2005). IGF, FGF, and PDGF are known to support chondrocytic activities, rather than promote chondrogenesis of MSCs. These findings, taken together with recent data showing the additive effects of combining TGF-ß3 with BMP-6 or IGF-1 for enhanced chondrogenesis of MSCs (Indrawattana et al. 2004), highlight the complexity of cartilage tissue engineering. The influence of mechanical loading, such as compression or hydrostatic pressure, on chondrocytes and MSCs to enhance the development of a cartilaginous phenotype has been reported by different groups, and indicates the functional importance of the mechano-biological component of cartilage tissue engineering (Carver and Heath 1999, Angele et al. 2003, Elder et al. 2005, Mauck et al. 2007).

Exciting research over the last two decades has pointed to the potential and challenge of cartilage tissue engineering. Development of biomimetic scaffolds, knowledge of chondrocyte and MSC biology, and application of chondroinductive stimuli are the key components to achieving tissue engineering-based improvement of articular cartilage regeneration strategies.

12.5 Conclusions

The limited repair capacity of articular cartilage, the potential progression of isolated articular cartilage defects into more severe stages of osteoarthritis, and a lack in sufficient therapeutic interventions to maintain or regenerate the unique composition, structure, and function of adult articular cartilage over a long period of time, underscore the need for the development of new technical approaches to improve current cartilage repair strategies. Tissue engineering is a particularly attractive approach to develop functional tissue substitutes for the treatment of cartilage defects. First stage clinical trials have already suggested the capability of matrix-based autologous chondrocyte transplantation for the treatment of isolated cartilage defects. The utilization of MSCs should address and overcome the shortcomings of these strategies, and represent a promising cell source for cartilage tissue engineering. Further challenges in functional tissue engineering are concerned with cell selection, scaffold design, and biological and mechano-biological stimulation. Sufficient regeneration or replacement of damaged or diseased cartilage over a long period as a future goal can only be achieved with increased understanding of cartilage physiology, and stem cell biology, technological development in engineering, and refinements of operative procedures.

Questions/Exercises

1. Describe articular cartilage formation during embryonic development. What characterizes this process?
2. What are the major groups of structural macromolecules in the articular cartilage? Give examples of each group.
3. How many zones can the articular cartilage be divided into? What are they? How are cells and fibers arranged in different zones?

4. What are the risk factors for development of osteoarthritis?
5. What enzymes are involved in cartilage degradation? Give two examples of factors regulating the expression of these enzymes.
6. What is the difference between chondral and osteochondral lesions? What are the natural outcomes of the two types of lesions?
7. What are the two major cell types used to repair cartilage defects? Discuss their limitations and advantages.
8. What potentials do mesenchymal stem cells (MSCs) have? Give examples of tissues MSCs can be isolated from.
9. What factors are needed to differentiate MSCs into chondrocytes in vitro? How do you know that MSCs have become cartilage?
10. What are the major components used in a project of tissue engineering? Give examples of materials used for cartilage tissue engineering.

References

Abramson SB, Attur M, Yazici Y (2006) Prospects for disease modification in osteoarthritis. Nat Clin Pract Rheumatol 2:304–312

Altaf FM, Hering, TM, Kazmi NH, Yoo JU, Johnstone B (2006) Ascorbate-enhanced chondrogenesis of ATDC5 cells. Eur Cell Mater 9:64–69

Andereya S, Maus U, Gavenis K, Muller-Rath R, Miltner O, Mumme T, Schneider U (2006) First clinical experiences with a novel 3D-collagen gel (CaReS) for the treatment of focal cartilage defects in the knee. Z Orthop Ihre Grenzgeb 144:272–280

Angele P, Yoo JU, Smith C, Mansour J, Jepsen KJ, Nerlich M, Johnstone B (2003) Cyclic hydrostatic pressure enhances the chondrogenic phenotype of human mesenchymal progenitor cells differentiated in vitro. J Orthop Res 21:451–457

Archer CW, Dowthwaite GP, F-W P (2003) Development of synovial joints. Birth Defects Res C Embryo Today 69:144–155

Arden N, Nevitt MC (2006) Osteoarthritis: epidemiology. Best Practice Res Clin Rheumatol 20:3–25

Baksh D, Song L, Tuan RS (2004) Adult mesenchymal stem cells: characterization, differentiation, and application in cell and gene therapy. J Cell Mol Med 8:301–316

Barry F, Boynton RE, Liu B, Murphy JM (2001) Chondrogenic differentiation of mesenchymal stem cells from bone marrow: differentiation-dependent gene expression of matrix components. Exp Cell Res 268:189–200

Bobic V (1999) Autologous osteo-chondral grafts in the management of articular cartilage lesions. Orthopade 28:19–25

Bramono DS, Richmond JC, Weitzel PP, Kaplan DL, Altman GH (2004) Matrix metalloproteinases and their clinical applications in orthopaedics. Clin Orthop Relat Res 428:272–285

Briggs MD, Chapman KL (2002) Pseudoachondroplasia and multiple epiphyseal dysplasia: mutation review, molecular interactions, and genotype to phenotype correlations. Hum Mutation 19:465–478

Brittberg M, Lindahl A, Nilsson A, Ohlsson C, Isaksson O, Peterson L (1994) Treatment of deep cartilage defects in the knee with autologous chondrocyte transplantation. New Engl J Med 331:889–895

Bruckner P, van der Rest M (1994) Structure and function of cartilage collagens. Microsc Res Tech 28:378–384

Buckwalter JA, Mankin HJ (1997) Articular cartilage. Part I: tissue design and chondrocyte–matrix interactions. J Bone Joint Surg 79A:600–611

Caplan AI, Dennis JE (2006) Mesenchymal stem cells as trophic mediators. J Cell Biochem 98:1076–1084

Carney SL, Muir H (1988) The structure and function of cartilage proteoglycans. Physiol Rev 68:858–910

Carranza-Bencano A, Perez-Tinao M, Ballesteros-Vazquez P, Armas-Padron JR, Hevia-Alonso A, Crespo FM (1999) Comparative study of the reconstruction of articular cartilage defects with free costal perichondrial grafts and free tibial periosteal grafts: an experimental study on rabbits. Calcified Tissue Int 65:402–407

Carter DR, Beaupré GS, Wong M, Smith RL, Andriacchi TP, Schurman DJ (2004) The mechanobiology of articular cartilage development and degeneration. Clin Orthop Relat Res 427 Suppl: 69–77. Review.

Carver SE, Heath CA (1999) Influence of intermittent pressure, fluid flow, and mixing on the regenerative properties of articular chondrocytes. Biotechnol Bioeng 65:274–281

Chambers MG, Kuffner T, Cowan SK, Cheah KS, Mason RM (2002) Expression of aggrecan and collagen genes in normal and osteoarthritic murine knee joints. Osteoathritis Cartilage 10:51–61

Chen FH, Rousche KT, Tuan RS (2006) Technology insight: adult stem cells in cartilage regeneration and tissue engineering. Nat Clin Pract Rheumatol 2:373–382

Chua KH, Aminuddin BS, Fuzina NH, Ruszymah BH (2004) Interaction between insulin-like growth factor-1 with other growth factors in serum depleted culture medium for human cartilage engineering. Med J Malaysia 59:7–8

Cima LG, Vacanti JP, Vacanti C, Ingber D, Mooney D, Langer R (1991) Tissue engineering by cell transplantation using degradable polymer substrates. J Biomech Eng 113:143–151

Craig FM, Bentley G, Archer CW (1987) The spatial and temporal pattern of collagens I and II and keratan sulphate in the developing chick metatarsophalangeal joint. Development 99:383–391

Cremer MA, Rosloniec EF, Kang AH (1998) The cartilage collagens: a review of their structure, organization, and role in the pathogenesis of experimental arthritis in animals and in human rheumatic disease. J Mol Med 76:275–288

Davidson D, Blanc A, Filion D, Wang H, Plut P, Pfeffer G, Buschmann MD, Henderson JE (2005) Fibroblast growth factor (FGF) 18 signals through FGF receptor 3 to promote chondrogenesis. J Biol Chem 280:20509–20515

Derfoul A, Miyoshi AD, Freeman DE, Tuan RS (2007) Glucosamine promotes chondrogenic phenotype in both chondrocytes and mesenchymal stem cells and inhibits MMP-13 expression and matrix degradation. Osteoarthritis Cartilage 15:646–655.

Derfoul A, Perkins GL, Hall DJ, Tuan RS (2006) Glucocorticoids promote chondrogenic differentiation of adult human mesenchymal stem cells by enhancing expression of cartilage extracellular matrix genes. Stem Cells 24:1487–1495

Djouad F, Fritz V, Apparailly F, Louis-Plence P, Bony C, Sany J, Jorgensen C, Noel D (2005) Reversal of the immunsupressive properties of mesenchymal stem cells by tumor necrosis factor alpha in collagen-induced arthritis. Arthritis Rheumatol 52:1595–1603

Domm C, Fay J, Schunke M, Kurz B (2000) Redifferentiation of dedifferentiated joint cartilage cells in alginate culture: effect of intermittent hydrostatic pressure and low oxygen partial pressure. Orthopäde 29:91–99

Donohue JM, Buss D, Oegema T, Thompson R (1983) The effects of indirect blunt trauma on adult canine articular cartilage. J Bone Joint Surg 65A:948–956

Dowthwaite GP, Flannery CR, Flannelly J, Lewthwaite JC, Archer CW, Pitsillides AA (2003) A mechanism underlying the movement requirement for synovial joint cavitation. Matrix Biol 22:311–322

Elder SH, Fulzele KS, McCulley WR (2005) Cyclic hydrostatic compression stimulates chondroinduction of C3H/10T1/2 cells. Biomech Model Mechanobiol 3:141–146

French MM, Gomes RR Jr, Timpl R, Hook M, Czymmek K, Farach-Carson MC, Carson DD (2002) Chondrogenic activity of the heparan sulfate proteoglycan perlecan maps to the N-terminal domain I. J Bone Mineral Res 17:48–55

Fuchs JR, Hannouche D, Terada S, Zand S, Vacanti JP, Fauza DO (2005) Cartilage engineering from ovine umbilical cord blood mesenchymal progenitor cells. Stem Cells 23:958–964

Fuchs JR, Nasseri BA, Vacanti JP (2001) Tissue engineering: a 21th century solution to surgical reconstruction. Ann Thorac Surg 72:577–591

Furukawa T, Eyre DR, Koide S, Glimcher MJ (1980) Biochemical studies on repair cartilage resurfacing experimental defects in the rabbit knee. J Bone Joint Surg 62A:79–89

Ge Z, Hu Y, Heng BC, Yang Z, Ouyang H, Lee EH, Cao T (2006) Osteoarthritis and therapy. Arthritis Rheumatol 55:493–500

Giordano A, Galderisi U, Marino IR (2007) From laboratory bench to patients's bedside: an update on clincal trials with mesenchymal stem cells. J Cell Physiol 211:27–35

Glasson SS, Askew R, Sheppard B, Carito B, Blanchet T, Ma HL, Flannery CR, Peluso D, Kanki K, Yang Z, Majumdar MK, Morris EA (2005) Deletion of active ADAMTS5 prevents cartilage degradation in a murine model of osteoarthritis. Nature 434:644–648

Glowacki J, Yates K, Maclean R, Mizuno S (2005) In vitro engineering of cartilage: effects of serum substitutes, TGF-beta, and IL-1alpha. Orthod Craniofac Res 8:200–208

Goldring SR, Goldring MB (2004) The role of cytokines in cartilage matrix degeneration in osteoarthritis. Clin Orthop Relat Res 427:S27–S36

Griffon DJ, Sedighi MR, Sendemir-Urkmez A, Stewart AA, Jamison R (2005) Evaluation of vacuum and dynamic cell seeding of polyglycolic acid and chitosan scaffolds for cartilage engineering. Am J Vet Res 66:599–605

Hangody L, Fules P (2003) Autologous osteochondral mosaicplasty for the treatment of full thickness defects of weight-bearing joints: ten years of experimental and clinical experience. J Bone Joint Surg (Am) 85A(Suppl 2):25–32

Hayes DWJ, Brower RL, John KJ (2001) Articular cartilage. Anatomy, injury, and repair. Clin Podiatr Med Surg 18:35–53

Hecht JT, Hayes E, Haynes R, Cole WG (2005) COMP mutations, chondrocyte function and cartilage matrix. Matrix Biol 23:525–533

Henrotin YE, Bruckner P, Pujol JP (2003) The role of reactive oxygen species in homeostasis and degradation of cartilage. Osteoarthritis Cartilage 11:747–755

Hildebrand A, Romaris M, Rasmussen LM, Heinegard D, Twardzik DR, Border WA, Ruoslahti E (1994) Interaction of the small interstitial proteoglycans biglycan, decorin and fibromodulin with transforming growth factor beta. Biochem J 302 (Pt 2):527–534

Holtzer H, Abbott J, Lash J, Holtzer S (1960) The loss of phenotypic trait by differentiated cells in vitro, I. dedifferentiation of cartilage cells. Proc Natl Acad Sci USA 46:1533–1542

Homandberg GA (2001) Cartilage damage by matrix degradation products: fibronectin fragments. Clin Orthop Relat Res 391(Suppl):S100–S107

Huard J, Li Y, Peng H, Fu FH (2003) Gene therapy and tissue engineering for sports medicine. J Gene Med 5:93–108

Hunt SA, Jazrawi LM, Sherman OH (2002) Arthroscopic management of osteoarthritis of the knee. J Am Acad Orthop Surg 10:356–363

Hunter W (1743) On the structure and disease of articulating cartilage. Philos Trans R Soc Lond 42b:514–521

Hunter CJ, Levenston ME (2004) Maturation and integration of tissue-engineered cartilages within an in vitro defect repair model. Tissue Eng 10:736–746

Hunziker EB (1999) Biologic repair of articular cartilage. Defect models in experimental animals and matrix requirements. Clin Orthop Relat Res 367(Suppl):S135–146

Hunziker EB (2002) Articular cartilage repair: basic science and clinical progress. A review of the current status and prospects. Osteoarthritis Cartilage 10:423–463

Hunziker EB, Quinn TM (2003) Surgical removal of articular cartilage leads to loss of chondrocytes from cartilage bordering the wound edge. J Bone Joint Surg (Am) 85A(Suppl 2):85–92

Indrawattana N, Chen G, Tadokoro M, Shann LH, Ohgushi H, Tateishi T, Tanaka J, Bunyaratvej A (2004) Growth factor combination for chondrogenic induction from human mesenchymal stem cell. Biochem Biophys Res Commun 320:914–919

Ito Y, Fitzsimmons J, Sanyal A, Mello M, Mukherjee N, O'Driscoll S (2001a) Localization of chondrocyte precursors in periosteum. Osteoarthritis Cartilage 9:215–223

Ito Y, Sanyal A, Fitzsimmons J, Mello M, O'Driscoll S (2001b) Histomorphological and proliferative characterization of developing periosteal neochondrocytes in-vitro. J Orthop Res 19:405–413

Jackson DW, Lalor PA, Aberman HM, Simon TM (2001) Spontaneous repair of full-thickness defects of articular cartilage in a goat model. A preliminary study. J Bone Joint Surg (Am) 83:53–64

Jakob RP, Franz T, Gautier E, Mainil-Varlet P (2002) Related articles, Autologous osteochondral grafting in the knee: indication, results, and reflections. Clin Orthop Relat Res 401:170–184

Johnson LL (1986) Arthroscopic abrasion arthroplasty. Historical and pathologic perspective present status. Arthroscopy 2:54–69

Johnson LL (2001) Arthroscopic abrasion arthroplasty: a review. Clin Orthop 391:306–317

Johnstone B, Hering TM, Caplan AI, Goldberg VM, Yoo JU (1998) In vitro chondrogenesis of bone marrow-derived mesenchymal progenitor cells. Exp Cell Res 238:265–272

Kiani C, Chen L, Wu YJ, Yee AJ, Yang BB (2002) Structure and function of aggrecan. Cell Res 12:19–32

Kim HKW, Moran ME, Salter RB (1991) The potential for regeneration of articular cartilage in defects created by chondral shaving and subchondral abrasion-an experimental investigation in rabbits. J Bone Joint Surg 73A:1301–1315

Kolf CM, Cho E, Tuan RS (2007) Mesenchymal stromal cells. Biology of adult mesenchymal stem cells: regulation of niche, self-renewal and differentiation. Arthritis Res Ther 9:204

Koyama E, Shibukawa Y, Rountree RB, Kingsley D, Enomoto-Iwamoto M, Iwamoto M, Pacifici M (2007) Limb synovial joints are produced by a distinct population of progenitor cells. In Orthopaedic research society 53rd annual meeting, San Diego, CA, p 41

Kruger T, Wohlrab D, Reichel H, Hein W (2000) The effect of arthroscopic joint debridement in advanced arthrosis of the knee joint. Zentralbl Chir 125:490–493

Kuo CK, Li WJ, Mauck RL, Tuan RS (2006) Cartilage tissue engineering: its potential and uses. Curr Opin Rheumatol 18:64–73

Kuo CK, Tuan RS (2003) Tissue engineering with mesenchymal stem cells. IEEE Eng Med Biol Mag 22:51–56

Langer F, Gross A, West M, Urovitz E (1978) The immunogenicity of allograft knee joint transplants. Clin Orthop Relat Res 132:155–162

Langer R, Vacanti JP (1993) Tissue engineering. Science 260:920–926

Laurencin CT, Ambrosio AMA, Borden MD, Cooper JAJ (1999) Tissue engineering: orthopedic applications. Annu Rev Biomed Eng 1:19–46

Lee JE, Kim KE, Kwon IC, Ahn HJ, Lee SH, Cho H, Kim HJ, Seong SC, Lee MC (2004) Effects of the controlled-released TGF-beta 1 from chitosan microspheres on chondrocytes cultured in a collagen/chitosan/glycosaminoglycan scaffold. Biomaterials 25:4163–4173

Li WJ, Danielson KG, Alexander PG, Tuan RS (2003) Biological response of chondrocytes cultured in three-dimensional nanofibrous poly (epsilon-caprolactone) scaffolds. J Biomed Mater Res 67A:1105–1114.

Li WJ, Tuli R, Okafor C, Derfoul A, Danielson KG, Hall DJ, Tuan RS (2005) A three-dimensional nanofibrous scaffold for cartilage tissue engineering using human mesenchymal stem cells. Biomaterials 26:599–609

Linschoten N, Johnson C (1997) Arthroscopic debridement of knee joint arthritis: effect of advancing articular degeneration. J South Orthop Assoc 6:25–36

Loening AM, James IE, Levenston ME (2000) Injurious mechanical compression of bovine articular cartilage induces chondrocyte apoptosis. Arch Biochem Biophys 381:205–212

Magnuson PB (1941) Joint debridement surgical treatment of degenerative arthritis. Surg Gynecol Obstet 73:1–9

Mahmoudifar N, Doran PM (2005) Tissue engineering of human cartilage in bioreactors using single and composite cell-seeded scaffolds. Biotechnol Bioeng 91:338–355

Malda J, Woodfield TB, van der Vloodt F, Wilson C, Martens DE, Tramper J, van Blitterswijk CA, Riesle J (2005) The effect of PEGT/PBT scaffold architecture on the composition of tissue engineered cartilage. Biomaterials 26:63–72

Mankin HJ (1982) The response of articular cartilage to mechanical injury. J Bone Joint Surg (Am) 64:460–466

Manning W, Bonner W (1967) Isolation and culture of chondrocytes from human adult articular cartilage. Arthritis Rheumatol 10:235–239

Marler JJ, Upton J, Langer R, Vacanti JP (1998) Transplantation of cells in matrices for tissue regeneration. Adv Drug Deliv Rev 33:65–182

Marlovits S, Zeller P, Singer P, Resinger C, Vecsei, V (2006) Cartilage repair: generations of autologous chondrocyte transplantation. Eur J Radiol 57:24–31

Martin JA, Buckwalter JA (2000) The role of chondrocyte–matrix interactions in maintaining and repairing articular cartilage. Biorheology 37:129–140

Mauck RL, Byers BA, Yuan X, Tuan RS (2007) Regulation of cartilagineous ECM gene transcription by chondrocytes and MSCs in 3D culture in response to dynamic loading. Biomech Model Mechanibiol 6:113–125

Mauck RL, Yuan X, Tuan RS (2006) Chondrogenic differentiation and functional maturation of bovine mesenchymal stem cells in long-term agarose culture. Osteoarthritis Cartilage 14:179–189

McGinley B, Cushner F, Scott W (1999) Debridement arthroscopy, 10-year followup. Clin Orthop 367:190–194

Merida-Velasco JA, Sanchez-Montesinos I, Espin-Ferra J, Rodriguez-Vazquez JF, Merida-Velasco JR, Jimenez-Collado J (1997) Development of the human knee joint. Anat Rec 248:269–278

Mitchell N, Shepard N (1987) Effect of patellar shaving in the rabbit. J Orthop Res 5:388–392

Mouw JK, Case ND, Guldberg RE, Plaas AH, Levenston ME (2005) Variations in matrix composition and GAG fine structure among scaffolds for cartilage tissue engineering. Osteoarthritis Cartilage 13:828–836

Mow VC, Rosenwasser MP (1988) Articular cartilage: biomechanics. American Academy of Orthopaedic Surgeons, Park Ridge, IL.

Neuhold LA, Killar L, Zhao W, Sung ML, Warner L, Kulik J, Turner J, Wu W, Billinghurst C, Meijers T, Poole AR, Babij P, DeGennaro LJ (2001) Postnatal expression in hyaline cartilage of constitutively active human collagenase-3 (MMP-13) induces osteoarthritis in mice. J Clin Invest 107:35–44

Noth U, Rackwitz L, Heymer A, Weber M, Baumann B, Steinert F, Schütze N, Jakob F, Eulert J (2007) Chondrogenic differentiation of human mesenchymal stem cells in a collagen type I hydrogel for articular cartilage repair. J Biomed Mater Res. A 83:626–635.

Outerbridge RE (1961) The etiology of chondromalacia patellae. J Bone Joint Surg (Br) 43:752–757

Park Y, Sugimoto M, Watrin A, Chiquet M, Hunziker EB (2005) BMP-2 induces the expression of chondrocyte-specific genes in bovine synovium-derived progenitor cells cultured in three-dimensional alginate hydrogel. Osteoarthritis Cartilage 13:527–536

Peach CA, Carr AJ, Loughlin J (2005) Recent advances in the genetic investigation of osteoarthritis. Trends Mol Med 11:186–191

Pelletier JP, Martel-Pelletier J, Raynauld JP (2006) Most recent developments in strategies to reduce the progression of structural changes in osteoarthritis: today and tomorrow. Arthritis Res Ther 8:206

Pridie KH (1959) A method of resurfacing osteoarthritic knee joints. J Bone Joint Surg 41:618–619

Rockett P, Wageck J (1997) The reparative response with abrasion arthroplasty. In ISAKOS 1997 Biennial Congress, Buenos Aires.

Roos EM (2005) Joint injury causes knee osteoarthritis in young adults. Curr Opin Rheumatol 17:195–200

Roughley PJ (2001) Articular cartilage and changes in arthritis: noncollagenous proteins and proteoglycans in the extracellular matrix of cartilage. Arthritis Res 3:342–347

Roughley PJ, Lee ER (1994) Cartilage proteoglycans: structure and potential functions. Microsc Res Tech 28:385–397

Roughley P, Martens D, Rantakokko J, Alini M, Mwale F, Antoniou J (2006) The involvement of aggrecan polymorphism in degeneration of human intervertebral disc and articular cartilage. Eur Cell Mater 11:1–7

Sangha O (2000) Epidemiology of rheumatic diseases. Rheumatology 39(Suppl 2):3–12

Schmidt H, Hasse E (1989) Arthroscopic surgical treatment of circumscribed cartilage damage with spongiolization or Pridie drilling. Beitr Orthop Traum 36:35–37

Schneider U, Andereya S (2003) First results of a prospective randomized clinical trial on traditional chondrocyte transplantation vs CaReS-Technology. Z Orthop Ihre Grenzgeb 141:496–497

Shapiro F, Koide S, Glimcher MJ (1993) Cell origin and differentiation in the repair of full-thickness defects of articular cartilage. J Bone Joint Surg (Am) 75A:532–553

Shum L, Coleman CM, Hatakeyama Y, Tuan RS (2003) Morphogenesis and dysmorphogenesis of the appendicular skeleton. Birth Defects Res C Embryo Today 69:102–122

Sledge S (2001) Microfracture techniques in the treatment of osteochondral injuries. Clin Sports Med 20:365–377

Song L, Baksh D, Tuan RS (2004) Mesenchymal stem cell-based cartilage tissue engineering: cells, scaffold and biology. Cytotherapy 6:596–601

Sprague N (1981) Arthroscopic debridement for degenerative knee joint disease. Clin Orthop 160:118–123

Srinivas GR, Barrach HJ, Chichester CO (1993) Quantitative immunoassays for type II collagen and its cyanogen bromide peptides. J Immunol Methods 159:53–62

Stanton H, Rogerson FM, East CJ, Golub SB, Lawlor KE, Meeker CT, Little CB, Last K, Farmer PJ, Campbell IK, Fourie AM, Fosang AJ (2005) ADAMTS5 is the major aggrecanase in mouse cartilage in vivo and in vitro. Nature 434:648–652

Steadman JR, Briggs KK, Rodrigo JJ, Kocher MS, Gill TJ, Rodkey WG (2003) Outcomes of microfracture for traumatic chondral defects of the knee: average 11-year follow-up. Arthroscopy 19:477–484

Steadman J, Rodkey W, Briggs K, Rodrigo J (1999) The microfracture technic in the management of complete cartilage defects in the knee joint. Orthopade 28:26–32

Steadman JR, Rodkey WG, Singelton SB, Briggs KK (1997) Microfracture technique for full thickness chondral defects: technique and clinical results. Op Tech Orthop 7:300–304

Stevenson S (1987) The immune response to osteochondral allografts in dogs. J Bone Joint Surg 69A:573–582

Stoltz JF, de Isla N, Huselstein C, Bensoussan D, Muller S, Decot V (2006) Mechanobiology and cartilage engineering: the underlying pathophysiological phenomena. Biorheology 43:171–180

Tew S, Kwan APL, Hann A, Thomson B, Archer CW (2000) The reactions of articular cartilage to experimental wounding: role of apoptosis. Arthritis Rheumatol 43:215–225

Thompson RC, Oegema TR, Lewis JL, Wallace L (1991) Osteoarthritic changes after acute transarticular load: an animal model. J Bone Joint Surg 73A:990–1001

Tuan RS (2004) Biology of developmental and regenerative skeletogenesis. Clin Orthop Relat Res 427(Suppl):S105–117

Tuan RS (2006) Stemming cartilage degeneration: adult mesenchymal stem cells as a cell source for articular cartilage tissue engineering. Arthritis Rheumatol 54:3075–3078

Tuan RS, Boland G, Tuli R (2003) Adult mesenchymal stem cells and cell-based tissue engineering. Arthritis Res Ther 5:32–45

Vacanti JP, Langer R (1999) Tissue engineering: the design and fabrication of living replacement devices for surgical reconstruction and transplantation. Lancet 354(Suppl 1):132–134

Vacanti CA, Langer R, Schloo B, Vacanti J (1991) Synthetic polymers seeded with chondrocytes provide a template for new cartilage formation. Plast Reconstr Surg 88:753–759

Vincent T, Hermansson M, Bolton M, Wait R, Saklatvala J (2002) Basic FGF mediates an immediate response of articular cartilage to mechanical injury. Proc Natl Acad Sci USA 99:8259–8264

Vincent T, Saklatvala J (2006) Basic fibroblast growth factor: an extracellular mechanotransducer in articular cartilage? Biochem Soc Trans 34:456–457

von der Mark K (1999) Structure, biosynthesis and gene regulation of collagens in cartilage and bone. Academic Press, Orlando.

Wagner H (1972) Möglichkeiten und klinische Erfahrungen mit der Knorpeltransplantation. Z Orthopädie 110:708–715

Wang X, Manner PA, Horner A, Shum L, Tuan RS, Nuckolls GH (2004) Regulation of MMP-13 expression by RUNX2 and FGF2 in osteoarthritic cartilage. Osteoarthritis Cartilage 12:963–973

Wong M, Wuethrich P, Buschmann MD, Eggli P, Hunziker E (1997) Chondrocyte biosynthesis correlates with local tissue strain in statically compressed adult articular cartilage. J Orthop Res 15:189–196

Wong M, Wuethrich P, Eggli P, Hunziker E (1996) Zone-specific cell biosynthetic activity in mature bovine articular cartilage: a new method using confocal microscopic stereology and quantitative autoradiography. J Orthop Res 14:424–432

Zang H, Vrahas MS, Baratta RV, Rosler DM (1999) Damage to rabbit femoral articular cartilage following direct impacts of uniform stresses: an in vitro study. Clin Biomech 14:543–548

Chapter 13
Basic Science and Clinical Strategies for Articular Cartilage Regeneration/Repair

Barry W. Oakes

Contents

13.1 Introduction 395
13.2 Articular Cartilage and its Repair 396
    13.2.1 Normal Articular Cartilage: Basic Science of Structure, Biology, and Function 397
    13.2.2 The Vital Role of Growth Factors in Normal Articular Cartilage Matrix Biosynthesis 398
    13.2.3 Optimal Chondrocyte Density in Cartilage 400
    13.2.4 Clinical Findings of Cartilage Structure Functionality 402
13.3 Hyaline Articular Cartilage Response to Injury 403
13.4 Treatment Options for Cartilage Injuries 405
13.5 Autologous Chondrocyte Implantation 408
    13.5.1 ACI Surgical Procedure and Technique 409
13.6 Key Questions About ACI 412
13.7 Conclusions: The Clinicians' View on Treatments for Articular Cartilage Repair 424
References 426

13.1 Introduction

In the last decade, much has been learnt regarding the regeneration and repair of tissues of the skeleton. Regeneration of tissue readily occurs in the embryo, is almost absent in neonates and is never observed in adults. This may be because of the relatively high proportion of undifferentiated progenitor cells found in embryos and their relative scarcity in adults, that is, 1/10,000 mesenchyme cells in newborn and $1/2 \times 10^6$ in an 80 year old adult (Haynesworth et al. 1994). Regeneration is a slow process and seems to recapitulate many steps during embryonic development. In contrast, repair is a much more rapid process probably designed for survival and involves the usual inflammatory cell cascade followed by matrix deposition and

B.W. Oakes (✉)
Mercy Tissue Engineering and Department of Anatomy and Cell Biology, Monash University Australia, Melbourne, Australia, 2 Reeve Court, Cheltenham, 3192, Australia
e-mail: barry.oakes@optusnet.com.au

M. Santin (ed.), *Strategies in Regenerative Medicine*,
DOI 10.1007/978-0-387-74660-9_13, © Springer Science+Business Media, LLC 2009

then a remodeling process which attempts to partially regenerate damaged tissues in the adult. Repair and remodeling process can take many months to occur in the musculoskeletal system and the resulting repaired tissues, although of inferior biomechanical properties, may aid patient's survival and be sufficient for adequate pain-free joint function (e.g., ligament/tendon repair in the adult may have a similar ultimate tensile strength but the strain performance of the repair tissue is inferior to that of normal adult ligament and tendon). This apparent paradox is thought to be due to a combination of factors such as poor neo-collagen fibril alignment in the actual repair region plus small collagen fibril diameters which have a density lower than intra-collagen fibril cross-links (Oakes 2003). It is from this now extensive knowledge-base of both repair and regeneration in "orthopedic tissues" that orthopedic tissue engineering has developed.

Tissue engineering involves the use of cells (either adult, mesenchymal, or embryonic) coupled with biological or artificial matrices or scaffolds which guide the cells during tissue repair / regeneration. These cells can be "driven" by specific bioactive molecules, ex vivo gene transfer and other physical factors within specially designed "bioreactors" to form neo-tissues in vitro for future re-implantation in vivo. Alternatively, the cells and special matrices which can include bioactive molecules such as growth factors can be combined in vivo to attempt to enhance tissue repair such as the one that has recently been achieved with human articular cartilage repair. In this approach, patient's own autologous chondrocytes are retrieved at the time of arthroscopy and then are expanded in vitro before their re-implantation into full thickness articular cartilage defects. The treated defects are finally covered with a sutured and fibrin-glued periosteal patch. Such articular cartilage repair has been shown to be clinically effective and durable up to seven years post-implantation (Peterson et al. 2000).

This chapter will discuss primarily the use of autologous chondrocyte implantation (ACI) for the repair / replacement of articular cartilage within the knee joint. The chapter will also compare this repair process with that of normal tissues and it will indicate briefly how tissue engineering concepts have enhanced orthopedic clinical outcomes for symptomatic patients with articular cartilage defects.

Gene-base technology has been deliberately omitted because this promising technology has as not yet been applied to articular cartilage repair in the human. However, many examples are available of the exciting use of genetic engineering in enhancing articular cartilage repair in animal models. Likewise, the use of embryonic stem cells will not be considered as it is still in its infancy and limited to animal models and in vitro systems.

13.2 Articular Cartilage and its Repair

For a full understanding of ACI, histological and physiological bases of articular cartilage have to be considered. Here, we provide an overview of articular cartilage basic science as related to its repair process.

13.2.1 Normal Articular Cartilage: Basic Science of Structure, Biology, and Function

Articular cartilage (AC) is a unique avascular, aneural, and alymphatic load bearing complex live tissue which is supported by the underlying subchondral bone plate which in turn is supported by trabecular bone (Ratcliffe and Mow 1996, Buckwalter and Mankin 1997a) (see schematic representations at www.pages.drexel.edu, www.orthobiomech.info). The superficial layer or "lamina splendens" is composed of tightly packed collagen fibrils parallel to the articular surface and helps to bind the deeper layers into one functional construct. AC has a unique structure as its extracellular matrix (ECM) is composed of a complex combination of type II collagen fibrils specifically arranged in arcades and vertically in the mid-zone. The AC ECM includes high molecular weight proteoglycans, the aggrecans which are able to entrap high water content. Because of their high negative charge, aggrecan molecules retain water avidly and the high osmotic pressure generated expands and places the collagen II fibril network under tension. It is this unique molecular combination that provides articular cartilage with resilience to withstand the intermittent high loads of daily life and also its durability which usually spans the subject's life time. The deep zone is calcified cartilage and is separated from the mid-zone by the "tidemark" which represents the non-calcified-calcified cartilage interface. The calcified cartilage in turn rests and is locked onto a complex undulating subchondral bone plate (Poole 2003). Chondrocytes which synthesize and turnover AC ECM occupy less than 10% of the tissue volume and are essentially "trapped" in the ECM (like osteoblasts) and do not (cannot) migrate to repair adjacent defects whether of traumatic or of other aetiology. Hence, this tissue suffers the problem of non-repair following any form of traumatic injury where AC is lost by shearing forces or other complex mechanically excessive loads.

In 1992, C.A. Poole emphasized the concept of the "chondron" in normal AC biology. Elegant studies using immuno-fluorescent labeling coupled with electron microscopy have demonstrated the tight type II/VI collagen matrix surrounding the chondrocyte and a tail inferiorly of type II/VI collagen which Poole suggests may be involved in acting like a transducer to monitor ECM loading and in stimulating the chondrocytes to respond appropriately by synthesizing more or less ECM.

It is also known that in vitro (de Witt et al. 1984) and in vivo AC chondrocytes respond to both loading and unloading. A dramatic loss of aggrecan is observed following chronic unloading in non-weight bearing conditions (Caterson and Lowther 1978), while aggrecan concentration increases under the repetitive loading of running (Kiviranta et al. 1987).

Buckwalter and Mankin (1997b) also demonstrated a marked loss of chondrocyte number per unit volume of tissue with age. Hence with age >20 years there are much fewer cells per unit volume available for any possible repair.

13.2.2 The Vital Role of Growth Factors in Normal Articular Cartilage Matrix Biosynthesis

In 1986, McQuillan et al. demonstrated the important role of IGF-1 in stimulating aggrecan synthesis in vitro and probably also in vivo. TGF-β is also an important driver of both type II collagen and aggrecan synthesis (Sporn et al. 1986). TGF-β is an important factor to induce chondrocyte differentiation from their parent undifferentiated mesenchymal stem cells during limb development and formation of the primary cartilage anlagen, the precursor tissue for future long bone development (Leonard et al. 1991). TGF-β is sequestered and lodged as an inactive precursor in the chondrocyte immediate ECM and appears to be able to maintain chondrocyte differentiation when in close proximity to the cell.

Once this adjacent ECM is disrupted (e.g., by enzymes for in vitro cell extraction), chondrocytes are dispersed without their surrounding ECM containing TGF-β, the cells de-differentiate into elongated fibroblast-like cells and lose their ability to synthesise both type II collagen and large MW aggrecan. Instead the fibroblast-like cells synthesise type I collagen and hyaluronan more typical of tendon and ligament and, in vivo, a fibrocartilage type ECM is formed. This process of chondrocyte de-differentiation unfortunately occurs with low density mono-layer plating of chondrocytes thus making chondrocyte expansion in vitro for ACI chondrocyte grafting a difficult objective to be obtained (Benya and Shaffer 1982). See Figures 13.1 and 13.2.

Serum IGF –I levels fall with age (Corpus et al. 1993) and may also be in part responsible for the decreased aggrecan chain length observed with age (Buck-walter et al. 1994).

Hunziker and Rosenberg (1996), Hunziker et al. (2001a,b), were the first to demonstrate that articular cartilage is capable of repair using special biological conditions. In elegantly designed quantitative experiments using a minipig femoral condyle model, Hunziker meticulously routed an entirely enchondral gutter or groove of known width and depth within the rabbit femoral condyle articular cartilage and examined the capacity of the use of TGF-β to induce repair of these surgically created defects by attracting synovial stem cells into the AC defect zone. In his early experiments, Hunziker examined the capacity of synovial cells to simply attach to the defect finding only few cells able to graft. In a further step, the investigator treated the defect with Chondroitinase ABC to remove the proteoglycans which were thought to be blocking cell attachment to the type II collagen matrix thus improving cell attachment in comparison with untreated defects. The author hypothesized that topical TGF-β may exert a chemotactic effect thus attracting synovial cells and maybe encouraging synovial stem cell differentiation towards the chondrocyte differentiation pathway. Indeed, ad hoc experiments showed that more cells were found within the defect compared to simply using chondroitinase ABC alone. Postulating that incoming synovial stem cells may require a matrix "instructing" them to fill the entire defect, the investigator treated some grooved defects with both chondroitinase

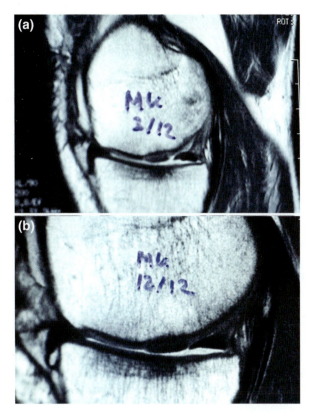

Fig. 13.1 (a, b) MRI saggital views of a lesion similar to that shown in Fig. 13.7a. **(a)** 2 months Post-ACI. Note fill and edges of defect can clearly be seen. **(b)** 12 months Post-ACI. Note overfill and edges of neo-repair can still clearly be seen. Note 'bone bruising' and a thinkened subchondral bone plate of the adjacent tibial surface

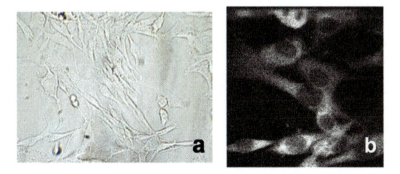

Fig. 13.2 In vitro de-differentiation of chondrocytes into fibroblast-like cells after two weeks in culture. **(a)** chondrocytes with an elongated morphology typical of fibroblasts, **(b)** chondrocytes resulting positive to collagen I rather than collagen II (J.Werkmeister *et al.* CSIRO, Melbourne)

ABC and a fibrin gel finding that synovial fibroblasts filled the defect and they were evenly dispersed throughout the fibrin gel. However, no chondrocytes were found. Liposomes-delivered with TGF-β dispersed throughout the fibrin gel induced differentiation of the synovial cells to chondrocytes.

This very elegant quantitative work demonstrated that, given the correct biological conditions, articular cartilage repair was feasible thus confirming the finding of Brittberg et al. (1994) using ACI.

Chaipinyo et al. (2002) were able to demonstrate that bovine chondrocytes seeded at low density in three-dimensional type I collagen gels proliferated and maintained their unique differentiated phenotype in serum free medium supplemented with a cocktail of three growth factors (3GF) ,TGF-β1,IGF-I, and β-FGF. These three growth factors enhanced the mitotic activity of bovine chondrocytes similar to 20% Foetal calf serum (FCS). In vitro, at day 21, chondrocytes proliferated 41 fold in gels-FCS and 37 fold in gels including 3GF. This study has also shown that a combination of three growth factors in serum-free medium has a similar effect on cell proliferation and rates of protein synthesis but a very different effect on the rate of proteoglycan synthesis and total matrix accumulation when compared to 20% FCS supplemented serum.

However, the matrix deposition observed showed histological differences. Smaller diameter collagen fibrils and a denser matrix were observed in FCS-enriched gels culture compared to 3GF-enriched gels where loosely arranged larger diameter fibrils (probably from the original type 1 collagen gel) were observed. This work was based on one of the first studies demonstrating the possibility of growing and maintaining chondrocytes in vitro in their differentiated state (Oakes et al. 1977, Handley et al. 1975).

13.2.3 Optimal Chondrocyte Density in Cartilage

Studies have also shown that the cell numbers required for ACI neo-cartilage repair must be calculated. Cell density in vitro, and most probably in vivo, is *very* important to maintain chondrocyte differentiation. Cell numbers for transplantation should be carefully considered to fill an AC defect of known volume. On average the mean thickness of human knee joint AC is 2.26 ± 0.49 mm (Stockwell 1971) whilst the maximum thickness is 3.8 ± 0.46 mm (Adam et al. 1998).The thickness of the non-meniscal cartilage of the medial tibial plateaux is 3.6 ± 0.34 mm (Kaab et al. 1998). For example, using Stockwell's data about the physiological density of human knee joint chondrocytes that is $14.1 \pm 3.2 \times 10^3$ cells/mm^3, a 2 cm^2 defect with a depth of 3.8 mm or an equal volume of 760 mm^3 would require 10.7×10^6 cells, whilst a 2 mm defect with the same surface area would require 5.6×10^6 cells.

NOTE: A valuable practical rule of thumb for surgeons using ACI is that for every 1 cm^2 of AC full thickness defect, $10 \times 10 \times 2.26$ mm dimension, 3.2×10^6 cells would be required using mean thickness of AC as 2.26 mm and Stockwell's normal knee joint AC density data. This means that human chondrocytes in

Table 13.1 Cell numbers required for successful ACI

| Size defect (cm²) | Cell numbers | |
	Required cell number ($\times 10^6$)*	Cell numbers obtained from 300 mg AC biopsy ($\times 10^6$)
1	3.2	5.3
2	6.4	10.6
5	16.0	26.5
10	32.0	53.0

*Calculated from normal adult femoral condyle AC cell density (Stockwell, 1971; Hunziker, 2002) and average AC thickness (Adam, 1998).

culture need to be expanded to at least 5.3 times of the original cell number obtained from a 300 mg biopsy (2000 cells/mg) for every 1 cm² of human AC defect area (Table 13.1).

Recent studies by Chaipinyo et al. (2004) indicate that debrided articular cartilage obtained at the time of the first arthroscopy from the surrounding AC of the defect (which is usually discarded) can be used as a source of healthy chondrocytes. This mean that less normal AC needs to be removed to obtain the necessary cell numbers described above for expansion in vitro. In brief, biopsies were retrieved from 12 patients who underwent debridment of AC (DAC). The age range of these subjects was 35–61 years. Normal healthy AC (NAC) was obtained from 2 autopsies aged 21 and 25 years. After 4 weeks in culture DAC chondrocytes in Type I collagen gels proliferated 18.34 ± 1.95 fold and was similar to NAC which proliferated 11.24 ± 1.02 fold. Synthesis of proteoglycan and collagen in DAC and NAC were similar. Newly synthesized matrices in gel cultures consisted of predominantly type II collagen as shown by immuno-labeling and SDS-PAGE followed by fluorography. Hence, adult chondrocytes from "debrided human AC". cultured at low density in type I collagen gels can be used for the ACI procedure as they provide sufficient viable cell numbers for ACI and maintain their chondrocyte phenotype as they synthesize a cartilage-like specific matrix.

Table 13.2 shows the expansion capability of human chondrocytes using type III collagen gels.

Table 13.2 Reseedings from human chondrocytes expanded in type I collagen gels after 28 and a further 28 days in type I gel culture (modified from Chaipinyo et al. 2004)

	Biopsy type (patient's age, gender, anatomical site)	1st Culture 28 days (A)	Reseeding 28 days (B)	Total 56 days 2 × 28 days (A × B)
		Cell proliferation		
DAC	37 years, male, MFC	21	8	168
	50 years, male, Trochlea	24	12	288
NAC	21 years, male, combined	8	28	224
	25 years, male, combined	12	19	228

Note: DAC = Debrided Articular Cartilage, MFC = Medial Femoral Condyle, NAC = Normal Articular Cartilage

13.2.4 Clinical Findings of Cartilage Structure Functionality

Articular cartilage damage is common and does not normally repair (Curl et al. 1997). The natural history of articular cartilage lesions remains elusive in vivo as there have been few long term follow-up studies of AC defects with critical imaging. Shelbourne et al. (2003) noted in 123 incidental chondral lesions observed at the time of more than 2700 ACL reconstructions caused patients to report lower (P<0.05) Noyes' subjective scores than did controls with normal articular cartilage after a mean of 8.7 years follow-up. Despite the absence of radiographic changes which are clearly a late stage of articular cartilage loss, lateral chondral lesions caused worse subjective scores compared to medial chondral lesions. In a review of 993 knee arthroscopies in patients with a mean age of 35 years, there was an 11% incidence of full- thickness AC lesions that may have benefited from early surgical treatment (Aroen et al. 2004). With the newer cartilage, specific MRI protocols, and a close correlation with chondral defects, clinical symptoms, and a likelihood of symptom progression have been established (Marlovits et al. 2005).

After partial meniscectomy (Cicuttini et al. 2002), up to 6.5% volumetric loss of AC per year has been demonstrated, indicating the important protective load-sharing role that the menisci play in the normal knee joint. The normal medial meniscus transmits 50% of the knee joint load in knee extension and 90% when the knee is in flexion (Walker and Erkman 1975). In vitro animal studies have demonstrated that loss of just 20% of a meniscus can lead to a 350% increase in contact forces and probable chondrocyte apoptosis in the long term (Seedhom and Hargreaves 1979).

Often young athletes and others are left with full thickness defects >1 cm diameter, which are symptomatic probably due to subchondral bone injury and oedema and seek pain relief. It is thought that the subchondral bone plate and the overlying complex articular cartilage act as a combined biological loading interface and transmit regular attenuated dynamic loading to the underlying trabecular bone which then eventually transmits loading to the more dense Haversian bone of the cortex of long bones. Indeed, when normal AC is lost on critical weight bearing areas of the femoral condyles, subchondral bone oedema is commonly found in symptomatic patients and is thought to be due to microfractures of the subchondral bone plate itself or to the underlying supporting trabeculae of the adjacent cancellous bone. These microfractures may be due to the loss of attenuation of bone loading guaranteed by both the articular cartilage and the menisci. Hence, it can be deduced that protection of both the subchondral bone plate and its supporting structures from critical point overload is one of the major functional roles of normal hyaline articular cartilage coupled with the menisci.

Cytokine activation especially IL-I and IL-6 has been shown to degrade aggrecan by activating aggrecanases (Tyler 1986, Ratcliffe et al. 1986). These enzymes can rapidly lead to a softening of the AC with loss of osmotic tension in the type II collagen framework. This leads to collagen framework over-load (with continued weight bearing on an injured knee joint with an associated hemarthrosis) which in turn leads to loss of the superficial tangentially arranged collagen fibrils by shearing wear. The consequence is the classic fibrillation of AC which results in a grade II damage by exposure of the predominantly vertically arranged large diameter collagen fibrils readily and regularly seen by orthopedic surgeons by arthroscopy (Buckwalter et al. 1998).

13.3 Hyaline Articular Cartilage Response to Injury

The avascular structure of articular cartilage is well known as well as the lack of lymphatic drainage and neural elements (Mankin 1982, Buckwalter and Mankin 1997). As a consequence, injury to this unique load-bearing layer is initially painless and, therefore, insidious for patients who often do not experi-ence symptoms until the cartilage lesion is of such a size that the unprotected subchondral bone-plate undergoes microfracture(s) and subsequent subchon-dral bone oedema; this is the so-called "bone bruising" that appears at MRI examination (Fig. 13.1a, b).

Articular chondrocytes are ineffective in responding to injury. Whilst there may be transient cell replication and increased local matrix synthesis in response to injury, it is not until the subchondral bone plate is penetrated that the usual vascular inflammatory response occurs, with cell recruitment from marrow elements. This transient response is usually incomplete and the final matrix is composed of largely Type I collagen with associated hyaluronan rather than Type II collagen and large MW aggrecan molecules found in normal AC (Figs. 13.2 and 13.3). In the long term, this neo-fibrocartilage repair fails to restore normal hyaline cartilage biomechanical compliance in relation to the subchondral bone plate. A major reason for the failure of AC to repair by

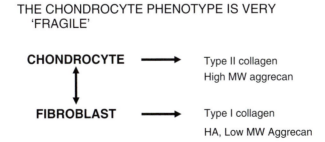

THE CHONDROCYTE PHENOTYPE

THE CHONDROCYTE PHENOTYPE IS VERY
'FRAGILE'

CHONDROCYTE ⟶ Type II collagen
High MW aggrecan

FIBROBLAST ⟶ Type I collagen
HA, Low MW Aggrecan

Fig. 13.3 The chondrocyte phenotype and its sensitivity to matrix removal and its reversible predisposition to de-differentiate and change its synthetic profile to a fibroblast profile

adjacent chondrocytes is probably the loss of growth factor stimulation that occurs with age. There is a decreased synthesis of growth factors such as GH and IGF-I (Corpus et al. 1993) and also TGF-β which are vital for stimulating mitogenesis and which drive cells to differentiate and to maintain their differentiation along the mature chondrocyte pathway and ECM synthesis.

Shapiro et al. (1993) studied the repair of 3 mm cylindrical full thickness defects over 48 weeks by a large detailed study using 122 rabbits. In the first few days, fibrinous arcades were established across the defects from surface edge to surface edge and served as a matrix to orientate bone derived mesenchymal cell ingrowth along the long axes. Synthesis of cartilage ECM was first evident at 10 days and at two weeks, cartilage was present beneath a surface of collagenous tissue which was rich in flattened fibrocartilaginous cells. At three weeks, in almost all the defects a well demarcated layer of cartilage containing chondrocytes was present and the defects were completely filled with a repopulation of cells at 6–12 weeks post-injury. A progressive differentiation of cells to chondroblasts, chondrocytes, and osteoblasts in their appropriate locations was observed. At 24 weeks, both the tidemark and the compact bone lamellar subchondral bone plate had been re-established. The cancellous woven bone in the depths of the defect was replaced by lamellar, course cancellous bone.

There was repair demonstrated from chondrocytes from the adjacent articular cartilage. The repair was mediated entirely by proliferation and differentiation of mesenchymal cells of the bone marrow.

However, traces of degeneration of the cartilage matrix was seen in many defects at 12–20 weeks, with prevalence of this degeneration increasing from 24–48 weeks. Polarized light microscopy demonstrated failure of the newly synthesized matrix to adhere and integrate with the host cartilage adjacent to the drill hole even when light microscopy had shown apparent continuity of the tissue. In many instances, a clear gap was seen between the repair and host cartilage. The authors attributed this lack of physical and chemical bonding of macromolecular components of the repair cartilage to the host "normal" cartilage which may allow vertical micromotion and macromotion under load maybe inducing further cartilage degeneration. They suggested this may be due to a circumferential cell death zone surrounding the drill hole (~100 μm) which prevented removal of the ECM or integration with the new repair ECM.

Buckwalter (1999) has documented the types of chondral defects taking place with age with osteochondral injuries occurring in the young where chondral injuries peaked at age 20–29 years and degenerative chondropenia occurred after age 40.

Mankin (1982) described the repair attempts of injured articular cartilage and classified such injuries into superficial endochondral or deep injuries penetrating the subchondral bone plate. With superficial injury no actual repair took place, where as in deep injuries the usual vascular repair mechanisms occurred with initial clot formation and then migration of bone marrow stem cells into the defect and described in more detail above by Shapiro's work.

The concept of local "bone-bruising" associated with local severe blunt trauma is probably associated with overlying chondrocyte damage and apoptosis; this process has not been studied very well in the human largely because of the inability to obtain access to such patients and also because of the reluctance of surgeons to biopsy such regions unless there is obvious AC damage such as cartilage flap formation due to shearing forces. Such a study could be done with careful planning with patient and Institutional ethics approval.

13.4 Treatment Options for Cartilage Injuries

Damage to articular cartilage is common (Curl et al. 1997) including the less common large area of loss of both subchondral bone and its overlying articular cartilage caused by osteochondritis dissecans (Fig. 13.4).

Osteochondritis dissecans is thought to be due to local avascular necrosis (AVN) causing the osteochondral defect which can lead to premature arthritis. While symptomatic elderly patients with low demand and large areas of grade IV AC loss can be successfully treated symptomatically and functionally by joint replacement, there are many symptomatic patients in the young and middle age groups for whom there is usually no acceptable or reliable treatment for areas of full thickness articular cartilage loss greater than 1 cm^2 (usually due to direct trauma). These defects are treated by the usual orthopedic debridement and lavage which is an attempt to smooth damaged AC by removing loose flaps of AC and smoothing remaining AC. The comparison of

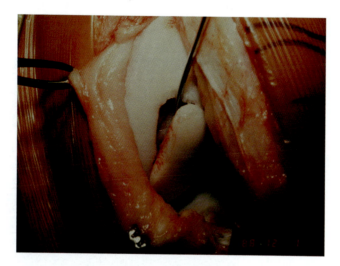

Fig. 13.4 A typical clinical view of osteochondritis dissecans

arthroscopic debridement and lavage was completed in 2005 and favors ACI long term when compared to simple destructive debridement and lavage (Fu et al. 2005).

Pharmacologic agents for intra-articular injection such as corticosteroids, hyaluronic acid and cross-linked hyaluronic acid (Synvisc ®) have been studied. However, none of these treatments have shown a positive effect beyond short term, symptomatic improvement equivalent to the use of oral anti-inflammatory medication (Adams et al. 1995, Wen 2000).

Correction of angular deformity and local treatment of articular cartilage disruption to minimise the progression of degeneration is recommended (Jones and Peterson 2006). Techniques to enhance the intrinsic but limited capacity of articular cartilage and subchondral bone to heal have been developed recently.

Pluripotential bone marrow stem cell recruitment by microfracture (Fig. 13.5) appears to be effective in terms of pain relief, but produces fibro-cartilage of limited durability (Steadman et al. 2002, Mithoefer et al. 2005c). This technique is called microfracture and although claims are made for its durability there has been little biopsy work to demonstrate that tissue other than fibrocartilage is produced following microfracture. Recent work by Mithoefer et al. (2005) clearly indicates that the fibrocartilage neo-repair of microfracture only lasts about two years and is directly related to body mass.

Localized articular cartilage replacement using osteochondral allografts and autografts have been proven to be surgically challenging, with incomplete incorporation and subsequent replacement of hyaline by fibrocartilage of limited efficacy (Hangody and Fules 2003, Bugbee 2004, 2005). The implantation of osteochondral autografts is defined as mosaicplasty (http://www.nice.org.uk/guidance/index.jsp?action = byID&r = true&o = 11211). The technique is based on osteochondral autograft harvesting and transplantation. Many small cylindrical osteochondral plugs are harvested from the less weight-bearing periphery of the patellofemoral area and are inserted into drill holes in the defective section of cartilage. It is believed that the limited repair mechanisms

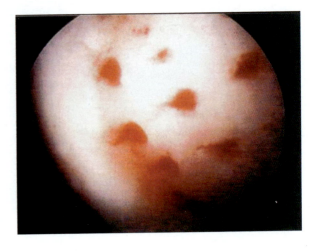

Fig. 13.5 A typical arthroscopic image of damaged cartilage after the microfracture procedure

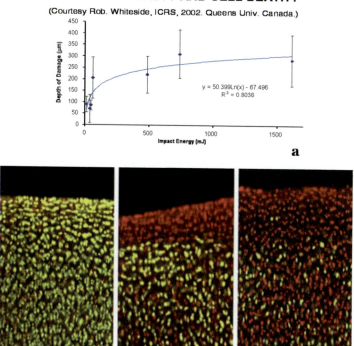

Fig. 13.6 (a) Cell death profiles following mosaicplasty-associated mechanical impact traumas. (a) Depth of impact cell damage as a function of impact energy. (b) Vital cell labelling indicating both cell survival and cell death with different types of commercial instrumentation for impacting the plug grafts. Note depth of cell death can be up to 250 μm (Courtesy R. Whiteside ICRs, Toronto, 2002)

produced by this technique may be due to the occurrence of cell necrosis in the autograft. Indeed, a relationship between the degree of tapping of the plug during implantation and the presence of relatively large zone of cell death in the superficial layers of the plug has been observed (Fig. 13.6a). This is one of the disadvantages associated with mosaicplasty that also suffers from limited tissue availability (Fig. 13.6b).

Mosaicplasty has advantages as it uses autologous tissue, it leads to good integration with the subchondral bone and it can be performed by single arthroscopy. For these reasons, this technique is usually recommended for symptomatic subjects younger than 40 years old and with AC lesions of femoral condylar cartilage.

The current preferred process is attempting articular cartilage repair using isolated autologous chondrocytes or whole tissue with chondrogenic potential such as perichondrium or periosteum (Brittberg et al. 1997). However, no long-term studies of successful periosteal grafting have thus far been published.

13.5 Autologous Chondrocyte Implantation

In 1994, Brittberg et al from Gotenberg used ACI for the first time to attempt of repair of full thickness symptomatic hyaline articular cartilage defects in the human knee (Brittberg et al. 1994).

The principle of using ACI technology is to synthesize and deposit a repair tissue that closely resembles normal AC and, hence, would have long-term durability and restor, near-to-physiological joint functionality. Since 1994, there has been a great proliferation of papers both at clinical and basic science level either verifying Brittberg et al's original and unique observations or attempting to improve on the technology by both surgical and biological modifications. It is not the role of this chapter to cover all aspects of articular cartilage repair but rather to present an experienced matrix biologist's and clinician's view about the salient and important papers which have clarified important unknown issues regarding the introduction of ACI technology. An excellent review on ACI has been published by Alford and Cole (2005) and a systematic review by Ruana-Ravina and Diaz (2006). This chapter will also review the extensive work in Melbourne at Mercy Tissue Engineering which has been recognized at an international level and it has answered many clinical and basic science questions. Indeed, clinical and basic science answers must underpin a diligent application of the current and well verified technology before any surgeon embarks on any further modification of the protocol established by the Gotenberg group (Lindahl et al. 2003).

It will no doubt surprise the non-orthopedic reader to know that after 12 years of ACI surgery, there has been no prospective, randomized controlled study performed for this technology. The author of this chapter believes that this type of study should be performed to clarify the value and efficacy of this technology. The reason for the overt failure of implementation of such an apparent obvious clinical trial is the unwillingness of surgeons around the globe to subject patients to a trial which they would feel may compromise their clinical outcome, at least in the short term, of those patients included in the control non-operated group. However, some important work has recently been carried out which will clarify this question of the lack of a control non-ACI group and some important clinical answers have recently been obtained.

Indeed, initial clinical and basic science studies performed in Sweden indicated for the first time in human clinical experience that it is possible to repair symptomatic damaged hyaline articular cartilage with a repair tissue which was very similar to normal hyaline articular cartilage. As this repair "neo-cartilage" had a similar morphological appearance and also similar stiffness parameters, this study indicated that such repair tissue may indeed be durable and suggested a possible bio-replacement for damaged grade IV articular cartilage loss in the knee joint and possibly other joints such as the ankle joint.

13.5.1 ACI Surgical Procedure and Technique

ACI surgical procedure has been well described in the literature (Brittberg et al. 1994, Alford and Cole 2005).

Briefly, an arthroscopy is performed on symptomatic patients with a known full-thickness cartilage defect(s) (Fig. 13.7a). The assessment of the lesion site, the general assessment of the status of the knee joint AC and especially the status and stiffness of the remaining AC within the joint are thus performed. If ACI is feasible a small cartilage biopsy is obtained usually from the intercondylar notch such that the wet weight of the AC obtained is ~300 mg.

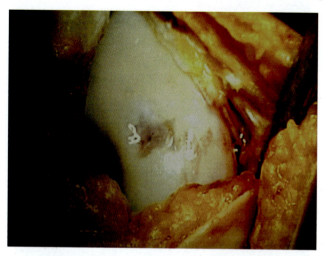

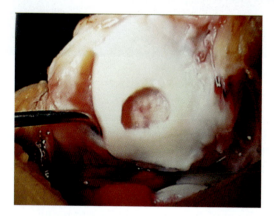

Fig. 13.7 Phases of autologous chondrocyte implantation procedure (**a**) full thickness symptomatic lesion. (**b**) Debridgement of damaged articular cartilage to the level of the calcified zone of the subchondral bone plate, (**c**) broaching the subchondral bone plate which allows bone marrow 'rogue' stemcells to enter the chondral compartment, (**d**) defect with periosteal patch sewn insitu and sealed with fibrin glue

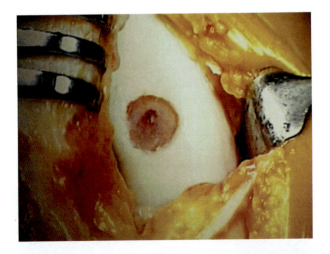

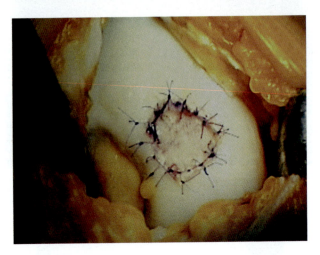

Fig. 13.7 (continued)

This cartilage is then sent to the laboratory for chondrocyte culture and expansion which usually takes about 3–4 weeks. The expanded cells are then implanted into the selected AC deficient site at an open arthrotomy. The side of the arthrotomy is determined by the side of the lesion. The defect is carefully debrided such that all loose and damaged cartilage is removed and a firm normal wall of AC is left. It is usual surgical practice to debride all the AC down to the subchondral bone plate and to remove any protruberant bone emerging from the subchondral bone plate or so called internal osteophyte or bone boss (Henderson and La Valette 2005) (Fig. 13.7b–c). This may cause bleeding and is treated by the use of adrenaline and sealed with fibrin glue to attempt to prevent bone marrow stem cells gaining access to the chondral compartment (Brittberg et al. 1997). After this basal debridement, a periosteal patch is obtained either from the medial of lateral femoral condyles or from the proximal tibia via a separate incision after careful synovial membrane removal. The size and shape of the AC defect is reproduced by a template and the periosteal patch is cut around this template to be a little larger than the peripheral margins of the defect. This is to take into account the marked shrinkage of the periosteal patch after its careful removal from the underlying bone because of its unusual high content of elastic fibers. These elastic fibers can be used as "markers" for the original periosteum in subsequent full thickness biopsy assessment of the ACI procedure. The periosteum is then carefully sewn in situ to the wall of the AC defect by use of ~1 mm spaced 9/0 PDS sutures with the suture knot lying on the apposed edge of the periosteum and of the AC defect. This interface is then sealed with fibrin glue except at the highest region of the defect where the cultured cells are introduced (Fig. 13.7d). If the area is larger than ~20 mm diameter, a central stay suture can be used with a small artery forcep to slightly "tent" the periosteum so that the volume of the AC defect can be measured. The edge of the AC defect is then carefully dried with swabs to allow for attachment of the fibrin glue. A 1 mm syringe is loaded with saline and a 19 guage plastic canula is then carefully introduced via the non-sealed region of the periosteum and some saline introduced into the cavity of the lesion. The surgeon looks for leakage of the sealed periosteal flap and further seals any leakage region with fibrin glue. The saline is then retrieved and the volume of the defect noted. The cultured and expanded cells from the laboratory are then placed in this volume of media and stirred within the microfuge tube to disperse the cell pellet which is then introduced into the defect through the same entrance site as for the previous saline. Once the lesion is "full", the final periosteal suture is tied and this last suture then finally sealed together with the small hole left after removal of the central tension suture. *At MTE, each lesion is customized, and the number of cells introduced is known with 10% more cells to account for any leakage, cell damage during syringe uptake and during dispersing in the microfuge tube (See Table 13.1).* The periosteal flap and lesion site is finally rehydrated with saline and the arthrotomy closed with a drain placed in the opposite femoral gutter. During closure of the arthrotomy it is important not to place pressure on the periosteal patch especially if the lesion is on the weight bearing area of the femoral condyles. In such a case, the knee is placed in slight flexion on the operating table. Similarly, when a full leg knee brace is applied after wound dressings are completed, it is important not to apply pressure to the repaired region to cause inadvertent periosteal leakage and hence loss of the introduced cultured cells.

The post-operative programme is also very important. Slow Continuous passive motion (CPM) is not permitted in the first 24 h post-operation for fear of damage to the periosteal patch and hence cell leakage could occur. Cells usually attach to their substrate within at least 12 h prior to proliferation and ECM synthesis. The knee brace is maintained for ~6 weeks and especially if the lesion is on the weight-bearing area of the femoral condyles non-weight bearing for the first six weeks is critical to allow the introduced cells to attach

AUTOLOGOUS CHONDROCYTE IMPLANTATION (ACI) :

(L.Peterson AAOS,1998)

– **STRENGTHS**
- First real application of tissue engineering
- Transplanted cells appear to grow an hyaline-Like AC
- Chondrocyte progenitor cells used
- Periosteum may add Growth Factors
- Integration of new & old cartilage is seamless
- May be durable
- Excellent results : **85%** patient improvement over **4 years F/Up**

Fig. 13.8 ACT strengths

AUTOLOGOUS CHONDROCYTE IMPLANTATION (ACI) :

(L.Peterson AAOS,1998)

– **WEAKNESSES**
- 2 Surgical Procedures: harvest and implantation
- Second procedure is an open arthrotomy
- Probably age dependent
- Demanding technique, not readily duplicated
- Very expensive, experts and tissue culture laboratory
- Not effective for Osteoarthritis
- If periosteal flap not sealed, cells will escape
- Cells used are de-differentiated chondrocytes = fibroblasts !
- Dog model failed with ACT, (Spector et al., JBJS,97).

Fig. 13.9 ACT weakness

and proliferate and synthesize their own ECM to fill the defect cavity with neo-cartilage. Then partial weight bearing in the 6–12 week phase is then gradually introduced.

See Figs. 13.8 and 13.9 for summary of strengths and weaknesses of ACI (ACT) surgical technology.

13.6 Key Questions About ACI

1. What is the quality of the repair neo-cartilage in term of its morphology, cell density and ECM composition and is this related to time post-ACI and to both short and long term clinical outcomes?

Although it is generally accepted that the "gold-standard" for AC neo-repair should be hyaline cartilage there are still very few studies with an adequate number of biopsies reported to make meaningful conclusions as to the quality of the "neo-repair" cartilage produced with ACI. Roberts et al. (2003a) reported on post-ACI biopsies and studied 23 full-depth cores of cartilage and subchondral bone obtained from 20 patients (mean age = 34.9 ± 9.2 yrs) who had undergone ACI between 9 and 34 months previously (mean = 14.8 ± 6.9 months). Six of these patients had been treated with ACI and mosaicplasty as a combined procedure, while the rest had ACI alone (17 patients). In the majority of patients, the femoral condyle was treated (11 medial and 6 lateral, 2 patella, and one talus). Frozen 7 μm-thick sections were used for histological analyses. Results indicated ∼2.5 mm-thick repair tissue with varying morphology ranging from predominantly hyaline in 22%, mixed in 48% through to predominantly fibrocartilage in 30%. Repair tissue was well integrated with the host tissue in all aspects viewed.

Oakes et al. (2004) reported on 58 post-ACI biopsies using plastic embedded 2–5 μm thick full thickness core biopsies.

58 patients with a mean age of 38 years at ACI implantation (range 14–62 years) had full thickness (including subchondral bone) central (C) and/or marginal (M) core biopsies using a Giebel needle of their repairing lesions at a mean of 12.2 months post-implantation (range 2.6–53.1 months) due to clinical symptoms requiring a "second–look" arthroscopy (Henderson et al. 2003). All patients had the classic Brittberg and Peterson ACI surgery using a sutured and fibrin-sealed periosteal patch. The biopsies were fixed without decalcification and embedded in Epon-Araldite and sectioned at 2–5 μm and stained with 1% Toluidine Blue and Azure A-Methylene Blue for histological classification. The biopsy tissues were classified as hyaline articular cartilage (HAC), "hyaline-like" articular cartilage (HLC), fibrocartilage (FC), periosteum and mixtures of the above, for example, HLC/FC.

Table 13.3 summarizes the results for detailed global knee joint and talar histological analyses.

Overall in the knee, 64% were either HAC or HLC. See Fig. 13.10 a–d

The above results are still the largest collection of full thickness core biopsies to be reported using the classic ACI technology with living sutured and fibrin-sealed autologous periosteum as the articular defect chondrocyte retainer and

Table 13.3 Histological classification from 58 ACI biopsies according to their anatomical retrieval site (modified from Oakes et al. 2004, ICRS symposium, Ghent)

Classification	Medial Femoral Condyle		Lateral Femoral Condyle		Trochlear		Patellar		Talus
Anatomical region	C	M	C	M	C	M	C	M	Dome
HAC	12	2	3		1	3	1		
HLC	4	1	1		2	1	2		3
FC	2	2			6				
HLC/ FC	5	1			1		1		1
HAC / HLC		1							
HAC / FC		1							
Periosteum									

Note: HAC = Hyaline Articular Cartilage, HLC = Hyaline Like Articular Cartilage, FC = Fibrocartilage

container at the initial ACI procedure. The 3 and 12 month clinical and MRI correlations have recently been reported (Henderson et al. 2003). In most of the biopsies examined at the 12 month interval post- ACI the periosteal layer was still intact and the gradation from periosteal fibrous tissue to HLC to HAC in the deep layer adjacent to the subchondral bone was observed. After 12 months, often the periosteal layer was absent and in four MFC patients periosteal remodeling was observed. Only in one patient was calcification observed. Of

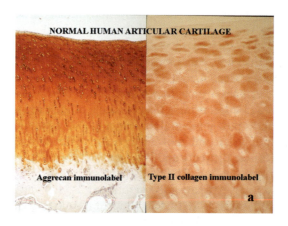

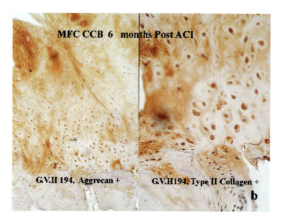

Fig. 13.10 Histological features of cartilage at different times after ACI using peroxidase immunolabelling with specific antibodies to both aggrecan and type II collagen with low power light microscopy, (**a**) physiological cartilage, note dense labelling of aggrecan and type II collagen throughout the section, (**b**) 6-months after ACI, Note, left: Medial femoral condyle, aggrecan labelling in the deep zone adjacent to the subchondral bone plate, right: Type II collagen labelling in the deep zone adjacent to the subchondral bone plate at bottom, (**c**) 12 months after ACI. Medial femoral condyle. Left: note even labelling in deep region. Right: note labelling in the deep zone, (**d**) 13 months after ACI, Medial femoral condyle: left, note even labelling in the superficial region, right, note labelling throughout the superficial zone. Left pictures = aggrecan staining, right pictures = collagen Type II staining

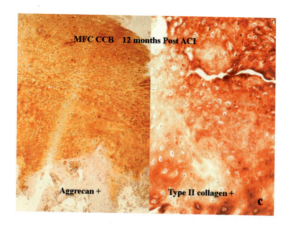

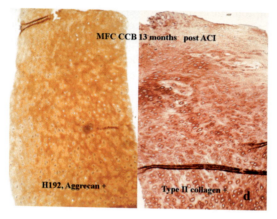

Fig. 13.10 (continued)

interest in this study was the large number of FC biopsies observed within the central trochlea patients (60%) compared to the central MFC biopsies (30%) suggesting perhaps an inadequate loading regimen during their rehabilitation program. Interface sections at the margins demonstrated "seamless" integration with normal HAC. These results were similar to those previously published by Knutsen et al. (2006) (n = 14) and more recently by Briggs et al. (2003) (n = 14) and Knutsen et al. (2004) (n = 40).

Pavesio et al. (2003) reported the use of hyaluronan-based scaffolds and the histology of 22 patients treated with this scaffold. It was observed that 14 of the 22 patients had HL cartilage whereas the rest had mixtures of HL/FC or FC. The mean follow-up time was 13.9 months for the second-look surgery-biopsy time. Remarkably 97% of patients reported a subjective improvement in knee function and symptoms. 86% of patients reported reduced pain and discomfort. There was no control group.

2. What is the stiffness of AC in the normal knee AC and also of the repair neo-cartilage when assessed by quantitative arthroscopic probing with a validated articular indentometer? Does the stiffness change with time post-ACI?

Lyyra et al. (1999) and Spring et al. (2001) reported the normal varying stiffness of the AC in the various anatomical regions of the normal knee joint in ~100 arthroscopies which were performed for reasons other than AC related symptoms.

This was important background information to be obtained before attempting any quantification of the stiffness of the neo-repairs following ACI. Table 13.4 shows the varying stiffness data of AC from the normal young adult human knee joint.

Both these studies used the Artscan 1000 which has been validated to measure AC stiffness and AC stiffness was demonstrated to have a linear relationship with aggrecan concentration within the AC (Lyyra 1997).

Vasara et al. (2005) reported the results from thirty patients with a mean age of 30 years with symptomatic chondral lesions in the knee treated with ACI (using a periosteal flap). Arthroscopy 1 year postoperatively was performed to evaluate tissue repair and stiffness using an indentation device (Artscan 1000). Clinical evaluations were also conducted using the Brittberg patients' evaluation scoring system, Lysholm functional score, Tegner activity score and the ICRS standard evaluation. Delayed Gadolinium-enhanced MRI was performed to estimate the relative proteoglycan concentration in cartilage in 4 patients.

Arthroscopic examination showed, in most cases, good filling of the lesion with repair tissue and integration with surrounding cartilage. The indentation stiffness of the repair tissue improved to 62% when compared to the adjacent normal cartilage. In 6 patients, the normalised stiffness was at least 80% of the adjacent cartilage, suggesting hyaline-like repair. 53% of patients graded their knee as excellent or good and 47% graded their knee as fair. Index values for gadolinium-enhanced MRI during follow-up of four patients was comparable to the values for the control tissue, indicating new proteoglycan deposition. The MRI findings correlated with the macroscopic findings; however, the same grafts

Table 13.4 Biomechanical properties of human articular cartilage obtained by use of Artscan 1000 (modified from Spring et al. 2001)

Standard sites	Stiffness (N)	Order of stiffness
Medial Femoral Condyle	3.71	1.28
Lateral Femoral Condyle	3.24	1.28
Medial Trochlea	3.15	1.43
Lateral Trochlea	2.69	1.52
Distal Trochlea	2.44	1.02
Medial Tibial Plateau	2.33	1.26
Lateral Tibial Plateau	2.27	1.19
Lateral Patella Facet	2.18	1.03
Proximal Trochlea	1.87	0.91
Medial Patella Facet	1.71	0.70

had very low stiffness values, indicating that the graft biomechanical properties do not necessarily reflect the proteoglycan concentration. In addition, a critical parameter such as the graft proteoglycan distribution was not considered.

This study can be compared with that of Henderson et al. (2006). In this study, the properties and qualities of autologous chondrocyte-driven repairs were evaluated in 66 autologous chondrocyte implantations in 57 patients. 55 out of 57 were assessed by histology, indentometry, and International Cartilage Repair Society (ICRS) repair scoring at reoperation for mechanical symptoms or pain. International Knee Documentation Committee (IKDC) scores were used to address the clinical outcome.

Maximum stiffness, normalized stiffness, and ICRS repair scoring were higher for hyaline articular cartilage repairs compared with fibrocartilage, with no difference in clinical outcome (Table 13.5).

Re-operations revealed 32 macroscopically abnormal repairs (Group B) and 23 knees with normal-looking repairs in which symptoms leading to arthroscopy were accounted for by other joint disorders (Group A). In Group A, 65% of repairs were either hyaline or hyaline-like cartilage compared with 28% in Group B. Autologous chondrocyte repairs composed of fibrocartilage showed more morphologic abnormalities and became symptomatic earlier than hyaline

Table 13.5 International cartilage repair society visual cartilage assessment scale

Scoring criteria	Repair points	Scores
Degree of defect repair	In level with surrounding cartilage	4
		3
	75% repair of defect depth	2
	50% repair of defect depth	1
	25% repair of defect depth	0
	0% repair of defect depth	
Integration to border zone	Complete	4
	Demarcating border <1 mm	3
	³/₄ of graft integrated, ¼ with a notable border >1 mm width	2
	½ of graft integrated, ½ with a notable border >1 mm	1
	From no contact to 1/4 of graft integrated with surrounding cartilage	0
Macroscopic appearance	Intact smooth surface	4
	Fibrillated surface	3
	Small, scattered fissures or	2
	Cracks	1
	Several, small or few but large fissures	0
	Total degeneration of grafted area	
Overall repair assessment	Grade I = normal	12
	Grade II = nearly normal	11–8
	Grade III = abnormal	7–4
	Grade IV = severely abnormal	3–1

Table 13.6 Patient Demographics and Scores (modified from Henderson et al. 2006)

Demographics	Study population	Prospective cohort
Patients' number	57	234
Gender	23 females, 34 males	73 females, 161 males
Mean age (range)	40.8 years (17–65)	38.5 years (14–64)
Mean follow-up (range)	28.1 years (8–54)	28.8 years (3–61)
Lesion site and number (size cm^2)	43 (4.3)	137 (4.2)
Medial femoral condyle	9 (4.2)	75 (4.3)
Lateral femoral condyle	9 (5.1)	52 (4.8)
Trochlea	5 (4.4)	43 (4.5)
Patella		
Scores		
Maximum stiffness (N ± SD)	3 ± 15	NA
Normalized stiffness (± SD)	1 ± 0.5	NA
IKDC (Maximum = 100)	39 ± 12	41 ± 17
Preoperatively (±SD)	54 ± 16†	61 ± 21*†
Postoperatively (±SD)	10 ± 1	NA
ICRS (maximum = 12)		

IKDC = International Knee Documentation Committee, ICRS = International Cartilage Repair Society Repair Scoring, SD = standard deviation, NA = not assessed.
*statistically significant difference between the prospective cohort and the study population at p<0.01,
† statistically significant difference between pre- and post-operative scores at p<0.0001.

or hyaline-like cartilage repairs. The hyaline articular cartilage repairs had biomechanical properties comparable to surrounding cartilage and superior to those associated with fibrocartilage repairs (Tables 13.6 and 13.7).

These two studies indicate favorable clinical outcomes and show that, following ACI, AC stiffness approaches normal cartilage and may even be comparable to normal cartilage. In unpublished immuno-labeling studies using specific EM antibodies to the G1 domain of aggrecan (gift from J.Mort, via A.Prof. Chris Little) and Types I, II, and III collagen, aggrecan was found within the periosteal layer at 12 and 13 months post ACI which may account for the normal stiffness of the ACI neo-repair observed in the recent Henderson et al. 2006 study.

Similar observations were reported by Roberts et al. (2003) using monoclonal antibodies against chondroitin 4 sulphate and keratan sulphate.

3. Is specialized AC semi-quantitative MRI reliable for assessing ACI clinical outcomes?

An excellent review of MR imaging and ACI is by James et al. (2005), provides detailed information to answer this question. In addition, it has to be mentioned that Roberts et al. (2003) were able to correlate MRI scores with histology scores. The MRI score was based on surface integrity, cartilage signal in the graft region, repair cartilage thickness, and changes in the underlying bone.

Table 13.7 Arthroscopy and histology results after second-look evaluation (modified from Henderson et al. 2006)

Evaluation criteria	Number of patients in ACI unrelated group (A)	Number of patients in ACI related group (B)
Histology (%)		
Hyaline cartilage	9 (39)	4 (13)
Hyaline-like cartilage	6 (26)	5 (16)
Hyaline-like cartilage/ fibrocartilage	4 (17)	8 (25)
Fibrocartilage	4 (17)	15 (47)
Number of lesion/site (size, cm^2)		
Medial femoral condyle	17 (4)	22 (4.1)
Lateral femoral condyle	3 (4.2)	4 (3.9)
Trochlea	3 (4.7)	6 (4.5)
Patella	0	0
Maximum stiffness (N ± SD)	3.4 ± 1.2	2.8 ± 1.2*
Normalized stiffness (N ± SD)	1.3 ± 0.5	1.0 ± 0.4†
IKDC score (maximum = 100 ± SD)		
Pre-operatively	39 ± 11	39 ± 12
Post-operatively	59 ± 17§	53 ± 16§
ICRS score (maximum = 12)	10.4 ± 0.59	9.7 ± 1.3‡
Previous surgeries		
Debridment	23	17
Microfracture	3	1
Subchondral drilling	1	0
Osteochondral grafting	0	1
ACI	0	2

IKDC = International Knee Documentation Committee, ICRS = International Cartilage Repair Society Repair Scoring, SD = standard deviation.
* Significant difference between group A and group B at $p<0.04$
† Significant difference between group A and group B at $p<0.01$
‡ Significant difference between group A and group B at $p<0.07$
§ Significant difference pre- and post-operatively in the same group at $p<0.0001$.

Henderson et al. (2003) were also able to demonstrate improvement in clinical status as measured by IKDC scores at 12 months using the MRI parameters of percentage fill, signal density, presence of a joint effusion and presence of subchondral bone oedema, in 78 lesions. However, no correlation was found between each individual MR criterion or the overall MR score and the IKDC clinical score or the subjective score.

Similar results have been reported by Takahashi et al. (2006) with 41 grafts available for MRI exam and comparison with clinical scores at 1 year.

O'Byrne et al. (2003) were also able to demonstrate changes in MR signal using delayed Gadolinium-enhanced T1 MRI of cartilage (dGEMRIC) when papain was injected into goat knee joints demonstrating the possibility of measuring proteoglycan loss by using quantitative MRI examination. Similarly, Samosky et al. (2005) have also demonstrated using dGEMRIC a correlation between GAG content and indentation of human tibial articular cartilage indicating quantitative specialized MRI can be used to measure mechanical parameters indirectly.

Burstein and Gray (2003) have reviewed recent developments in MRI technology to image cartilage ECM which currently are used only in MRI research labs and hopefully may become clinically available in the near future. The ability to obtain maps of the biochemical composition of cartilage may also provide a non-invasive means of determining its mechanical properties so vital for long-term cartilage function and durability. Work has already begun to examine the relationships between local mechanical properties and MRI parameters, with good correlations noted between dGEMRIC and site-matched indentation stiffness measurements (Nieminen et al. 2001).

4. Is ACI cost-effective?

Derrett et al. (2005) have assessed the costs and health status outcomes following ACI and mosaicplasty in a cross-sectional study that was conducted at the Royal National Orthopaedic Hospital between 1997 and 2001. 53 ACI, 20 mosaicplasty, and 22 patients waiting for ACI were taken into account. Secondary-care resource use was collected to two years postoperatively using a resource collection proforma. Participants responded to postal questions about socio-demographic characteristics and knee-related (Modified Cincinnati Knee Rating System) and general health status (EQ-5D). The average cost per patient was higher for ACI (£10,600: 95% confidence interval [CI], £10,036–£11,214) than mosaicplasty (£7,948: 95% CI, £6,957–£9,243). Postoperatively, ACI and mosaicplasty patients (combined) experienced better health status than those waiting for ACI. ACI patients tended to have better health status outcomes than mosaicplasty patients, although this was not shown to be statistically significant. Estimated average EQ-5D social tariff improvements for quality-adjusted life year (QALY) calculations were 0.23 (ACI) and 0.06 (mosaicplasty). Average costs per QALY were £23,043 (ACI) and £66,233 (mosaicplasty). The incremental cost effectiveness ratio (ICER) for providing ACI over mosaicplasty was £16,349. The average costs were higher for ACI than mosaicplasty up to two years following surgery, however, both the estimated cost per QALY and ICER for providing ACI over mosaicplasty fell beneath an implicit English funding threshold of £30,000 per QALY.

5. How does ACI compare with other purported cartilage repair procedures?

In alternative cartilage repair treatments using microfracturing or drilling the clinical outcome is usually satisfactory for the first two years but then gradually the clinical results declines and the long-term results are poor. (ICRS newsletter 1998, Lindahl et al. 2003). Section 13.7 offers a more detailed discussion of this aspect of ACI.

6. Are mesenchymal stem cells (MSC) coupled with specified growth factors such as TGF-β and IGF-I a viable source of precursor chondrogenic cells?

Because of their multipotent properties MSC can be isolated by standard techniques, expanded through many generations whilst retaining their ability to differentiate when exposed to appropriate signals (Barry 2003). Conditions for the differentiation of these cells in vitro have been well described. MSC cultured into high- density three-dimensional cultures (e.g., alginate beads) in the presence of TGF-β rapidly progress from undifferentiated cells with a fibroblastic morphology to a mature chondrocyte phenotype with abundant ECM production. It is important to notice that MSC's derived from patients with OA showed reduced proliferation in vitro and lower rates of differentiation along the chondrogenic pathway (Martin and Buckwalter 2003). Barry injected MSC's transduced with green fluorescent protein and delivered the cells as a suspension into previously medial-menisctomized and ACL-transected goat joints. He was able to demonstrate the formation of a neo-medial meniscus stimulated by these stem cells in the posterior compartment of these goat knee joints. The neo-meniscus tissue had a hyaline-like appearance with focal areas of type II collagen. These observations suggested a possible role for injected MSC's in the treatment of osteoarthritic joints.

7. Are animal models of cartilage repair useful for predicting ACI performances in clinics?

The very thin AC of small-size animal models as well as of goats and sheep are relatively much thinner when compared to human tissue. Hence, the mimicking of human AC has been attempted by equine models which are, however, limited because of their relatively high costs. Hunziker (2002) shows that animal models are problematic as the relatively thin cartilage makes suturing and long-term adherence of the periosteal flap doubtful. This may also explain the failure of dog models (Breinan et al. 1997).

8. Does in vitro chondrocyte de-differentiation limit ACI?

There is no doubt that chondrocytes de-differentiate once cultured in vitro at low density in monolayer cultures as applied in currently commercially available technology (Benya et al. 1977). The excellent study by Binette et al. (1998) demonstrated clearly that human articular chondrocytes change phenotype in monolayer culture and have a fibroblast morphology with concomitant cell proliferation. Increased proliferation was associated with downregulation of the synthesis of cartilage-specific aggrecan, inhibition of type-II collagen expression, and upregulation of type-I collagen and versican. Chondrocyte de-differentiation was shown to be reversible in about 4–6 weeks after cell transfer into either suspension or alginate bead 3D culture. It was also demonstrated that type X collagen and osteopontin markers of chondrocyte hypertrophy were not expressed by adult articular chondrocytes even after five months indicating these chondrocytes are special and do not attempt to proceed down the endchondral ossification pathway. Similar studies have been reported by others using human adult cartilage cells which indicate that cells isolated from younger human donors have a greater chondrogenic potential than cells from older donors especially in monolayers culture systems (Huckle et al. 2003, Novartis Foundation).

Hardingham (2004) has recently emphasized the need for cell biology in tissue engineering and also has emphasized the importance of chondrocyte phenotype in delivering successful clinical outcomes for cartilage engineering. Hardingham et al. (2006) have investigated the epigenetic events leading to chondrocyte differentiation and de-differentiation and have used the transcription factor SOX9, growth factor stimulation and hypoxic conditions to significantly improve the response of chondrocytes in 3D culture systems. Human mesenchymal cells have also been differentiated into chondrocytes and molecular events leading to this phenotypic change studied.

9. Is implanted cell density important to the repair quality and clinical outcome in full thickness human defects?

In vitro and in vivo works from Chaipinyo et al. (2002, 2004) and from other authors (Tables 13.1–13.4) substantiate the hypothesis that implanted chondrocyte density plays a key role in cartilage repair and clinical outcome. Oakes et al. (2004), Roberts et al. (2003), and Knutsen et al. (2006) have been collecting substantial histologic data over a two year (up to five years) follow-up period, which are able to provide a full assessment of ACI technology on more than 80 patients. Recent work has also drawn excellent correlations between histological features and arthroscopic indentometry data (Henderson et al. 2006). Oakes et al have provided histological data showing that ACI free-cell technology is delivering approximately a 50% acceptable repair rate (Table 13.7). When compared with Vasara et al. (2005), the best indentation stiffness was 62% (Table 13.7, Henderson et al. 2006).

There are contrasting opinions among surgeons about the importance of cell number upon implantation. It is the personal evidence-based view of the author of this chapter that different cell numbers can yield different degrees of repair. This view seems to be supported by a recent review by Jones and Peterson (2006).

The author of this chapter has placed considerable importance and effort to customize the cell number to match that of the normal cell number per unit volume of the defect (+10% for cell losses during transfer, etc.) and to minimize chondrocyte de-differentiation in vivo. The results so far achieved are the best published to date in terms of histological features and indentometry thus confirming the importance of adequate chondrocyte cell density in ACI.

10. Can biological or artificial biomaterial scaffolds improve ACI?

An elegant work by Vunjak-Novakovic et al. (1999) have demonstrated the formation of thick (1–5 mm) cartilage constructs in vitro by chondrocytes grown on PGA scaffolds for six weeks in a continuously perfused and rotated Synthecon bioreactor (Houston, USA). In comparison, static culture yielded small and fragile constructs. Turbulent flow in mixed flasks generated constructs with fibrous outer capsules. Both these environments resulted in constructs with poor mechanical properties. Conversely, the constructs that were freely suspended in a dynamic laminar flow field in perfused rotating vessels formed the largest and continuous cartilage-like ECM; the highest fractions of glycosaminoglycans and collagen resulting in significantly improved mechanical properties.

Marlovits et al. (2003) have demonstrated that aged human articular cartilage can form tissues in a scaffold-free bioreactor. Articular chondrocytes isolated from 10 aged patients (median age, 84 years) were expanded in monolayer culture. A single-cell suspension from de-differentiated chondrocytes was inoculated in a rotating wall perfused vessel without the use of any scaffold or supporting gel material. After 90 days of perfusion cultivation, a three dimensional cartilage-like tissue was formed that was encapsulated in a fibrous tissue resembling a perichondrial membrane. Morphology revealed differentiated chondrocytes ordered in clusters within a continuous dense cartilaginous matrix demonstrating strong positive staining with monoclonal antibodies to Type II collagen and articular proteoglycans. This work also demonstrated that the absence of a biomaterial scaffold delays cartilage formation in vitro. Conversely, the use of scaffold reduces the culturing time, supports the chondrocytes close juxta-position at high density and retains ECM precursors.

Ochi et al. (2002) have reported the use of autologous chondrocytes cultured in an Atelocollagen gel (Type I collagen) which was implanted into 28 knees (26 patients) with the use of a periosteal flap and followed for a minimum period of 25 months. Arthroscopic assessment indicated 26 knees (93%) had good or excellent outcomes. Biomechanical testing revealed the transplants had a stiffness similar to the surrounding cartilage. However, only two patients were examined histologically.

11. Are current International Knee scoring systems suitable and adequate for measuring clinical outcomes in ACI and other cartilage repair procedures?

There is still no wide agreement about the best instrument to measure cartilage repair. This is essentially because the ICRS outcomes scoring system for chondral injuries has yet to be clinically validated. The Cincinnati Knee Rating System is favored in the USA whereas other countries favor other scoring systems such as the ICRS, Tegner, Lysholm, IKDC, KOOS.

12. Is knee joint debridement and lavage a useful procedure and is ACI superior to it in the long term clinical outcome?

Fu et al. (2005) have compared ACI with joint debridement and lavage. They compared 54 ACI patients with 42 debridement patients. Eighty one percent of the ACI patients and 60% of the debridement patients reported median improvements of five points and two points, respectively, in the overall condition score over the three years of the study. ACI patients also reported greater improvements in the median pain and swelling scores than did debridement patients. The treatment failure rate was the same for each group of patients. Eighteen ACI patients and one debridement patient had a subsequent operation. The strengths of this study was the inclusion and exclusion criteria, the use of the same outcome instruments at baseline and follow-up, the multi-center design of the study, and the sample size ensuring ample statistical power for comparing outcomes between the two study groups. There was a small potential bias in the response rate for the ACI group that was higher than the debridement group (72% vs. 93%). The study was also not randomized. Group analyses showed that debridement was minimally effective in patients with chronic lesions with respect to all outcome parameters measured, whereas ACI was effective in patients regardless of the type of lesion treated (chronic or acute). This initial three year data suggests ACI is more efficacious than debridement in carefully selected patients.

13. Is there an age limit to the surgical use of ACI?

It appears that there are age limits affecting chondrocyte viability and capacity to synthesize a cartilage-like ECM. Barbero et al. (ICRS, 2002) have noted age-related changes in human chondrocyte biology. Thirty five human AC biopsies were classified into age groups and cell yields were calculated per tissue wet weight.

Cells were expanded in either control medium with 10% FCS (CTR) or in medium supplemented with TGFβ1, FGF-2, and PDGF-BB (TFP). Cell doublings were calculated and cell redifferentiation determined histologically and by GAGs and Collagen II RT-PCR. Results demonstrated that cell yields were lower in cells isolated from subjects older than 40. Cell proliferation in CTR was decreased when cells were isolated from subjects younger than 20 and was significantly higher in TFP medium with no age-dependant decrease. TFP-cultured cells synthesized more GAG than CTR cells. TFP-cultured cells also showed significantly higher cell proliferation at any age and greater chondrogenic capacity when isolated from donors younger than 55.

Martin and Buckwalter (2003) have elegantly demonstrated that age induced changes in cell senescence markers (i.e., β-galactosidase expression, mitotic activity, telomere length, mitochondrial DNA, membrane potential, and numerical density) in human articular chondrocytes from 27 donors from 1 to 87 years. They found that β-galactosidase expression increased with age, while mitotic activity and telomere length decreased. Also as population doublings increased, mitochondrial DNA was degraded, membrane potential was lost and their number per cell also decreased.

The authors related these findings to the known age-related decreases in chondrocyte synthetic activity, mitotic activity, and decreased responsiveness to anabolic cytokines and mechanical stimuli. They also suggested that in vivo chondrocyte senescence is important in the age-related increase of osteoarthritis incidence and decrease in the efficacy of cartilage repair.

13.7 Conclusions: The Clinicians' View on Treatments for Articular Cartilage Repair

Verdonk et al. (2006) have recently used a web-based survey of 285 orthopedic surgeons from 46 countries world wide where 2/3 were experienced clinicians performing at least 10 cartilage repair interventions per year. The 19-question survey covered different aspects of cartilage repair spanning from patient selection to treatment follow-up, from physician-pursued treatment goals to evidence-based results, and regulatory agents impacting on future clinical practice.

Despite the availability of several treatment options, the survey clearly revealed a high unmet need in the field of knee cartilage injuries, especially when the long-term outcome goals were considered. Seventy percent of surgeons agreed on performing cartilage repair in symptomatic young adults with knee cartilage injuries who were at considerable risk from osteoarthritis later in life. According to the survey, these young adults should be targeted for primary prevention of premature OA as well as to improve short- and long-term pain and symptoms. The quality of the repair tissue was also regarded as very important and the rebuilding of hyaline cartilage instead of fibrous tissue was highly valued.

Monitoring of the success of the repair procedures was performed by the majority of the surgeons (87%) both through a range of clinical outcome scores and through objective MRI assessments in post-operative follow-up. In a recent study (Marlovits et al. 2006) showed good correlations between the clinical outcome scores and MRI variables.

Surgeons also gave a higher score to ACI technology when compared to microfracture; the former resulting in more satisfactory long-term results. Unfortunately, the number of head-to-head comparisons of the different cartilage repair surgical techniques remains limited in terms of both methodology and long-term results >2 years with conflicting results between well-performed case studies. For example, Knutsen et al. (2004) reported similar clinical outcome studies for ACI and microfracture in 80 patients after two years. Conversely, Brown et al. (2004) found that ACI was superior to microfracture in 112 patients with a mean follow-up of 13 and 15 months respectively. Mithoefer's et al. (2005) prospective study concluded that microfracture only provided good short term results with small defects and that clinical improvement waned after two years. This observation was very recently confirmed by Kreuz et al. (2006) who demonstrated that the deterioration of the results of microfracture commences about 18 months post-surgery, with young patients having the best prognosis. These results confirm that probably fibrocartilage is produced in microfracture with incomplete defect filling which deteriorates over time. Hence, it has been suggested that microfracture should be perhaps limited to small defects <2.5 cm^2. The Knutsen's et al. (2006) study did not demonstrate deterioration in clinical results between 2 and 5 years post-surgery.

Therefore, the authors of the survey recommend more and better designed studies to clarify the conflicting results originating from the various investigations. To favor the collection of more comparable data, they suggested six specific guidelines that included

1. Randomized controlled trial with clear endpoints, adequate randomization, and power analysis
2. Clearly defined inclusion and exclusion criteria

3. Validated outcome measures
4. Outcomes assessed by an independent investigator
5. Clearly stated time of the outcome assessment (ideally >2 years)

Acknowledgments The author would like to thank Ms Maria Kouvalaris from Mercy Tissue Engineering for her great help in the preparation of this manuscript. Also Mrs Sue Connell for preparation and staining of the histology sections. Also Associate Professor Chris Little for the immuno-labeling studies presented here on selected ACI biopsies.

References

Adam C, Eckstein F, Milz S, Putz R. (1998) The distribution of cartilage thickness within joints of the lower limb of elderly individuals. J Anat 193:203–214

Adams ME, Atkinson MH, Lussier AJ, Schulz JI, Siminovitch KA, Wade JP, et al. (1995) The role of viscosupplementation with hylan G-F 20 (Synvisc) in the treatment of osteoarthritis of the knee: a Canadian multicenter trial comparing hylan G-F 20 alone, hylan G-F 20 with non-steroidal anti-inflammatory drugs (NSAIDs) and NSAIDs alone. Osteoarthritis Cart 3:213–25

Alford JW, Cole BJ (2005) Cartilage restoration, Part 1 basic science, historical perpspective, patient evaluation, and treatment options. Am J Sports Med 33:295–306

Aroen A, Loken S, Heir S et al. (2004) Articular cartilage lesions in 993 consecutive knee arthroscopies. Am J Sports Med 32:211–215

Barbero A, Grogan SP, Ploegert S, et al. (2002) Age related changes in chondrocyte biology. In: Proceeding of Fourth International Cartilage Repair Society Symposium (ICRS), Toronto, Canada

Barry FP (2003) Mesenchymal stem cell therapy in joint disease. In: Tissue engineering of cartilage and bone. Novartis foundation symposium 249. J Wiley and Sons UK, pp 86–102

Benya PD, Shaffer JD (1982) Dedifferentiated chondrocytes re-express the differentiated phenotype when cultured in agarose gels. Cell 30:215–224

Benya PD, Padilla SR, Nimni ME (1977) The progeny of rabbit articular chondrocytes synthesize collagen types I and III and type I trimer, but not type II. Verifications by cyanogen bromide peptide analysis. Biochemistry 16:865–872

Binette F, McQuaid DP, Haudenschild DR et al. (1998) Expression of a stable articular phenotype without evidence of hypertrophy by adult human articular chondrocytes in vitro. J Orthop Res 16:207–216

Breinan H, Minas T, Hsu H, Nehrer S, Sledge C, Spector M (1997) Effect of cultured autologous chondrocytes on repair of chondral defects in a canine model. J Bone Joint Surg 79A:1439–1451

Briggs TWR, Mahroof S, David LA, Flannelly J, Pringle J, Bayliss M (2003) Histological evaluation of chondral defects after autologous chondrocyte implantation on the knee. J. Bone and Joint Surg 85-B:1077–83

Brittberg M, Lindhal A, Nilsson et al. (1994) Treatment of deep cartilage defects in the knee with autologous chondrocyte transplantation. NEJM 331:889–895

Brittberg M, Sjogren-Jansson E, Lindahl A, Peterson L (1997) Influence of fibrin sealant (Tisseel) on osteochondral defect repair in the rabbit knee. Biomaterials 18:235–242

Brown WE, Potter HG, Marx RG, Wickiewicz TL, Warren R (2004) Magnetic resonance imaging appearance of cartilage repair in the knee. Clin Orthop Relat Res, 422:214–223

Buckwalter JA (1999) Evaluating methods of restoring cartilage articular surfaces. Clin Orthop Rel Research 367S:S224–S238

Buckwalter JA, Mankin H (1997a) Articular cartilage. Part I: tissue design and chondrocyte-matrix interactions. Instructional course lectures, The American Academy of Orthopaedic Surgeons. J Bone Joint Surg Am 79:600–611

Buckwalter JA, Mankin H (1997b) Articular cartilage. Part II: degeneration and osteoarthrosis, repair, regeneration and transplantation. Instructional course lectures, The American Academy of Orthopaedic Surgeons. J Bone Joint Surg Am 79:612–632

Buckwalter JA, Hunziker E, Rosenberg L et al. (1998) Articular cartilage: composition and structure. In: Woo SLY, Buckwalter JA (eds) Injury and repair of the musculoskeletal tissues. American Academy of Orthopaedic Surgeons, Park Ridge, IL, pp 405–425

Buckwalter JA, Pita JC, Muller FJ, Nessler J (1994) Structural differences between two populations of articular cartilage proteoglycan aggregates. J Orthop Res 12:144–148

Bugbee W (2005) Fresh osteochondral allografts. Instructional lecture, American Academy Orthopaedic Surgeons, Washington, DC Feb 2005

Bugbee W (2004) Fresh Osteochondral Allografts for the Knee. Tech Knee Surg 3:1–9

Burstein D, Gray M (2003) New MRI techniques for imaging cartilage. J. Bone Joint Surg Am 85:70–77

Caterson B, Lowther D (1978) Changes in the metabolism of the proteoglycans from sheep articular cartilage in response to mechanical stress. Biochim Biophys Acta 540:412–422

Chaipinyo K, Oakes BW, van Damme MPI (2002). Effects of growth factors on cell proliferation and matrix synthesis of low-density, primary bovine chondrocytes cultured in collagen I gels. J. Orthopaedic Research 20: 1070–1078.

Chaipinyo K, Oakes BW, van Damme MPI (2004) The use of debrided human articular cartilage for autologous chondrocyte implantation: maintenance of chondrocyte differentiation and proliferation in type I collagen gels. J Orthop Res 22:446–455

Cicuttini FM, Forbes A, Yuanyuan W, Rush G, Stuckey SL (2002) Rate of knee cartilage loss after partial meniscectomy. J Rheumatol 29:1954–1956

Corpus E, Harman M, Blackman M (1993) Human growth hormone and ageing. Endocr Rev 14:20–39

Curl WW, Krome J, Gordon ES, et al. (1997) Cartilage injuries: a review of 31,516 knee arthroscopies. Arthroscopy 13:456–460

Derrett S, Stokes ES, James M, Bartlett W, Bentley G (2005) Cost and health analysis after autologous chondrocyte implantation and mosaicplasty: a retrospective comparison. Int J Technol Assess Health Care 21:358–367

de Witt MT, Handley CJ, Oakes BW, Lowther DA (1984) In vitro response of chondrocytes to mechanical loading. The effect of short term mechanical tension. Connect Tissue Res 12:97–109

Fu FH, Zurakowski D, Browne JE, Mandelbaum B, Erggelet C, Moseley JB Jr, Anderson AF, Micheli LJ (2005) Autologous chondrocyte implantation versus debridment for treatment of full thickness chondral defects of the knee. An observational cohort with 3-year follow-up. Am J Sports Med 33:1658–1666

Hangody L, Fules P (2003) Autologous osteochondral mosaicplasty for treatment of full-thickness defects of weight-bearing joints: ten years experimental and clinical experience. J Bone Joint Surg Am 15(Suppl 2):25–32

Hangody L (1998) Mosaicplasty in the treatment of the focal chondral and osteochondral Defects. ICRS Newsletter, Spring, 1998, p11

Handley CJ, Bateman JF, Oakes BW, Lowther DA (1975) Characterization of the collagen synthesizied by cultured cartilage cells. Biochem Biophys Acta 386:444

Hardingham TE, Oldershaw RA, Tew S (2006) Cartilage, SOX9 and Notch signals in chondrogenesis. J Anat 209:469–480

Hardingham T (2004) Tissue engineering – view from a small island. Tissue Eng 9:1063–1064

Haynesworth SE, Goldberg VM, Caplan AI (1994) Diminution of the number of mesenchyme stem cells as a cause for skeletal ageing. In: Buckwalter JA, Goldberg VM, Woo SL-Y (eds)

Musculoskeletal soft-tissue ageing: impact on mobility. American Academy of Orthopaedic Surgeons, Rosemont, IL, pp 79–87

Henderson IJP, Tuy B, Connell D, Oakes B, Hettwer WH (2003) Prospective clinical study of autologous chondrocyte implantation and correlation with MRI at three and 12 months. J Bone Joint Surg 85-B:1060–1066

Henderson IJP, La Valette DP (2005) Subchondral bone overgrowth in the presence of full-thickness cartilage defects in the knee. Knee 12:435–440

Henderson IJP, Lavigne P, Valenzuela H, Oakes B (2007) Autologous chondrocyte implantation: superior biologic properties of hyaline cartilage repairs. Clin Orthop Relat Res, 455:253–261

Hunziker EB (2002) Articular cartilage repair: basic science and clinical progress. A review of the current status and prospects. Osteoarthr Cartil Jun 10:432–63

Hunziker EB, Driesang MK, Morris EA (2001) Chondrogenesis in Cartilage repair is induced by members of the transforming growth factor-beta superfamily. Clin Orthop Relat Res 391S:S171–S181

Hunziker EB, Rosenberg LC (1996) Repair of partial-thickness defects in articular cartilage: cell recruitment from the synovial membrane. J Bone Joint Surg 78A:721–733

Hunziker EB (2001a) Growth-factor-induced healing of partial-thickness defects in adult articular cartilage. Osteoarthr Cartil 9:22–32

Hunziker EB (2001b) Growth-factor-induced healing of partial-thickness defects in adult articular cartilage. J Orthop Res 20:1070–1078

Huckle J, Dootson G, Medcalf N, Mc Taggart S, Wright E, Carter A, Schreiber R, Kirby B, Dunkelman N, Stevenson S, Riley S, Davisson T, Ratcliffe A (2003) Differentiated chondrocytes for cartilage tissue engineering. In: Tissue engineering of cartilage and bone. Novartis foundation symposium 249. J Wiley and Sons, UK, pp 103–117

James SLJ, Connell DA, Saifuddin A, Skinner JA, Briggs TWR (2005) MR imaging of autologous chondrocyte implantation of the knee. Eur Radiol 16:1022–1030

Jones DG, Peterson L (2006) Autologous chondrocyte implantation. Selected instructional course lectures. J Bone Joint Surg Am 88:2501–2501

Kaab M, Gwyn A, Notzli HP (1998) Collagen fibre arrangement in the tibial plateaux articular cartilage of man and other mammalian species. J Anat 193:23–34

Kiviranta I, Jurvelin J, Tammi M, Säämänen A-M, Helminen HJ (1987) Weight bearing controls glycosaminoglycan concentration and thickness of articular cartilage in the knee joints of young beagle dogs. Arthritis Rheum 30:801–808

Knutsen G, Engebretson L, Ludvigsen TC, Drogset JO, Grontvedt T, Solheim E, Strand T, Roberts S, Isaksen V, Johansen O (2004) Autologous chondrocyte implantation compared with microfracture in the knee: a randomised trial. J. Bone and Joint Surg 86-A:455–464

Knutsen G, Drogset JO, Engebretsen L, et al. (2006) Autologous chondrocyte implantation compared with microfracture in the knee. (Five year follow-up) Proceedings ICRS symposium San Diego (abstract)

Kreuz PC, Steinwachs MR, Ergggelet C et al. (2006) Results after microfracture of full-thickness chondral defects in different compartments of the knee. Osteoarthr Cartil 14:1119–1125

Leonard CM, Fuld HM, Frenz DA et al. (1991) Role of transforming growth factor – beta in chondrogenic pattern formation in the embryonic limb: stimulation of mesenchymal condensation and fibronectin gene expression by exogenous TGF-beta and evidence for endogenous TGF-beta like activity. Dev Biol 145:99–109

Lindahl A, Brittberg M, Peterson L (2003) Cartilage repair with chondrocytes : clinical and cellular aspects. In: Tissue engineering of cartilage and bone. Novartis foundation symposium 249. J. Wiley and Sons Ltd. UK, pp 175–189

Lyyra T (1997) Development, validation and clinical application of indentation technique for arthroscopic measurement of cartilage stiffness. University of Kupio, Kuopio, Finland: 1–99

Lyyra T et al. (1999) In vivo charactertiyation of indentation stiffness of articular cartilage in the normal human knee. J Biomed Mater Res. 48:482–487

Mankin HJ (1982) Current concepts review. The response of articular cartilage to mechanical injury. J Bone Joint Surg March 64A:460–466

Marlovits S, Tichy B, Truppe M, Gruber D, Vecsei V (2003) Chondrogenesis of aged human articular cartilage in a scaffold-free bioreactor. Tissue Eng 9:1215–1226

Marlovits S, Singer P, Zeller P, Mandl I, Haller J, Trattnig S (2006) Magnetic resonance observation of cartilage repair tissue (MOCART) for the evaluation of autologous chondrocyte transplantation: determination of interobserver variability and correlation to clinical outcome after 2 years. Eur J Radiol 57:16–23

Marlovits S, Striessnig G, Kutscha-Lissberg F, Resinger C, Aldrian SM, Vecsei V, Trattnig S (2005) Early post-operative adherence of matrix-induced autologous chondrocyte implantation for the treatment of full-thickness cartilage defects of the femoral condyle. Knee Surg Sports Traumatol Arthrosc 13:451–457

Martin JA, Buckwalter J (2003) The role of chondrocyte senescence in the pathogenesis of osteoarthritis and in limiting cartilage repair. JBJS 85-A:S106-S110

McQuillan DJ, Handley CJ, Campbell MA, Bolis S, Milway VE, Herington AC (1986) Stimulation of proteoglycan biosynthesis by serum and insulin-like growth factor-I in cultured bovine articular cartilage. Biochem J 240:423–430

Mithoefer K et al. (2005a) Articular cartilage repair in soccer players with autologous chondrocyte transplantation. Am J Sports Med 33:1639–1646

Mithoefer K et al. (2005b) Functional outcome of knee articular cartilage repair in adolescent athletes. Am Orthop Soc Sports Med 33:1147–1153

Mithoefer K, Williams RJ III, Warren RF, Potter H, Spock CR, Jones EC, Wickiewics TL, Marx RG (2005c) The microfracture technique for the treatment of articular cartilage lesions of the knee. A prospective cohort study. J Bone Joint Surg Am 87:1911–1920

Nieminen MT, Toyras J, Laasanan MS, Rieppo J, Silvennoinen J, Helminen HJ, Jurvelin JS (2001) MRI quantitation of proteoglycans predicts cartilage stiffness in bovine humeral head. Proceedings American Orthopaedic Society Feb 25–28, San Francisco, California, USA

Oakes BW, Connell S, Henderson I (2004) Histological classification of 58 patients with plastic – embedded full thickness core biopsies 2.6–53 months following Autolgous Chondrocyte Implantation (ACI) . Proceedings ICRS meeting Ghent, 2004. (Abstract).

Oakes BW (2003) Tissue healing and repair: tendons and ligaments. In: Frontera WR (ed) Rehabilitation of sports injuries. Scientific basis. The Encylopaedia of Sports Medicine. An IOC Medical Commission Publication. Blackwell Publishing, UK, Australia & Germany. pp 56–98

Oakes BW, Handley CJ, Lisner F, Lowther DA (1977) An ultrastructural and biochemical study of the high density primary culture of embryonic chick chondrocytes. J Embryol Exp Morph 38:239–263

O'Byrne E, Pellas T, Laurent D (2003) Qualitative and quantitative in vivo assessment of articular cartilage using magnetic resonance imaging. In: Tissue engineering of cartilage and bone. Novartis foundation symposium 249. J Wiley and Sons Ltd. UK pp 190–202

Ochi M, Uchio Y, Kawasaki K, Wakitani S, Iwasa J (2002) Transplantation of cartilage-like tissue made by tissue engineering in the treatment of cartilage defects of the knee. J Bone Joint Surg 84B:571–578

Pavesio A, Abatangelo G, Borrione A, et al. (2003) Hyaluronan-based scaffoids (Hyalogoraft) in the treatment of knee cartilage defects: preliminary clinical findings. In: Tissue Engineering of Cartilage and Bone. Novartis Foundation Symposium 249. J Wiley and Sons Ltd. pp 203–217.

Peterson L, Minas T, BrittbergM, Nilsson A, Sjogren-Jansson E, Lindahl A (2000) Two- to – 9 year outcome after autologous chondrocyte transplantation of the knee. Clin Orthop Relat Res 374:212–234

Poole AR (2003) What type of cartilage repair are we attempting to attain. JBJS 85A: 40–44

Poole CA, Ayad S, Gilbert RT (1992) Chondrons from articular cartilage: V. Immunohisto-chemicql evaluation of type VI collagen in isolated chondrons by light, confocal and electron microscopy. J Cell Sci 103:1101–1110

Ratcliffe A, Mow VC (1996) Articular cartilage. In: Wayne D Comper (ed) Extracellular matrix, volume 1, tissue function, pp 234–302. Harwood Academic Publishers, The Netherlands.

Ratcliffe A, Tyler JA, Hardingham TE (1986) Articuolar cartilage cultured with interleukin-1: increased release of link protein, hyaluronate-binding region, and other proteoglycan fragments. Biochemi J 238:571–580

Roberts S, McCall IW, Darby AJ, Menage J, Evans H, Harrison PE, Richardson JB (2003a) Autologous chondrocyte implantation for cartilage repair: monitoring its success by magnetic resonance imaging and histology. Arthritis Res Ther 5:R60-R73

Roberts S, Hollander AP, Caterson B, Menage J, Richardson JB (2003b) Matrix turnover in human cartilage repair tissue in autologous chondrocyte implantation. Arthritis and Rheum 44:2586–2598

Ruana-Ravina A, Diaz MJ (2006) Autologous chondrocyte implantation: a systematic review. Osteoarthr Cartil 14:47–51

Samosky JT, Burstein D, Grimson WE, Howe R, Martin S, Gray M (2005) Spatially-loclized correlation of dGEMRIC-measured GAG distribution and mechanical stiffness in the human tibial plateau. J Orthop Res 23:93–101

Seedhom BB, Hargreaves DJ (1979) Transmission of load in the knee joint with special reference to the role of the menisci, Part II: experimental results, discussions and conclusions. Eng Med 8:220–228

Shapiro F, Koide S, Glimcher MJ (1993) Cell origin and differentiation in the repair of full-thickness defects of articular cartilage. J. Bone Joint Surg Am 75:532–553

Shelbourne KD, Jari S, Gray T (2003) Outcome of untreated articular cartilage defects of the knee. A natural history study. J Bone Joint Surg 85A:S8-S16

Sporn M, Roberts AB, Wakefield LM, Assoian RK (1986) Transforming growth factor-beta: biological function and chemical structure. Science 233:532–534

Spring BJ, Staudacher HM, Henderson IJP (2001) Biomechanical evaluation of normal articular cartilage in the human knee. (Abstract). Proceedings of the Asia Pacific Orthopaedic Association, Adelaide, April 1–6 p 134

Steadman JR, Rodkey WG, Briggs KK (2002) Microfracture to treat full-thickness chondral defects: surgical technique, rehabilitation, and outcomes. J Knee Surg 15:170–1176

Stockwell R (1971) The interrelationship of cell density and cartilage thickness in mammalian cartilage. J Anat 109:411–421

Takahashi T, Tins B, McCall IW, Richardson JB, Takagi K, Ashton K (2006) MR appearance of autologous chondrocyte implantation in the knee: correlation with the knee features and clinical outcome. Skeletal Radiol 35:16–26

Tyler JA (1986) Chondrocyte-mediated depletion of articular cartilage proteoglycans in vitro. Biochemi J 225:493–507

Vasara AI, Nieminen MT, Jurvelin JS, Peterson L, Lindahl A, Kivianta I (2005) Indentation stiffness of repair tissue after autologous chondrocyte transplantation. Clin Orthop and Relat Res 433:233–242

Verdonk R, Steinwachs M, vanlauwe J, Engebretsen L (2006) Trends in cartilage repair. Current opinion of orthopaedic surgeons on cartilage repair: the results of a web-based survey. 3–14. From ICRS web site www.cartilage.org

Vunjak-Novakovic G, Martin I, Obradovich B, Treppo S, Grodzinsky AJ, Langer R, Freed LE (1999) Bioreactor cultivation conditions modulate the composition and mechanical properties of tissue-engineered cartilage. J Orthop Res 17:130–138

Walker PS, Erkman MJ (1975) The role of the menisci in force transmission across the knee. Clin Orthop 109:184–192.

Wen DY (2000) Intra-articular hyaluronic acid injections for knee osteoarthritis. Am Fam Physician 62:565–570, 572

Chapter 14
Bone Biology: Development and Regeneration Mechanisms in Physiological and Pathological Conditions

Hideki Yoshikawa, Noriyuki Tsumaki, and Akira Myoui

Contents

14.1   Introduction   431
14.2   Macroscopic and Microscopic Structure of Bone   432
14.3   Differentiation and Functional Activities of Osteoblastic Cells.   433
14.4   Bone Remodeling   434
14.5   Bone Development and Growth.   435
14.6   Bone Repair and Regeneration   436
14.7   Bone Morphogenetic Proteins and Transforming Growth Factor-βs   438
14.8   Bone Tissue Engineering by BMP   441
14.9   Angiogenesis and Bone Regeneration   443
14.10   Bone Formation in Pathologic Conditions   443
        14.10.1   Bone Tumors.   443
        14.10.2   Myositis Ossificans and Fibrodysplasia Ossificans Progressiva
                  (FOP).   444
        14.10.3   Ossification of the Spinal Ligaments.   445
14.11   Conclusions   445
Questions   446
References   446

14.1 Introduction

Bone is a highly specialized connective tissue which serves various functions essential for life, such as protection for vital organs including bone marrow, mechanical support as a locomotive organ, metabolic regulation for the minerals, and maintenance of serum homeostasis. Bone development, repair and regeneration are complicatedly regulated by cells, matrix, and local and systemic hormones or cytokines. Understanding the mechanisms of bone development and regeneration is essential for the bone regenerative medicine and bone tissue engineering.

H. Yoshikawa (✉)
Department of Orthopaedic Surgery, Osaka University Graduate School of Medicine,
2-2 Yamadaoka, Suita 565-0871, Japan
e-mail: yhideki@ort.med.osaka-u.ac.jp

M. Santin (ed.), *Strategies in Regenerative Medicine*,
DOI 10.1007/978-0-387-74660-9_14, © Springer Science+Business Media, LLC 2009

14.2 Macroscopic and Microscopic Structure of Bone

There are five types of bones in the human body: long, short, flat, irregular, and sesamoid. Long bones consist of a long shaft (the diaphysis) and two articular surfaces, called epiphyses. The ends of the diaphysis are metaphyses, located in contact with growth plates. Short bones are roughly cube-shaped and have only a thin cortical bone. The bones of the wrist and ankle are short bones. Flat bones are thin and curved with two parallel layers of cortical bones. Skull, scapula, and sternum are flat bones. Irregular bones are not categorized into the above types. Their shapes are irregular and complicated. The bones of pelvis and spine are irregular bones. Sesamoid bones such as patella and pisiform are embedded in tendons. They act to hold the tendon and play a role in the joint movement.

Bone consists of compact and cancellous bones. The former is also referred to as cortical bone, which is the hard outer layer of bones with minimal gaps and cavities. The latter is also called trabecular bone, which is composed of a network of needle-like or flat bone and allows room for bone marrow. The exterior of bones is covered by the periosteum with an external fibrous layer and an internal osteogenic layer.

Osteoblasts are mononuclear bone-forming cells derived from osteoprogenitor cells. They are located on the surface of osteoid and produce a matrix protein as osteoid, which mineralizes to become bone (Fig. 14.1). Osteoid is

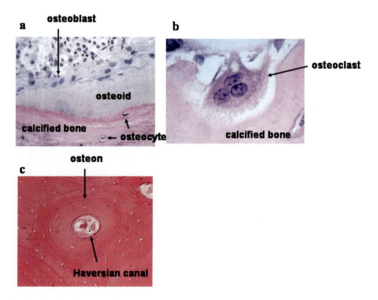

Fig. 14.1 Histologic pictures of bone tissues. Bone space formation site. Osteoblasts, osteoid, and calcified bone with osteocytes can be seen. (hematoxylin and eosin staining, × 200). Bone resorption site. Osteoclast with the ruffled border and calcified bone can be seen. (hematoxylin and eosin staining, × 400). Cross section of cortical bone. An osteon with a Haversian canal can be seen (hematoxylin and eosin staining, × 100)

mainly composed of Type I collagen. The major inorganic component of mineralized bone matrix is calcium hydroxyapatite. Osteocytes originate from osteoblasts which are trapped and surrounded by bone matrix. The spaces which they occupy are so called lacunae. Osteocytes have many processes (canaliculi) and maintain calcium homeostatis, and possibly act as mechano-receptors, regulating the response to mechanical stress (Nijweide et al. 1996).

Osteon (Haversian system) is a minimal unit of structures in lamellar or compact bone. In the center of the osteon is a central canal, called Haversian canal (Fig. 14.1).

Osteoclasts are large, multinucleated cells located on bone surfaces. At the site of active bone resorption, the osteoclast forms "ruffled border", which contacts with bone surface and facilitates demineralization and removal of bone matrix (Fig. 14.1). Osteoclast releases hydrogen ions, acidifying and dissolving the mineralized bone matrix. In addition, they produce hydrolytic enzymes such as cathepsin K and collagenase, digesting the organic compo-nents of the bone matrix (Vannanen 1996).

14.3 Differentiation and Functional Activities of Osteoblastic Cells

The principle cells that mediate bone formation or repair are following: osteo-progenitor cells that contribute the maintaining the osteoblast population; osteoblasts that produce bone matrix on bone surfaces; osteocyte that support bone structure and regulate the homeostasis of minerals. Studies on the mole-cular and cellular mechanisms regulating the osteoblastic differentiation and function have been focused on various signaling pathways, growth factors, transcription factors, cytokines, and hormones (Table 14.1). Osteoblast origi-nates from a local mesenchymal stem cell under the influence of local factors such as bone morphogenetic protein (BMP)/TGF-β superfamily (Urist 1965, Wozney et al. 1988), parathyroid hormone (PTH), and PTH-related protein (PTHrP) (Karaplis and Goltzman 2000), fibroblast growth factor (FGF) (Ornitz and Marie 2002), insulin-like growth factor (IGF), Indian hedgehog, and the Wnt proteins (Hill et al. 2005). These precursors undergo proliferation and differentiation into osteoprogenitors and then into mature osteoblasts. Osteoblasts are always lining on the surface of bone matrix that they are producing. The plasma membrane of the osteoblast is characteristically rich in alkaline phosphatase, which is an index of bone formation. Osteoblasts express receptors for parathyroid hormone, prostaglandins, estrogen, and vita-minD_3. They also express receptor activator of NF-κB ligand (RANKL) which can activate osteoclasto- genesis in a local and paracrine manner and osteopro-tegerin, a decoy RANK receptor inhibiting osteoclast formation (Simonet and Lacey 2003). After the secreting period, the osteoblasts become either an osteocytes or a flat lining cell.

Table 14.1 Regulation of Bone Development and Regeneration

Regulatory Factors	Functions
Growth factors	
BMP (bone morphogenetic protein)2/4/7	Osteoblastic differentiation in vitro, ectopic bone formation in vivo
TGF-β (transforming growth factor)	Periosteal caltilage/bone formation
IGF-I/IGF-II (insulin-like growth factor)	Anabolic regulators of bone formation
PDGF (platelet-derived growth factor)	Osteoblastic proliferation, inhibition of osteoblastic differentiation
EGF (epidermal growth factor)	Differentiation of mesenchymal stem cell
Signaling molecules	
Wnt (wingless and INT-1)	Regulator of chondrogenesis and osteogenesis
β catenin	Regulator of canonical Wnt signaling
Tob	Inhibitors of BMP2/TGFβ signaling
Noggin	Soluble BMP antagonist
Ihh (Indian hedgehog)	Promotion of osteogenic phenotype
Sclerostin	BMP antagonist from osteocyte
Leptin	Antiosteogenic hypothalamic factor in vivo
SDF-1 (stromal derived factor-1)	Promotion of growth and survival of marrow stromal cells
IL-18	Mitogenic cytokine for chondrocytes and osteoblasts
Transcriptional regulators	
Runx2 (Cbfa1)	Essential molecule for initial stage of bone formation
Osterix	Essential molecule for late stage of bone formation
NF-κB	Negative regulator of osteoblastic differentiation
ATF4 (activating transcription factors4)	Maintenance of bone mass through regulation of osteoblast genes
MSX1/2 (homeobox proteins)	Negative regulator of genes in mature osteoblasts
AP-1 (activating protein1)	Osteoblastic regulator respond to growth and mechanical stimuli
Hormones and vitamin	
PTH (parathyroid hormone)	Anabolic effects via regulation of many osteoblastic genes
Estrogen	Osteoprogenitor proliferation and collagen synthesis
1,25(OH)2D3	Upregulation of many osteoblastic genes in bone matrix
Glucocorticoid	Osteoprogenitor cell differentiation, but apoptosis of mature osteoblasts
Growth hormone	Stimulator of proliferation and differentiation of osteoblasts

14.4 Bone Remodeling

Bone remodeling is the process of resorption of old bone followed by replacement of new bone with little change in bone volume. This turnover of bone is performed by the coordinated actions of osteoclasts and osteoblasts on trabecular surfaces and in Haversian systems. Activation of osteoclasts is the initial step in the remodeling sequence, although the mechanism responsible for the initiation of remodeling is unknown. After bone resorption by osteoclasts,

osteoblasts produce new bone matrix to fill the defects. Osteoclasts and osteoblasts are coupled together via various paracrine cell signalings, and these processes occur in tandem at site-specific locations. The main biochemical signaling
necessary to regulate the coupling is RANKL/RANK interactions (Theill et al.
2002). Osteoblasts express RANKL on their cell surface, whereas RANKL
expression in bone marrow stromal cells is largely inducible by various cytokines
and hormones. Interaction of RANKL with its receptor RANK expressed on
osteoclast precursors promotes osteoclast differentiation, and the interaction of
the RANKL with RANK on mature osteoclasts results in their activation and
prolonged survival. OPG in the bone microenvironment, is secreted primarily by
osteoblasts/bone marrow stromal cells, blocks the interaction of RANKL with its
receptor RANK on osteoclast precursors as well as on mature osteoclasts, thus
acting as a main physiological regulator of bone remodeling. On the other hand,
bone growth factors, such as BMPs, TGF-β, IGFs, PDGF, and FGFs, could be
also involved in the coupling phenomenon (Canalis et al. 1989). These factors
seem to be released locally from bone matrix resolved by osteoclasts. They may
act in a sequential manner to induce cellular events required for bone formation.
Among them TGF-β superfamily may be important in the coupling that links
bone formation to bone resorption.

14.5 Bone Development and Growth

Two types of processes are involved in bone development: endochondral ossification (long bones) (Fig. 14.2) and intra-membranous ossification (flat bone)
(Baron 2003). The critical difference between them is the appearance of cartilaginous step in the former. In the area where bone is formed, mesenchymal cells

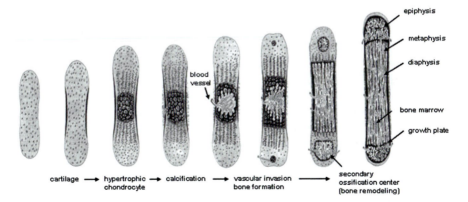

Fig. 14.2 Bone development. The schematic diagram shows endochondral ossification. Con-
densed mesenchymal cells form a cartilage model of the bone to be formed. After chondrocyte
hypertrophy and calcification, vascularization and chondroclastic/osteoclastic activity result
in formation of the primary and secondary ossification centers. Mature bone with bone
marrow is regulated by bone remodeling with bone marrow

condense and form mass of high cell density that represent outlines of future skeletal elements. Mesenchymal cells within condensations can either differentiate into bone-forming cells (osteoblasts), or into chondrocytes. Direct differentiation into osteoblasts occurs in areas of intra-membranous ossification, such as in the skull, the clavicle, the maxilla, and mandible, and in the subperiosteal bone-forming layer of long bones, while differentiation into chondrocytes occurs in the remaining skeleton. The selection of mesenchymal cells to differentiate into osteoblasts or chondrocytes is regulated by a canonical Wnt (wingless and INT-1) signaling (http://www.stanford.edu/~rnusse/wntwindow.html) (Hill et al. 2005). In areas of membranous ossification, Wnt signaling results in high level of β-catenin (http://www.stanford.edu/~rnusse/wntwindow.html) in mesenchymal cells. This induces the gene expressions that are required for osteoblastic differentiation and inhibits transcription of genes needed for chondrocytic differentiation. The transcription factors, RUNX2 (runt related transcription factor 2) (Komori et al. 1997) and osterix (OSX) (Nakashima et al. 2002) are also critical for differentiation of mesenchymal cells to osteoblasts.

On the other hand, differentiation into chondrocytes is regulated by the transcription factor SOX9 (SRY (sex determining region Y)-box 9) and other SOX members (Akiyama et al. 2002). Chondrocytes then proliferate and produce extracellular matrix to form the primordial cartilage, and then proliferating chondrocytes in the central region of the cartilage undergo differentiation into hypertrophic chondrocytes (Fig. 14.2). Hypertrophic chondrocytes exit the cell cycle and synthesize an extracellular matrix that has a different composition than that of proliferating cartilage. The hypertrophic cartilage is invaded by blood vessels, osteoblasts, osteoclasts, and hematopoietic cells, resulting in formation of primary ossification centers (Baron 2003). Within these centers, the hypertrophic cartilage matrix is degraded, the hypertrophic chondrocytes die, and bone replaces the disappearing cartilage. These processes occur later in growth plate located between the epiphysis and metaphysis of long bones. Elongation of long bones is the result of replacement of cartilage with bone at growth plates. At the metaphyseal end of the cartilage, hypertrophic chondrocytes die and calcified matrix is partly removed by osteoclasts/chondroclasts carried by blood vessels. Osteoblasts carried by blood vessels form woven bone on the remnant of the calcified cartilage matrix, modeling the trabeculae of the primary spongiosa. Trabecular bone then undergoes remodeling consisting of resorption by osteoclasts and apposition of newly formed bone matrix by osteoblasts, creating secondary spongiosa (Fig. 14.2). Osteoclastic bone resorption and osteoblastic bone formation are coupled on the surface of trabeculae of secondary spongiosa (Baron 2003).

14.6 Bone Repair and Regeneration

Bone regeneration after fracture or surgical resection is a unique repair process in which the original structure and biochemical integrity is largely restored (Fig. 14.3). The process of fracture healing consists of several phases:

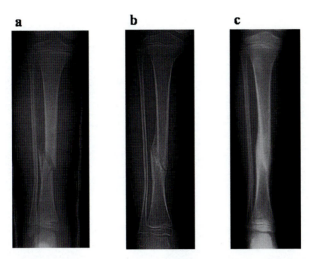

Fig. 14.3 Radiologic pictures of a fracture healing process (the right tibia of an 8-year-old boy). (**a**) The boy sustained a closed tibial fracture by a bicycle accident. The cast fixation was performed just after fracture.(**b**) Vigorous callus formation can be seen at four weeks after fracture. (**c**): Bony union has been completed at eight weeks after fracture

(1) inflammatory response including bleeding, (2) recruitment and differentiation of mesenchymal cells, (3) endochondral bone formation and primary bone formation, and (4) secondary bone formation and remodeling where the newly formed callus is shaped to restore the anatomical structure and mechanical strength (Fig. 14.4) (Gerstenfeld and Einhorn 2006). It is generally accepted that the processes of fracture healing most recapitulated those that work during the embryological development of the skeleton. Namely, two types of ossification contribute to fracture healing: one around the gap region of the fracture (endochondral) and one at the peripheral and adjacent regions to the cortices (intra-membranous). The molecular mechanisms that regulate fracture healing have been studied extensively. Immediately after fracture, various inflammatory factors such as prostaglandins (PGs), interleukin-1 (IL-1), IL-6, IL-11, TNF-α, IFN-γ, RANKL, oeteoprotegerin, macrophage-colony stimulating factor (M-CSF), and cyclooxygenase-2 (COX-2) are induced with bleeding and soft tissue injury, and play some roles in the initiation of the repair process (Gerstenfeld and Shapiro 1996 and Gerstenfeld and Einhorn 2006).

The second group of factors is quite specific in action to enhance cellular recruitment, proliferation, and differentiation. TGF-β superfamily, which includes bone morphogenetic proteins (BMPs), TGF-β(1–3), and growth and development factors (GDFs), seems to be involved in the recruitment, proliferation, and differentiation of osteogenic cells (Tsumaki and Yoshikawa 2005). Among them, BMPs are the most important from a viewpoint of the biological activity and one of the most extensively studied proteins in the field of bone repair and regeneration. Nakase et al. first reported the temporal and special distribution of BMP-4 during fracture repair (Nakase et al. 1994 and Nakase

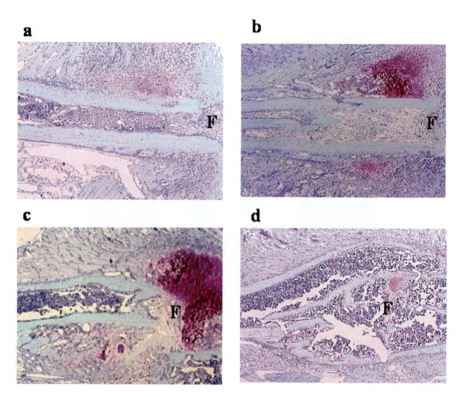

Fig. 14.4 Histological picture of a fracture healing process of the mouse rib (F: fracture site).
(**a**) Mesenchymal cells recruitment and a slight petiosteal reaction can be seen at four days after
fracture. (**b**) Periosteal chondrogenic differentiation can be seen at eight days after fracture.
(**c**) Vigorous chondrocyte proliferation and hypertrophy of the cartilage can be seen at 12 days
after fracture, (**d**) Bone formation and the coupled remodeling can be seen at 20 days after fracture

and Yoshikawa 2006) (Fig. 14.5), and Cho et al. showed that some members of
TGF-β superfamily including BMPs may play roles at the various stages of
endochondral and intra-membranous bone formation during fracture healing
(Cho et al. 2003).

14.7 Bone Morphogenetic Proteins and Transforming Growth Factor-βs

BMPs were originally identified as proteins capable of inducing ectopic cartilage and bone formation when implanted subcutaneously or in muscle pouches
(Urist 1965) (Fig. 14.6). This ectopic cartilage/bone formation recapitulates the
entire sequence of events that occurs during endochondral bone development in
limb buds, where a sequential cascade of events: chemotaxis of mesenchymal
cells, mesenchymal cell condensation, and proliferation and their differentiation into chondrocytes that produce cartilage matrix, angiogenesis and vascular

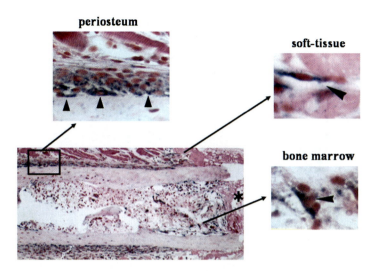

Fig. 14.5 Expression and localization of bone morphogenetic protein-4 (BMP-4) mRNA during the process of fracture healing in mice (in situ hybridization). Before fracture, BMP-4 mRNA is not detected. At 2–4 days after fracture, BMP-4 mRNA is expressed in the cells at the periosteum, para-fracture soft-tissue, and intramedullary cavity. (arrow heads: BMP-4 mRNA signals, * fracture site)

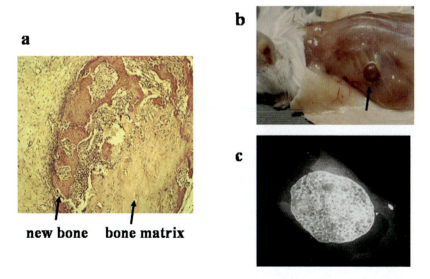

Fig. 14.6 Ectopic bone formation induced by bone morphogenetic protein. (**a**) Histologic picture of an ectopic bone formation in a rat induced by decalcified bone matrix (hematoxylin and eosin staining, × 100). Macroscopic picture of an ectopic bone formation in a mouse induced by 5 μg of rhBMP-2. Radiologic picture of an ectopic bone formation in a mouse induced by 5 μg of rhBMP-2 (soft X-ray picture)

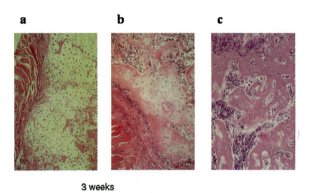

3 weeks

Fig. 14.7 Sequential histologic pictures of an ectopic bone formation induced by 5 µg of rhBMP-2 (hematoxylin and eosin staining, × 100). Differentiation into chondrocytes can be seen at seven days after implantation. Angiogenesis and endochondral bone formation can be seen at 14 days after implantation. Replacement by bone and bone marrow formation can be seen at 21 days after implantation

invasion, absorption of cartilage with appearance of osteoblasts that deposit bone matrix, and mineralization and remodeling (Fig. 14.7). In 1988, Wozney et al. (Wozney et al. 1988) cloned the BMP-1-4 genes and subsequent studies have revealed that BMPs compose a large subfamily of the TGF-β superfamily. BMPs bind to BMP receptors on the cell surface, and these signals are transduced intracellularly by Smad proteins (Derynck et al. 1998). Namely, the signaling is transduced through two types of serine/threonine kinase receptors: types I and II (Miyazono et al. 2005). Smads (http://www.grt.kyushu-u.ac.jp/spad/account/tf/smad.htmlare) the major downstream targets of type I receptors of TGF-β/BMP superfamily proteins (Attisano and Wrana 1998, Heldin et al. 1997). Smads 1, 5 and, 8 are R-Smads that transduce BMP signals, which interact with transcriptional factors and bind directly or indirectly to specific DNA sequences to activate gene transcription (Fig. 14.8). In addition, recent studies of tissue-specific activation and inactivation of BMP signals have revealed that BMP signals control proliferation and differentiation of chondrocytes, differentiation of osteoblasts and bone quality (Tsumaki and Yoshikawa 2005, Okamoto et al. 2006). BMP activity can be suppressed by a variety of mechanisms; (1) extracellular specific antagonist BMP binding proteins such as noggin, chordin, follistatin, and sclerostin (Yanagita 2005), (2) inhibitory Smads 6 and 7 (Murakami et al. 2003), (3) ubiquitination and degradation of Smads by Smurf 1 and 2 (Murakami et al. 2003), and (4) intracellular Smad binding proteins, Tob (Yoshida et al. 2000) and Ski (Canalis et al. 2003). BMPs are now used in regeneration of bone in fracture healing and spine fusions and in dental tissue engineering. BMPs 2 (Jones et al. 2006) and 7 (Vaccaro et al. 2005) are currently approved by the Food and Drug Administration (FDA) for clinical use.

On the other hand, TGF-βs were identified as a group of molecules to mediate many key events in normal development and growth of diverse tissues,

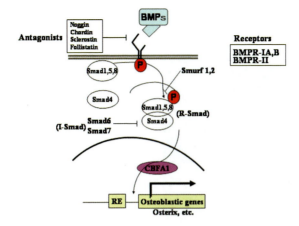

Fig. 14.8 A scheme of BMP signal transduction. Upon ligand binding, types I and II receptors form multimers. Type II receptors phosphorylate type I receptors. Then, type I receptors phosphorylate downstream targets, such as Smads. Smads 1, 5, and 8 are R-Smads that transduce BMP signals. Phosphorylated R-Smads form heteromers with Smad4, which is a common-partner Smad (Co-Smad), and translocate into the nucleus. There, they interact with transcription factors and bind directly or indirectly to specific DNA sequences to activate osteoblastic gene transcription. Several inhibitory mechanisms regulate these BMP signaling pathways. Extracellularly, antagonists bind to BMPs and prevent them from interacting with their receptors. Intracellularly, inhibitory Smads, such as Smads 6 and 7 and Smurfs inhibit Smad signaling

including cartilage and bone (Joyce et al. 1990). Daily injections of TGF-β onto the periosteum of parietal bones or long bones of neonatal rats resulted in localized cartilage and bone formation (Noda and Camilliere 1989). TGF-βs seem to stimulate new cartilage and bone formation by the periosteal chondro-progenitor and osteoprogenitor cells.

14.8 Bone Tissue Engineering by BMP

Bone morphogenetic protein (BMP) is a biologically active molecule capable of inducing new bone formation, and thus applicable for bone tissue engineering (Myoui et al. 2004, Kaito et al. 2005, Yoshikawa and Myoui 2005). We have developed the interconnected porous hydroxyapatite ceramics (IP-CHA) as a scaffold for bone tissue engineering (Tamai et al. 2002) (Fig. 14.9), and analyzed the efficacy as a delivery system for recombinant human BMP-2 (rhBMP-2). We combined two biomaterials to construct a carrier/scaffold system for rhBMP-2: IP-CHA and a synthetic biodegradable polymer poly D, L-lactic acid-polyethy-leneglycol block co-polymer (PLA-PEG) (Saito et al. 2001). We used a rabbit radii model to evaluate the bone-regenerating activity of rhBMP-2/PLA-PEG/IP-CHA composite. At eight weeks after implantation, all bone defects in groups treated with 5–20 μg of rhBMP-2 were completely repaired with sufficient strength (Akita et al. 2004, Kaito et al. 2005) (Fig. 14.10). Using this carrier scaffold system, we

a   b

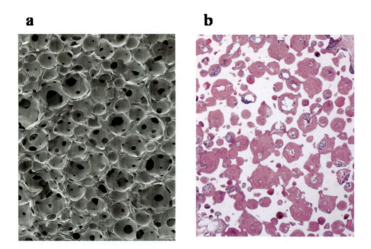

Fig. 14.9 Bone tissue engineering by a ceramic scaffold and bone marrow mesenchymal cells. (a) A scanning electronic microscopic picture of the interconnected porous hydroxyapatite ceramic (IP-CHA). Spherical pores (100–200 μm in diameter) were divided by thin walls and interconnected by interpores (10–80 μm in diameter). (b) New bone formation within the IP-CHA with bone marrow-derived mesenchymal stem cells in a rat. The IP-CHA was recovered from rat subcutaneous tissue at six weeks after implantation. Most of the pores were filled with newly formed bone (hematoxylin and eosin staining, × 50)

reduced the amount of rhBMP-2 necessary for such results to about a tenth of the amount needed in previous studies. Enhancement of bone formation is probably due to the superior osteoconduction ability of IP-CHA and the optimal drug delivery system provided by PLA-PEG. The synthetic biodegradable polymer (PLA-PEG) /IP-CHA composite is an excellent carrier/scaffold delivery system for rhBMP-2, and strongly encourages the clinical effects of rhBMP-2 in bone tissue regeneration (Myoui et al. 2004, Yoshikawa et al. 2005).

a   b   c   d

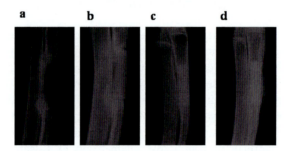

Fig. 14.10 Soft X-ray photographs of the bone regeneration at the bone defect (15 mm) of rabbit radius. (a) The bony defect failed to heal without implantation. (b) The IP-CHA alone group at eight weeks after implantation. Radiolucent lines were clearly visible between the IP-CHA and host bone, and the radiodensity of IP-CHA did not increase. (c) The rhBMP-2 (5 μg)/IP-CHA group at eight weeks after implantation. Bony unions were observed at the junction sites and the radiodensity of IP-CHA increased. (d) The rhBMP-2 (20 μg)/IP-CHA group at eight weeks after implantation. Bony unions and extensive bone formation were observed

14.9 Angiogenesis and Bone Regeneration

Angiogenesis of the growth plate or fracture callus is a crucial event that couples chondrogenesis and osteogenesis (Gerstenfeld and Einhorn 2006).

The terminal stages of chondrogenesis consist of chondrocyte hypertrophy, apoptosis, resorption of calicified cartilage matrix by chondroclasts allowing for the primary bone formation by osteoblasts. Chondrocyte apoptosis occurs just after invasion of vascular endothelial cells and death of chondrocyte is induced by soluble factors from the vasculature or hematopoietic components during angiogenesis (Gerstenfeld and Shapiro 1996).

The importance of angiogenesis during fracture healing or bone regeneration has been shown by several recent studies that treatment of fracture with vascular endothelial growth factor (VEGF) enhances healing, and that angiogenic inhibitors completely prevent fracture healing, callus formation, and periosteal bone formation (Street et al. 2002). Clinically, bone grafting with vascular pedicles are useful and successful for bone repair and regeneration, even at the site with an unfavorable condition such as infection or large bony defect. Therefore, introduction of vasculature into synthetic bone substitutes is another important aspect in regeneration for larger bone defects. We investigated the possibility of integrating IP-CHA with a capillary vessel network via insertion of a vascular pedicle, and to determine whether this procedure enhances new bone formation in tissue engineering. This kind of approach to support blood supply generation in porous biomaterials has also been reported by others using subcutaneous tissue and muscle flaps. IP-CHA loaded with 10 μg of rhBMP-2 was implanted subcutaneously into rat groin with insertion of superficial inferior epigastric vessels. At three weeks, IP-CHA/BMP composite with vascular insertion exhibited abundant new bone formation in the pores of the deep portion close to the inserted vessels (Akita et al. 2004, Yoshikawa et al. 2005). This novel system of integrating a vascular network with IP-CHA is considered as a useful technique for bone tissue engineering.

14.10 Bone Formation in Pathologic Conditions

Pathological ossification including heterotopic ossification can be detected in the various clinical situations such as bone tumors, myositis ossificans, or ossification of the ligaments.

14.10.1 Bone Tumors

In 1926, Phemister demonstrated that osteosarcomas contain normal non-tumorous bone formed by normal osteoblasts in addition to tumorous bone formation produced by osteosarcoma cells (Phemister 1926). Because BMPs are potent inducers of normal osteoblasts, it has been hypothesized that BMPs may be involved in the pathogenesis of the reactive bone formation observed in

osteosarcoma. Clinically, BMP-positive human osteosarcomas were characterized by more sclerotic radiologic features (Yoshikawa et al. 1994b and 1997),
and experimentally, recombinant murine BMP-4 elicited periosteal new bone
formation in vivo (Yoshikawa et al. 1994a). Thus, the periosteal reaction in
osteosarcoma may be produced by normal osteoblasts induced by BMPs.
Osteoid osteoma and osteoblastoma are believed to be manifestations of a
benign osteoblastic process characterized by the osteoid nidus and by a distinctive surrounding zone of reactive bone formation. Urist et al. has speculated
that the nidus may release a bone-stimulating agent such as BMP, from the
observation of a transplanted osteoid osteoma into nude mice (Urist et al.
1979). In addition, ectopic bone formation in soft tissue associated with these
tumors has been described in the literature, in which ectopic bone was detected
at the ligamentous tissue adjacent to the tumor (Takaoka et al. 1994, Okuda
et al. 2001). Immuno-histochemical assays demonstrated that BMP-2/4 was
expressed in the cytoplasms of the osteoblastic cells in osteoid osteoma (Okuda
et al. 2001), raising a possibility that BMPs secreted from the tumor cells
triggered the reactive bone formation and the ectopic bone formation.

14.10.2 Myositis Ossificans and Fibrodysplasia Ossificans
         Progressiva (FOP)

The pathogensesis of myosistis ossificans remains unknown, but locally increased
production of BMP-like proteins has been proposed as the cause of the pathologic condition. Heterotopic bone formation occasionally developed at the periar-
ticular regions in the patients with a spinal cord injury or a severe brain damage
(Garland 1988). Inhibitory signals through the central nervous against bone
morphogens system may be disrupted after nerve injury, resulting in heterotopic
bone formation. Recently, it has been reported that fibrodysplasia ossificans
progressiva, a systemic type of heterotopic bone formation, may be caused by
overexpression of BMP-4 (Shafritz et al. 1996). Fibrodysplasia ossificans progressiva (FOP) (http://www.familyvillage.wisc.edu/lib_fop.htm) is an extremely
rare, and about 700 cases have been reported so far in the literature. Typically, the
age of onset is around 3–4 years with episodes of acute soft-tissue swelling (flare-
ups) in the neck or upper spine. Large painful tumors of highly vascularized
growing tissue develop spontaneously or after minor trauma and become to
heterotopic bone formations. Gradually, bone masses develop at various sites
and immobilize the joints and lead to ankylosis. Ambulation is classically lost in
the twenties or thirties due to ankylosis of the hip. Major complications arise
from rigidity of the rib cage and ankylosis of the jaw causing restrictive lung
disease and nutritional impairment. BMP-4 mRNA was expressed in lymphoblastoid cell lines from 26 of 32 patients with fibrodysplasia ossificans progressiva
but from only 1 of 12 normal subjects. BMP-4 and its mRNA were detected in the
lymphoblastoid cell lines from a man with fibrodysplasia ossificans progressiva
and his three affected children (two girls and a boy), but not from the children's

unaffected mother. No other bone morphogenetic proteins were detected. In any case, it is strongly suggested that heterotopic bone formation may be induced by BMP-like proteins which are locally overexpressed at the pathologic conditions.

14.10.3 Ossification of the Spinal Ligaments

Ossification of the posterior longitudinal ligament (OPLL) and the ligamentum flavum (OLF) is a pathological condition in the spinal ligament, with hetero-topic bone mainly through endochondral ossification. Most reports on OPLL have originated from Japan, with only a few reports from Western countries. The incidence of OPLL among Japanese people is reported to be about 3%, which is higher than the incidence reported for Chinese (0.2–1.8%), Koreans (0.95%), Americans (0.12%), or Germans (0.1%). The natural course of ossi-fication in 94 patients who underwent surgery was followed. There were 75 men and 19 women, whose age ranged from 23 to 79 years (mean 54.8 years). Progression of ossification was recognized at the site of increased strain in the intervertebral disc. The prognosis of patients with OPLL has been disappoint-ing, depending upon the degree of myelopathy (Matsunaga and Sakou 2006).

BMPs and transforming growth factor-β might be causative factors in patho-genesis of OPLL and OLF. Possible mechanisms are as follows: (1) Systemic overexpression of BMPs/TGF-βs and/or their receptors in the patients, such as BMP-4 overexpression in fibrodysplasia ossificans progressiva (Shafritz et al. 1996), (2) local overexpression of BMPs/TGF-βs and/or their receptors around or in the spinal ligaments, and (3) enhancement of responsiveness to BMPs/TGF-βs in the mesenchymal cells around or in the ligaments. Systemic overexpression of s in OPLL or OLF patients has never been reported, but there have been several evidences on local overexpression of BMPs/TGF-βs. A couple of reports demonstrated that local implantation of BMPs could induce ossification of the ligamentum flavum in animals (Miyamoto et al. 1992, Saito et al. 1992). Miya-moto et al. reported that ossification of the ligamentum flavum and secondary spinal-cord compression were induced in mice by implantation of partially purified BMP (Miyamoto et al. 1992). The study indicates that OLF can be experimentally induced by BMP and that mesenchymal cells which can respond to BMP and differentiate into chondrocytes/osteoblasts do exist in or around the ligamentum flavum of mice. These data suggest that BMPs/TGF-β may play some significant roles in the pathogenesis of OPLL and OLF.

14.11 Conclusions

In this chapter, we have described bone biology: development and regeneration mechanisms of bone. We have also reviewed pathological bone formation as well as physiological bone formation because knowledge obtained from the pathologic bone formation must contribute to the solution of not only the pathomechanism of diseases but also the molecular mechanisms underlying normal bone development or bone repair. For example, bone regeneration induced by BMPs recapitulates the entire sequence of events that occurs during endochondral bone development in limb buds. Therefore, understanding of bone biology, such as endochondral bone formation, fracture healing, angiogenesis during bone formation, or bone remodeling, is essential for bone regenerative medicine and bone tissue engineering. Recently, various scaffolds, bioactive factors, and stem cells are available and have been used for regeneration of bony defects, but in conclusion, we believe the most important is that based on the fundamental bone biology, novel biomaterials, or technologies should be clinically applied and utilized.

Questions

1. Highlight structural features of compact and trabecular bone and link them to their mechanical properties.
2. What are the hormones and cell types involved in bone remodeling?
3. Which is the cell lineage the osteoclasts derive from?
4. Provide examples of growth factors playing a major role in bone regeneration and describe their mechanism of action on bone cells.
5. Discuss the role of main proteins of the bone extracellular matrix and their role in its formation.
6. What are the main biological phases following bone fracture and what is their role in bone regeneration?
7. Analyze the molecular and cell mechanisms leading to eteropic calcification.
8. Describe stem cell location in bony tissues and their role in bone regeneration.
9. Present angiogenesis in bone and its importance in tissue physiology.

References

Akita S et al. (2004) Capillary vessel network integration by inserting a vascular pedicle enhances bone formation in tissue-engineered bone using interconnected porous hydroxyapatite ceramics. Tissue Eng 10:789–795

Akiyama H et al. (2002) The transcription factor Sox9 has essential roles in successive steps of the chondrocyte differentiation pathway and is required for expression of Sox5 and Sox6. Genes Dev 16:2813–2828

Attisano L, Wrana JL (1998) Mads and Smads in TGFβ signaling. Curr Opin Cell Biol 10:188–194

Baron R (2003) General principles of bone biology. In: Primer on the metabolic bone diseases and disorders of mineral metabolism, 5th edn. The American Society of Bone and Mineral Research, Washington DC, pp 1–8

Canalis E et al. (1989) Growth factors and the regulation of bone remodeling. J Clin Invest 81:277–281

Canalis E et al. (2003) Bone morphogenetic proteins, their antagonists and the skeleton. Endocr Rev 24:218–235

Cho TJ et al. (2003) Differential temporal expression of members of the transforming growth factor beta superfamily during murine fracture reapir. J Bone Miner Res 17:513–520

Derynck, R et al. (1998) Smads: transcriptional activators of TGF-β responses. Cell 95:737–740

Garland DE (1988) Clinical observations on fractures and heterotopic ossification in the spinal cord and traumatic brain injured populations. Clin Orthop 233:86–101

Gerstenfeld LC, Einhorn TA (2006) Fracture healing: the biology of bone repair and regeneration In: Primer on the metabolic bone diseases and disorders of mineral metabolism, 6th edn, The American Society of Bone and Mineral Research, Washington, DC, pp 42–48

Gerstenfeld LC, Shapiro FD (1996) Expression of bone-specific genes by hypertrophic chondrocytes: Implication of the complex functions of the hypertrophic chondrocyte during endochondral bone development. J Cell Biochem 62:1–9

Heldin CH et al. (1997) TGF-β signaling from cell membrane to nucleus through SMAD proteins. Nature 390:465–471

Hill TP et al. (2005) Canonical Wnt/beta-catenin signaling prevents osteoblasts from differentiating into chondrocytes. Dev Cell 8:727–738

Jones AL et al. (2006) Recombinant human BMP-2 and allograft compared with autogenous bone graft for reconstruction of diaphyseal tibial fractures with cortical defects. A randomized, controlled trial. J Bone Joint Surg 88A:1431–1441

Joyce ME et al. (1990) Transforming growth factor-beta and the initiation of chondrogenesis and osteogenesis in the rat femur. J Cell Biol 110:2195–2207

Kaito T et al. (2005) Potentiation of the activity of bone morphogenetic protein-2 in bone regeneration by a PLA-PEG/hydroxyapatite composite. Biomaterials 26:73–79

Karaplis AC, Goltzman D (2000) PTH and PTHrP effects on the skeleton. Rev Endocr Metab Disord 1:331–341

Komori T et al. (1997) Targeted disruption of Cbfa1 results in a complete lack of bone formation owing to maturational arrest of osteoblasts. Cell 89:755–764

Matsunaga S, Sakou T (2006) OPLL: disease entity, incidence, literature search, and prognosis. In: Yonenobu K et al. (eds) Ossification of the posterior longitudinal ligament, 2nd edn. Springer Tokyo, pp 11–17

Miyamoto S et al. (1992) Ossification of the ligamentum flavum induced by bone morphogenetic protein. An experimental study in mice. J Bone Joint Surg 74B:279–283

Miyazono K et al. (2005) BMP receptor signaling: Transcriptional targets, regulation of signals, and signaling cross-talk. Cytokine Growth Factor Rev 16:251–263

Murakami G et al. (2003) Cooperative inhibition of bone morphogenetic protein signaling by Smurf1 and inhibitory Smads. Mol Biol Cell 14:2809–2817

Myoui A et al. (2004) Three-dimensionally engineered hydroxyapatite ceramics with interconnected pores as a bone substitute and tissue engineering scaffold. In: Yaszemski MJ et al. (eds) Biomaterials in Orthopedics. Marcel Dekker, New York, pp 287–300

Nakase T, Yoshikawa H (2006) Potential roles of bone morphogenetic proteins (BMPs) in skeletal repair and regeneration. J Bone Miner Metab 24:425–433

Nakase T et al. (1994) Transient and localized expression of bone morphogenetic protein 4 messenger RNA during fracture healing. J Bone Miner Res 9:651–659

Nakashima K et al. (2002) The novel zinc finger-containing transcription factor osterix is required for osteoblast differentiation and bone formation. Cell 108:17–29

Nijweide PJ et al. (1996) The osteocyte. In: Bilezikian JP et al. (eds) Principles of bone biology. Academic Press, New York, p 115–126

Noda M, Camilliere JJ (1989) In vivo stimulation of bone formation by transforming growth factor-β. Endocrinology 124:2991–2994

Okamoto M et al. (2006) Bone morphogenetic proteins in bone stimulate osteoclasts and osteoblasts during bone development. J Bone Miner Res 21:1022–1033

Okuda S et al. (2001) Ossification of the ligamentum flavum associated with osteoblastoma: a report of three cases. Skeletal Radiol 30:402–406

Ornitz DM, Marie PJ (2002) FGF signaling pathways in endochondral and intramembranous bone development and human genetic disease. Genes Dev 16:1446–1465

Phemister DB. (1926) A study of the ossification in bone sarcoma. Radiology 7:17–23

Saito H et al. (1992) Histopathologic and morphometric study of spinal cord lesion in a chronic cord compression model using bone morphogenetic protein in rabbits. Spine 17:1368–1374

Saito N et al. (2001) Biodegradable poly, -lactic acid–polyethylene glycol block copolymers as a BMP delivery system for inducing bone. J Bone Joint Surg 83A:S92–98

Shafritz AB et al. (1996) Overexpression of an osteogenic morphogen in fibrodysplasia ossificans progressiva. N Engl J Med 335:555–561

Simonet WS, Lacey DL (2003) Osteoclast differentiation and activation. Nature 423:337–342

Street J et al. (2002) Vascular endothelial growth factor stimulates bone repair by promoting angiogenesis and bone turnover. Proc Natl Acad Sci U S A 99:9656–9661

Takaoka K et al. (1994) Ectopic ossification associated with osteoid osteoma in the acetabulum. A case report. Clin Orthop 299:209–211

Tamai N et al. (2002) Novel hydroxyapatite ceramics with an interconnective porous structure exhibit superior osteoconduction in vivo. J Biomed Mater Res 59:110–117

Theill LE et al. (2002) RANKL and RANK: T cells, bone loss, and mammalian evolution. Annu Rev Immunol 20:795–823

Tsumaki N, Yoshikawa H (2005) The role of bone morphogenetic proteins in endochondral bone formation. Cytokine Growth Factor Rev 16: 279–285

Urist MR (1965) Bone: formation by autoinduction. Science 150:893–899

Urist MR et al. (1979) Growth of osteoid osteoma transplanted into athymic nude mice. Clin Orthop 141:275–280

Vaccaro AR et al. (2005) Comparison of OP-1 Putty (rhBMP-7) to iliac crest autograft for posterolateral lumbar arthrodesis: a minimum 2-year follow-up pilot study. Spine 30:2709–2716

Vannanen K (1996) Osteoclast function: biology and mechanisms. In: Bilezikian JP et al. (eds) Principles of bone biology. Academic Press, New York, p 103–113

Wozney JM et al. (1988) Novel regulators of bone formation: molecular clones and activities. Science 242:1528–1534

Yanagita M (2005) BMP antagonists: their roles in development and involvement in pathophysiology. Cytokine Growth Factor Rev 16:309–317

Yoshida Y et al. (2000) Negative regulation of BMP/Smad signaling by Tob in osteoblasts. Cell 103:1085–1097

Yoshikawa H, Myoui A (2005) Bone tissue engineering with porous hydroxyapatite ceramics. J Artif Org 8:131–136

Yoshikawa H et al. (1994a) Periosteal sunburst spiculation in osteosarcoma: a possible role for bone morphogenetic protein. Clin Orthop 308:213–219

Yoshikawa H et al. (1994b) Expression of bone morphogenetic proteins in human osteosarcoma: Immunohistochemical detection with monoclonal antibody. Cancer 73:85–91

Yoshikawa H et al. (1997) Bone morphogenetic proteins (BMPs) in musculoskeletal oncology. J Musculoskel Res 1:1–12

Chapter 15
Clinical Applications of Bone Tissue Engineering

Silvia Scaglione and Rodolfo Quarto

Contents

15.1   Introduction   449
15.2   The Clinical Problem   452
15.3   Stem Cells in Orthopedic Therapy   453
15.4   Biomaterials as Biomimetic Scaffolds   456
15.5   Clinical Applications in Orthopedics   459
15.6   Conclusions   460
Questions/Exercises.   462
References   462

15.1 Introduction

Several major progresses and improvements have been introduced in the field of bone regenerative medicine during the last few years, as innovative alternatives to current therapies which still present many limitations. Natural processes of bone repair are sufficient to restore the skeletal integrity for most fractures. However, the auto-regenerative potential of bone cannot handle large size "critical" lesion. Therefore, manipulation of natural healing mechanisms to regenerate larger bone segments is often required in reconstructive surgery.

Tissue Engineering (TE) has represented surely the most innovative and attractive approach for potentially solving many of the problems in orthopedic surgery.

Tissue Engineering, a relatively new field in medicine, has to be considered as a sequence of phases going from the research project to the design and production of bioactive matrices and ideally of a living tissue substitute. This main objective can be approached in different ways. In this chapter, we will limit our discussion to the use of stem/progenitor cells and biomaterials in the TE scenario, with a particular focus on regeneration of bone tissue through cell

S. Scaglione (✉)
Advanced Biotechnology Center, Genova, Italy; Department of Communication,
Computer and System Sciences, University of Genova, Italy
e-mail: silvia.scaglione@unige.it

M. Santin (ed.), *Strategies in Regenerative Medicine*,
DOI 10.1007/978-0-387-74660-9_15, © Springer Science+Business Media, LLC 2009

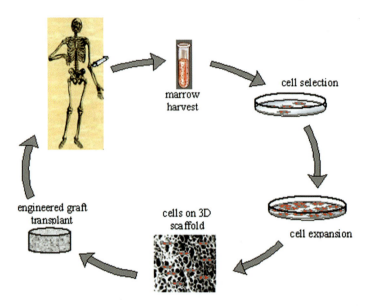

Fig. 15.1 Concept of cell therapy: this approach requires a source of stem cells which, being usually scarce in the tissue of origin, needs to be selected, and expanded in vitro. After choosing the required biomaterial, stem cells will be allowed to adhere to it to be finally delivered at the lesion site

therapy. Cell therapy is the branch of TE which is mainly dealing with cells considered as a key element to achieve the regeneration of the target tissue (Fig. 15.1). In this context and in a reductive view, biomaterials are intended as mere cell delivery vehicles. However, biomaterials are not only simple bio-inert "cell carriers", but they should have the difficult task of driving tissue regeneration in vivo and, most importantly, of being "informative" for cells, which means to direct cell differentiative fate providing a correct microenvironment and physical-chemical inputs (Fig. 15.2).

Furthermore, modern biomaterials should start a regenerative process and properly drive it. At the same time, they should also be resorbed in vivo proportionally to tissue regeneration gradually loosing their original mechanical performance. They should finally disappear as soon the neo tissue formed is able to fully substitute the graft.

In general, the TE philosophy relies on providing the region to be reconstructed with initial biomechanical properties, encouraging progenitor cells to form new bone, and then degrading the synthetic carrier to allow the new bone to remodel and to gradually restore the required biomechanical support function. Basic requirements of these biomaterials include the development of biomimetic scaffolds with an internal architecture able to favor cell migration and in vivo vascularization, and a chemical composition conducive for cell attachment and maintenance of cellular functions.

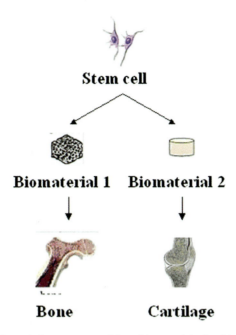

Stem cell

Biomaterial 1   Biomaterial 2

Bone   Cartilage

Fig. 15.2 Biomaterials can influence stem cell fate. Biomaterials should not be thought as bio-inert cell delivery vehicles, but they have to be considered as informative scaffolds able to provide the cell with a correct microenvironment and physical-chemical inputs. As an example, in the figure a stem cell population is directed towards two different differentiation pathways by two chemically distinct biomaterials

Stem cell therapy of skeletal tissues has proven to be a potentially successful strategy by providing a novel therapeutic approach that responds to pathological conditions of patients suffering large bone losses because of injuries or diseases. Among different sources of osteoprogenitor cells which have been proposed, mesenchymal stem cells (MSC) represent in most cases the first choice, due to their ability to proliferate in culture and to give rise to specialized cells. Their use hold promises of wide spread applications particularly in areas of spinal cord injury, non-unions, critical bone defects, spinal fusions, augmentation of ligament reconstructions, cartilage repair, and degenerative disc disorders (Garbossa et al. 2006, Ge et al. 2005, Jorgensen et al. 2001, Klein and Svendsen 2005, Krampera et al. 2006, Leung et al. 2006, Marcacci et al. 2003, Noth et al. 2005, Sakai et al. 2005, Sykova et al. 2006). Although they may be routinely isolated and expanded in culture, their use for therapeutic strategies requires technologies not yet perfected. In this context, different strategies have been reported for using MSC in the regeneration of bone tissue in vivo. MSC can be "driven" by specific bioactive molecules, ex vivo gene transfer, and other physical factors to form mineralized matrix in vitro for future re-implantation in vivo (Caterson et al. 2001, Luskey et al. 1990, Partridge et al. 2002).

Alternatively, these cells are combined with special matrices, which can carry bioactive molecules such as growth factors, and implanted in the attempt to enhance bone regeneration (Caterson et al. 2001, Luskey et al. 1990, Partridge et al. 2002).

Although the interest of this approach is very high and many studies have been performed in vivo in both small and large animal systems, still only two pilot clinical studies have been reported (Quarto et al. 2001, Vacanti et al. 2001).

How should we interpret this apparent failure? Potentially several elements may have limited the experimental applications and a wider diffusion of this fascinating approach: (a) the field of stem cells still needs time and knowledge before being considered a standard approach in regenerative medicine, (b) some "caveat" should induce scientists to explore side effects of stem cells as therapeutic agent; (c) biomaterials, although improving fast, are still far from being optimal for a TE use; (d) guidelines for use of cells in therapy are becoming more and more rigorous and stringent.

15.2 The Clinical Problem

Bone lesions/defects occur in a wide variety of clinical situations, and their reconstruction to provide mechanical and functional integrity is a necessary step in the patient's rehabilitation. Moreover, a bone graft or a bone substitute is often required in maxillo-facial and orthopedic surgery to assist healing in the repair of osseous congenital deformities, or in the repair of defects due either to trauma or to surgical excision of bone pathological lesions exceeding a certain size.

Most of the therapeutic approaches to repair large bone defects include graft transplant (auto-, allo-, and xeno- grafts; different biomaterial implants).

Bone auto-grafts, both non-vascularized and vascularized, are considered at the moment the optimal choice for osteogenic bone replacement in osseous defects (Gazdag et al. 1995, Hokugo et al. 2006, Kneser et al. 2006, Movassaghi et al. 2006, Zarate-Kalfopulos and Reyes-Sanchez 2006). Presently vascularized grafts are the most widely used, being considered as the most successful. Autologous bone grafts reliably fill substance deficits and induce bone tissue formation at the defect site following transplantation. Chips, larger pieces, and even blocks of several centimeters in size could be harvested. The success rate is high indeed, but complications, such as infections, bleeding, hematoma, non unions, etc. are very frequent especially in large shaft reconstructions (Kumta et al. 1997, Norman-Taylor et al. 1997, Sutherland et al. 1997). Furthermore, large reconstructions with autologous bone require significant harvests of healthy tissue with important donor site morbidity (Banwart et al. 1995, Kurz et al. 1989, Seiler and Johnson 2000). Therefore, this approach is limited by definition.

Many of the problems presented by the autografts, in particular the immediate availability of tissue for transplant, could be solved by allo- and possibly xeno-grafts, which are commonly used for repair of osseous defects when autologous transplantation is not applicable. Although the initial properties of allo- or xeno- grafts resemble those of autologous bone in terms of biomechanical stability and elasticity, host immune response, and potential risk of infectious diseases transmission raise several concerns limiting their applications.

Novel materials, cellular transplantation, and bioactive molecule delivery are being explored alone and in various combinations to address the problem of bone regeneration and repair. The aim of these strategies is to exploit the body's natural ability to repair injured bone with new bone tissue, and later to remodel the newly formed bone in response to the local stresses it experiences. In general, the strategies discussed in this paper attempt to provide the reconstructed region with appropriate initial mechanical properties, encourage new bone to form, and then gradually degrade to allow the newly formed bone to remodel and acquire the proper mechanical support function.

15.3 Stem Cells in Orthopedic Therapy

The use of stem cells in therapy is today a consolidated procedure at least in some clinical settings (Bianco and Gehron Robey 2000, Caterson et al. 2001, Jorgensen et al. 2001, Krampera et al. 2006, Quarto et al. 2001, Rando 2006). Bone marrow transplant is possibly the oldest and most established example of the clinical use of stem cells. The progresses of cell biology and the advancement of the knowledge of stem cell biology together with the most recent technological innovations in cell cultures have made possible to manipulate in vitro stem cell derived from virtually any tissue (Fig. 15.3) (Bianco and Gehron Robey 2000, Bosch et al. 2000, Caterson et al. 2001, Dyce et al. 2004, Guasch 2006, Johnston 2004, Olszewski 2004, Otteson and Hitchcock 2003, Reh and Fischer 2001).

Among stem cells reservoir, bone marrow is the most commonly used. Bone marrow derived mesenchymal stem cells (MSCs) are multi-potent cells capable of giving rise to virtually all connective tissue cells. Sources of these cells, as well as other cells with similar features, have been described to include embryos, umbilical cord, and certain sites in adults such as periostium, bone marrow, spleen, thymus, skeletal muscle, adipose tissue, skin, and retina (Bosch et al. 2000, Chacko et al. 2003, Chailakhan and Lalykina 1969, Doherty et al. 1998, Friedenstein and Lalykina 1972, Friedenstein et al. 1966, Huang et al. 2002, Kim et al. 2004, Levy et al. 2001, Schantz et al. 2002, Zuk et al. 2002, Zuk et al. 2001). MSCs isolated from different tissues share many phenotypical and functional characteristics. More interestingly, MSC present a common differentiative potential also with

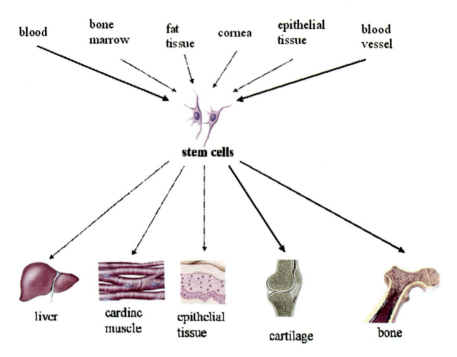

blood   bone marrow   fat tissue   cornea   epithelial tissue   blood vessel

stem cells

liver   cardiac muscle   epithelial tissue   cartilage   bone

Fig. 15.3 Stem cells origin: stem cells with similar features can be derived from different tissues and in turn can give rise to a variety of cell phenotypes and eventually of different tissues

pericytes, perivascular cells with established osteogenic potential. Therefore, vascular pericytes may be regarded as cells strictly related to MSC, which may suggest a common link between angiogenesis and bone formation (Collett and Canfield, 2005).

Recently cells with features close or comparable to MSC have been isolated from peripheral blood, representing a previously unrecognized circulatory component to the process of bone formation (Kuznetsov et al. 2001).

The ideal source of stem cells should be a tissue easily accessible and easy to be harvested with the minimal possible inconvenience for the donor. Peripheral blood is of course the ideal tissue, but still little is known about circulating cells with mesenchymal potential, their isolation, culture conditions, and differentiation potential. On the other hand, the harvesting of small amounts of bone marrow is a relatively simple procedure performed routinely in virtually any medical center. Furthermore, protocols for isolation and culture of marrow derived mesenchymal stem cells are well known and established. Therefore, the preferred source of skeletal stem cells so far is the bone marrow as it contains adult stem cells that can be easily driven towards a bone phenotype. In adult individuals of multiple vertebrate species, MSCs can be readily isolated and expanded in culture from a small bone marrow aspirate, displaying in vitro the capability of differentiating along several lineage pathways such as bone, cartilage, fat, muscle, tendon, liver, kidney, heart, and even brain cell phenotypes (Bianco and Gehron Robey 2000).

The use of osteogenic stem cells or osteoprogenitors to regenerate skeletal tissues is a popular area of research investigation with high potential for successful use of tissue-engineering principles in orthopedics. Recent studies demonstrate the successful uses of osteoprogenitors in the treatment of large bone defects (Bruder et al. 1998, Kadiyala et al. 1997, Kon et al. 2000, Petite et al. 2000). When implanted in vivo in fact, MSC are capable of reconstituting an organoid composed of bone tissue originated from transplanted MSC, and bone marrow originated from host hemopoietic progenitors. A significant improvement in this field has been represented by the association between MSC and porous bioceramics, where the intrinsic osteoconductive properties of the bioceramics are integrated with the biological properties of the osteoprogenitor cells allowing the formation of vital, well vascularized and biomechanically functional bone tissue.

A critical aspect to be clarified is the putative/controversial capability of transplanted MSC to exert a long lasting suppressant activity on the immune system of recipient organisms (Aggarwal and Pittenger 2005, Le Blanc 2003, Le Blanc et al. 2004, Liu et al. 2006, Rasmusson 2006, Sudres et al. 2006). If true, such an activity would support the possibility that administration of MSCs may also represent a novel therapeutic strategy for immune-mediated disorders. On the other hand, because of the potential consequences of a long lasting suppressant activity, the use of MSC in therapy should be carefully considered and evaluated.

Another point deserving attention is the fact that MSC effective dosage (i.e., number of MSC sufficient to provide a therapeutic effect) has not been established yet. Although bone marrow (BM) represents an easily retrievable source of stem cells, BM aspirates alone are not consistent or rich enough in MSC to use unmodified as a primary source. Moreover, only recently few surface markers have been described to specifically identify this cell population from bone marrow, and a consensus on the validity of such markers has not been reached yet (Alison et al. 2006, Boheler 2004). Bone marrow derived MSC are, therefore, selectively expanded in monolayer (2D), due to their capacity to adhere to a plastic surface, and cultured until reaching a number sufficient to colonize the overall volume of three-dimensional (3D) scaffold to be implanted. However, cell density (i.e., the number of MSC per surface unit) does not necessary guarantee a positive outcome of the clinical application. Concerning this, several other biological aspects have to be considered, such as cell differentiative stage, their number of doublings (i.e., age in culture), donor age and health status, and further key points not yet well established (Banfi et al. 2002, Banfi et al. 2000, Javazon et al. 2004).

A part from these considerations, other key points to be considered for cell therapy application in the clinical routine are the need for standardization, process automation, time- and personnel- costs, legal, and ethical implications. The possibility of selectively expanding MSC directly within 3D scaffolds in a safe and reproducible process may open new opportunities to the simplified and streamlined production of osteoinductive grafts for the cell therapy in orthopedic.

These few examples can just give a light flavor of the amount of studies that need to be performed before these cells could be admitted in the clinical routine.

15.4 Biomaterials as Biomimetic Scaffolds

Although material science technology has resulted in clear improvements in the field of regenerative medicine, no ideal bone substitute has been developed yet and hence large bone defects still represent a major challenge for orthopedic and reconstructive surgeons.

A number of bone substitute biomaterials are easily available. The intended clinical use defines the desired properties of engineered bone substitutes. Anatomical defects in load bearing long bones, for instance, require devices with high mechanic stability whereas for craniofacial applications, initially injectable or moldable constructs are favorable.

Scaffold chemical composition is of crucial importance for the osteoconductive properties and the resorbability of the material. In addition, scaffolds should have an internal structure permissive for vascular invasion. Porous bioceramics (hydroxyapatite, tricalcium phosphate) are very promising candidates in bone substitution, because of their bone-like chemical composition, osteoconductivity and mechanical properties (Boyde et al. 1999, Dong et al. 2001, Dong et al. 2000, Flautre et al. 1999, Gauthier et al. 1998, Gauthier et al. 2005, Kon et al. 2000, Livingston et al. 2003, Lu et al. 1998, Marcacci et al. 1999). They are also particularly advantageous for bone tissue engineering application as they induce neither an immune nor an inflammatory response in recipient organisms (El-Ghannam 2005, Erbe et al. 2001, Livingston et al. 2002) (for more details see chapter 3).

The most intriguing concept in modern biomaterials is obtaining materials able to mimic a specific eventually pre-existing environment and, therefore, inducing cells to differentiate in a predetermined manner and to regenerate by themselves the desired tissue according to physiological pathways. This approach should allow to avoid biomaterials requirements of long-term stability, mechanical loading, host reactions, since implanted grafts should be gradually resorbed and substituted by neo-autologous bone tissue.

Therefore, one of the elements required to mimic the morphogenic process leading to bone repair and delivering osteoprogenitor cells is the use of a proper scaffold. Biomaterial scaffolds have to be considered a critical component of the experimental design, since they have to promote cell adhesion, proliferation and differentiation as well as to encourage vascular invasion and ultimately new bone formation, mimicking the microenvironment of a regenerating bone.

The initial clinical studies in animal models and in pilot clinical studies have been encouraging and suggests that stem cells combined with proper scaffolds could be a promising method for bone reconstruction (Marcacci et al. 1999,

Muraglia et al. 1998, Petite et al. 2000, Quarto et al. 2001, Vacanti et al. 2001). In contrast to classic biomaterial approach, TE is based on the understanding of tissue formation and regeneration, and it aims to induce new functional tissues, rather than just to implant new "spare parts". In this approach, the ceramic performs as a mechanical support, an osteomimetic surface and as a template for the newly formed bone tissue. On the other hand, the MSC recognize the bioceramic surface as pre-existing bone (osteo-mimesis) and differentiate into osteoblasts depositing bone extracellular matrix. The system has been widely used for studies aimed at biotechnological applications, biomaterial testing and regenerative medicine.

Besides their chemical composition, the other critical parameter to improve the efficiency of biomaterials to be used in bone tissue engineering is the overall structure and architecture (Chang et al, 2000, De Oliveira et al. 2003, Gauthier et al. 1998, Karageorgiou and Kaplan 2005, Lu et al. 1999, Mankani et al. 2001, Navarro et al. 2004). Indeed an important advancement in this field has been represented by the introduction of synthetic porous scaffolds. In these cases in fact the internal architecture can be intelligently designed and the density, pore shape, pore size and pore interconnection pathway of the material can be predetermined. The result is that the surface available for tissue regeneration as well as for cell delivery can be very high. The increase in surface has also effects on scaffold resorbability. Moreover, high porosity levels are necessary for the in vivo bone tissue ingrowth, since it allows migration and proliferation of osteoblasts and mesenchymal cells, and matrix deposition in the empty spaces (Fig. 15.4). Macroporosity has a strong impact on osteogenic outcomes

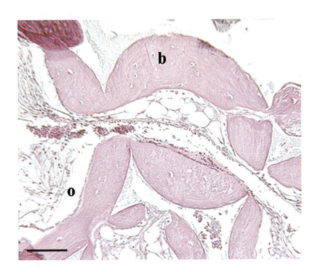

Fig. 15.4 MSC generate bone when implanted in experimental animals: a histological section of a construct implanted ectopically in nude mice shows lining osteoblasts (**o**) depositing neo-bone matrix tissue (**b**), stained in pink, starting from the ceramic (**c**) surface, gradually filling the pore cavities. Blood vessel (**v**) running through an interconnection joining two adjacent pores of the scaffold is displayed. Scale bar:100 μ

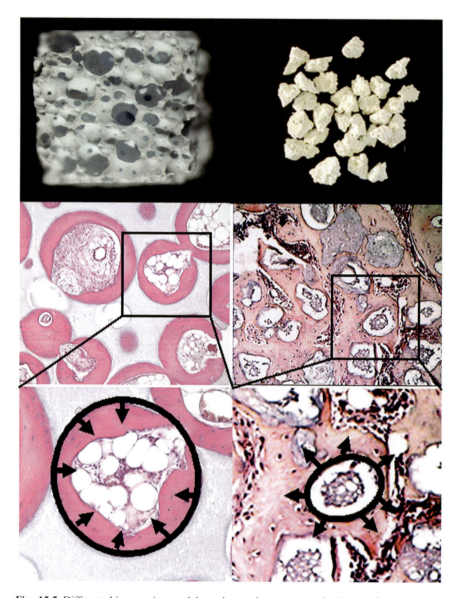

Fig. 15.5 Different bioceramics models to be used as osteoconductive grafts. At the left column, a representative highly porous ceramic foam is displayed, while at the right column ceramic based granules are displayed, both in terms of their macrostructure (*first line*) and in terms of bone forming efficiency (*second line*). The histological sections of the two constructs implanted ectopically in nude mice show a different approach of in vivo bone matrix deposition (*stained in pink*) due to their different internal architecture: on the left side, bone matrix has been deposed within pore cavities, starting from the ceramic external surface, on the right, bone tissue was deposited from the external surface of granules, gradually filling the overall structure of the implanted graft (*third line*)

and pore interconnection pathway plays an important role as well. An incomplete pore interconnection or a limiting calibre of the interconnections could represent an important constraint to the overall biological system by limiting blood vessels invasion (Mastrogiacomo et al. 2006). Therefore, bioceramics with high porosity and appropriate interconnection pathway should allow the tissue to infiltrate and to fill the scaffold.

Following these considerations, completely "open structure" based scaffolds with no physical constraints (i.e., pore size, interconnection size, total porosity) might be modelled and developed in order to provide a time- and space-unlimited in vivo blood vessel invasion and neo-bone tissue ingrowth within the scaffold. Although highly porous bioceramics still represents a promising approach for generating osteoconductive grafts (Boyde et al. 1999, Chang et al. 2000, Gauthier et al. 1998, Karageorgiou and Kaplan 2005, Mankani et al. 2001), their small pore interconnection size and internal pores architecture represent a physical constraint for blood vessel size and in vivo bone ingrowth within pores cavities (Mastrogiacomo et al. 2006).

A possible alternative approach might be the development of granules based-grafts, where no physical limits are present and, therefore, the in vivo bone forming efficiency might be significantly higher (Fig. 15.5). Computer assisted design (CAD) and rapid prototyping (RP) techniques may also be used to generate intelligent scaffolds of defined architecture.

15.5 Clinical Applications in Orthopedics

Large bone defects still represent a major problem in orthopedics. Bone-repair treatments rely on two main approaches: the bone transport (Ilizarov technology) and the graft transplant (autologous and allogenic bone grafts and biomaterial grafts). Thus far, none of these strategies have proven to be always resolving.

Regeneration of skeletal tissues involves slow replacement of tissues with identical tissue. In contrast, repair is a more rapid process and is probably designed for survival. It involves the usual inflammatory cell cascade, followed by matrix deposition and then a remodeling process which attempts, in part, to regenerate damaged tissues in the adult. A now extensive knowledge base of both repair and regeneration in "skeletal tissues" has enabled the development of orthopedic tissue engineering.

Ideal skeletal reconstruction depends on regeneration of normal tissues that result from initiation of progenitor cell activity. However, knowledge of the origins and phenotypic characteristics of these progenitors and of the controlling factors governing bone formation and remodeling is still limited.

Bone Marrow Stromal Cells (BMSC) ability to form bone when locally transplanted in small animals was originally demonstrated in an ectopic model of bone formation by Friedenstein and co-workers (Friedenstein et al. 1966). Only several years later, this ability of BMSC has been transposed to preclinical models in large size animals to verify the feasibility of this cell therapy approach in larger scale lesions (Bruder et al. 1998, Kadiyala et al. 1997, Kon et al. 2000, Petite et al. 2000). In all these studies, the basic conceptual approach has been similar. Autologous marrow samples were harvested and the osteoprogenitors were isolated and expanded in culture. A critical size segmental defect was surgically created in a long bone. The surgical lesion was filled with biomaterials carrying autologous in vitro expanded osteogenic progenitors. Radiographic and histological analysis of the retrieved specimens revealed: (a) excellent integration of the host bone/implants, (b) amount of neo-formed bone significantly higher in the scaffolds loaded with osteoprogenitors than in acellular control grafts. The results of all these studies were in good agreement suggesting an important advantage in bone formation and, therefore, in the healing of the segmental defect when marrow derived osteoprogenitors were delivered together with the biomaterial scaffolds.

It is surprising that after the initial enthusiasm demonstrated by the flourishing of very encouraging large animal studies, only two pilot clinical studies have been performed (Quarto et al. 2001, Vacanti et al. 2001). In 2001, Quarto and co-workers have transposed the cell-based tissue engineering approach, verified in large animals, in the clinical setting to treat patients with substantial (4–7 cm) long bone defects. All patients recovered limb function between 6 and 15 months. Again in the 2001, Vacanti and co-workers have applied the same basic concept to regenerate a phalanx. Results in this case were less encouraging since only little bone neo-formation was observed with a consequent limited functional recovery.

15.6 Conclusions

In this chapter, we have described the combination of stem cells and biomaterials as a powerful potential therapeutic tool for the regeneration and repair of large skeletal lesions. Several experimental preclinical models applying this approach have provided promising results of high interest. In 2001, two encouraging although preliminary, clinical studies have been reported. What is puzzling, based on this premises, is the lack of a more extensive use in the orthopedic practice of this fascinating and promising approach. No other clinical reports have been described in the last five years. How can we explain that?

For sure the few clinical results existing, although interesting, are far from being optimal and several problems have still to be solved before considering cell therapy in orthopedics a real alternative. Major concerns in this approach can be pointed out in both cells and biomaterials, but problems are also represented by stringent guidelines, lack of internationally recognized standards, lack of proper quality controls, and perhaps a weak crosstalk between medical and basic sciences.

Stem cells represent indeed an important novel tool in regenerative medicine and surgery and they will surely find striking applications in the therapy of a variety of pathologies. Still we miss the full knowledge of their biology and the standardization level required for clinical applications.

Cell cultures should be highly standardized and culture media and each component used should be certified for this kind of use. Which cell population will perform better and which culture technology (traditional plastic, bioreactors, etc.) will provide us with the best cell preparation at the lowest cost and at the highest qualitative standards are the points that cell biologists and biotechnologists will have to solve in the next future. Convenient quality criteria and controls will have to be established.

Among the standardization criteria to be established, the effective number of MSC sufficient to provide a therapeutic effect represents a point of primary importance. The problem involves the heterogeneity in differentiation stages of most MSC preparations due to non-standardized culture conditions. The risk is to use a significantly higher number of cells whereas only a fraction might be the one needed. A consensus on the definition of surface markers identifying the cell phenotype responsible of the therapeutic effect may help solving this problem. Recently few surface markers have been reported which could identify MSC phenotypes, but a wide consensus on the validity of these markers has not been reached yet.

MSC have been reported to be responsible of exerting a long lasting suppressant activity on the immune system of the recipient organism upon engraftment. This controversial capability could have important applications in the treatment of immune-mediated disorders. On the other hand, it could also have negative consequences in terms of suppression of the antineoplastic surveillance of the immune system, thus potentially favoring the onset of neoplasias in the transplanted host (Djouad et al. 2003). Further studies are indeed needed to assess the real importance of this point. Still because of the potential consequences of this critical and previously unrecognised aspect, MSC use in therapy should be carefully evaluated.

Scaffold also will have to evolve to provide higher standards of efficiency and efficacy; possibly future biomaterials will need to have geometry specifically designed for hosting cell and allow their survival and a better vascularization once delivered at the implant site in vivo. Scaffolds will have to improve in order to have resorption rates adequate to the specific application and possibly to became at least at some extent more easily workable and allow some movement (flexion, torsion, compression, etc.).

Some issues remain at the forefront of the controversy involving stem cell research – legislation, ethics and public opinion, cost, and production methods (See Chapter 16). As is true with any new technology, the generated enthusiasm, having potential to influence virtually every orthopedic case management, must be balanced by subjecting it to stringent clinical and basic research investigations and the respect of the current guidelines. The scientific and clinical challenge remains: to perfect cell-based tissue-engineering protocols in order to utilize the body's own rejuvenation capabilities by managing surgical implantations of scaffolds, bioactive factors, and reparative cells to regenerate damaged or diseased skeletal tissues.

Questions/Exercises

1. Can the chemical composition of the scaffolds influence the cell adhesion mechanisms?
2. Which parameters may influence the vascularization process of the implanted graft?
3. Which scaffold parameter should be well controlled during its planning and development?
4. Which is the best porosity of engineered grafts able to guarantee an efficient in vivo bone formation?
5. Can suitable mechanical properties requirements influence the choice of the scaffold to be implanted, in terms of its chemical composition?
6. Which parameters may influence the in vivo degradation process of the implanted graft?
7. Can the internal architecture of the scaffolds influence the cell migration mechanisms and the in vivo matrix deposition kinetic?
8. Can different clinical settings (maxillo-facial, orthopedic surgery, etc.) influence the choice of the bone graft to be implanted?
9. Can bone defect size influence the choice of the bone graft to be implanted, in terms of its chemical composition and/or internal architecture?
10. Has the use of osteogenic stem cells still displayed some limitations for their wide use to regenerate skeletal tissues?

References

Aggarwal S, Pittenger MF (2005) Human mesenchymal stem cells modulate allogeneic immune cell responses. Blood 105:1815–22

Alison MR, Brittan M, Lovell MJ, Wright NA (2006) Markers of adult tissue-based stem cells. Handb Exp Pharmacol 174:185–227. Review

Banfi A, Muraglia A, Dozin B, Mastrogiacomo M, Cancedda R, Quarto R (2000) Proliferation kinetics and differentiation potential of ex vivo expanded human bone marrow stromal cells: Implications for their use in cell therapy. Exp Hematol 28:707–15

Banfi A, Bianchi G, Notaro R, Luzzatto L, Cancedda R, Quarto R (2002) Replicative aging and gene expression in long-term cultures of human bone marrow stromal cells. Tissue Eng 8:901–10

Banwart JC, Asher MA, Hassanein RS (1995) Iliac crest bone graft harvest donor site morbidity. A statistical evaluation. Spine 20:1055–60

Bianco P, Gehron Robey P (2000) Marrow stromal stem cells. J Clin Invest 105:1663–8

Boheler KR (2004) Functional markers and the "homogeneity" of human mesenchymal stem cells. J Physiol 554(Pt3):592

Bosch P, Musgrave DS, Lee JY, Cummins J, Shuler T, Ghivizzani TC, Evans T, Robbins TD, Huard (2000) Osteoprogenitor cells within skeletal muscle. J Orthop Res 18:933–44

Boyde A, Corsl A, Quarto R, Cancedda R, Bianco P (1999) Osteoconduction in large macroporous hydroxyapatite ceramic implants: evidence for a complementary integration and disintegration mechanism. Bone 24:579–89

Bruder SP, Kraus KH, Goldberg VM, Kadiyala S (1998) The effect of implants loaded with autologous mesenchymal stem cells on the healing of canine segmental bone defects. J Bone Joint Surg Am 80:985–96

Caterson EJ, Nesti LJ, Albert T, Danielson K, Tuan R (2001) Application of mesenchymal stem cells in the regeneration of musculoskeletal tissues. Medscape General. Medicine 3(1), E1. Review

Chacko DM, Das AV, Zhao X, James J, Bhattacharya S, Ahmad I (2003) Transplantation of ocular stem cells: the role of injury in incorporation and differentiation of grafted cells in the retina. Vision Res. 43:937–46

Chailakhan RK, Lalykina KS (1969) Spontaneous and induced differentiation of osseous tissue in a population of fibroblast-like cells obtained from long-term monolayer cultures of bone marrow and spleen. Dokl Akad Nauk SSSR. 187:473–5

Chang BS, Lee CK, Hong KS, Youn HJ, Ryu HS, Chung SS, Park KW (2000) Osteoconduction at porous hydroxyapatite with various pore configurations. Biomaterials. 21:1291–8

Collett GD, Canfield AE (2005) Angiogenesis and pericytes in the initiation of ectopic calcification. Circ Res. 96:930–8

De Oliveira JF, De Aguiar PF, Rossi AM, Soares GA (2003) Effect of process parameters on the characteristics of porous calcium phosphate ceramics for bone tissue scaffolds. Artif Organs. 27:406–11

Djouad F, Plence P, Bony C, Tropel P, Apparailly F, Sany J, Noel D, Jorgensen C (2003) Immunosuppressive effect of mesenchymal stem cells favors tumor growth in allogeneic animals. Blood 102:3837–44

Doherty MJ, Ashton BA, Walsh S, Beresford JN, Grant ME, Canfield AE (1998) Vascular pericytes express osteogenic potential in vitro and in vivo. J Bone Miner Res 13:828–38

Dong J, Kojima H, Uemura T, Kikuchi M, Tateishi T, Tanaka J (2001) In vivo evaluation of a novel porous hydroxyapatite to sustain osteogenesis of transplanted bone marrow-derived osteoblastic cells. J Biomed Mater Res 57:208–16

Dong J, Uemura T, Shirasaki Y, Tateishi T (2002) Promotion of bone formation using highly pure porous beta-TCP combined with bone marrow-derived osteoprogenitor cells. Biomaterials 23:4493–502

Dyce PW, Zhu H, Craig J, Li J (2004) Stem cells with multilineage potential derived from porcine skin. Biochem Biophys Res Commun 316:651–8

El-Ghannam A (2005) Bone reconstruction: from bioceramics to tissue engineering. Expert Rev Med Devices 2:87–101

Erbe EM, Marx JG, Clineff TD, Bellincampi LD (2001) Potential of an ultraporous beta-tricalcium phosphate synthetic cancellous bone void filler and bone marrow aspirate composite graft. Eur Spine J (10 Suppl) 2:S141–6

Flautre B, Anselme K, Delecourt C, Lu J, Hardouin P, Descamps M (1999) Histological aspects in bone regeneration of an association with porous hydroxyapatite and bone marrow cells. J Mater Sci Mater Med 10:811–4

Friedenstein AJ, Lalykina KS (1972) Thymus cells are inducible to osteogenesis. Eur J Immunol 2:602–3

Friedenstein AJ, Piatetzky S II, Petrakova KV (1966) Osteogenesis in transplants of bone marrow cells. J Embryol Exp Morphol 16:381–90

Garbossa D, Fontanella M, Fronda C, Benevello C, Muraca G, Ducati A, Vercelli A (2006) New strategies for repairing the injured spinal cord: the role of stem cells. Neurol Res 28:500–4

Gauthier O, Bouler JM, Aguado E, Pilet P, Daculsi G (1998) Macroporous biphasic calcium phosphate ceramics: influence of macropore diameter and macroporosity percentage on bone ingrowth. Biomaterials 19:133–9

Gauthier O, Muller R, von Stechow D, Lamy B, Weiss P, Bouler JM, Aguado E, Daculsi G (2005) In vivo bone regeneration with injectable calcium phosphate biomaterial: a three-dimensional micro-computed tomographic, biomechanical and SEM study. Biomaterials 26:5444–53

Gazdag AR, Lane JM, Glaser D, Forster RA (1995) Alternatives to autogenous bone graft: efficacy and indications. J Am Acad Orthop Surg 3:1–8

Ge Z, Goh JC, Lee EH (2005) The effects of bone marrow-derived mesenchymal stem cells and fascia wrap application to anterior cruciate ligament tissue engineering. Cell Transplant 14:763–73

Guasch G (2006) Epithelial stem cells in the skin. Med Sci (Paris) 22:710–2

Hokugo A, Sawada Y, Sugimoto K, Fukuda A, Mushimoto K, Morita S, Tabata Y (2006) Preparation of prefabricated vascularized bone graft with neoangiogenesis by combination of autologous tissue and biodegradable materials. Int J Oral Maxillofac Surg 35(11):1034–40

Huang JI, Beanes SR, Zhu M, Lorenz HP, Hedrick MH, Benhaim P (2002) Rat extramedullary adipose tissue as a source of osteochondrogenic progenitor cells. Plast Reconstr Surg 109:1033–41; discussion 1042–3

Javazon EH, Beggs KJ, Flake AW (2004) Mesenchymal stem cells: paradoxes of passaging. Exp Hematol 32:414–25.

Johnston, N (2004) Skin stem cells. Drug Discov Today 9:994

Jorgensen C, Noel D, Apparailly F, Sany J (2001) Stem cells for repair of cartilage and bone: the next challenge in osteoarthritis and rheumatoid arthritis. Ann Rheum Dis 60:305–9

Kadiyala S, Young RG, Thiede MA, Bruder SP (1997) Culture expanded canine mesenchymal stem cells possess osteochondrogenic potential in vivo and in vitro. Cell Transplant 6:125–34

Karageorgiou V, Kaplan D (2005) Porosity of 3D biomaterial scaffolds and osteogenesis. Biomaterials 26:5474–91

Kim DS, Cho HJ, Choi HR, Kwon SB, Park KC (2004) Isolation of human epidermal stem cells by adherence and the reconstruction of skin equivalents. Cell Mol Life Sci 61:2774–81

Klein S, Svendsen CN (2005) Stem cells in the injured spinal cord: reducing the pain and increasing the gain. Nat Neurosci 8:259–60

Kneser U, Schaefer DJ, Polykandriotis E, Horch RE (2006) Tissue engineering of bone: the reconstructive surgeon's point of view. J Cell Mol Med 10:7–19

Kon E, Muraglia A, Corsi A, Bianco P, Marcacci M, Martin I, Boyde A, Ruspantini I, Chistolini P, Rocca M, Giardino R, Cancedda R, Quarto R (2000) Autologous bone marrow stromal cells loaded onto porous hydroxyapatite ceramic accelerate bone repair in critical-size defects of sheep long bones. J Biomed Mater Res 49:328–37

Krampera M, Pizzolo G, Aprili G, Franchini M (2006) Mesenchymal stem cells for bone, cartilage, tendon and skeletal muscle repair. Bone 39:678–83

Kumta SM, Kendal N, Lee YL, Panozzo A, Leung PC, Chow TC (1997) Bacterial colonization of bone allografts related to increased interval between death and procurement: an experimental study in rats. Arch Orthop Trauma Surg 116:496–7

Kurz LT, Garfin SR, Booth RE Jr (1989) Harvesting autogenous iliac bone grafts. A review of complications and techniques. Spine 14:1324–31

Kuznetsov SA, Mankani MH, Gronthos S, Satomura K, Bianco P, Robey PG (2001) Circulating skeletal stem cells. J Cell Biol 153:1133–40

Le Blanc K (2003) Immunomodulatory effects of fetal and adult mesenchymal stem cells. Cytotherapy 5:485–9

Le Blanc K, Rasmusson I, Sundberg B, Gotherstrom C, Hassan M, Uzunel M, Ringden O (2004) Treatment of severe acute graft-versus-host disease with third party haploidentical mesenchymal stem cells. Lancet 363:1439–41

Leung VY, Chan D, Cheung KM (2006) Regeneration of intervertebral disc by mesenchymal stem cells: potentials, limitations, and future direction. Eur Spine J (15 Suppl) 15:406–13

Levy MM, Joyner CJ, Virdi AS, Reed A, Triffitt JT, Simpson AH, Kenwright J, Stein H, Francis MJ (2001) Osteoprogenitor cells of mature human skeletal muscle tissue: an in vitro study. Bone 29:317–22

Liu H, Kemeny DM, Heng BC, Ouyang HW, Melendez AJ, Cao T (2006) The immunogenicity and immunomodulatory function of osteogenic cells differentiated from mesenchymal stem cells. J Immunol 176:2864–71

Livingston T, Ducheyne P, Garino J (2002) In vivo evaluation of a bioactive scaffold for bone tissue engineering. J Biomed Mater Res 62:1–13

Livingston TL, Gordon S, Archambault M, Kadiyala S, McIntosh K, Smith A, Peter SJ (2003) Mesenchymal stem cells combined with biphasic calcium phosphate ceramics promote bone regeneration. J Mater Sci Mater Med 14:211–8

Lu JX, Gallur A, Flautre B, Anselme K, Descamps M, Thierry B, Hardouin P (1998) Comparative study of tissue reactions to calcium phosphate ceramics among cancellous, cortical, and medullar bone sites in rabbits. J Biomed Mater Res 42:357–67

Lu JX, Flautre B, Anselme K, Hardouin P, Gallur A, Descamps M, Thierry B (1999) Role of interconnections in porous bioceramics on bone recolonization in vitro and in vivo. J Mater Sci Mater Med 10:111–20

Luskey BD, Lim B, Apperley JF, Orkin SH, Williams DA (1990) Gene transfer into murine hematopoietic stem cells and bone marrow stromal cells. Ann N Y Acad Sci 612:398–406

Mankani MH, Kuznetsov SA, Fowler B, Kingman A, Robey PG (2001) In vivo bone formation by human bone marrow stromal cells: effect of carrier particle size and shape. Biotechnol Bioeng 72:96–107

Marcacci M, Kon E, Zaffagnini S, Giardino R, Rocca M, Corsi A, Benvenuti A, Bianco P, Quarto R, Martin I, Muraglia A, Cancedda R (1999) Reconstruction of extensive long-bone defects in sheep using porous hydroxyapatite sponges. Calcif Tissue Int 64:83–90

Marcacci M, Kon E, Zaffagnini S, Vascellari A, Neri MP, Iacono F (2003) New cell-based technologies in bone and cartilage tissue engineering. I. Bone reconstruction. Chir Organi Mov 88:33–42

Mastrogiacomo M, Scaglione S, Martinetti R, Dolcini L, Beltrame F, Cancedda R, Quarto R (2006) Role of scaffold internal structure on in vivo bone formation in macroporous calcium phosphate bioceramics. Biomaterials 27:3230–7

Movassaghi K, Ver Halen J, Ganchi P, Amin-Hanjani S, Mesa J, Yaremchuk MJ (2006) Cranioplasty with subcutaneously preserved autologous bone grafts. Plast Reconstr Surg 117:202–6

Muraglia A, Martin I, Cancedda R, Quarto R (1998) A nude mouse model for human bone formation in unloaded conditions. Bone 22:131S–134S

Navarro M, del Valle S, Martinez S, Zeppetelli S, Ambrosio L, Planell JA, Ginebra MP (2004) New macroporous calcium phosphate glass ceramic for guided bone regeneration. Biomaterials 25:4233–41

Norman-Taylor FH, Santori N, Villar RN (1997) The trouble with bone allograft. BMJ 315:498

Noth U, Schupp K, Heymer A, Kall S, Jakob F, Schutze N, Baumann B, Barthel T, Eulert J, Hendrich C (2005) Anterior cruciate ligament constructs fabricated from human mesenchymal stem cells in a collagen type I hydrogel. Cytotherapy 7:447–55

Olszewski WL (2004) Stem cells of the human skin epithelium–can they be isolated and resume function as single-cells transplanted into recipient skin defects? Ann Transplant. 9:34–6

Otteson DC, Hitchcock PF (2003) Stem cells in the teleost retina: persistent neurogenesis and injury-induced regeneration. Vision Res 43:927–36

Partridge K, Yang X, Clarke NM, Okubo Y, Bessho K, Sebald W, Howdle SM, Shakesheff KM, Oreffo RO (2002) Adenoviral BMP-2 gene transfer in mesenchymal stem cells: in vitro and in vivo bone formation on biodegradable polymer scaffolds. Biochem Biophys Res Commun 292:144–52

Petite H, Viateau V, Bensaid W, Meunier A, de Pollak C, Bourguignon M, Oudina K, Sedel L, Guillemin G (2000) Tissue-engineered bone regeneration. Nat Biotechnol 18:959–63

Quarto R, Mastrogiacomo M, Cancedda R, Kutepov SM, Mukhachev V, Lavroukov A, Kon E, Marcacci M (2001) Repair of large bone defects with the use of autologous bone marrow stromal cells. N Engl J Med 344:385–6

Rando TA (2006) Stem cells, ageing and the quest for immortality. Nature 441:1080–6

Rasmusson I (2006) Immune modulation by mesenchymal stem cells. Exp Cell Res 312:2169–79

Reh TA, Fischer (2001) Stem cells in the vertebrate retina. Brain Behav Evol 58:296–305

Sakai D, Mochida J, Iwashina T, Watanabe T, Nakai T, Ando K, Hotta T (2005) Differentiation of mesenchymal stem cells transplanted to a rabbit degenerative disc model: potential and limitations for stem cell therapy in disc regeneration. Spine 30:2379–87

Schantz JT, Hutmacher DW, Chim H, Ng KW, Lim TC, Teoh SH (2002) Induction of ectopic bone formation by using human periosteal cells in combination with a novel scaffold technology. Cell Transplant 11:125–38

Seiler JG 3rd, Johnson J (2000) Iliac crest autogenous bone grafting: donor site complications. J South Orthop Assoc 9:91–7

Sudres M, Norol F, Trenado A, Gregoire S, Charlotte F, Levacher B, Lataillade JJ, Bourin P, Holy X, Vernant JP, Klatzmann D, Cohen JL (2006) Bone marrow mesenchymal stem cells suppress lymphocyte proliferation in vitro but fail to prevent graft-versus-host disease in mice. J Immunol 176:7761–7

Sutherland AG, Raafat A, Yates P, Hutchison JD (1997) Infection associated with the use of allograft bone from the north east Scotland Bone Bank. J Hosp Infect 35:215–22

Sykova E, Jendelova P, Urdzikova L, Lesny P, Hejcl A (2006) Bone marrow stem cells and polymer hydrogels-two strategies for spinal cord injury repair. Cell Mol Neurobiol 26(7–8):1113–29

Vacanti CA, Bonassar LJ, Vacanti MP, Shufflebarger J (2001) Replacement of an avulsed phalanx with tissue-engineered bone. N Engl J Med 344:1511–4

Zarate-Kalfopulos B, Reyes-Sanchez A (2006) Bone grafts in orthopedic surgery. Cir Cir 74:217–22

Zuk PA, Zhu M, Mizuno H, Huang J, Futrell JW, Katz AJ, Benhaim P, Lorenz HP, Hedrick MH (2001) Multilineage cells from human adipose tissue: implications for cell-based therapies. Tissue Eng 7:211–28

Zuk PA, Zhu M, Ashjian P, De Ugarte DA, Huang JI, Mizuno H, Alfonso ZC, Fraser JK, Benhaim P, Hedrick MH (2002) Human adipose tissue is a source of multipotent stem cells. Mol Biol Cell 13:4279–95

Chapter 16
Conclusions: Towards High-Performance and Industrially Sustainable Tissue Engineering Products

Matteo Santin

Contents

16.1 Introduction 467
16.2 From Interfacial to Complete Tissue Regeneration 468
    16.2.1 Development of Non-Invasive Implantation Procedures 470
    16.2.2 Controlling and Exploiting the Inflammatory Response 473
    16.2.3 Biochemical Cues Analogues 476
    16.2.4 Biostructural Cues Analogues 481
16.3 Complete Tissue and Organ Regeneration 484
    16.3.1 Technical Issues in Tissue and Organ Regeneration 485
16.4 Regulatory Issues 487
    16.4.1 The American Federal Law............................... 487
    16.4.2 The EC Consultation Process and Regulation 488
16.5 Towards the Exploitation of Tissue Engineering Products 488
    16.5.1 Bioactive Biomaterials Market 489
    16.5.2 Stem Cells Market..................................... 489
16.6 Overall Conclusions 490
Acknowledgments... 490
References 490

16.1 Introduction

Since the early 1990s, the research in biomedical field has been dominated by a paradigm shift whereby the concept of replacing damaged tissues and organs with medical devices has been overcome by the goal of their partial or complete regeneration. Although traditional implants (i.e., orthopedic, dental, cardiovascular, and ocular) are still widely used to ameliorate and save patients' life, it is now possible to engineer materials, bioactive components, and cells able to stimulate the regeneration of tissues damaged by either disease or trauma (Sipe 2002). It has to be outlined that this paradigm shift neither emerged independently nor aims to supersede completely the traditional tissue replacement by

M. Santin (✉)
School of Pharmacy & Biomolecular Sciences, University of Brighton, Cockcroft
Building Lewes Road, Brighton BN2 4GJ, UK
e-mail: m.santin@brighton.ac.uk

M. Santin (ed.), *Strategies in Regenerative Medicine*,
DOI 10.1007/978-0-387-74660-9_16 © Springer Science+Business Media, LLC 2009

implants or by pharmaceutical therapies. Rather, lessons have been learnt from the development of traditional implants and drugs which can now be integrated in the development of new strategies for regenerative medicine.

This chapter summarizes the principal aspects of regenerative medicine which have been presented in the previous chapters and highlights common denominators which may lead to the consolidation of medical protocols and to standardization of regulations. Furthermore, the progresses made so far in this field and the future ambitions will be weighed against commercial sustainability and ethical issues.

16.2 From Interfacial to Complete Tissue Regeneration

In the majority of the cases, the failure of the traditional biomedical implants is caused by their lack of integration with the surrounding tissue (Anderson 2001). A typical example is the failure of metal implants for orthopedic (e.g., hip and knee implants) and dental applications (Steflik et al. 1998). At long term, the absence of a tight bonding between the implant metal surface and the bone mineral phase leads to mechanical instability and, as a consequence, to the fracture of the surrounding bone (Steflik et al. 1998). In these cases, instead of a mineralized bony tissue, a new soft tissue is formed at the interface which is mainly composed by collagen and proteglycans (Steflik et al. 1998). Similarly, in the case of soft tissues, natural and synthetic grafts are not always able to form a structural continuum with the surrounding tissue and they are often encapsulated by a fibrotic tissue which impedes their complete integration and functionality (Santin et al. 1996). It is widely accepted that these interposed tissues are formed as a consequence of the inflammatory response which is triggered by the traumatic character of the surgical intervention and it is protracted by the presence of the foreign body that is the implant (Anderson 2001). Since the early phases of implantation, inflammatory and tissue cells compete for adhering on the implant surface (Fig. 16.1 a and b); only a balanced activity of the different cell types will determine the formation of new physiological tissue and, therefore, the degree of integration of the implant (Fig. 16.2 a–d).

To avoid the foreign body response, implant surfaces have been modified to be recognized as "self" by the body. Aim of these modifications is facilitating surface colonization by tissue cells; ultimate goal is promoting physiological tissue regeneration at the interface and, as a consequence, implant integration. The physico-chemical properties and biocompatibility of conventional biomaterials presented in Chapters 2 and 3 as well as those of biomimetic biomaterials discussed in Chapter 4 clearly elucidate how the mimicking of components of the tissue ECM (either soft or mineralized) has significantly shifted the interest of scientists towards specific biorecognition. At molecular and cellular levels, four main factors are believed to favor interfacial tissue regeneration:

(a)   (b)

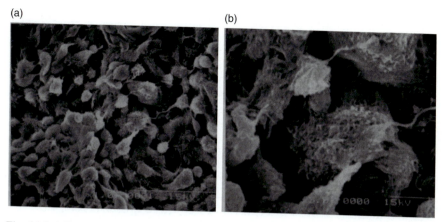

Fig. 16.1 Adhesion of different cell types on a PHEMA/gelatine interpenetrated polymer network hydrogel implanted subcutaneously in rats (modified from Santin M et al. 1996). (**a**) cell population and (**b**) cell morphology at the interface between PHEMA/gelatine implant and fibrotic capsule

(i) Preservation of the native conformation in proteins adsorbing the implant surface
(ii) Modulation of the inflammatory response to control tissue regeneration
(iii) Adequate adhesion of tissue cells to make them able to produce and deposit new ECM on the implant surface
(iv) Bonding between components of the newly formed tissue and the implant surface.

Indeed, it is known that the ability of cells to differentiate or to maintain their differentiated phenotype depends on their ability to interact with the surrounding environment through specific interactions established between bioligands (e.g., ECM structural proteins) and cell receptors (e.g., integrins). These interactions can be guaranteed only by the preservation of the bioligand structural conformation. Similarly, once deposited by the cells, the ECM components need to form chemical bonds with those of the tissue as well as with the biomaterial surface thus ensuring the structural continuum required for restoring and sustaining tissue functionality. Similarly, bone hydroxyapatite can be covalently bound to surfaces of implants exposing specific functional groups (e.g., phosphates, phosphoserine, Santin et al. 2006) and coatings (synthered hydroxyapatite, calcium phosphate, Bioglass®, Merolli et al. 2004a, b).

The development of tissue engineering constructs where specific cell types are seeded onto biomaterial scaffolds prior to implantation has enormously benefited from the progresses made in the field of surface engineering as applied to traditional biomedical implants. Molecularly tailored biomaterials for implant manufacture which are able to promote appropriate cell adhesion and functionality have been later adopted to design 3D tissue engineering scaffolds (see Chapter 4). This approach is particularly attractive in those applications where

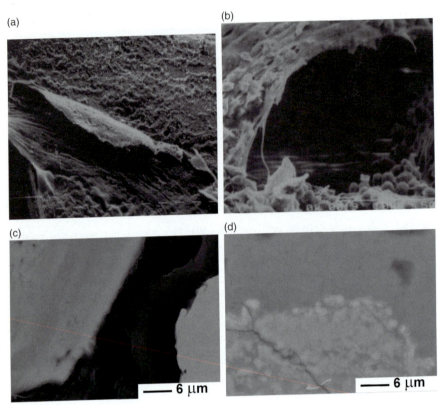

Fig. 16.2 Different degrees of integrations of implants within tissues. (**a**) tightly integrated fibrotic capsule surrounding a PHEMA/gelatine interpenetrated polymer network hydrogel implanted subcutaneously in rats (modified from Santin M et al. 1996), (**b**) loosely associated fibrotic capsule surrounding a PHEMA (no gelatine) hydrogel implanted subcutaneously in rats (modified from Santin M et al. 1996) (**c**) interfacial gap between a titanium implant and the surrounding bone (Merolli A et al. 2006), (**d**) complete osteointegration of a hydroxyapatite-coated titanium implant (Merolli A et al. 2006)

the formation of new tissue is pursued in vitro by the means of integrated systems such as the bioreactors (see Section 16.3.1.1). Contrariwise, in the case of tissue engineering constructs which are implanted without waiting for the in vitro consolidation of a mature tissue, the biomolecular cues and the cells used for the engineering of the construct can be significantly compromised by the harsh conditions of implantation (e.g., inflammatory response) and be prevented from fulfilling their function.

16.2.1 Development of Non-Invasive Implantation Procedures

Implantation of medical devices is often accompanied by invasive surgical procedures which lead to patients' discomfort and to longer recovery times.

Furthermore, it is widely accepted that, despite the biocompatibility of the implant, an early inflammatory response is always triggered by the surgical procedure. In those circumstances where surgery is particular invasive, this inflammatory response is exacerbated and the implant integration process may be significantly compromised. Therefore, alongside with more biocompatible biomaterials, scientists, and clinicians have been trying to develop non-invasive implantation procedures. Typical examples of strategies proposed to minimize surgical traumas include the use of: (i) injectable biomaterials and tissue engineering constructs, (ii) self-assembling biomacromolecules, (iii) nanostructured biomaterials. A large number of biomaterials belonging to these three classes have been developed and some of them have reached the market. Typical examples are provided in the following sections of this chapter.

16.2.1.1 Injectable Biomaterials

Injectable biomaterials mainly include polymers and composites comprising polymeric and ceramic materials. Injectable polymers have been developed which are able to preserve their stability over relatively long period of time or able to set at body temperature. Among the many polymers and composites so far developed, few have already reached the market (Zaffe et al. 2005). Injectable PLA/PGA-based hydrogels such as Fisiograft (Ghimas, Italy, www.ghimas.it) have been made available for periodontal applications. Polytetrafluoroethylene (PTFE) pastes have been commercialized under the registered names of Polytef® and Urethrin® and proposed as biomaterials for urinary tract tissue augmentation (Tsai et al. 2006, Zaffe et al. 2005). However, the biocompatibility of all these products still suffers of significant limitations. Contradictory findings have been published about the tissue regeneration potential of PLA/PGA-based hydrogels in bone applications. The early regeneration of bony tissue in bone defects treated with these biomaterials has been shown to be followed by adverse reaction to the polymer biodegradation by-products leading to bone resorption (Hedberg et al. 2005). PTFE-based pastes have been shown to induce a foreign body response, fibrotic tissue formation around them when implanted in the urinary tract, and even migration in other body districts (Tsai et al. 2006). In addition, the relative rigidity of the spheres composing these pastes makes their injectability difficult. For these reasons, FDA has not approved the use of Polytef and Urethrin for urological applications. Macroplastique® (www.uroplasty.com) is a biomaterial made of poly (dimethylsiloxane) elastomeric beads mixed with poly (vinylpirrolydone) gel for soft tissue regeneration. As for the other pastes, these biomaterials have also shown clear sign of foreign body reaction when used for the regeneration of soft tissues. Hyaluronic acid has also been provided as injectable biomaterials (Tsai et al. 2006). Several formulations of hyaluronic acid and its derivatives such has Hyal-System, Hyalo gyn, Hyaff have been commercialized by Fidia Advanced Biopolymers, Italy (www.fidiapharma.it) for several tissue regeneration applications including intra-dermal administration, vaginal bioequilibrant gel, and cell encapsulation for cartilage regeneration. The proven biocompatibility of the hyaluronic acid makes these products preferred by many clinicians. Alongside the many injectable collagen formulations for cosmetic purposes, Contigen® (Bard, www.bardurological.com), a glutaraldehyde-crosslinked collagen Type I has been used for urological applications and shown to induce a significant foreign body response. Among the composite biomaterials, Norian SRS is an injectable calcium phosphate cement commercialized by Synthes (www.synthes.com). Although its injectable properties and setting at body temperature make this cement very attractive to orthopedic surgeons, studies have reported painful soft tissue reactions in patients treated with this material (Welkerling et al. 2003).

16.2.1.2 Self-Assembling Biomolecules

Self-assembling peptides are among the most common biomolecules that, in recent years, have been considered for biomedical purposes (Bell et al. 2006). Peptides forming supra-molecular structures have been developed at the MIT to stop bleeding, while other beta sheets-forming peptides have been synthesized to form 3D fibrillar scaffolds with a bioactive surface capable of inducing re-mineralization of damaged or diseased dental enamel (Heller and Wei 2006, Kirckham et al. 2007). Many research groups have focused their investigation in synthesizing self-assembling peptides mimicking the ECM structure and their potential application in myocardial tissue regeneration has been shown (Ramachandran and Yu 2006). Among them, it is worth mentioning self assembling peptides such as the $-(RADA)_n$ (Horii et al. 2007) and the $-(KLD)_{12}$ (Kisiday et al. 2002) sequences to which many types of cell adhesion peptides have been grafted (See Chapter 4). Peptides have also been developed that can mimic the rheological properties of hyaluronan and their use as joint lubricants for osteoarthritis has been suggested (Bell et al. 2006). As for the injectable biomaterials, self-assembling composites have also been produced combining peptide scaffolds with hydroxyapatite powder (Ignjatovic et al. 2007).

16.2.1.3 Nanostructured Biomaterials

Nanostructured biomaterials have become the focus of many research groups worldwide and public and private investments have been made in recent years to support research in this field (Freitas 2005). Nanostructured biomaterials can be framed in the wider context of nanomedicine which has been defined as 'the process of diagnosing, treating and preventing disease and traumatic injury, relieving pain, and persevering and improving human health using molecular tools and molecular knowledge of the human body' (Freitas 2005). Although nanomedicine is a discipline that has been established only recently, its concept stems from the vision of the former Nobel laureate in physics, Richard P. Feynman. In the late 50s, Feynman postulated the manufacture of small machines which can be permanently implanted in the body to support the function of organs and that these machines could interact with the body at cellular level. Indeed, it is now widely recognized that cells respond to nanostructures and that their behavior can be finely tuned by changing the underlying substrate at nano-scale level. The use of nanostructured biomaterials such as nanoparticles/nanoshells, dendrimers, and semi-dendrimers has been proposed for the targeted and controlled delivery of drugs, growth factors, and genes. The interaction of these nanostructured biomaterials with cells has been improved by their functionalization with specific molecules facilitating biorecognition. More information about the synthesis of dendrimers and their potential to enhance tissue regeneration will be given in Section 16.2.4.

As for injectable biomaterials and self-assembling structures, nanostructured materials offer the opportunity of implantation of highly performing biomaterials by minimally invasive surgery. Their solubility in aqueous environment or the ease of suspension in physiological solutions allows their in situ implantation by significantly reducing the surgical procedure.

16.2.2 Controlling and Exploiting the Inflammatory Response

The widely accepted link between the implantation of a foreign material and the insurgence of a chronic inflammatory response has for many years misled biomaterial scientists in their efforts to produce highly performing biomaterials. For years, the inflammatory response and the consequent development of a granulation tissue has been considered as an adverse reaction to the material "tout court" and their role in tissue regeneration overlooked. As a consequence, the search for inert materials has been in the majority of the cases sterile and the same definition of inertness is controversial. As proteins and other biomolecules recognizable by the cells adsorb onto the surface of an implant during the early phases of implantation, the interaction of biological components with the implant is inevitable (Norde 1986). Moreover, as these molecules tend to loose their native conformation upon contact with the material surface, the triggering of a chronic inflammatory response is also difficult to rule out (Anderson 2001). As mentioned above, many research and development projects have been based on the wrong premise that the inflammatory response is an adverse effect; only recently it has been recognized that the inflammatory response is a key step of the tissue regeneration process (Martin and Leibovich 2005). Autocrine and paracrine biochemical signaling released by inflammatory cells such as the monocytes and their differentiated form the macrophages is fundamental to induce tissue regeneration. A recent review paper has emphasized that there are different types of inflammatory cell activation and some of them are fundamental to a physiological regeneration of a damaged tissue (Martin and Leibovich 2005). The presence of different sub-populations of monocytes/macrophages and their biomaterial-induced activation has also been demonstrated (Martin and Leibovich 2005). Evidences have emerged showing the different states of activation of inflammatory cells isolated from control and diabetic subjects when in contact with biomaterials; the cells from

Type 1 and Type 2 diabetes show a basal activity higher than those of control subjects which predispose them to biomaterial-induced activation (Harrison et al. 2007).

Studies have also demonstrated that monocytes/macrophages, or a hemo-progenitor cell sub-set that is included in the mononuclear cell population, can change their phenotype into fibroblasts or endothelial cells or osteoblasts when exposed to specific biochemical signaling or into myofibroblasts when adhering on stainless steel surfaces (Leu et al. 1988, Stewart et al. 2008). Although this transformation may contribute to the formation of the fibrotic capsule often surrounding implants, more accurate studies need to be performed to understand whether the phenotype plasticity of this type of cells can be exploited to control tissue formation.

16.2.2.1 Control of Macrophage Activation

Monocytes/macrophages, the cells with a key role in the inflammatory response, exist in our body as cells circulating in our vascular system or as resident in the tissues. These cells acquire an activated state when tissue trauma or foreign body invasion takes place (Anderson 2001). This activated state confers to the cell the ability of destroying, engulfing (phagocytosis, Fig. 16.3 a) or walling off (giant cell formation, Fig. 16.3 b) tissue fragments and invading foreign bodies (e.g., bacteria, viruses, biomedical implants, or any foreign body).

Later in the tissue repair process, macrophages secrete biochemical signaling (e.g., cytokines and growth factors) which are capable of regulating their own activity (autocrine signaling) and that of the tissue cells (paracrine signaling). The development of the next generation of biomaterials for medical implants and tissue engineering constructs will need to take into account such a dynamic character of the inflammatory response and make sure that interactions with

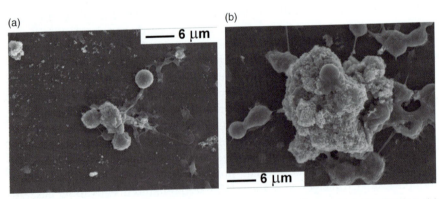

Fig. 16.3 Micrographs of monocytes/macrophages in their different state of activation. (**a**) cells engulfing metal nano-particles, (**b**) cells attacking a foreign body and fusing into giant cell to wall it off

the host tissue will be tuned with each phase of healing. Biomimicking structures and bioactive cues are, therefore, required to reduce the host response and to guide healing towards a physiological pathway.

Chapter 4 has discussed the advantages related to presenting cells with biomimetic biomaterials and surfaces and emphasized the importance of mimicking the physico-chemical and biochemical characteristics of the ECM. In the same chapter, it has been pointed out that in ECM biomacromolecules interact without loosing their native conformation (a problem taking place upon contact with implants) and cells activity is finely regulated to ensure either a regular tissue remodeling or, in the case of damage, its repair. It is widely accepted that the native conformation of the ECM structural components and the correct phenotype of cells is ensured by the hydrophilic character of the ECM of both soft tissues and cartilage (see Chapters 5, 12, and 13). As a consequence, the use of hydrogels has been privileged to design fillers and tissue engineering constructs. Although the biocompatibility of these hydrogels is satisfactory in many surgical applications, their ability to mimic biological structures relies on the integration of bioligands (See Chapter 4). The scale-up of these bioligands for industrial production is possible, but their relatively high costs and their relative instability arise problems of commercial sustainability and storage conditions. It can be speculated that the synthesis of new hydrogels may avoid or minimize the use of these bioligands. For example, recent studies show that, when compared to traditional hydrogels such as PHEMA and PVA, hydrogels exposing zwitterionic groups promote an organization of the water molecules which is more favorable to preserve protein conformation upon adsorption and to reduce inflammatory cell activation (Santin unpublished data). Hydrogels with zwitterionic character such as poly(phosphatidyl choline) have already been commercially exploited to produce contact lenses and coatings for cardiovascular stents (Patel et al. 2005). The rationale that led to the synthesis of these classes of polymers was the abundance of a phospholipid, the phosphatidylcholine in the membrane of blood red cells; this phospholipid renders red blood cells hemocompatible. However, the same physico-chemical properties reduce cell adhesion thus making this biomaterial not suitable for tissue regeneration applications. It is envisaged that the integration of biochemical cues (e.g., bioligands) within the structure of these types of hydrogels may lead to novel hydrogels closely mimicking the cell membrane or the ECM in their structures and biofunctionality.

16.2.2.2 Exploitation of the Host Response

The early exposure either of an implant or of a tissue engineering construct to the biological environment of a traumatized or diseased tissue may alter its properties, limit its integration, and impair its biofunctionality. In Section 16.2.2, this type of failure has been attributed to the host response. However, in the same section it has been outlined that the cell playing a central role in the host response, the macrophages, have also a plastic phenotype capable of transforming these cells into tissue cells. Knowledge about the causes leading to macrophage trans-differentiation is still in its infancy especially if related to its dependence from substrate characteristics (Stewart et al. 2008). The ability of both tissue cells and macrophages to recognize bioligands such as the –RGD- amino acid sequence present in several ECM structural proteins may orient the design of new substrates able to exploit macrophages as a source of tissue cells in situ. Rationalized studies are also required to link specific microenvironment characteristics to the differentiation of macrophages into their post-inflammatory phenotype capable of secreting anti-inflammatory growth factors rather than pro-inflammatory cytokines (Martin and Leibovich 2005).

16.2.3 *Biochemical Cues Analogues*

The induction of cell proliferation or differentiation in regenerative medicine is achieved by using two main classes of biochemical cues:

(i) Bioligands which are recognized by cell receptors
(ii) Growth factors

As pointed out in Section 16.2, biochemical cues are relatively expensive to be widely employed at commercial level and their relative instability at room temperature requires refrigerated storage, an undesired practice in hospital units. A typical example is the bone filler commercialized by Stryker, which is based on demineralized bone powder enriched by the bone morphogenetic protein-7 (BMP-7) also called osteopontin-1 (OP-1) http://www.stryker.com. In addition, bioactive molecules such as growth factors are purified from either animal or human sources or they are of recombinant origin with related risks of transmittable diseases, batch-to-batch variations and they require laborious purification procedures. For these reasons, several companies have been developing tissue culture media where the addition of animal serum or purified growth factors is not needed (Table 16.1).

In this section, typical examples of synthetic molecules able to act as biochemical cues for cells will be presented in the attempt of demonstrating that in the future new biochemical cues analogues will be exploited to lead to highly controlled, commercially sustainable, and clinically safe tissue engineering constructs.

16.2.3.1 Bioligands Analogues

A large number of papers have been published showing the role of the amino acid –RGD- sequence in recognizing cell membrane integrins and, consequently, favoring cell adhesion (See Chapter 4). This sequence can be

Table 16.1 Websites of companies producing specialized media and reagents, animal reagent-free tissue culture media, and relative equipment

Company	Website	Product
BD Biosciences	www.bdbiosciences.com	Hepatocytes culture media
HyClone	www.hyclone.com	Media preparation vessel
Irvine Scientific	www.irvinesci.com	Medium for CHO cell lines
Mediatech	www.cellgro.com	Low levels serum media for embryonic stem cells
Millipore	www.millipore.com	Serum-free media for embryonic mouse stem cells
Novozymes	www.novozymes.com	Recombinant human transferring and albumin
PAA	www.paa.com	Stem cell growth and differentiation media
PromoCell	www.promocell.com	Cell-type specific markers for cell isolation and purification
R&D Systems	www.rndsystems.com	Serum-free media specific for stem cells
StemCell Technologies	www.stemcell.com	Stem cell (hematopoietic) standardized media and reagents

synthesized in a relatively easy manner and several lateral domains have been proposed to ensure the best exposure of the relevant amino acids to the cell approaching the biomaterial surface (Guler et al. 2006). In Chapter 4, a list of alternative peptide sequences supporting cell adhesion is reported (Table 4.6).

These methods are mainly dependent on traditional solid-phase peptide protocols which are highly reproducible and lead to relatively pure products. A drawback to these efficient synthetic methods is the relatively high cost of the reagents (i.e., amino acid with blocked functional groups).

An alternative synthetic molecule has, therefore, been proposed which is relatively simple to synthesize and more economical; agmantine (Franchini et al. 2006). Agmantine has a molecular structure relatively similar to the –RGD- amino acid sequence and it has been shown to promote cell adhesion. More studies of this nature are required to synthesize analogues of other bioligands relevant to cell adhesion, migration and proliferation.

16.2.3.2 Growth Factors Analogues

As for bioligands, most growth factors or, more broadly, cell activity regulators are peptides. Although the native molecule of a typical growth factor is relatively large, several studies have identified within the molecule the sequences which are relevant to biointeractions and bioactivity thus paving the way towards the production of completely synthetic growth factor analogues (D'Andrea et al. 2006). Typical examples are the restricted sequences of the OP-1 and of the vascular endothelial growth factor (VEGF, Table 16.2).

Table 16.2 Typical growth factor analogue peptides with a potential use in cell culture and tissue engineering

Peptide Sequence/Name	Function	Reference
SSRS	Insulin Growth Factor-1 analogue/ corneal regeneration	Yamada et al. (2006)
RKIEIVRKK	Platelet Derived Growth Factor-BB analogue/angiogenesis	Brennand et al. (1997)
QPWLEQAYYSTF	Vascular Endothelial Growth Factor analogues/angiogenesis	Hardy et al. (2007)
YPHIDSLGHWRR		
LLADTTHHRPWT		
AHGTSTGVPWP		
VPWMEPAYQRFL		
TLPWLEESYWRP		
KLTWQELYQLKYKGI	Vascular Endothelial Growth Factor analogue/angiogenesis	D'Andrea et al. (2006)
NSVNSKIPKACCVPTELSAI	Bone Morphogenetic Protein-2 analogue/bone regeneration	Suzuki et al. (2000)

Table 16.3 Typical stem cell homing peptides with a potential use in cell culture and tissue engineering

Peptide Sequence/Name	Function	Reference
-GGSKPPGTSS-CONH$_2$	Bone marrow homing/ cell differentiation	Horii et al. (2007)
-GGPPSSTKT-CONH$_2$	Bone marrow homing/ cell differentiation	Horii et al. (2007)
fMLP	Prototypic N-formyl peptide/ stimulates cell adhesion to ECM proteins	Viswanathan et al. (2007)
CTCE-0214	SDF-1 peptide analogue/ survival of cord blood hemoprogenitor cells	Li K et al. (2006)

In addition to these analogues, it is worth mentioning other peptides which are involved in vivo in the homing of bone marrow and hemoprogenitor stem cells. These peptides can be exploited for cell culture and tissue engineering purposes (Table 16.3).

16.2.3.3 Drugs and Bioactive Natural Molecules

Pharmaceutical compounds which induce cell phenotypes and functions have been developed throughout the years for many types of therapies. For the needs of regenerative medicine applications, molecules with a known activity on cell proliferation or differentiation can be integrated in tissue engineering constructs to regulate the formation of new tissue in situ or in bioreactor systems. A plethora of molecules can be used which have been widely exploited by in vitro cell culturing. Typical examples are ascorbic acid and β-glycerophosphate which are used in combination to maintain the phenotype of osteoblasts in culture. Similarly, for stimulating the differentiation of stem cells into osteoblasts, dexamethasone is commonly added to the growth medium. For culturing vascular smooth muscle cells, the presence of insulin is indispensable. An analogue of the neurotransmitter γ-aminobutirric acid (GABA), the muscimol has been used to tether the surface of biomaterial surfaces by an avidin/biotin recognition system (Vu et al. 2005). The muscimol-tethered biomaterial surfaces can thus control the activity of post-synaptic neuron receptors.

The use of these compounds in tissue engineering constructs can support tissue regeneration in many clinical applications. The synthetic origin of these molecules also rules out problems related to transmittable diseases and minimizes batch-to-batch variations. However, although their dosage in vitro has been optimized, no rationalized study has yet been performed to establish the required amounts and delivery kinetics in 3D tissue engineering constructs. Protocols need to be made available where the optimal concentrations of relevant molecules per number of cell encapsulated in the scaffold will be established. However, this objective is not trivial if considered that the effect on the cells can still be affected by the delivery kinetics as determined by the drug diffusion throughout the scaffolds, cell spatial distribution, and co-culturing (or invasion) by different cell types.

Medicinal plants could also been exploited for regenerative medicine purposes. The study of the plant-derived bioactive compounds is a very old and successive discipline that has not been yet fully exploited in tissue engineering. One of the few options is the use of isoflavones. Isoflavones belong to the class of plant flavonoids, molecules with a well ascertained activity on immunocompetent and tissue cells (Middleton et al. 2000). These molecules are present in many plants, but their concentration and types are particularly rich in soybean. It is now known that isoflavones exert a differentiation role on osteoblasts by penetrating the cell membrane and interacting preferentially with the estrogen receptor β present on the nuclear membrane of the cell. The ligand/receptor complex inhibits the topoisomerase II, a nuclear enzyme responsible for cell replication (Shao et al. 1998). As a consequence, in a certain range of concentration, isoflavones such as daidzein and genistein inhibit cell proliferation, but they induce osteoblast differentiation measured as alkaline phosphatase activity (Morris et al. 2006). It is believed that the ability of these molecules to bind the estrogen receptor β is due by the similarity of their molecular structure with synthetic estrogens (Fig. 16.4).

Furthermore, it is believed that isoflavones inhibit immunocompetent cells through a different biochemical pathway. Studies have demonstrated that these molecules inhibit tyrosine kinase pathways by interacting with cell membrane receptors. Although these receptors have not been identified, it has been speculated that isoflavones can bind the TGF-β receptors. If proven, this finding could open an important role for the soybean isoflavones and, in general, for

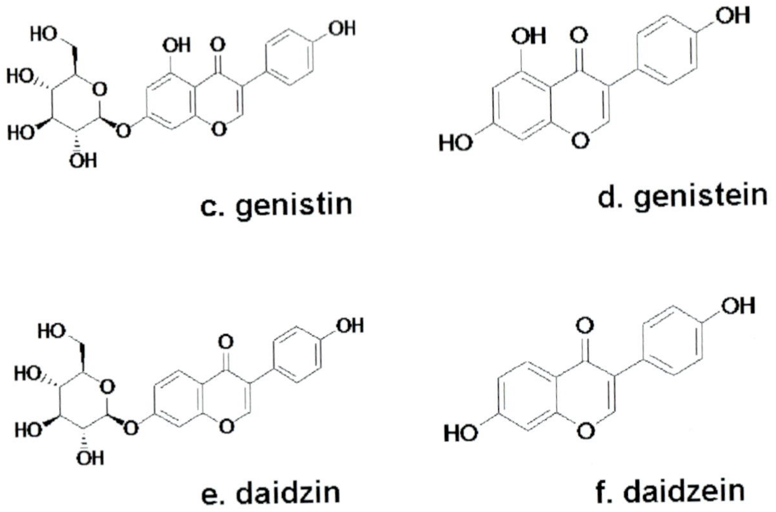

c. genistin   d. genistein

e. daidzin   f. daidzein

Fig. 16.4 Molecular structure of the soybean isoflavones daidzein (**a**) and genistein (**b**) and their glycosylated forms daidzin (**c**) and genistin (**d**)

plant flavonoids in controlling macrophage phenotype in regenerative medicine applications (see Section 16.2.2.1).

By binding this receptor isoflavones may compete with other growth factors important to wound. Urokinase-type plasminogen activator (uPA) regulates ECM turnover thus affecting wound healing and genistein has been found to have an inhibitory effect on epidermal growth factor-dependent uPA upregulation in human gingival fibroblasts (Smith et al. 2004). This isoflavone also downregulates hepatocyte growth factor expression (HGF) (Okada et al. 2000). HGF is a mesenchymal cell-derived pleiotropic cytokine and plays an important role in cell migration and proliferation (Forte et al. 2006). Genistein also upregulates matrix metalloproteinase-1 expression in human dermal fibroblasts thus favoring ECM degradation and cell migration during healing and tissue remodeling (Varani et al. 2004).

Although many clinicians consider granulation tissue formation favorable for split-thickness skin graft application (for details see Chapter 5), overexpression of fibrovascular tissue may cause scarring problems at later stages of wound healing. Therefore, control over granulation tissue formation via tyrosine kinase inhibition, as well as the anti-inflammatory effect of isoflavones (Middleton et al. 2000), may result in functionally and cosmetically more acceptable outcomes of deep burns treatment and, more broadly, to physiological tissue regeneration. Indeed, genistein has been used in vivo in combination with ointment to treat burns in rats showing reduced scar formation upon healing (Hwang et al. 2001). In the genistein-treated wounds, collagen deposition was more physiological and its packing typical of scar avoided.

In burns treatment, nitric oxide (NO) and inducible NO synthase (iNOS) may cause macrophage dysfunction after thermal injury (Luo et al. 2005).

Macrophages are known to produce cytokines such as transforming growth factor α, basic fibroblast growth factor, transforming growth factors $\beta 1$, $\beta 2$, and $\beta 2$, platelet-derived growth factor, vascular endothelial growth factor, and others which are important for wound healing. Indeed, genistein has also been found to inhibit macrophage activation (Middleton et al. 2000).

Additional advantages in using isoflavones are the possibility of synthesizing rather than extracting them from plants as well as the opportunity of exploiting their glycosylated forms (Fig. 16.4). Indeed, the glycosylated forms daidzin and genistin have no activity on cells (Zhang et al. 2007), but their sugar moiety is easily hydrolyzed in biological environments thus releasing the bioactive compound. This transformation of the glycosylated non-active molecules into non-glycosylated active ones can be exploited to prolong the effect of these molecules in vivo. The prolonged activity could also been ensured by the relatively hydrophobic character of these molecules which can establish van der Waals interactions with the biomaterial employed to make the scaffolds and, therefore, gradually released during implantation especially in those cases where the scaffold is biodegradable. Thus far, the effect of isoflavones on inflammatory and tissue cells has been exploited by the preparation of novel soybean-based biomaterials. As soybean is mainly composed of proteins (38%), carbohydrates (42%) and oil (18%) fractions, biodegradable biomaterials have been prepared by crosslinking defatted soybean curd or soybean extracts (Santin et al. 2002; 2007a, Santin and Ambrosio 2008). The crosslinking of these biodegradable biomaterials can be obtained either by thermosetting or by ionic or by covalent crosslinking of the protein and carbohydrate fractions (Santin et al. 2007). When tested in vitro on osteoblasts or implanted in vivo in rabbit bone, these biomaterials have shown to stimulate alkaline phosphatase activity and collagen production by the osteoblasts and to regenerate bone (Santin and Ambrosio 2008).

More recently, both the soybean-based biomaterials and purified isoflavones have shown to induce mesenchymal stem cell differentiation into osteoblasts with or without the addition of other osteogenic compound such as β-glycerophosphate and dexamethasone (Illsley et al. 2007). These findings are particularly relevant in the field of stem cell culturing standardization as they may pave the way towards approaches alternative to the use of growth factors and feeders cells of animal origin.

16.2.4 Biostructural Cues Analogues

In the previous sections of this book and, indeed of this chapter, it has been emphasized that the behavior of a cell is tightly dependent on the contacts it establishes with the surrounding environment and with other cells. In the natural ECM of both soft and hard tissues, cells are completely surrounded by structural components of protein and polysaccharide nature. These components are synthesized and gradually secreted by the cell itself which, therefore, provide its own right environment. The appropriate environment is established depending on the cell status and needs. Migrating cells produce a fluid ECM coat around themselves which is mainly based of polysaccharides such as hyaluronan (Sugahara et al. 2003). Static cells engulf themselves in a collagen Type I matrix which, in the case of bony and cartilage tissue, is gradually mineralized. Vascular SMC in their contractile phenotype produce proteins such as elastin that ensure mechanical pliability to the vessel wall (Sjolund et al. 1986). Epithelial and endothelial cells sit on a basement membrane which includes collagen Type IV, a collagen with a reticulated structure (LeBleu et al. 2007). The biofunctionality of these types of ECM is not only due to the biological properties of each single macromolecule, but it is also linked to their ability to assemble with other macromolecules of different kind to form a structural continuum with completely new properties (e.g., biomechanical properties). In other words, the ECM biofunctionality is achieved by a bottom-up approach where single biomacromolecules interact to form a new, more complex structure bearing new characteristics.

The advent of new disciplines such as the nanotechnology has opened new frontiers in the engineering of new synthetic macromolecules capable of mimicking at molecular level the structure of components of the ECM and bringing in their structures chemical groups favoring their assembly into ordered macrostructures. The combination of these nanostructured materials with the biochemical cues described above can result into unique microenvironments capable of tuning finely cell behaviour and, therefore, tissue regeneration. For example, dendrimers and semi-dendrimers have been synthesized by Michael addition reaction or by peptide synthesis methods (Hobson and Feast 1999). Dendrimers (from the ancient Greek δενδρον = tree) are hyperbranched polymers with a highly ordered nanostructure. These macromolecules can be synthesized by either liquid phase or solid phase synthesis to give either spherical star shaped nanostructures (dendrimers) or tree-like structures (semi-dendrimers, Fig. 16.5) of different branching generation (G_x) up to nine (G_9) .

Dendrimers have been used as drug and gene delivery systems as they can be internalized by the cells (Tang et al. 1996). Later their potential in other biomedical applications has been postulated and, recently, they have been used to increase the density of bioligands available to the cell and to functionalize the surface of biomaterials to improve their tissue integration (Monaghan 2001, Santin et al. 2007). In these approaches, the last branching generation of either the dendrimers or the semi-dendrimers has been decorated with specific amino acids or peptides able to promote biointeractions. Similarly, sugar-based or sugar-decorated dendrimers and semi-dendrimers can be used to expose other moieties important in biorecognition (Dubber and Lindhorst 2001). Bioligands recognized by cell integrins as well as phosphoserine groups able to facilitate the formation of a bone-like mineral phase have been used to make these hyperbranched polymers useful in regenerative medicine (Monaghan 2001, Santin et al. 2007). Functionalities can also be introduced in these polymers to favor their assembly into reticulated macrostructures bearing features similar to those of

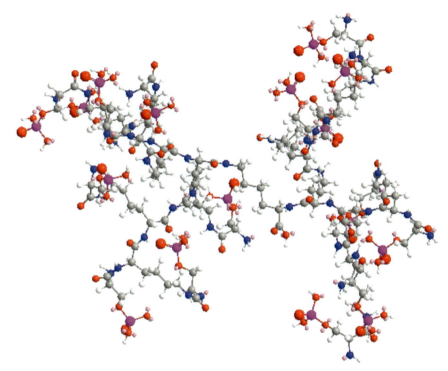

Fig. 16.5 Typical molecular structures of a semi-dendrimer with a phosphoserine decoration able to stimulate calcium phosphate crystal formation

specific ECM types (e.g., basement membrane). Together with self-assembling peptides (See Chapters 3 and 4), dendrimers, and semi-dendrimers may be integrated in macroporous scaffolds to provide to the cells a true 3D environment. This would be a major step forward if considered that the majority of the scaffolds for tissue engineering have been designed with macroporosity to ensure molecular diffusion, cell migration and, ultimately, tissue ingrowth which are not recognized as 3D by the cells (Fig. 16.6).

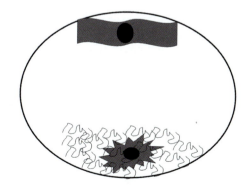

Fig. 16.6 Schematic representation of a scaffold macroporous where cells can recognize the wall as a 2D substrate (*top*) or encapsulated into a true 3D environment due to the presence of self-assembling nanomaterials (*bottom*)

After an injury, in the attempt to regenerate, axons have to penetrate a complex 3D environment made of components of ECM as well as of inflammatory and tissue cells. Microgrooves have been shown to promote axon directional growth in nerve cells and that once synapses are established the axon becomes suspended and thus independent from the microgrooves surface (Goldner et al. 2006). It has been hypothesized that neurons can produce forces which override the effect of the microgrooved substrate. It can be envisaged that nanostructures can be engineered capable of triggering the axon regeneration process while minimizing the presence of a foreign material at the site of injury. The positive effect of insoluble analogues of the ECM structure as well as of polyamide filaments on axon growth may play in favor of such approach (Yannas and Hill 2004).

Artificial materials able to mimic the structure of the ECM surrounding hepatocytes in livers and β-cells islets in the pancreas may help to control the difficult objective of preserving the phenotype and functionality of these types of cells. Similarly, the study of the different histological areas where stem cells can be found may lead to the development of new biostructure analogues capable of preserving the potential of this cell to differentiate or to direct them towards a specific phenotype.

16.3 Complete Tissue and Organ Regeneration

The development of new biocompetent biomaterials for implant manufacturing has paved the way towards complete tissue and organ regeneration. This book has clearly illustrated the state-of-the-art in the various research fields and clinical applications and has highlighted the opportunity provided by the combined use of these biomaterials when implanted as such or as carriers in cell therapy. The isolation of cells from animal (xenogenic) or human (allogenic) donors or from the same patient (autogenic) and their combination with biomaterials aim at reproducing the biological events routinely occurring in the human body to preserve its physiology. Tissue engineering aims at mimicking at a relatively smaller scale the process of division and differentiation that hundreds of trillions of cells undergo at any time in the human body and their ability to form new tissues. Through tissue engineering a possibility is offered to clinicians to implant differentiated (tissue-competent) cells or toti-/pluri-potent stem cells (from embryos or adult donors) into either the damaged or diseased tissue or other optimal therapeutic locations (e.g., immuno-privileged sites, highly vascularized tissues) with the goal of in situ tissue regeneration. Alternatively, the tissue engineering constructs can be incubated in bioreactor systems under specific conditions (e.g., temperature, pH, mechanical stimuli) until a mature tissue is formed and available for implantation. Both strategies offer advantages and suffer of limitations. The encapsulation of therapeutic cells with injectable biomaterials offers the chance of a minimally invasive surgery as described in the Section 16.2.1, while facing the challenge of regenerating large tissue damaged in presence of an inflammatory response or diseased conditions. In the majority of the applications, tissue regeneration in bioreactor systems is expected to be associated to conventional surgical procedures with a relatively high degree of invasiveness. However, the integration of the implanted new tissue or organ can be favored by an *ad hoc* surgical protocol and by an inflammatory reaction mainly circumscribed to the interface between the surrounding tissue and the tissue engineering construct. For both approaches, the use of cells has been so far limited by technical and ethical issues which need to be solved to satisfy the regulatory and commercial needs linked to the use of tissue engineering products.

16.3.1 Technical Issues in Tissue and Organ Regeneration

The in-depth analysis of the technical issues affecting the optimization and standardization of future tissue engineering products is out of the scope of this chapter and the reader is encouraged to read three excellent review papers by Metcalfe and Ferguson (2007). Here, it is worth giving an overview of these issues and show their tight links.

16.3.1.1 Biomaterials and Bioreactors

Although conventional polymeric and ceramic biomaterials have been widely proposed as suitable scaffolds and carriers for cell-based therapy, it is widely accepted that a fine modulation of the cell functions will tightly depend on the formulation of new biomimetic, bioresponsive biomaterials. In addition, in the case of tissue and organ regeneration in vitro, the future tissue engineering product will rely upon the availability of appropriate bioreactors. Therefore, a big and coordinated effort among cell biologists, material scientists, and engineers is required to reach the holy grail of complete tissue and organ regeneration.

Technical Issues Associated to the Use of Biomimetic and Bioresponsive Biomaterials

From an analysis of the many biomimetic and bioresponsive biomaterials presented in this book and of the specific clinical applications, it emerges that specific formulations will need to be synthesized and engineered for each clinical application. The relative instability associated to most bioresponsive and biomimetic biomaterials will require comprehensive studies showing their functionality after sterilization and at different storage conditions. Furthermore, rationalized studies are necessary to make sure that their formulation will be tuned to the needs of the cell types that they are destined to carry.

Technical Issues Associated to the Use of Bioreactors

A number of bioreactors have been developed where cells can be cultured in conditions resembling the physiological conditions. In particular, in recognition of the effect induced by the mechanical stimuli on tissue regeneration, efforts have been paid to optimize bioreactors able to generate compression or peristaltic movements and to transmit them to the growing cells. It is now proven that cells sense these stimuli and are directed in their morphological, proliferative, and differentiation status. Issues are still to be solved about the optimal stimuli to be given to the cells under different circumstances and how these have to vary according to the rate of degradation of the scaffold and to new tissue deposition. Dynamic chambers have been developed by research groups and companies in the attempt of addressing these issues, but so far no system is able to integrate all the required functions and parameters. Among them, it is worth mentioning the dynamic system developed by a Bose Electroforce (http://www.bose-electroforce.com) that is made available in different configurations able to simulate the mechanical stimulation of the main tissues (i.e., bone, cartilage, and vascular system). The size and configuration of most of dynamic bioreactors are still not suitable for the on-line monitoring of the cells (e.g., microscopy). Although ports have been made available to access the culture chamber, no complete specification has been optimized to control all the main cell culturing parameters. It is envisaged that sensors and biosensors will need to be included in the systems to sense changes of pH, temperatures, products of cell metabolism.

Technical Issues Associated to the Use of Cells

Cell therapy is widely recognized as the most promising approach in regenerative medicine. This book has provided specific examples of tissue regeneration where the use of stem/progenitor cells and differentiated cells/tissues is shown. There are clinical protocols where autologous cells are harvested from the patient during the surgery and immediately re-implanted without requiring any manipulation. Although this type of approach has been shown to be successful in several types of surgery, there are cases where their number needs to be expanded in culture without loss of their characteristics. In particular, when stem cells are cultured they need to:

- Preserve a stable phenotype overtime with no loss of pluri- or multi-potency in the case of stem cells
- Differentiate into the desired tissue when tissue regeneration is obtained in vitro (e.g., in bioreactor systems)
- Avoid any contamination with potentially hazardous, animal-derived factors, or other cell types

Technical hurdles are still to be overcome to make sure that these requirements will be satisfied. For example, it is known that when in contact with culturing substrates, stem cells may loose their phenotype. The same is true for chondrocytes the phenotype of which is very difficult to preserve in culture. Bone marrow mesenchymal stem cells which are cultured to produce osteoblasts always yield a percentage of undesired fibroblast-like cells. The cells also need to be manipulated in an animal-free and well chemically defined culture conditions which can reduce the risks of transmittable diseases and batch-to-batch variations. Research still needs to be conducted to identify growth factors specific for human embryonic stem cells and to control their differentiation into specific differentiated phenotypes. The need of mouse feeder cells in supporting stem cell culture is also cause of concerns as cells fusion may lead to genotypic alteration thus arising serious safety and ethical issues. Another challenge is represented by their validation in animal models. To provide evidences of safety, efficacy of the stem cells clinically reflective animals will be required. Finally, recognition of cell types as therapeutic-grade lines will need to be achieved through their manipulation in specialized GMP facilities by qualified technical staff.

16.4 Regulatory Issues

In the past, the clinical use and commercial exploitation of tissue engineering suffer of a regulatory vacuum especially in relation with the use of cells. As a consequence, in recent years, both the European Commission and the FDA have been trying to regulate the field.

16.4.1 The American Federal Law

Under the American federal law, a human medical product can be classified as drug, biologic, device or combination product. The American Food and Drug Administration (FDA) regulate tissue engineering, regenerative medicine, and cell therapy products mainly as combination products. FDA oversees this regulation through three centers dedicated to human medicinal products: the Centre of Drug Evaluation and Research, the Centre for Biologics Evaluation and Research and the Center for Device and Radiological Health. The jurisdiction of the product evaluation is given to any of these centers depending on the product primary mode of action. Since combination products have more than one mode of action, the most important one determines the allocation to any of the three centers. Therefore, a tissue engineering product will be examined as biologics if its mode of action falls into the definition of biologic product that is 'a virus, therapeutic serum, toxin, antitoxin, vaccine, blood, blood component or derivative, allergenic product or analogous product applicable to the prevention, treatment, or cure of diseases or injuries of man' [42USC262(a)]. These products require an approval following a Biological Licence Application that demonstrates the safety and effectiveness of the product. Similarly, the drug definition applies when the product is ' an article intended for use in the diagnosis, cure, mitigation, treatment, or prevention of disease in humans or other animals, and an article (other than food) and other articles intended to affect the structure or any function of the body of humans or other animals' [21USC321(g)]. In those cases, where the approval as device is applicable (i.e., biomaterial implants), a Pre-Market Application approval has to be sought. A device is defined as 'an instrument, apparatus, implant, in vitro reagent or other similar or related article which is intended for use in diagnosis of disease or conditions, or in the cure, mitigation, treatment or prevention of disease, in humans or other animals, or intended to affect the structure or any function of the body . . . and which does not achieve any of its principal intended purposes through chemical action within or on the body . . . , and which is not dependent on being metabolized for the achievement of any of its principal intended purposes' [21USC201(h)]. Alternatively, a pre-market notification (510 k) can be obtained by demonstrating properties equivalent to a pre-existing approved device.

This regulatory framework is complemented by the Current Good Practice Final Rule that was published in 2004 and that became effective since 2005.

16.4.2 The EC Consultation Process and Regulation

Within the EC, a consultation process has been initiated since 2005. Gene therapy, somatic cell therapy, and tissue engineering products are already considered by the EC as advanced therapies as they are based on complex and highly innovative manufacturing processes where the properties of biological components (gene and cells) are somehow modified to the purpose of the clinical application. Gene and somatic cell therapy products have been already classified as medicinal products and, as such, regulated by the EC. Contrariwise, tissue engineering products have for long time fallen outside any legislative framework thus resulting in the adoption of different policies by the Member States. There have been a series of consultation instigated by the EC in the attempt to address this problem and to find a legislative framework for tissue engineering. As a result, a decision has been reached whereby tissue engineering products have been regulated as medicinal products.

16.5 Towards the Exploitation of Tissue Engineering Products

While the technical and regulatory hurdles will be addressed, public and private investments in multi-disciplinary science, engineering and medical research, and development projects need to be strategically planned and implemented to support the commercialization of clinically performing regenerative medicine products. A mentality switch is required by both research bodies and institutions which will need to base their research strategies and projects on sound market analysis and demonstrated clinical needs. Product development plans need to be shaped to fulfil the needs for financial support, the regulatory requirements and the reimbursement policies. For example, taking into account the clinicians' view and needs in the specific medical applications and how these are linked to the market projections as well as understanding the product regulatory requirement and the reimbursement rates will allow companies to draw their commercialization strategy. Hereinafter, a brief market analysis overview is provided for two of the sectors more relevant to regenerative medicine and tissue engineering: the bioactive biomaterials market and the stem cell market.

16.5.1 Bioactive Biomaterials Market

Global Industry Analysts, Inc has estimated that the worldwide market for biocompatible materials was valued at US$28.7 billion in 2006 and it is forecast to be worth US$31.1 billion in 2007. The market is further projected to rise to US$ 39.1 billion in 2010, corresponding to a compound annual growth rate (CAGR) of 7.8% over the period 2000–2010. North America leads the global market with revenues estimated at US$11.9 billion in 2006, and projected to be US$12.8 billion in 2007. Japan is the fastest growing market its growth projected at an annual rate of 8.15% for the period 2000–2010, it is expected to reach US$4.8 billion by 2010.

Among the three main classes of materials (polymers, metals, and ceramics) for biomedical implant manufacturing, polymeric materials represent the largest segment of the global biocompatible materials market with an estimated share of 85.6% in 2006. In the same year, the revenues from this segment were estimated at US$ 24.6 billion and they are projected to reach US$26.7 billion by the end of 2007.

Some of the largest manufacturers of polymers are trying to develop new bioactive and biodegradable biomaterials, among them DuPont, Dow Chemical and BASF. Key drivers of innovation and development in this sector is the goal of significantly reducing patients'morbidity, recovery time, and overall costs for the healthcare system.

16.5.2 Stem Cells Market

Navigant Consulting has estimated that in 2005 the worldwide overall market for cell culture has generated revenues in excess of US$ 1billion. The predicted annual growth rate is of 12.7% and revenues should reach US41.8 billion by 2010. Even more optimistic figures have been provided by analysts at the Business Communications Company who have estimated revenues of US$1.7 by 2008. The market is currently dominated by firms such as Invitrogen (34%),

Fisher-Scientific (20%), JRH Bioscience (14%), Serological (9%), Cambrex (7%), and Sigma-Aldrich (6%). Invitrogen is also the main supplier of stem cell culture products by including in their catalogue more than 1200 products. Recent acquisitions have been made such as the one of JRH Bioscience by Sigma-Aldrich for about US$ 342.5 million. These companies have been attracted by the stem cell research market that in 2005 was worth US$820 million. This market has been estimated by Navigant Consulting to have a CAGR of 8.97% to reach US$1.95 in 2015.

A significant public funding has also been made available in America and Europe. For example, in the US the state of California has allocated $3 billion to stem cell research over the next 10 years. EU has allocated a budget of €51 billion for the period 2007–2013.

16.6 Overall Conclusions

The pages of this book have shown how the vision and the scientific rigor of many scientists and clinicians have led to pioneering and multi-disciplinary projects in the field of regenerative medicine and tissue engineering. We are now at a turning point in the history of medicine where the role of cell-based therapy will progressively gain a prevalent role. As it happened in the 20th century for the research in biomedical implant, funding and regulatory bodies, as well as biomedical companies need to put their trust in tissue engineering and work together to develop a regulatory and financial framework where new discoveries and products can be achieved.

It is hoped that, by reviewing the scientific knowledge and clinical targets so far achieved in regenerative medicine, this book may give a humble contribution to the experts in the field and it may catch the interest and enthusiasm of young scientists from various disciplines. It is with deep gratitude that we should collect the great inheritance left to us by so many valid scientists and clinicians and it is with great excitement that we should look forward to the work of the new generations.

Acknowledgments The author is grateful to Mr. Mike Helias for its scanning electron microscopy technical support and to Dr. Steve Meikle for providing the chemical structures.

References

Anderson JM (2001) Biological response to materials. Ann Rev Mater Res 31:81–110
Bell CJ, Carrick LM, Katta J, Jin ZM, Ingham E, Aggeli A, Boden N, Waigh TA, Fisher J (2006) Self-assembling peptides as injectable lubricants for osteoarthritis. J Biomed Mater Res Part A 78A:236–246
Brennand DM, Dennehy U, Ellis V, Scully MF, Tripathi P, Kakkar VV, Patel G (1997) Identification of a cyclic peptide inhibitor of platelet-derived growth factor-BB

receptor-binding and mitogen-induced DNA synthesis in human fibroblasts. FEBS Letters 413:70–74

D'Andrea LD, Del Gatto A, Pedone C, Benedetti E (2006) Peptide-based molecules in angiogenesis. Chem Biol Drug Design 67:115–126

Dubber M, Lindhorst TK (2001) Trehalose-based octopus glycosides for the synthesis of carbohydrate-centered PAMAM Dendrimers and thiourea-bridged glycoclusters. Organic Letters 3:4019–4022

Forte G, Minieri M, Cossa P, Antenucci D, Sala M, Gnocchi V, Fiaccavento R, Carotenuto F, De Vito P, Baldini PM, Prat M, Di Nardo P (2006) Hepatocyte growth factor effects on mesenchymal stem cells: proliferation, migration and differentiation. Stem Cells 24:23–33

Franchini J, Ranucci E, Ferruti P (2006) Synthesis, physicochemical properties, and preliminary biological characterizations of a novel amphoteric agmatine-based poly(amidoamine) with RGD-like repeating units Biomacromolecules 7:1215–1222

Freitas RA Jr, (2005) What is nanomedicine? Nanomedicine 1:2–9

Goldner JS, Bruder JM, Li G, Gazzola D, Hoffman-Kim D (2006) Neurite bridging across micropatterned grooves. Biomaterials 27:460–472

Guler MO, Hsu L, Soukasene S, Harrington DA, Hulvat JF, Stupp SI. (2006) Presentation of RGDS epitopes on self-assembled nanofibers of branched peptide amphiphiles. Biomacromolecules 7:1855–1863

Hardy B, Raiter A, Weiss C, Kaplan B, Tenenbaum A, Battler A (2007) Angiogenesis induced by novel peptides selected from a phage display library by screening human vascular endothelial cells under different physiological conditions. Peptides 28:691–701

Harrison M, Siddiq A, Guildford A, Bone A, Santin M (2007) Stent material surface and glucose activate mononuclear cells of control, type 1 and type 2 diabetes subjects. J Biomed Mater Res Part A 83A:52–57

Hedberg EL, Kroese-Deutman HC, Shih CK, Crowther RS, Carney DH, Mikos AG, Jansen JA (2005) In vivo degradation of porous poly(propylene fumarate)/poly(DL-lactic-co-glycolic acid) composite scaffolds. Biomaterials 26:4616–4623

Heller M, Wei C (2006) Self-assembly peptide prevents blood loss Nanomedicine 2:216

Hobson LJ, Feast WJ (1999) Poly(amidoamine) hyperbranched systems: synthesis, structure and characterization. Polymer 40:1279–1297

Horii A, Wang X, Gelain F, Zhang S (2007) Biological designer self-assembling peptide nanofibre scaffolds significantly enhance osteoblast proliferation, differentiation and 3-D migration. PLoS ONE 2(2):E190 doi:10.1371/journal.pone.0000190

Hwang K, Chung RS, Schmitt JM, Buck D, Winn SR, Hollinger JO (2001) The effect of topical genistein on soft tissue wound healing in rats. J Histotechnol.24:95–99

Ignjatovic NL, Liu CZ, Czernuszka JT, Uskoković DP (2007) Micro- and nano-injectable composite biomaterials containing calcium phosphate coated with poly(DL-lactide-co-glycolide). Acta Biomaterialia 3:927–935

Ikada Y (2007) Challenges in tissue engineering. J Roy Soc Interface 4:589–601

Illsley M, James E, James S, Santin M (2007) The dose-dependant increase in alkaline phosphatase activation in response to the phytoestrogen genistein in cultured multipotent mesenchymal stromal cells. Proceedings of the 21st European Society for Biomaterials Conference.

Kirckham J, Firth A, Vernals D, Boden N, Robinson C, Shore RC, Brookes SJ, Aggeli A (2007) Self-assembling peptide scaffolds promote enamel remineralization. J Dent Res 86:426–430

Kisiday J, Jin M, Kurz B, Hung H, Semino C, Zhang S, Grodzinsky AJ (2002) Self-assembling peptide hydrogel fosters chondrocyte extracellular matrix production and cell division: implications for cartilage tissue repair. Proc Nat Ac Sci 99:9996–10001

LeBleu VS, MacDonald B, Kalluri R (2007) Structure and function of basement membranes. Exp Biol Med 232:1121–1127

Leu HJ, Feigl W, Susani M, Odermatt B (1988) Differentiation of mononuclear blood cells into macrophages, fibroblasts and endothelial cells in thrombus organization. Exp Cell Biol 56:201–210

Li K, Chuen CKY, Lee SM, Law P, Fok TF, Ng PC, Li CK, Wong D, Merzouk A, Salari H, Gu GJS, Yuen PMP (2006) Small peptide analogue of SDF-1 alpha supports survival of cord blood CD34(+) cells in synergy with other cytokines and enhances their ex vivo expansion and engraftment into nonobese diabetic/severe combined immunodeficient mice. Stem Cells 24:55–64

Luo GX, Peng DZ, Zheng JS, Chen XW, Wu J, Elster E, Takadi D (2005) The role of NO in macrophage dysfunction at early stage after burn injury. Burns 31:138–144

Martin P, Leibovich SJ. (2005) Inflammatory cells during wound, repair: the good, the bad and the ugly. Trends Cell Biol 15:599–607

Merolli A, Gabbi C, Santin M, Locardi B, Tranquilli Leali P (2004a) Bioactive glass coatings of Ti6Al4V Promote the tight apposition of newly-formed bone in-vivo. Key Eng Mater 254:789–792

Merolli A, Gabbi C, Santin M, Locardi B, Tranquilli Leali P (2004b) Bioactive glass coatings on Ti6Al4V promote the tight apposition of newly-formed bone in-vivo. Key Eng Mater 254–256:789–792

Merolli A, Giannotta L, Tranquilli Leali P, Lloyd AW, Denyer SP, Rhys-Williams W, Love WG, Gabbi C, Cacchioli A, Bosetti M, Cannas M, Santin M (2006) In vivo assessment of the osteointegrative potential of phospatidylserine-based coatings. J Mater Sci Mater Med 17:789–794

Metcalfe AD, Ferguson MWJ (2007) Tissue engineering of replacement skin: the crossroad of biomaterials, wound healing, embryonic development, stem cells and regeneration. J Roy Soc Interface 4:413–437

Middleton E, Kandaswami C, Theoharides TC (2000) The effect of plant flavonoids on mammalian cells: Implications for inflammation, heart disease, and cancer. Pharmacol Rev 52:673–751

Monaghan S. (2001) Solid-phase synthesis of peptide-dendrimer conjugates for an investigation of integrin binding. ARKIVOC 46–53

Morris C, Thorpe J, Ambrosio L, Santin M (2006) The soybean isoflavone genistein induces differentiation of MG63 human osteosarcoma osteoblasts. J Nutr 136:1–5

Norde W (1986) Adsorption of proteins from solution at the solid-liquid interface. Adv Colloid Interface Sci 25:267–340

Okada M, Hoio Y, Ikeda U, Takahashia M, Takizawa T, Morishita R, Shimada K (2000) Interaction between monocytes and vascular smooth muscle cells induces expression of hepatocyte growth factor. J Hypert 18:1825–1831

Patel JD, Iwasaki Y, Ishihara K, Anderson JM (2005) Phospholipid polymer surfaces reduce bacteria and leukocyte adhesion under dynamic flow conditions. J Biomed Mater Res Part A 73A:359–366

Ramachandran S, Yu YB (2006) Peptide-based viscoelastic matrices for drug delivery and tissue repair. Biodrugs 20:263–269

Santin M, Ambrosio L (2008) Soybean-based biomaterials: preparation methods, physico-chemical properties and tissue regeneration potential. Exp Rev Med Dev, 53: 349–358.

Santin M, Huang SJ, Iannace S, Nicolais L, Ambrosio L, Peluso G (1996) Synthesis and characterization of poly(2-hydroxyethylmethacrylate)-gelatin interpenetrating polymer network. Biomaterials, 17:1459–1467

Santin M, Morris C, Standen G, Nicolais L, Ambrosio L (2007a) A new class of bioactive and biodegradable soybean-based bone fillers. Biomacromol 8:2706–2711

Santin M, Nicolais L, Ambrosio L (2002) Soybean-based thermoplastics as biomaterials. Patent n. WO/2002/010261

Santin M, Oliver G, Standen G, Meikle S, Lloyd A (2007b) Solid Surface Functionalisation by Bi-Functional Semi-Dendrimers GB patent application

Santin M, Rhys-Williams W, O'Reilly J, Davies MC, Shakesheff K, Love WG, Lloyd AW, Denyer SP (2006) Calcium-binding phospholipids as a coating material for implant osteointegration. J R Soc Interface 3:277–281

Shao ZM, Alpaugh ML, Fontana JA, Barsky SH (1998) Genistein inhibits proliferation similarly in estrogen receptor-positive and negative human breast carcinoma cell lines characterized by P21(WAF1/CIP1) induction, G(2)/M arrest, and apoptosis. J Cell Biochem 69:44–54

Sipe JD (2002) Tissue engineering and reparative medicine. Ann NY Acad Sci 961:1–9

Sjolund M, Madsen K, Vondermark K, Thyberg J (1986) Phenotype modulation in primary cultures of smooth-muscle cells from rat aorta – synthesis of collagen and elastin. Differentiation 32:172–180

Smith PC, Santibanez JF, Morales JP, Martinez J (2004) Epidermal growth factor stimulates urokinase-type plasminogen activator expression in human gingival fibroblasts. Possible modulation by genistein and curcumin. J Periodont Res 39:380–387

Steflik DE, Corpe RS, Lake FT, Young RT Sisk AL, Parr GR, Hanes PJ, Berkery DJ (1998) Ultrastructural analyses of the attachment (bonding) zone between bone and implanted biomaterials. J Biomed Mater Res 39:611–620

Stewart H, Guildford AL, Lawrenson-Watt D, Santin M (2008) Substrate-induced phenotypical change of monocytes/macrophages into myofibroblast-like cells: a new insight into the mechanism of in-stent restenosis. J Biomed Mater Res, doi 10.1002/jbm.a.32100

Sugahara KN, Murai T, Nishinakamura H, Kawashima H, Saya H, Miyasaka M (2003) Hyaluronan oligosaccharides induce CD44 cleavage and promote cell migration in CD44-expressing tumor cells. J Biol Chem 278:32259–32265

Suzuki Y, Tanihara M, Suzuki K, Saitou A, Sufan W, Nishimura Y (2000) Alginate hydrogel linked with synthetic oligopeptide derived from BMP-2 allows ectopic osteoinduction in vivo. J Biomed Mater Res 50:405–409

Tang MX, Redemann CT, Szoka FC (1996) In vitro gene delivery by degraded polyamidoamine dendrimers. Bioconjug Chem 7:703–714

Tsai CC, Lin V, Tang L (2006) Injectable biomaterials for incontinence and vesico-ureteral reflux: current status and future promise. J Biomed Mater Res Part B Appl Biomater 77B:171–178

Varani J, Kelley EA, Perone P, Lateef H (2004) Retinoid-induced epidermal hyperplasia in human skin organ culture: inhibition with soy extract and soy isoflavones. Exp Mol Pathol 77:176–183

Viswanathan A, Painter RG, Lanson NA, Wang GS (2007) Functional expression of N-formyl peptide receptors in human bone marrow-derived mesenchymal stem cells. Stem Cells 25:1263–1269

Vu TQ, Chowdhury S, Muni NJ, Qian HH, Standaert RF, Pepperberg DR (2005) Activation of membrane receptors by a neurotransmitter conjugate designed for surface attachment. Biomaterials 26:1895–1903

Welkerling H, Raith J, Kastner N, Marschall C, Windhager R (2003) Painful soft-tissue reaction to injectable Norian SRS calcium phosphate cement after curettage of enchondromas J Bone Joint Surg 85B:238–239.

Yamada N, Yanai R, Kawamoto K, Takashi Nagano T, Nakamura M, Inui M, Nishida T (2006) Promotion of corneal epithelial wound healing by a Tetrapeptide (SSSR) derived from IGF-1. Invest Ophthalmol Vis Sci 47:3286–3292

Yannas IV (2007) Similarities and differences between induced organ regeneration in adults and early foetal regeneration. J Roy Soc Interface 4:403–417

Yannas IV, Hill BJ (2004) Selection of biomaterials for peripheral nerve regeneration using data from the nerve chamber model. Biomaterials 25:1593–1600

Zaffe D, Leghissa GC, Pradelli J, Botticelli AR (2005) Histological study on sinus lift grafting by Fisiograft and Bio-Oss. J Mater Scie Mater Med 16:789–793

Zhang EJ, Kim NG, Luo KQ (2007) Extraction and purification of isoflavones from soybeans and characterization of their estrogenic activities. J Agr Food Chem 55:6940–6950

Index

A

Abrasion arthroplasty, 378
Achilles' tendon, 19
Adhesion peptide sequences in biomimetic
    materials, 123–124
Adipogenesis, 205–206
    See also Angiogenesis
Aggrecan, 369–370
Agmantine, 477
Albumin usage in synthetic grafts, 44
Alginate as natural biodegradable scaffolds,
    66–67
Alloderm, 41
    See also Skin
American federal law for human medical
    product, 487–488
Angiogenesis, 164, 190–191
    acetylated low-density lipoprotein
        (AcLDL) in, 220
    activators and inhibitors, 196
    in adult life, 191–193
    cellular therapy, 218
    molecular regulators of, 195
        angiopoietins and Tie receptors,
            199–200
        hypoxia-inducible factor 1, 198–199
        platelet derived growth factor
            (PDGF) family, 200–201
        vascular endothelial growth factor
            (VEGF), 197–198
    in regeneration, 216–219
        cell-based pro-angiogenic therapy,
            220–222
        growth factors based pro-angiogenic
            therapy, 219–220
    steps during, 194
    and tissue development
        adipose tissue, 204–206
        liver, 202–203

    nervous systems, 203–204
    ocular neovascularization, 213
    pancreas, 202
    in pathological conditions, 206–214
    psoriasis, 213
    in tumors, 207–212
    tools for study, 214–216
        chamber models, 215–216
        cornea pocket assay, 215–216
        matrigel tube formation assay, 215
    VEGF therapy in, 219
Angiopoietins (Ang) and Tie receptors,
    199–200
Animal vertebra compression assay, 86
Anterior cruciate ligament (ACL) of knee
    joint, 34–35
    biomechanics of, 36
    ruptures and, 37–38
Apligraf, 40
    See also Skin
Arteries
    composition and structure, 25–27
    mechanical properties of, 27–28
Articular cartilage (AC), 29, 368–371,
    396–403
    artificial biomaterial scaffolds
        improvement, 422–423
    autologous chondrocyte implantation
        (ACI), 408–412
        incremental cost effectiveness ratio
            (ICER), 420
        surgical procedure and, 409–412
    cartilage injuries, treatment options,
        405–407
    cartilage oligomeric matrix protein
        (COMP), 370
    cell-matrix assembly in, 31
    charged groups on, 32
    composition, collagen types, 369–371

Articular cartilage (AC) (*cont.*)
    and compressive stresses, 31
    coupled phenomena, 33
    damage and factors, 373
    defects and, 375–377
    development, 368–369
    electromechanical behavior of, 32
    hyaline response to injury, 403–405
    localized replacement, osteochondral and
        autograft, 406
    as macrocontinuum of molecules, 32–33
    mathematical models for, 33
    mesenchymal stem cells (MSCs), 379
    operative strategies for, 377–379
    osteoarthritis (OA), therapies
        arthritis, 372
        pathology, 372–375
        pathophysiology of, 374
        pharmacological therapies, 375
        risk factors for, 373–374
    patients, 423
    proteoglycans, 30, 369–371
    and repair, 396–403
        clinicians' view on treatments for,
            424–426
    repair and regeneration
        AC defects and spontaneous repair,
            375–377
    small-size animal models of, 421
    structure, 371
    surgical, age limit in, 424
    tissue engineering, 381–382
        experimental approaches for, 382–386
    in vitro chondrocyte de-differentiation
        limit, 421–422
Atherosclerosis, 26, 213–214
Atmospheric plasma spraying, 70
Autografts, 35
    conjunctival limbal autograft
        (CLAU), 358
    cultured epithelial autografts (CEAs),
        183–185
    ostechondral autograft implantation
        procedure, 406
Autologous chondrocyte implantation
        (ACI), 408–412
    arthroscopy and histology results, 419
    biological/artificial biomaterial scaffolds,
        422–423
    chondrocytes de-differentiate, 421
    clinical outcome, 422
    costs and health status, 420
    free-cell technology, 422

International cartilage repair society visual
        cartilage assessment scale, 417
    knee scoring systems, 423
    MR imaging and, 418
    MSC and, 421
    patient demographics and scores, 418
    post-ACI biopsies, 413
    stiffness of neo-repairs, 416
    surgical procedure and technique, 409–412
    weakness of, 412

B
Basement membrane (BM), 157
Basic fibroblast growth factor (bFGF), 139
Biglycan, 369–370
Biobrane, 40
    See also Skin
Biocarb, 35
    See also Carbon fibers based prostheses
Biodegradable polymers
    synthetic
        poly (ε-caprolactone) PCL, 63
        polyethilene glycol-based hydrogels
            (PEG) with RGD sequences, 64
        poly (lactic acid) (PLA) and poly
            (glycolic acid) (PGA), 61–62
        polypropylene fumarate (PPF), 63
Bioglass®, 82
Biomaterials, 98–99
    American society for biomaterials, 5
    bioligands analogues, 476–477
    degrees of integrations of implants within
        tissues, 470
    European society for biomaterials, 6
    injectable biomaterials, 471–472
    isoflavones, plant flavonoids, 479–480
    mechanical properties of instructive, 131
    muscimol-tethered biomaterial
        surfaces, 479
    nanostructured biomaterials, 472–473
    roles as artificial cellular matrices, 110
    self-assembling
        nanomaterials, 483
        peptides, 472
    and stem cell, 451
        bioceramics models used as
            osteoconductive grafts, 459
    technical issues, 485–487
    worldwide market for, 489
Biomimetic materials
    biomimetic rationale and design
        principles, 99–102

biomimetic tissue regeneration
matrices and tissue conduction,
110–111
strategies for, 109–115
biomimicking biomaterials science
bioactive biomaterials, synthesis of, 5
pioneers of, 4–6
cell adhesion domains, 122–131
conjugation of cell-binding motifs,
125–126
nano-fiber materials, 127
parameters and, 128–131
reactive macromers, 127–128
enzymatically degradability, 133–136
growth factor activity
conjugation, 139–141
encapsulation of, 137–139
heparin and heparin-binding
incorporation, 141–143
incorporation and, 137
release profiles, 143
hybrid and composite materials, 118
natural biopolymers, 115–117
starting materials, 115
substrate mechanics, 131–133
synthesis
conjugation methods, 119
free radical polymerization, 120
macromers and, 119
molecular self-assembly, 122
physical association, 121–122
in situ material synthesis, 120
step-growth polymerization, 120–121
synthetic polymers, 117–118
technical issues, 485–487
Bioreactor systems, 284
Bio-stable materials
hydroxyapatite (HAP), natural, 69–71
bioactive ceramics, 71
bioceramic coatings, 70
synthetic materials
metallic, 72–75
polymers, 75–77
Bisphenol A-glycidylmethacrylate (Bis-
GMA), 76–77
Blastocysts, pre-implantation embryos, 346
Blood precursor cells, 191
Blood vessel
and blood vasculature, 195
tumor vessels, 210
normal morphology of, 208
permeability of, 211
structure, 201–202

Bone, 431–446
angiogenesis and, 443
bone morphogenetic protein (BMP)
BMP-2 derived peptide
sequence, 140
ectopic bone formation, 439
expression and localization, 439
for growth, 10
positive human osteosarcomas, 444
recombinant human BMP-2
(rhBMP-2), delivery system, 441
signal transduction, 441
tissue engineering by, 441–442
transforming growth factor-βs and,
438–441
bruising, 403
development and, 435–436
engineering, 56–60
cell sourcing, 57–58
growth factors (GFs), 58–59
scaffolds, 59–60
formation in pathologic conditions
bone tumors, 443–444
myositis ossificans and fibrodysplasia
ossificans progressiva (FOP),
444–445
spinal ligaments, ossification
of, 445
fracture healing process, 437
function and structure, 56
macroscopic and microscopic structure
of, 432–433
osteoprogenitor cells, 433
pathological ossification, 443–445
regeneration and, 436–438
remodeling process of, 434–435
repair biodegradable scaffolds
bioactive glass ceramics, 64–65
bio-stable materials, 69–77
natural polymers, 66–67, 69
synthetic materials, 60–64
Bone tissue engineering, 449–461
biomimetic scaffolds and, 456–459
clinical problem, 452–453
orthopedic therapy
clinical applications in, 459–460
stem cells in, 453–456

C
Cadaveric allograft, 182
Cadaveric keratolimbal allograft
(KLAL), 358

Calcium phosphates as cell scaffolds, 64–65
    calcium phosphate glasses (CPG), 81
    injectable as cement for bone
        regeneration, 71
    molecular structures of crystal
        formation, 483
Carbodiimide (EDC) and succinimide
    (sulfo-NHS), 126
Carbon fibers based prostheses, 35
Cardiovascular tissue engineering centre, 8
Carticell®, 10
Cartilage replacement
    cartilage
        composition and structure, 28–31
        mechanical properties, 31–34
        joint resurfacing and biological
            autograft, 45
        tissue engineered constructs
            agarose/chondrocyte constructs, 47
            Pluronic F-127, 46–47
            poly glycolic acid (PGA), 48–49
            porous scaffold, 49
        total joint replacement, 46
CEA- seeded Integra device, 41–42
    clinical outcome, 184
    patient treated with, 185
    See also Skin
Cells
    bidirectional feedback between
        environment and, 107
    cell-binding motifs
        conjugation, 125–126
        parameters and, 128–131
        reactive macromers containing,
            127–128
        self-assembling nanofibers, 127
    cellular adhesion, 123–124
        PHEMA/gelatine interpenetrated
            polymer network hydrogel, 469
    enzymatic degradability, 135–136
    fusion, 297–299
    and growth factors, 112
    interconversion/transdifferentiation,
        273–274
    local physical and chemical
        environment, 103
    migration and tissue healing, 108
    mitotic cells, categories of, 344
    NeoHep cells, 305–306
    pentoxyresorufin-O-dealkylase activity
        (PROD) by phenobarbital,
        induction of, 303
    sourcing and preservation, 114

technical issues associated to use of,
    486–487
therapy
    concept of, 450
    legal aspects of, 285
    and tissue neogenesis, 113–115
    in vitro cell expansion, 181
Central nervous system (CNS) and blood
    brain barrier (BBB), 195
Chamber models, 215–216
Characterization methods
    adsorbed protein amount, 86
    dynamic and static contact angles
        measurement, 85
    induced coupled plasma-mass
        spectroscopy (ICP-MS) analysis, 86
    mechanical stimulus, 85
    porosity and microstructure, 83–84
    statistical study, 86
    surface properties, 84
    Z potential and IP, 85
Chemotaxis, 38–39
Chitosan as natural biodegradable scaffolds,
    66–69
Chondroadherin, 370
Chondrogenesis, 443
Cicatricial pemphigoid, 358
Cincinnati knee rating system, 423
Coated scaffolds, 82
Collagen
    collagen-cement composites, 81
    fibers in ECM, 104
    matrix, 327
    type I collagen as natural biodegradable
        scaffolds, 66
    See also Periphral nerve system (PNS)
College of Medicine of University of Illinois
    Urbana Champaign, 26
Comb-like graft co-polymers, 130
Composite materials
    3D scaffolds preparation methods, 77–78
    porous cement 3D scaffold, 80
    surface modification methods, 79
Connective tissues, extracellular matrix
    (ECM), 16
Contigen®, 472
Cornea pocket assay, 215
Cre recombinase (cre)-based reporter
    bone marrow cells, 298
    expressing recipients, 299
    Harris and Alvarez-Dolado
        experiments, 299
    systems for cell fusion, 297

Crigler-Najjar Syndrome Type I, 289–290
Cultured skin substitutes (CSS), 42
    See also skin
Curis Inc, 89
Cysteine capped adhesion peptides, 135

D
Dacron© grafts, 43
Decorin, 369–370
Delta-like ligand 4 (Dll4)/Notch signaling
        and angiogenesis, 193
Dendrimers as drug and gene delivery
        systems, 482
Dermis, 21–22, 157–158
    cellular component of, 158–159
Detonation-gun spraying, 70
Diabetes mellitus, 262
    β-cell
        death mechanism, 265–269
        dysfunction in, 264–265
        GLP-1 and dipeptidyl peptidase IV
            (DPP-IV) inhibitors, 268–269
        thiazolidinediones (TZDs), 267–268
    type 1 and type 2, 262
Diacerein, 375
Dip coating, 70
Disease-modifying OA drugs
        (DMOADs), 375

E
Elastic cartilage, 29
    See also Cartilage
μ-Emulsion routes, 70
    See also Bio-stable materials
Endothelial cell
    endothelialization, 44
    specialization, 193–195
Epidermis, 21–22, 157
    See also Skin
Epiphyseal dysplasia, 373
Epiphyses, 432
    See also Bone
Epithelialization, 164
Extra cellular matrix (ECM)
    artificial
        formation, materials and methods, 110
        starting components of, 115
    biofunctionality of, 482
    cell-mediated release, 113
    components and structure, 157–158
    degradation and tissue dynamics, 108

glycoproteins of, 158
proteoglycans and hyaluronic acid
        (HA), 158
Eye, 342
    age related macular degeneration
        (ARMD), 359
    aniridia, 351
    corneal limbal stem cell (LSC) deficiency
        and blindness, 352
    cytokines and MMP activation, tissue
        damage, 355–356
    embryonic stem cells (ES), 345
    epithelium and retina, 344
    morphogenesis and *Pax6* gene, 350
    ocular stem cells, 350–351
        corneal epithelial stem cells, 352
        endothelial and retinal, 353
        in new blood vessels, 353–354
    ocular tissue regeneration, 357–360
        ex vivo limbal cell expansion, 359
        keratolimbal transplantation, 358–359
        retina regeneration of, 359–360
    retinal stem cell (RSC) and, 360
    stem cells
        adult and umbilical cord blood, 347
        amniotic membrane and, 359
        disorders of, 354
        embryonic, blastocysts, 345–346
        fate of, 347–348
        genes controlling embryonic stem
            cells, 350
        genetic regulations of, 349–352
        hierarchy of, 345
        limbal stem cells, 352
        markers of, 349
        metaplasia and transdifferentiation,
            351–352
        micro-environment and, 348
        ocular regeneration with, 357–360
        severe deficiency and chronic
            inflammation, 349

F
Fat tissue vascularity, 204–206
Fibrin
    degradation, 141
    gels, 143
    sealants as natural biodegradable
        scaffolds, 66
Fibrocartilge, 29
Fibrodysplasia ossificans progressiva (FOP),
        444–445

Fibromodulin, 369–370
Fibronectin as biomaterial surfaces, 76,
        105, 370
Fisiograft, injectable PLA/PGA-based
        hydrogels, 471
Frit enameling, 70
Fusogenic cell therapy, 295

G
Genistein and ECM degradation, 480–481
Genzyme, 9–10
Glass polyalkenoate cements, 75
GLP-1 and dipeptidyl peptidase IV
        (DPP-IV) inhibitors, 268–269
Glucose homeostasis, 263
Glycogen storage disease type 1a, 290
Glycosaminoglycans (GAGs), 103
    in ECM, 104–105
    glucosamine sulfate and chondroitin
        sulfate, 375
Goat skin sample, cyclic loading and
        unloading on, 25
Gore Tex ligament, 35
    microporous PTFE fibers, 36
Grafting method and ligand
        bioavailability, 132
Growth differentiation factor-5 (GDF-5),
        bone morphogenetic protein
        (BMP) family, 369
Growth factors
    analogues, 477–478
    and cellular induction, 111–113

H
Haversian system, *see* Osteon
Hedgehog (Hh) pathway and bone
        morphogenic proteins
        (BMPs), 89
Hemangioblast, 191
Hematopoietic stem cells (HSCs)
    hematopoiesis and angiogenesis, 191
Hemochromatosis, 295
Hemostasis, 163–164
Heparin
    and heparin-binding incorporation,
        141–143
    usage in synthetic grafts, 44
Hepatocyte
    2-acetylaminofluorene (2-AAF) and
        proliferation, 294
    acute liver failure, 289

Cre/loxP recombinase-based reporter
        system, 298
    drug development, metabolism and
        toxicology, 291–292
    FAH$^{-/-}$mouse model, 296–297
    human hepatocyte specific antigen, 301
    isoflavone and, 480
    isolation and culture, 285–288
    isolation process, perfusion buffers used
        in, 288
    LacZ reporter gene and, 297
    markers of differentiation and cell
        sources, 309–311
    metabolic capacity of stem cell derived
        cells with, 300
    2-(2-nitro-4-trifluoro-methylbenzyol)-1,
        3-cyclohexanedione (NTBC) for
        deterioration, 294
    RNA/protein analysis, 308
    stem cell technology, 292–312
        cell fusion and, 295–296
        extrahepatic cells, transdifferentiation
            of, 294–295
        extrahepatic human stem cells,
            302–306
        fusogenic cells in mouse bone marrow,
            296–297
        human stem and precursor cells in
            livers of mice, 300–302
        quantitative analyses, 306–308
        real differentiation, 297–299
    transplantation, 289–291
    urokinase-type plasminogen activator
        (uPA)/recombinant activation
        gene-2 (RAG-2) mice, study, 300
Hot-isostatic pressing, 70
Human liver
    acute liver failure, 289
    cell therapy of end-stage disease, 290
    hepatocyte-like cells and, 293
    human albumin positive cells and, 301
    inherited factor VII deficiency and
        hepatocyte transplantation, 290
    phase II metabolism, 291
    working with resected liver tissue,
        286–287
    xenobiotics and environmental
        pollutions, 291
Hyaff commercial biopolymers, 471
Hyaline cartilage, 29
    See also Cartilage
Hyalo gyn and Hyal-system commercial
        biopolymers, 471

Hyaluronate as natural biodegradable
    scaffolds, 67
Hyaluronic acid-collagen, 81–82
Hydrogels, 134
    and Michael-type addition reaction, 127
    osteopontin-mimetic, 129
Hydroxyapatite (HAP), 65
    See also Calcium phosphates as cell
        scaffolds
Hydroxyethyl methacrylate (HEMA), 76–77
Hyperbilirubinaemia, 289
Hypertrophic scarring, 166
Hypodermis, 22
Hypoxia-inducible factor 1 (HIF-1) and
    physiological oxygen tension,
    198–199

I

Indian hedgehog, 433
Inflammation, 163–164
Integra, 41
    Integra dermal regeneration template, 9
    See also Skin
Integraft, 35
    See also Autografts
Integrins, 105–106
    a disintegrin and metalloproteinase with
        thrombospondin motifs
        (ADAMTS), 374
    integrin-binding peptides, 126–127
    integrin-binding RGD peptide
        sequences, 126
    integrin binding site clustering and
        gradients, 130
International cartilage repair society (ICRS)
    visual cartilage assessment scale, 417
International knee documentation
    committee (IKDC), 417
Intima, 26
    See also Arteries
Intussusception, 192
    See also Angiogenesis
Invitrogen, 489–490
Ion-beam assisted deposition (IBAD), 70
Ion beam sputtering, 70
Ischemia, 190
Islet transplantation therapy, 269–270

J

Jenkins, 35
    See also Carbon fibers based prostheses

K

Kennedy ligament agumentation device
    (LAD), 35
Keratinocytes, 21
Knitted fabrics, 36

L

Labile cells, 344
Lacunae, 433
Laminin, 105
Langerhans cells, 21
Leeds-Keio and ligament advanced
    reinforcement system
    (LARS), 35
Leukemia inhibitory factor (LIF), 350
    and β cells, 272
    See also Pluripotency of stem cells
LifeCell product, 10
Ligaments, 16
    anterior cruciate ligament (ACL), 19
    composition and structure, 17–18
    ligamentum flavum (OLF)
        ossification, 445
    prostheses, 34–38
    stress–strain diagram of, 19–20
    time-dependent mechanical
        properties, 20
Ligand clustering, 130
Living-related conjunctival limbal allograft
    (lr-CLAL), 358

M

Macrophages
    activation control, 474–475
    and inflammation, 163–164
    macrophage colony-stimulating growth
        factor (M-CSF) and fibroblast-like
        cells, 305
    monocytes/macrophages, micrographs
        of, 474
    trans-differentiation of, 476
Macroplastique®, 471
Mammalian tissue and natural ECM as
    design guides
    composition and structure,
        104–105
        types and function, 102–103
    receptors, 105–106
Matrigel™, 133
Matrigel tube formation assay, 215
Matrilins, 370

Matrix metalloproteinase (MMP)
    activation and initiation of cascade, 125
    MMP-degradable peptides, 135
McIndoe's techniques, 2–3
Mechanoreceptors, 35
    See also Anterior cruciate ligament
        (ACL) of knee joint
Medtronic Sofamor Danek, 89
Melanocytes, 21
Mesenchymal stem cells (MSCs), 421
    bone tissue engineering and, 457–458
    chondrogenic differentiation of, 379–381
Metaplasia, 351–352
Microtomography, 84
    See also Characterization methods
Mosaicplasty, 407
    MosaicPlasty^{TM}, 378
    See also Autografts

N
Nano-fibers, 111
Nanog gene in ES, 350
Neurorrhaphy, 324–325
Neurotization, 326
N-Hydroxysuccinimide (NHS), 126
Ni free shape memory alloys, 75
Non-invasive prosthetic techniques, 334
    capacitive interfacing with transistors
        and nanowires, 334
    current based interfacing, 334–335
    partial regeneration through sieve
        electrodes, 335

O
Oct3/4 gene in ES, 350
Ocular tissue, 342
    evolution of, 343–344
    gene therapy, 360–361
    healing and, 343
    inflammation and tissue damage, 354–356
    mitotic cells, categories, 344
    NF-κB role in, 356–357
    ocular neovascularization, 213
    regeneration of, 345
Organogenesis product, 9–10
Ossification, 369
Osteoarthritis (OA), 372–375
    pathology, 372
    pathophysiology of, 374
    pharmacological therapies, 375
    risk factors for, 373–374

Osteoblasts
    mononuclear bone-forming cells, 432
    osteoblastoma, 444
Osteochondral Autologous
        Transplantation^{TM} (OATS^{TM}), 378
Osteochondritis dissecans, 405
Osteocytes, 433
Osteoid osteoma, 444
Osteon, 433
Osteopontin-1 (OP-1), 476
    See also Bone
Osteoporosis, 87

P
Pancreas
    acinar cells, 272
    β-cell transcription factors, co-expression
        of, 271
    islets of Langerhans, 263
Papillary dermis, 24
Pax6 gene
    as candidate gene for aniridia, 350
    heterozygosity for deficiency, 351
PEGylation, 127
Peptide-amphiphiles, 127
Periphral nerve system (PNS), 321–322
    electric field effects
        aid and guide nerve growth, 328–329
        and cellular machinery, 329–330
    myelinated nerve fiber and groups of
        unmyelinated nerve fibers,
        323–324
    nerve injuries
        clinical treatment of, 323
        epineurial repair, 324
        fascicular repair, 325
        grafting, 325
        grouped fascicular repair, 324–325
        guidance tube, 326–328
    nerve transfers, 326
    spinal cord repair and, 330–331
    stem cells, use, 331–335
    transection of, 322
Perlecan, 369–370
Permanent cells, 344
Permanent prostheses, 35
    See also Ligaments
Pioglitazone treatment in T2D, 268
Pittsburgh tissue engineering initiative, 8
Plasmin serine protease, 108
Platelet derived growth factor (PDGF) family
        as chemotactic factors, 200–201

Pluripotency of stem cells, 350
Pluripotential bone marrow stem cell, 406
Polymers
    conjugation of cell-binding motifs, 125–126
    enzymatic substrate peptide sequences, 134
    matrix composites applications, 82
    Michaelis-Menten degradation kinetics, 134
    peptide cross-linkers, 135–136
    PHSRN FNIII-9 domain with peptide spacer arm, 127
    poly (glycolic acid) degradable sutures, 75
    poly (glycolic-co-lactic acid) bone screws, 75
    polyhydroxyalkanoates (PHA) as natural biodegradable scaffolds, 67
    polyhyroxyethilmethacrylare (pHEMA) and composite ligament, 37
    poly (lactic acid) (PLA) and poly (glycolic acid) (PGA), 61
        degradation time of, 63
        synthetic pathway for, 62
        See also Bone
    poly (methyl methacrylate) (PMMA)
        bone cement, 75
        S. aureus adhesion and, 76
    polytetrafluoroethylene (PTFE) grafts, 43
    polyurethanes (PU) grafts, 43–44
    poly (vinyl siloxane) dental impression materials, 75
    proteoglycans (PGs), 30
    tissue infiltration and, 136
Polymorphnuclear (PMN) cells and clean-up of wound site, 163
Polytef for urological applications, 471
Porosimetry, 83
    See also Characterization methods
Porosity, 80
Porous tantalum scaffold, osteointegration in, 73
Posterior longitudinal ligament (OPLL) ossification, 445
Pridiedrilling, 378
Primary healing, 162
    See also Wound
Prolast composite ligament, 36
Psoriasis, 213
PTH-related protein (PTHrP), 433
Pulsed laser deposition (PLD), 70

R
Receptor activator of NF-κ B ligand (RANKL), 433
Reconstitution, 343
Refsum disease, 290
Regenerative medicine, 8
    and biomaterials, 114
    clinical applications of, 9
    commercialization and, 9–11
Resin modified glass ionomer, 76
RF magnetron sputtering, 70
Rheumatoid arthritis (RA), 372
ROCK transcription factor and cross-talking signaling networks, 109
Rosiglitazone treatment in T2D, 268

S
SaluMedica, 327
    See also Periphral nerve system (PNS)
3D Scaffold composites, 68
Schwann cells proliferation, 323
Secondary healing, 162
    See also Wound
Self assembly nanofibers for biomineralization, 82–83
Sesamoid bones, 432
Sessil drop method, 85
    See also Characterization methods
Sex determining region Y-box 9 (SRY), 436
Shape memory alloys (SMA), 74–75
Skin, 155–186
    commercially available/marketed
        dermal constructs, 178–180
        dermo-epidermal skin constructs, 176
        epidermal constructs, 177
    composition and structure, 20–22, 156
        basement membrane (BM), 157
        dermis, 157–158
        epidermis, 157
    European skin bank, 182
    extensibility and structural heterogeneities, 24
    GAG-collagen interactions, 24
    gold standard replacement skin, 167
    loss mechanisms
        pathological damage, 159–160
        surgical damage, 160–161
    preconditioning of, 24–25
    regeneration
        cell spraying on burn, 172

Skin (*cont.*)
    clinical application of skin substitutes,
      175–185
    combined skin mesh and sprayed cell
      treatment in burn patient,
      172–174
    commercially available skin
      substitutes, 174–175
    four Ps, requirements, 175
    surgical use of cultured heratinocytes
      in burns patients, 169–174
    replacements, 38–42
      substitutes for wound cover, 40
      wound closure for, 41–42
    small intestinal submucosa (SIS) for
      remodeling of skin, 10
    split-thickness skin graft, 168
    uniaxial stress–strain plot, 23
Smad signaling and Smurfs, 441
Soft tissue replacements, 34
Sol–gel coating, 70
Somatic cell nuclear transfer techniques
    (SCNT), 346
Somatostatin, 263
SOX9 transcription factor, 436
Soybean isoflavones, 479–480
Sprouting, 192
    See also Angiogenesis
Stable cells, 344
Stem cells, 114, 331–335
    aiding repair
      micro-electrode arrays and
        microchips techniques, 333–334
      trophic factors, controlled delivery,
        332–333
    market, 489–490
    mechanical guidance and microchip
      technologies, 332
    non-invasive prosthetic techniques
      micromachined sieve structures, 335
      microneedle arrays, 334–335
      transistors and nanowires, capacitive
        interfacing, 334
    origin of, 454
    therapy, 452
Steven-Johnson's syndrome, 358
Stress relaxation test for AC, 34
Structural proteins and mechanical integrity
    in ECM, 104
Stryker-Dacron prostheses, PET based
    device, 35–36
Subchondral bone and bone marrow
    release, 45

Succinimidyl-4-[N-
    maleimidomethyl]cyclohexane-
    1-carboxylate (SMCC linker), 126
Synthetic matrices, 111

T
Tantalum foam surface in biomedical devices
    and components, 73
Tendons, 16
    Achilles' tendon, 19
    composition and structure, 17–18
    stress–strain diagram of, 19–20
    time-dependent mechanical properties, 20
Tensional homeostasis, 109
Tertiary healing, 162
    See also Wound
Thermoresponsive polymers and
    networks, 126
Thermo-responsive semi-interpenetrating
    poly(NIPAAm-co-AAc)
    (poly(N-isopropylacrylamideco-
    acrylic acid), 126
Thiazolidinediones (TZDs), 267–268
Ti–6Al–4 V alloys as metallic biomaterials, 73
Tie (Tek) family of receptors and Ang,
    199–200
Tip cells
    and sprouting, 192–193
    VEGF role in migration, 193
Tissue engineering
    bioreactors, technical issues, 486
    bone, 450–452
      biomaterials as biomimetic scaffolds,
        456–459
      clinical applications in orthopedics,
        459–460
      clinical problem, 452–453
      stem cells in orthopedic therapy,
        453–456
    business
      bone fractures, incidence of, 87
      business market on, 87–89
    β-cell replacement
      adult stem cells and, 271–273
      bone marrow, 274–275
      embryonic stem cells (ESC), 275–277
      engineering of, 270–271
      islet transplantation, 269–270
      liver cells, 273–274
    Directive 2001/83/EC and Regulation
      (EC) No 726/2004 of European
      Community, 284

drugs and bioactive natural molecules,
478–481
growth factor analogue peptides, 478
high-performance and industrially
sustainable
American federal law, 487–488
biochemical and biostructural cues
analogues, 476–484
complete tissue and organ
regeneration, 484–487
EC consultation process and, 488
inflammatory response, controlling
and exploiting, 473–476
interfacial and complete regeneration,
468–470
non-invasive implantation
procedures, development, 470–473
products, exploitation, 488–490
technical issues affecting, 485–487
ideal vascular replacements and
architecture and characteristics,
238–240
large-diameter vessels-*in vitro* and *in vivo*
studies, 250–251
nanotechnology and, 482
pioneers of, 2–8
plastic surgery and, 3–4
Sir Archibald's pioneering surgical
procedures, 2
small-diameter vessels-*in vitro* and *in vivo*
studies, 247–250
stem cell homing peptides, 478
tissue development *in vitro* and
remodeling processes, 253
and tissue dynamics
and ECM metabolism, 106–108
enzymatic catabolism, 108–109
Tissue Engineering Regenerative
Medicine Society International
(TERMIS), 8
tissue infiltration, 136
vascular, landmarks of, 237
Websites of companies, 477
Tissue inhibitors of matrix proteases
(TIMPs) and MMP activity, 108
Titanium dental implant, 72
Titanium hip prosthesis, 74
Trabecular bone, 432
Transcyte, 40
See also Skin
Transdifferentiation and gene expression,
351–352
Trevira Hochfest, 35

β-Tricalcium phosphate (β-TCP), 65
See also Calcium phosphates as cell
scaffolds
Troglitazone treatment in T2D, 268
Tropocollagen, 164
Tunica media, 26–28
See also Arteries
Type 1 and type 2 diabetes, 263–267
Type 316L stainless steels, cobalt–chromium
alloys, 73

U
Urethrin for urological applications, 471
Urokinase-type plasminogen activator
(uPA) and ECM, 480

V
Vacuum plasma spraying, 70
Vascular endothelial growth factor (VEGF)
anti-VEGF therapy in cancer patients,
studies, 198
with cross-linked collagen, 139
expression level of, 198
heparin-bound, 143
isoforms, 197
lyophilized, 143
in and outside of vascular system, 197
peptide-amphiphile heparin gels loaded
with FGF-2 and, 142–143
signaling and adipogenesis, 206
variants, 140–141
Vascular system
atherosclerotic vascular diseases, 231–232
ideal vascular replacement, requirements
for, 236–238
synthetic vascular grafts
Dacron (PET), 233
expanded Polytetrafluorethylene
(ePTFE), 233–234
polyurethanes (PU), 234
vascular prostheses, 232–233
clinical data, 252
clinical experience with tissue-
engineered, 251–252
graft transplant, 42–45
limitations of, 234–236
mechanical requirements and,
246–247
patency rates, 235–236
tissue engineering of, 240–246
vasculogenesis, 190–191

Vascular system (*cont.*)
  in vitro and *in vivo* studies
    large-diameter vessels, 250–251
    small-diameter vessels, 247–250
    See also Angiogenesis
Vegf gene transfer, 219

W
Wallerian degeneration, 322
White cartilage. *See* Fibrocartilge
Wilson's disease, 295
Wnt proteins, 433
  Wnt (wingless and INT-1) signaling, 436
World Technology Evaluation Center
    (WTEC), 8
Wound
  closure with Integra™, 183–185
  contraction, 165

depth
  intermediate and full thickness
    injuries, 160
  partial thickness injuries, 159
healing process of
  clinical approaches, 166–169
  hemostasis and inflammation,
    163–164
  problems in, 165–166
  proliferative phase, 164–165
  scar maturation/remodeling
    phase, 165
  wound closure categories, 161–162
skin flap, 167
Woven ligaments, 36

Z
Zinc polycarboxylate, 75